Air Pollution, Global Change and Forests in the New Millennium

ELSEVIER Ltd.
The Boulevard, Langford Lane
Kidlington, Oxford OX5 1GB, UK

First edition 2003

Library of Congress Cataloging in Publication Data
A catalog record from the Library of Congress has been applied for.

British Library Cataloguing in Publication Data
A catalogue record from the British Library has been applied for.

ISBN: 0-08-044317-6
ISSN: 1474-8177

⊗ The paper used in this publication meets the requirements of ANSI/NISO Z39.48-1992 (Permanence of Paper).
Printed in The Netherlands.

Air Pollution, Global Change and Forests in the New Millennium

Edited by

D.F. Karnosky[a]
K.E. Percy[b]
A.H. Chappelka[c]
C. Simpson[b]
J. Pikkarainen[a]

[a] *School of Forestry and Wood Products, Michigan Technology University, Houghton, MI 49931-1295, USA*
[b] *Natural Resources Canada, Canadian Forest Service, Atlantic Forestry Centre, New Brunswick E3B 5P7, Canada*
[c] *School of Forestry, Auburn University, Auburn, AL 36849-5418, USA*

2003

ELSEVIER

Amsterdam – Boston – Heidelberg – London – New York – Oxford
Paris – San Diego – San Francisco – Singapore – Sydney – Tokyo

Dedication

Dedicated to our friend, colleague and mentor Dr. David Shriner, who died much too soon.

Dr. David Shriner had a lifelong interest in the effects of air pollution and climate change on forest ecosystems. While at Oak Ridge National Lab, David was one of the principal authors of the US National Acid Precipitation Assessment Program State of Science Report, and he served as a member of the National Science and Technology Council under President Clinton.

While serving as Assistant Director of the USDA Forest Service's North Central Research Station from 1998–2003, David was an enthusiastic and tireless promoter of the Aspen FACE project. He presented the lead keynote address at the May, 2000 IUFRO meeting leading to this book's development.

CONTENTS

Acidic Deposition

Interaction of Air Pollutants and Climate Change

Conclusions and Research Needs

Developments in Environmental Science

Introduction to the Book Series

Environmental pollution has played a critical role in human lives since the early history of the nomadic tribes. During the last millennium industrial revolution, increased population growth and urbanization have been the major determinants in shaping our environmental quality.

Initially primary air pollutants such as sulfur dioxide and particulate matter were of concern. For example, the killer fog of London in 1952 resulted in significant numbers of human fatality leading to major air pollution control measures. During the 1950s, scientists also began to understand the cause and atmospheric mechanisms for the formation of the Los Angeles photochemical smog. We now know that surface level ozone and photochemical smog are a worldwide problem at regional and continental scales, with specific geographic areas of agriculture, forestry and natural resources, including their biological diversity at risk. As studies continue on the atmospheric photochemical processes, air pollutant transport, their atmospheric transformation and removal mechanisms, so is the effort to control the emissions of primary pollutants (sulfur dioxide, oxides of nitrogen, hydrocarbons and carbon monoxide), mainly produced by fossil fuel combustion.

During mid-1970s environmental concerns regarding the occurrence of "acidic precipitation" began to emerge to the forefront. Since then, our knowledge of the adverse effects of air pollutants on human health and welfare (terrestrial and aquatic ecosystems and materials) has begun to rise substantially. Similarly, studies have been directed to improve our understanding of the accumulation of persistent inorganic (heavy metals) and organic (polyaromatic hydrocarbons, polychlorinated biphenyls) chemicals in the environment and their impacts on sensitive receptors, including human beings. Use of fertilizers (excess nutrient loading) and herbicides and pesticides in both agriculture and forestry and the related aspects of their atmospheric transport, fate and deposition; their direct runoff through the soil and impacts on ground and surface water quality and environmental toxicology have become issues of much concern.

In the recent times environmental literacy has become an increasingly important factor in our lives, particularly in the so-called developed nations. Currently the scientific, public and political communities are much concerned with the increasing global scale air pollution and the consequent global climate change. There are efforts being made to totally ban the use of chlorofluorocarbon and organo-bromine compounds at the global scale. However, during this millennium many developing nations will become major forces governing environmental health as their populations and industrialization grow at a rapid pace. There is an on-going international debate regarding policies and the mitigation strategies to be adopted to address the critical issue of climate change. Human health and environmental impacts and risk assessment and the associated cost–benefit analyses, including global economy are germane to this controversy.

An approach to understanding environmental issues in general and in most cases, mitigation of the related problems requires a systems analysis and a multi- and inter-disciplinary philosophy. There is an increasing scientific awareness to integrate environmental processes and their products in evaluating the overall impacts on various receptors. As momentum is gained, this approach constitutes a challenging future direction for our scientific and technical efforts.

The objective of the book series *"Developments in Environmental Science"* is to facilitate the publication of scholarly works that address any of the described topics, as well as those that are related. In addition to edited or single- and multi-authored books, the series also considers conference proceedings and paperback computer-software packages for publication. The emphasis of the series is on the importance of the subject topic, the scientific and technical quality of the content and timeliness of the work.

Sagar V. Krupa
Chief Editor, Book Series

List of Contributors

Alonso, R. Ecotoxicology of Air Pollution, CIEMAT-DIAE, Ed. 70. Avda. Complutense 22, Madrid-28040, Spain.

Amores, G. Department of Chemistry and Soil Science, University of Navarre, Irunlarrea s/n, 31080 Iruña-Pamplona, Spain.

Ballarin-Denti, A. Department of Mathematics and Physics, Catholic University of Brescia, via Musei 41, 25121 Brescia, Italy

Beer, W. Wisconsin Department of Natural Resources, Bureau of Air Management (AM/7), Box 7921, Madison, WI 53707-7921, USA.

Bermejo, V. Ecotoxicology of Air Pollution, CIEMAT-DIAE, Ed. 70. Avda. Complutense 22, Madrid-28040, Spain.

Birdsey, R.A. USDA Forest Service, Northern Global Change Program, 11 Campus Boulevard, Suite 200, Newtown Square, PA 19073, USA.

Bucher, J. Swiss Federal Institute for Forest, Snow & Landscape Research, Zürcherstrasse 111, CH-8903 Birmensdorf, Switzerland.

Bussotti, F. Dipartimento di Biologia Vegetale, Università di Firenze, Piazzale delle Cascine 28, 50144 Firenze, Italy.

Bytnerowicz, A. USDA Forest Service, Pacific Southwest Research Station, 4955 Canyon Crest Drive, Riverside, CA 92507, USA.

Callan, B. Natural Resources Canada, Canadian Forest Service, Pacific Forestry Centre, 506 West Burnside Rd., Victoria, British Columbia V8Z 1M5, Canada.

Cavero, R. Department of Botany, University of Navarre, Irunlarrea s/n, 31080 Pamplona, Spain.

Castillo, F.J. Departamento Ciencias del Medio Natural, Universidad Pública de Navarra, Pamplona-31006, Spain.

Chappelka, A.H. Auburn University, School of Forestry & Wildlife Sciences, 206 M. White-Smith Hall, Auburn, AL 36849-5418, USA.

Cozzi, A. Linnaea-ambiente, Via Sirtori 37, 50137 Firenze, Italy.

Dickson, R.E. USDA Forest Service, North Central Forest Experiment Station, Forestry Sciences Laboratory, 5985 Highway K, Rhinelander, WI 54501, USA.

Ederra, A. Department of Botany, University of Navarre, Irunlarrea s/n, 31080 Pamplona, Spain.

Elvira, S. Ecotoxicology of Air Pollution, CIEMAT-DIAE, Ed. 70. Avda. Complutense 22, Madrid-28040, Spain.

Ferretti, M. Linnaea-ambiente, Via Sirtori 37, 50137 Firenze, Italy.

Garrigó, J. Department of Chemistry and Soil Science, University of Navarre, Irunlarrea s/n, 31080 Pamplona, Spain.

Gerosa, G. DMF, Department of Mathematics and Physics, Università Catolica del Sacro Cuore, Via Musei 2, 25121 Brescia, Italy.

Gimeno, B.S. Ecotoxicology of Air Pollution, CIEMAT-DIAE, Ed. 70. Avda. Complutense 22, Madrid-28040, Spain.

Gravano, E. Dipartimento di Biologia Vegetale, Università di Firenze, Piazzale delle Cascine 28, 50144 Firenze, Italy.

Grossoni, P. Dipartimento di Biologia Vegetale, Università di Firenze, Piazzale delle Cascine 28, I-50144 Firenze, Italy.

Hansen, K. Department of Forest Ecology, Danish Forest and Landscape Research Institute, Hørsholm, Kongevej 11, 2970 Hørsholm, Denmark.

Hendrey, G.R. Brookhaven National Laboratory, 1 South Technology Street, Upton, NY 11973, USA.

Hom J. USDA Forest Service, Northern Global Change Program, 11 Campus Boulevard, Suite 200, Newtown Square, PA 19073, USA.

Honrath Jr., R.E. Department of Civil and Environmental Engineering, Michigan Technological University, 1400 Townsend Drive, Houghton, MI 49931, USA.

Hopkin, A. Natural Resources Canada, Canadian Forest Service, Great Lakes Forestry Centre, 1219 Queen Street, Sault Ste. Marie, Ontario P6A 5E2, Canada.

Host, G. Natural Resources Research Institute, University of Minnesota, 5013 Miller Trunk Highway, Duluth, MN 55811, USA.

Huttunen, S. Department of Biology, University of Oulu, P.O. Box 3000, FIN-90014 Oulu, Finland.

Inclán, R. Ecotoxicology of Air Pollution, CIEMAT-DIAE, Ed. 70. Avda. Complutense 22, Madrid-28040, Spain.

Isebrands, J.G. Environmental Forestry Consultants, LLC, P.O. Box 54, E7323 Hwy 54, New London, WI 54961, USA.

Jepsen, E. Wisconsin Department of Natural Resources, 101 S. Webster, P.O. Box 7921, Madison, WI 53707, USA.

Karnosky, D.F. School of Forest Resources and Environmental Science, Michigan Technological University, 101 U.J. Noblet Forestry Building, 1400 Townsend Drive, Houghton, MI 49931, USA.

Kieliszewska-Rokicka, B. Institute of Dendrology, Polish Academy of Sciences, 5 Parkowa Str. 62-035, Kórnik, Poland.

King, J. School of Forest Resources & Environmental Science, Michigan Technological University, 101 U.J. Noblet Forestry Building, 1400 Townsend Drive, Houghton, MI 49931, USA.

Kinouchi, M. School of Forest Resources & Environmental Science, Michigan Technological University, 101 U.J. Noblet Forestry Building, 1400 Townsend Drive, Houghton, MI 49931, USA.

Kolb, T.E. School of Forestry, Northern Arizona University, Box 15018, Flagstaff, AZ 86011-5018, USA.

Kruger, E. University of Wisconsin-Madison, Department of Forest Ecology, 1630 Linden Dr., Madison, WI 53706, USA.

Krupa, S.V. University of Minnesota, Plant Pathology Department, 420 Stak H, 1519 Gortner Avenue, St. Paul, MN 55108, USA.

Kubiesa, P. Institute for Ecology of Industrial Areas, 6 Kossutha Str. 40-832 Katowice, Poland.

Kubiske, M.E. USDA Forest Service, North Central Research Station, Forestry Sciences Laboratory, 5985 Highway K, Rhinelander, WI 54501, USA.

Kurki, S. Department of Biology, University of Oulu, P.O. Box 3000, FIN-90014 Oulu, Finland.

Legge, A.H. Biosphere Solutions, 1601 11th Ave. NW, Calgary, Alberta T2N 1H1, Canada.

Lenz, K.E. Department of Mathematics and Statistics, University of Minnesota, University Drive, Duluth, MN 55812, USA.

Leski, T. Institute of Dendrology, Polish Academy of Sciences, 5 Parkowa Str. 62-035, Kórnik, Poland.

Luchetta, L. Ecole Nationale Supérieure de Chimie de Toulouse, 118 Route de Narbonne, 31077 Toulouse, France.

Madotz, N. Department of Botany, University of Navarre, Irunlarrea s/n, 31080 Pamplona, Spain.

Manes, F. Dipartimento di Biologia Vegetale, Università di Roma "La Sapienza", Piazzale Aldo Moro 5, I-00185 Rome, Italy.

Mankovska, B. Forest Research Institute, T.G. Masarykova Street 2195, 960 92 Zvolen, Slovakia.

Manninen, S. Department of Ecology and Systematics, University of Helsinki, P.O. Box 65 (Viikinkaari 1), FIN-0014 Helsinki, Finland;
Botanical Institute, University of Gothenburg, P.O. Box 461, SE-40530 Gothenburg, Sweden.

Martin, M.J. Natural Resources Research Institute, University of Minnesota, 5013 Miller Trunk Highway, Duluth, MN 55811, USA.

Matyssek, R. Lehrstuhl für Forstbotanik, Technische Universität München, Am Hochanger 13, D-85354 Freising, Germany.

Mazzali, C. Fondazione Lombardia per l'Ambiente, Piazza Diaz 7, 20123 Milano, Italy.

McDonald, E.P. USDA Forest Service, Forestry Sciences Laboratory, 5985 Highway K, Rhinelander, WI 54501, USA.

McNulty, S.G. USDA Forest Service, Southern Global Change Program, 920 Main Campus Drive, Venture Center II, Suite 300, Raleigh, NC 27606, USA.

Mickler, R.A. ManTech Environmental Technology Inc., Southern Global Change Program, 920 Main Campus Drive, Venture Center II, Suite 300, Raleigh, NC 27606, USA.

Mori, B. Dipartimento di Biologia Vegetale, Università di Firenze, Piazzale delle Cascine 28, I-50144 Firenze, Italy.

Nellemann, C. Department of Biology and Nature Conservation, Agricultural University of Norway, Box 5014, 1432 Ås, Norway.

Noormets, A. The University of Toledo, Department EEES, LEES Lab, Mail Stop 604, Toledo, OH 43606, USA.

Oksanen, E. University of Kuopio, Department of Ecology and Environmental Science, POB 1627, 70211 Kuopio, Finland.

Paoletti, E. IPAF-CNR, Piazzale delle Cascine 28, I-50144 Florence, Italy.

Percy, K.E. Natural Resources Canada, Canadian Forest Service, Atlantic Forestry Centre, P.O. Box 4000, Fredericton, New Brunswick E3B 5P7, Canada.

Pirttiniemi, N. Department of Biology, University of Oulu, P.O. Box 3000, FIN-90014 Oulu, Finland.

Podila, G.K. University of Alabama, Department of Biological Sciences, 301 Sparkman Drive, WH142, Huntsville, AL 35899, USA.

Pregitzer, K.S. School of Forest Resources and Environmental Science, Michigan Technological University, Houghton, MI 49931, USA.

Prichard, T. Wisconsin Department of Natural Resources, 101 S. Webster, P.O. Box 7921, Madison, WI 53707, USA.

Prus-Glowacki, W. Adam Mickiewicz University, Faculty of Biology, Department of Genetics, Miedzychodzka 5, PL-60 371 Poznan, Poland.

Roth, J. Wisconsin Department of Natural Resources, Bureau of Air Management (AM/7), Box 7921, Madison, WI 53707-7921, USA.

Rudawska, M. Institute of Dendrology, Polish Academy of Sciences, 5 Parkowa Str. 62-035, Kórnik, Poland.

Santamaría, J.M. Department of Chemistry and Soil Science, University of Navarre, Irunlarrea s/n, 31080 Iruña-Pamplona, Spain.

Sharma, P. School of Forest Resources and Environmental Science, Michigan Technological University, 101 U.J. Noblet Forestry Building, 1400 Townsend Drive, Houghton, MI 49931, USA.

Shriner, D.S. USDA Forest Service, North Central Research Station, 1992 Folwell Avenue, St. Paul, MN 55108, USA.

Sober, J. School of Forest Resources and Environmental Science, 101 U.J. Noblet Forestry Building, 1400 Townsend Drive, Michigan Technological University, Houghton, MI 49931, USA.

Söderberg, U. Department of Forest Resource Management and Geomatics, Swedish University of Agricultural Sciences, SE-901 83 Umeå, Sweden.

Sorjamaa, R. Department of Applied Physics, University of Kuopio, P.O. Box 1627, FIN-70211 Kuopio, Finland;
Department of Physics (Biophysics), University of Oulu, P.O. Box 3000, FIN-90014 Oulu, Finland.

Staszewski, T. Institute for Ecology of Industrial Areas, 6 Kossutha Str. 40-832 Katowice, Poland.

Thakur, R.C. School of Forest Resources and Environmental Science, Michigan Technological University, 101 U.J. Noblet Forestry Building, 1400 Townsend Drive, Houghton, MI 49931, USA.

Tani, C. Dipartimento di Biologia Vegetale, Università di Firenze, Piazzale delle Cascine 28, I-50144 Firenze, Italy.

Thomsen, M.G. Norwegian Institute of Land Inventory, Raveien 9, 1432 Ås, Norway.

Wustman, B.A. Amicus Therapeutics Inc., Commercialization Center for Innovative Technologies, 675 US Highway One, North Brunswick, NJ 08902, USA.

Foreword and Acknowledgments

This book is the outgrowth of the International Union of Forestry Research Organizations' (IUFRO) May 28–31, 2000, meeting of 7.04.00 "Impacts of Air Pollution on Forest Ecosystems". The meeting was sponsored by the USDA Forest Service Global Change Program, Canadian Forest Service, Arthur Ross Foundation, National Council of the Paper Industry for Air and Stream Improvement, and Michigan Technological University.

The editors appreciate the assistance of over 40 reviewers who reviewed and provided constructive critiques of the manuscripts. The editors also acknowledge the group discussion sessions that led to the final chapter on knowledge gaps and research needs. Finally, the editors thank undergraduate students Dave Karnosky and Caroline Reed, graduate students Michiko Kinouchi and Pooja Sharma, and post-doctoral fellow Ramesh Thakur for their help in proofreading of all manuscripts.

Preface

Our atmosphere bears an increasing load of pollutants: carbon dioxide, ozone, oxides of nitrogen and sulfur, particulates, and heavy metals. These pollutants are deposited on leaf and soil surfaces, enter leaves through stomatal openings, and affect climate. The physiological responses of trees to these pollutants and to climate change are complex and only partially understood. Ecosystem processes such as nutrient cycling are also affected yet only beginning to be elucidated. Several decades of research have yielded important advances in understanding forest responses to single factors and certain combinations of factors. Complex, time-varying responses to multiple factors and their interactions are yet poorly understood. We remain incapable of extrapolating results from experiments and observations in space and time, with sufficient confidence to anticipate how forest vegetation in different regions will respond to air pollution and climate change in the 21st century.

Understanding past and current vegetation responses to air pollution and climate change is the critical first step toward improving our ability to anticipate future vegetation responses, and if necessary, initiate adaptive management responses and mitigation activities. Land managers require a solid scientific foundation for developing a strategy to adapt to anticipated 21st century air quality and climate. Policy makers need information about the location and magnitude of vegetation responses to various levels of air pollutants so that concentrations and exposure standards can be established with consideration of the effects on forest vegetation and the many services provided by healthy forest ecosystems.

The chapters in this book present a snapshot of the state of knowledge of air pollution effects at the beginning of the 21st century. From their different disciplines, a distinguished collection of authors document their understanding of how leaves, trees, and forests respond to air pollutants and climate change. Scenarios of global change and air pollution are described. Authors describe responses of forests to climate variability, tropospheric ozone, rising atmospheric CO_2, the combination of CO_2 and ozone, and deposition of acidic compounds and heavy metals. The responses to ozone receive particular attention because of increasing concern about its damaging effects and increasing concentrations in rural areas. Scaling issues are addressed—from leaves to trees, from juvenile trees to mature trees, from short-term responses to long-term responses, and from small-scale experiments and observations to large-scale forest ecosystems.

This book is one major product of a conference sponsored by the International Union of Forestry Research Organizations, the USDA Forest Service Global Change Northern Stations Program, the Arthur Ross Foundation,

NCASI, the Canadian Forest Service, and Michigan Technological University. The conference, held in May 2000 in Houghton, Michigan, USA, was appropriately titled "Air Pollution, Global Change, and Forests in the New Millennium". The editors, David Karnosky, Kevin Percy, Art Chappelka, Caroline Simpson, and Janet Pikkarainen performed a great service by organizing the conference and editing this book. All of the authors, editors, and peer reviewers deserve congratulations for taking this important step in the journey toward sustaining healthy forests in a changing atmospheric environment.

Richard A. Birdsey
USDA Forest Service
Global Change Research Program

Air Pollution, Global Change and Forests in the New Millennium
D.F. Karnosky et al., editors

1

Chapter 1

Air pollution and global change:
A double challenge to forest ecosystems

D.F. Karnosky*

*School of Forest Resources and Environmental Science, Michigan Technological University,
101 U.J. Noblet Forestry Building, 1400 Townsend Drive, Houghton, MI 49931, USA
E-mail: karnosky@mtu.edu*

K.E. Percy

*Natural Resources Canada, Canadian Forest Service-Atlantic Forestry Centre, P.O. Box 4000,
Fredericton, New Brunswick, E3B 5P7 Canada*

R.C. Thakur

*School of Forest Resources and Environmental Science, Michigan Technological University,
101 U.J. Noblet Forestry Building, 1400 Townsend Drive, Houghton, MI 49931, USA*

R.E. Honrath Jr.

*Department of Civil and Environmental Engineering, Michigan Technological University,
1400 Townsend Drive, Houghton, MI 49931, USA*

Abstract

The world's forests provide a host of wood products, and non-wood resources, and they are critically important in conserving plant, animal, insect and microbial diversity, maintaining soil and water resources, and providing opportunities for employment and recreation. Only recently have we started to value forests for their ability to sequester carbon from the atmosphere. The rapidly changing atmospheric environment with its mix of increasing anthropogenic emissions means that the future world's forests will be faced with unprecedented levels of carbon dioxide and other greenhouse gases, and rising temperatures due to the trapping of radiative heating by the greenhouse gases. In addition, large expanses of these ecosystems will be concurrently exposed to elevated levels of tropospheric ozone, particulates, nitrogen oxides, and acidic rainfall or other air pollutants. Finally, increasing demand for forest products and expanding development pressures from our rapidly growing world population will mean continued land use change and forest habitat loss. Thus, it is very difficult to predict the condition or productivity of forests in this century. In this book, a number of

---
*Corresponding author.

DOI:10.1016/S1474-8177(03)03001-8

forest and atmospheric scientists summarize what is known on the impacts of air
pollution and climate change on forest ecosystems.

1. Introduction

Forests cover 3.87 billion ha worldwide or 30% of the Earth's land area
(Fig. 1). Besides providing annually about 3.3 billion cubic meters of round-
wood and nearly 3.0 billion cubic meters of fuelwood, forests are important
for many non-wood forest products as well as for soil and water conservation,
biological diversity conservation, support of agricultural systems, employment
generation, provision of recreational opportunities, and protection of natural
and cultural heritage (FAO, 2001). Nearly 70% of the water vapor passes
through the stomata of forest trees and forests hold about 50% of the world's
carbon stocks. It is estimated that forests sequester some 2.0 Pg annually of
carbon emitted that would otherwise end up in the atmosphere, contributing to
global warming (Houghton, 2001). Furthermore, forests account for ~70% of
the carbon exchange between land and the atmosphere (Schlesinger, 1997).

2. How will the world's forests respond to elevated CO_2, warming climate, and increasing air pollution loading?

It is well known that atmospheric carbon dioxide (CO_2) is rising globally
(Keeling et al., 1995) and that much of the increase in atmospheric CO_2 is
due to elevated anthropogenic emissions (IPCC, 2001) and degradation of
tropical forests that would otherwise be larger CO_2 sinks (O'Brien, 2000).
Concurrently, other greenhouse gases, such as methane and nitrous oxide, are
also increasing (Fig. 2). Together these greenhouse gases are trapping considerable radiant energy near the Earth's surface, resulting in the so-called
"greenhouse effect" of warming climate (Fig. 3). Simultaneously, the atmospheric concentration of tropospheric ozone (O_3) is increasing (Fig. 4)
downwind of major metropolitan regions around the world such that nearly
50% of the world's forests are expected to be at risk from levels of O_3 over
60 ppb by the year 2100 (Fowler et al., 1999). In addition, particularly in
developing countries where industrialization and urbanization are expanding at a rapid rate, levels of acidic deposition from sulfur and nitrogen oxides emitted into the atmosphere are also increasing (Streets et al., 2000;
Streets and Waldhoff, 2000). Thus, our world's forests will be exposed to a
combination of air pollutant stresses and rapidly changing climate over the
next century making it difficult to predict how forest ecosystems will respond.

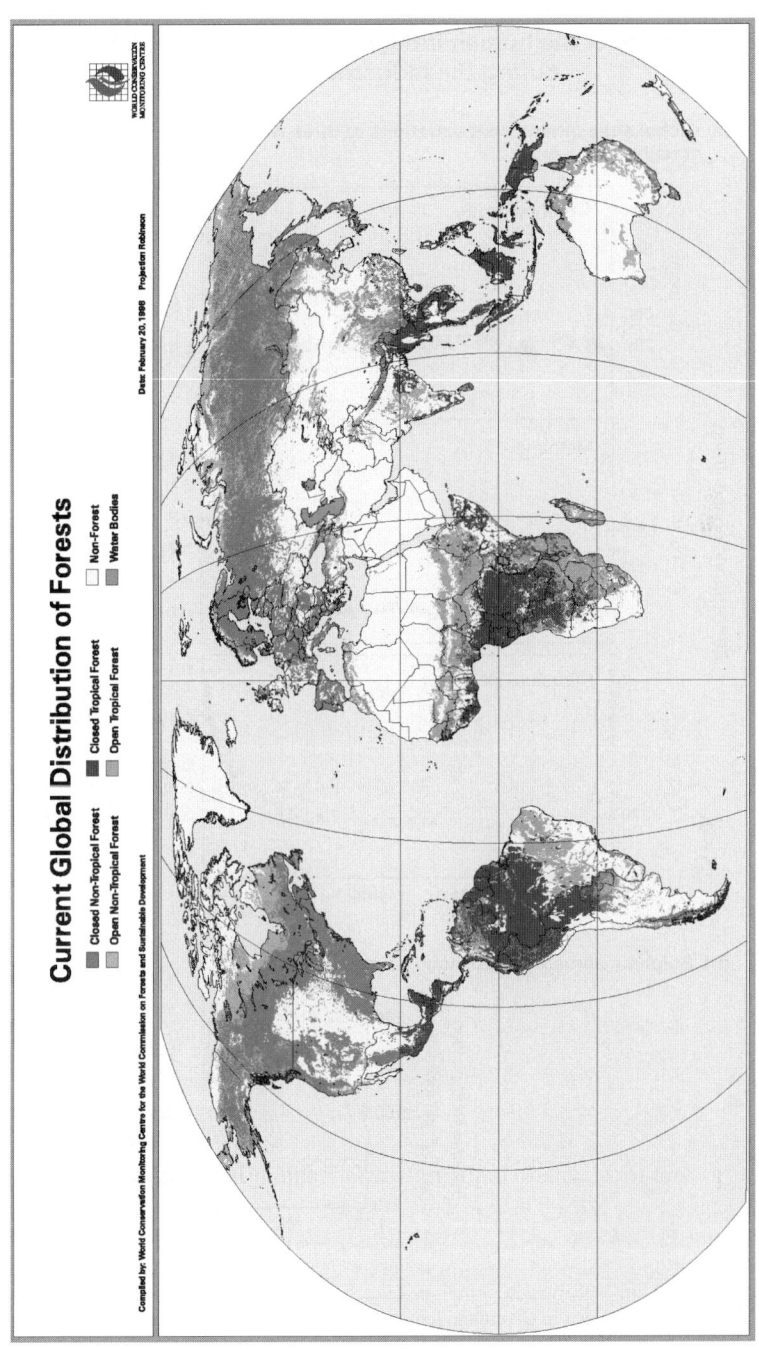

Figure 1. Global distribution of the world's forests (from FAO, 2001).

Indicators of the human influence on the atmosphere during the Industrial Era

(a) Global atmospheric concentrations of three well mixed greenhouse gases

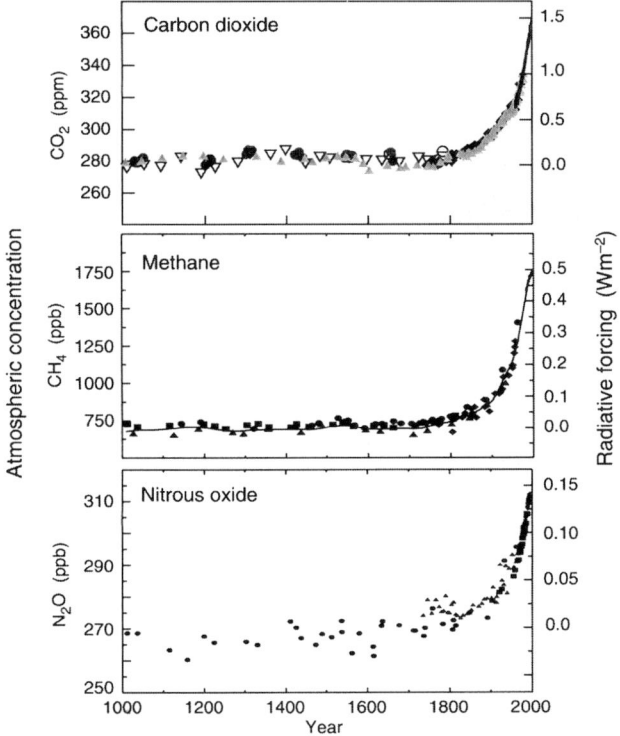

(b) Sulphate aerosols deposited in Greenland ice

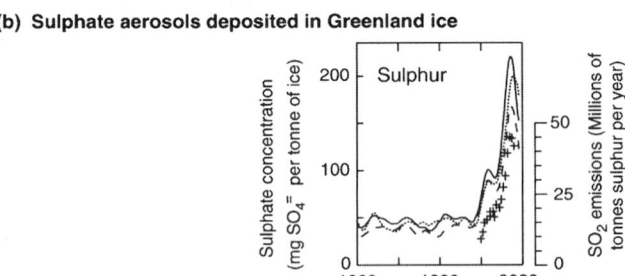

Figure 2. Historical atmospheric concentrations of carbon dioxide, methane, and nitrous oxides, and sulphate aerosols deposited in Greenland ice (from IPCC, 2001).

Variations of the Earth's surface temperature for:

(a) the past 140 years

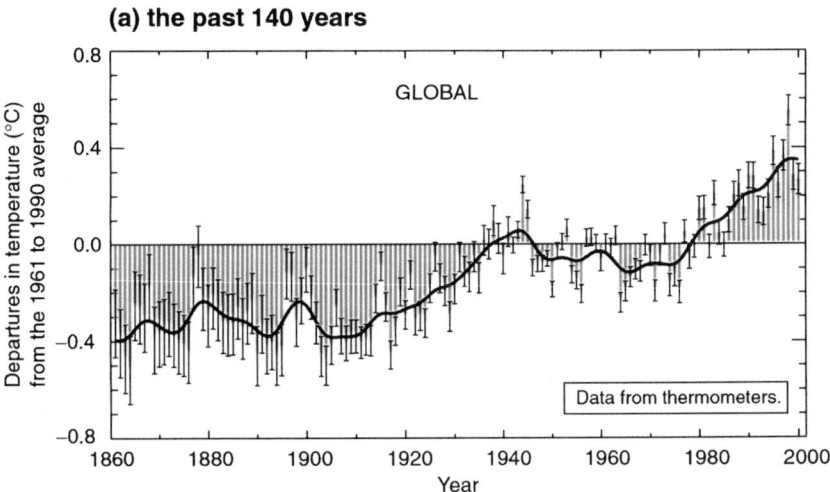

(b) the past 1,000 years

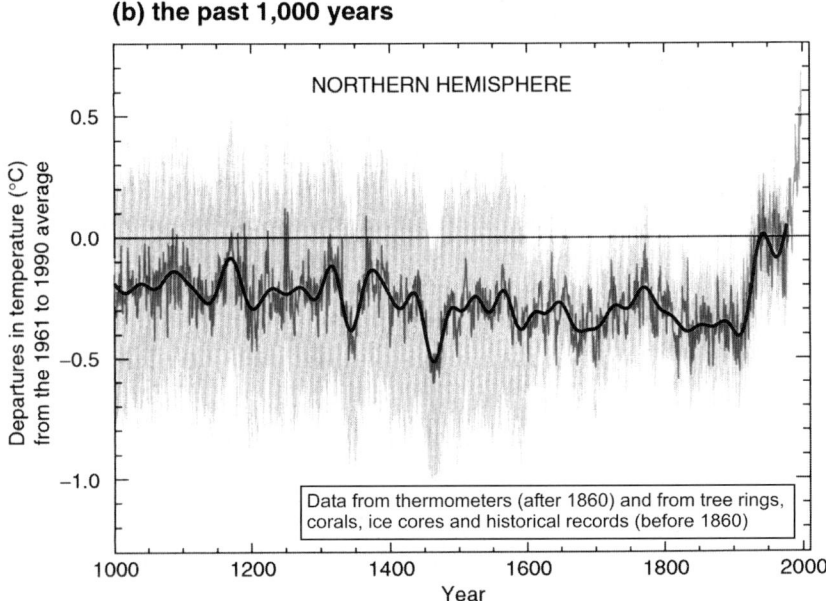

Figure 3. Variations of the Earth's surface temperature for (a) the past 140 years and (b) the past 1000 years (from IPCC, 2001).

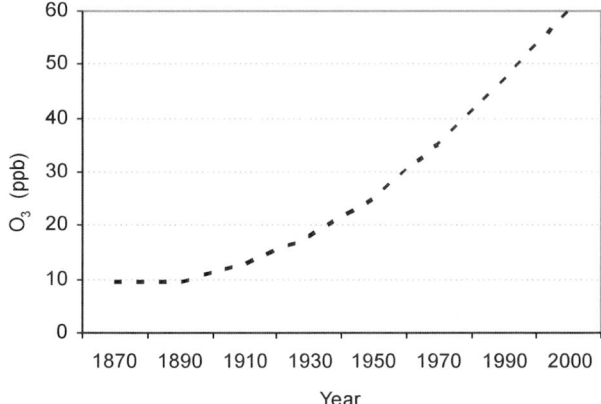

Figure 4. Historical trends in tropospheric O_3 (adapted from Marenco et al., 1994).

3. CO₂

The rise in atmospheric CO_2 has been described as an "all-you-can-eat-buffet" for forest trees as CO_2 is the basic building block of photosynthesis and as it rises, trees have higher photosynthetic rates. On average, trees grown under elevated CO_2 photosynthesize at about a 60% higher rate than under background CO_2 levels (Norby et al., 1999). In addition, trees growing under elevated CO_2 generally have lower stomatal conductance and improved water use efficiency.

Short-term growth responses under elevated CO_2 have predictably followed the same trends as photosynthetic enhancement with average growth enhancement being about 27% (Norby et al., 1999). However, it is difficult to extrapolate the results of the growth studies to growth trends for trees over their life-times or for forest stands over their rotation as most growth studies have been conducted:

- for a relatively short time (from less than one year to a few years) considering that forest trees have life times or rotation ages from decades to centuries;
- for the most part, using small seedlings whose responses may or may not be indicative of older and larger trees;
- in laboratory growth chambers, greenhouses, or open-top chambers with trees grown in pots or with different environmental conditions (temperature, light and humidity, for example) than trees would receive in the forests;
- free of weed competition; and
- with pest control.

Table 1. Summary of growth responses for forest trees exposed to elevated CO_2 in free-air CO_2 exposure (FACE) experiments

FACE experiment	Species	Soil nutrients	Tree age (yrs)	Growth enhancement	Growth acclimation	References
POPFACE	Hybrid poplars	Moderate	2	Yes (+10 to 11%)[a]	–	Gielen and Ceulemans, 2001
FACTS II	Trembling aspen	Moderate	5	Yes (+12% to +13%)[b]	No	Isebrands et al., 2001; Percy et al., 2002
	Paper birch	Moderate	5	Yes (+24 to +25%)[c]	No	Karnosky et al., 2003
	Sugar maple	Moderate	5	No[c]	–	Karnosky et al., 2003
Oak Ridge	Sweetgum	Low	20	Yes (+15 to +33%)[d]	No	Norby et al. 2001, 2002
FACTS I	Loblolly pine	Low	20	Yes (~26%)[d]	Yes	DeLucia et al., 1999; Oren et al., 2001; Hamilton et al., 2002

[a] Height.
[b] Volume.
[c] Heights and diameters.
[d] Basal area.

D.F. Karnosky et al.

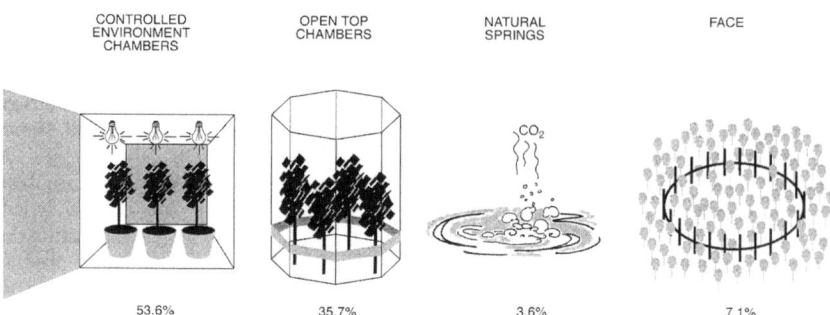

CONTROLLED ENVIRONMENT CHAMBERS   OPEN TOP CHAMBERS   NATURAL SPRINGS   FACE

53.6%   35.7%   3.6%   7.1%

Figure 5. The most common types of facilities used in CO_2-enrichment studies (from Gielen and Ceulemans, 2001). In addition, branch chambers have been useful for examining gas exchange parameters of large trees in-situ (Teskey et al., 1991; Vann and Johnson, 1995).

Research on forest trees has evolved from controlled environment chambers to open-top chambers and then on to studies utilizing natural CO_2 springs or free-air CO_2 exposure (FACE) facilities (Fig. 5). Long-term growth studies around CO_2 vents and two of the former FACE experiments (the FACTS I loblolly pine study and the Oak Ridge sweetgum study) (Table 1) suggest that growth enhancement under elevated CO_2 may be rather limited and that nutrient status of the soils may drive the response. The two poplar experiments are being conducted on soils higher in nutrients and growth enhancement has not diminished through two years (POPFACE) or 5 years (FACTS II).

The effects of elevated CO_2 on forest ecosystems are still being actively studied. However, from the standpoint of individual trees, we know that elevated CO_2 stimulates photosynthesis (Tjoelker et al., 1998; Noormets et al. 2001a, 2001b), impacts foliar senescence in autumn (Karnosky et al., 2003), and stimulates aboveground (Norby et al. 1999, 2002) and belowground (King et al., 2001; Kubiske and Godbold, 2001) growth. Trees grown under elevated CO_2 generally have lower nitrogen concentrations in their foliage, lower Rubisco concentrations (Moore et al., 1999), and altered defense compounds (Lindroth et al. 1993, 1997) and altered levels of antioxidants (Polle et al. 1993, 1997; Wustman et al., 2001). See Chapter 3 for more information on CO_2 effects on forest ecosystems.

4. O_3

Evidence indicates that our emissions of nitrogen oxides ($NO_x = NO + NO_2$) and volatile organic compounds (VOCs) have significantly increased levels of O_3 over large regions of the globe (e.g., Crutzen, 1988; Marenco et al., 1994;

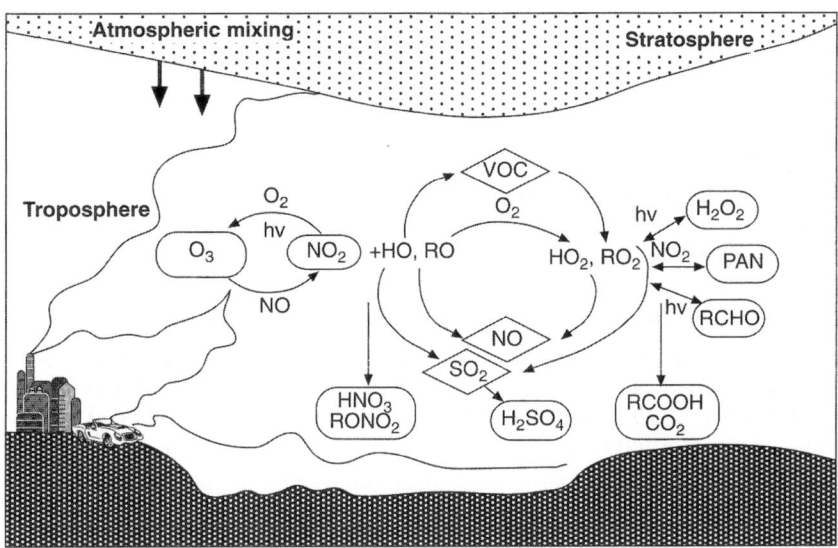

Figure 6. Schematic representation of gas-phase chemistry resulting in the generation of ozone and other by-products in polluted air. Primary pollutants, emitted from anthropogenic sources, are shown in diamond-shaped boxes; secondary pollutants, formed as a result of atmospheric reactions, are shown in circular boxes. PAN, peroxyacyl nitrate; VOC, volatile organic compounds (from Barnes and Wellburn, 1998).

Yienger et al., 2000). Anthropogenically driven increases in tropospheric O_3 form a large fraction (~20%) of the estimated greenhouse effect (Hauglustaine et al., 1994; Marenco et al., 1994; Kiehl et al., 1999; Berntsen et al., 2000). In addition, increasing levels of "background" O_3 are expected to affect strategies for attainment of air quality standards in urban areas in the future (see below) and increase the size of regions over which crop production is reduced due to O_3 damage (Chameides et al., 1994; Fowler et al., 1999). Finally, O_3 is a primary dfeterminant of the oxidizing strength of the troposphere, through its photolysis in the presence of water vapor to form HO radicals, and changing tropospheric O_3 levels result in alteration of HO concentrations, impacting the lifetimes of most potential pollutants in the troposphere (Thompson, 1992) (Fig. 6).

Impacts on the tropospheric O_3 budget on a global scale occur through two mechanisms: (1) the production of O_3 over regions of O_3 precursor emissions followed by export to the global atmosphere of a fraction of the O_3 so produced, and (2) the export of O_3 precursors followed by production of O_3 in regions remote from sources. Ozone production efficiency is non-linear with respect to NO_x concentration (Liu et al., 1987) and is NO_x-limited in most of the non-urban troposphere (Crutzen, 1988). As a result, in-situ production of O_3 in remote regions as the result of exported nitrogen oxides is expected to be more efficient than is production in source regions followed by export of O_3. Global model simulations indicate a significant impact of long-range transport of peroxyacetyl nitrate (PAN) and its analogs upon NO_x levels in remote regions (Moxim et al., 1996; Horowitz and Jacob, 1999; Levy II et al., 1999), and calculate that photochemical production is by far the dominant source term in the O_3 budget throughout the troposphere (Wang et al., 1998). Impacts of precursor emissions from North America, Europe, and Asia upon air quality in Europe, Asia, and North America, respectively, are sufficient to potentially affect the ability of nations in the downwind regions to attain air quality standards for O_3, currently or in the future (Jacob et al., 1999; Berntsen et al., 1999; Lin et al., 2000; Yienger et al., 2000; Lelieveld et al., 2002). Impacts of long-range transport from Asia upon O_3 in air reaching North America have been observed (e.g., at the Cheeka Peak site (Jaffe et al., 1999)) and in northern California (Parrish et al., 1992), as has transport from North America carrying elevated O_3 to Europe (e.g., Stohl and Trickl, 1999). Indeed, modeling analyses indicate that 20% of the violations of the European Council O_3 standard that occurred in the summer of 1997 would not have occurred in the absence of North American emissions (Li et al., 2003). However, emissions of nitrogen oxides worldwide are changing rapidly. Globally, emissions increases are expected to significantly enhance export, particularly from Asia; in contrast, it is likely that an increasing emphasis upon NO_x reductions to decrease O_3 standard violations in the United States and Europe will result in declining nitrogen oxides export from North America in the future (e.g., Jacob et al., 1999; Jonson et al., 2001).

4.1. Worldwide O_3 trends

While peak values of O_3 around major metropolitan areas in the US have generally decreased over the past 20 years (Lin et al., 2001), background base levels continue to increase worldwide (Fowler et al., 1998; Collins et al., 2000; Derwent et al., 2002). Particularly noteworthy is the rapid increase in O_3 levels near major cities in developing countries in Asia (Aunan et al., 2000; Cheung and Wang, 2001; Gupta et al., 2002), Central America (Raga and Raga, 2000; Skiba and Davydova-Belitskaya, 2002), and South America (Romero et al., 1999). Probably the highest O_3 concentrations in the world now occur in the vicinity of Mexico City which is faced with ideal conditions for photochemical oxidant production (high elevation, high incident radiation that does not vary significantly during the year, high daily temperatures, and high VOC and NO_x

emissions by mobile and fixed sources in the rapidly growing city (Raga and Raga, 2000)).

4.2. Ozone and forests

Ozone is a highly reactive oxidative stressor that enters the plant through the stomates and is highly reactive with cell walls and membranes in the cells surrounding the stomatal cavity. Ozone causes degradation of chlorophyll (Keller, 1988) and rubisco, adversely affecting the important machinery for photosynthesis (Coleman et al., 1995a). Ozone also induces premature leaf abscission (Keller, 1988; Karnosky et al., 1996) and can affect leaf size (Oksanen et al., 2001) and carbon allocation to roots (Coleman et al., 1995b; Coleman et al., 1996). Ozone has also been implicated in weakening trees such that they succumb to insect (Cobb and Stark, 1970; Percy et al., 2002) or disease (Karnosky et al., 2002) attacks. For agricultural crops in the US, it is estimated that 90% of the crop loss caused by air pollution is the result of O_3, either alone or in combination with other pollutants (Heck et al., 1982). We suspect a similar statement could be made for O_3 and forest trees in North America (McLaughlin and Percy, 1999).

The impacts of O_3 on forest tree populations have been studied in considerable detail. One of the first such problems to be diagnosed was the oxidant damage to ponderosa pine (*Pinus ponderosa*) over large areas of the San Bernardino Mountains in southern California (Miller et al., 1963). Community changes related to natural succession caused by interspecific variability in response to oxidants were initially described in this region by Miller (1973). He noted that mixed forests of ponderosa pine, sugar pine (*Pinus lambertiana*) and white fir (*Abies concolor*) were changing to predominantly fir because of the greater sensitivity of the pines to oxidants. Similar results have been described for Jeffrey pine (*Pinus jeffreyi*) and ponderosa pine at several locations along the western slope of the Sierra Nevadas (Peterson et al., 1989; Miller et al., 1996; Kurpius et al., 2002).

More recently, similar O_3-induced population impacts have been noted in the mountain pine forests surrounding Mexico City where O_3 remains at exceedingly high levels (100–200 ppb peaks or more) throughout the year (Miller, 1993; Miller and Tejeda, 1994). In this area, *Pinus hartwegii* appears to be the most highly sensitive to O_3 and has been severely impacted since the 1970s (Hall et al., 1996) with widespread dieback and decline resulting in its replacement in an extensive forest area surrounding Mexico City.

In the eastern United States, O_3 has been linked to visible foliar injury and growth decrease (Dochinger and Seliskar, 1970), decreased reproduction (Benoit et al., 1983), and increased mortality rates (Karnosky, 1981) for eastern white pine (*Pinus strobus*). Since the responses of eastern white pine appear

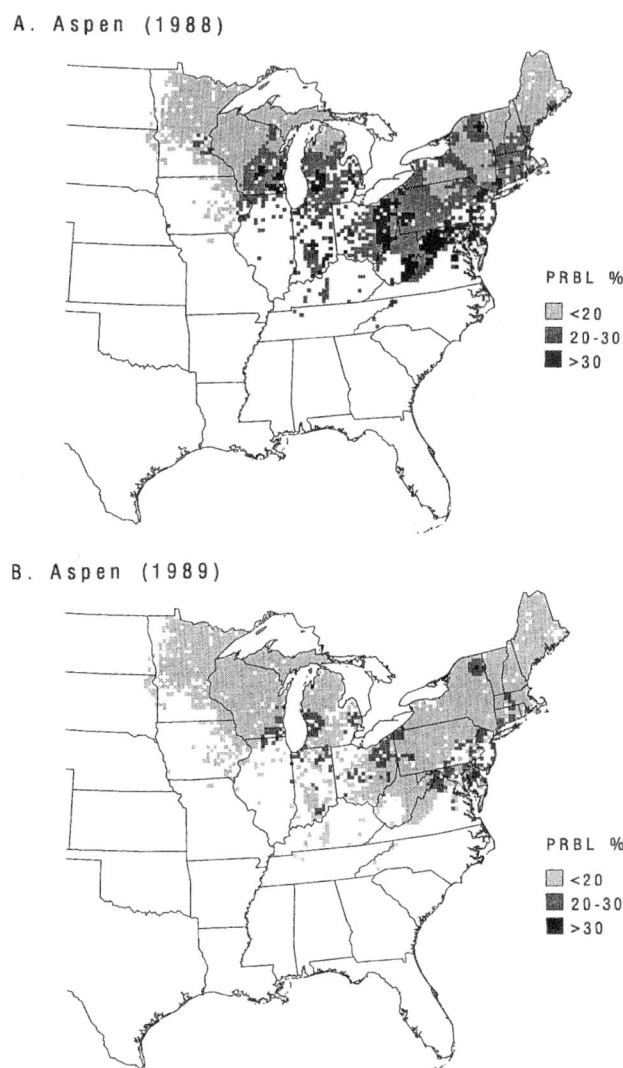

Figure 7. Variation in aspen seedling biomass loss with year-to-year exposure variation: 1988 (A) and 1989 (B) estimated exposures. PRBL calculated for each 20-km cell based on estimated ozone exposure value (three-month SUM06) and Weibull parameters for each species' response function (from Hogsett et al., 1997).

to be highly heritable, the components are in place for Phase I of natural selection, that is the elimination of sensitive genotypes. Since O_3 sensitive genotypes make up a relatively small portion of natural eastern white pine stands and the selection pressure conveyed by O_3 is rather low (Taylor and Pitelka,

1991), the question has been raised: "Does it really matter if we lose these sensitive genotypes?" Surely, this remains an openly debated and important research topic.

Similar responses to O_3 for sensitive genotypes of trembling aspen are expected. Evidence for population changes induced by O_3 in trembling aspen (*Populus tremuloides*) in the eastern United States are the studies by Berrang et al. (1986, 1989, 1991) which have shown a strong positive correlation between O_3 concentration at the population origin and the mean O_3 tolerance of the population. Populations from more heavily polluted areas tended to be more tolerant of O_3 than did populations from relatively pristine areas. As in eastern white pine, O_3 responses in aspen are highly heritable (Karnosky, 1977). Ozone has been shown to decrease aboveground biomass accumulation by 20 to 40% or more for sensitive genotypes (Wang et al., 1986; Karnosky et al., 1996) (Fig. 7) and 10 to 20% for more tolerant genotypes of aspen (Karnosky et al., 1996; Isebrands et al., 2001).

Ozone can also affect the relative abundance of understory vegetation in forests. Barbo et al. (1998) showed that O_3 exposures can cause shifts in the competitive interactions between plant species, thereby altering community structure. These understory plant interactions could also influence the ability of forest trees to naturally regenerate, grow and reproduce.

Not all tree species are susceptible to current levels of O_3. For example, Taylor (1994) has suggested that growth reductions for loblolly pine (*Pinus taeda*) in the southeastern US are not occurring at current O_3 levels. At the other extreme of O_3 effects is the Mexico City area where hundreds of thousands of pine (*Pinus* spp.) trees are dying due to prolonged exposures to very high levels of O_3 (Hall et al., 1996). For the eastern United States, Chappelka and Samuelson (1998) estimate that growth losses average about 0 to 10%. Worldwide, Fowler et al. (1999) estimate that some 24% of the world's forests are currently exposed to damaging concentrations and that this number will increase to 50% of the world's forests by the year 2100.

There remain many unanswered questions about the effects of O_3 on forest trees. The reader is referred to Chapters 4–13, 22 and 23 in this book for additional research findings on O_3 effects on forest trees.

5. Global warming

The global average surface temperature has increased since 1861. Over the 20th century the increase has been $0.6 \pm 0.2\,°C$ (Fig. 3). The 1990s was the warmest decade and 1998 was the warmest year on record (IPCC, 2001). Furthermore, the average nighttime temperature is increasing about $0.2\,°C$ per decade, twice as fast as daytime temperature increases.

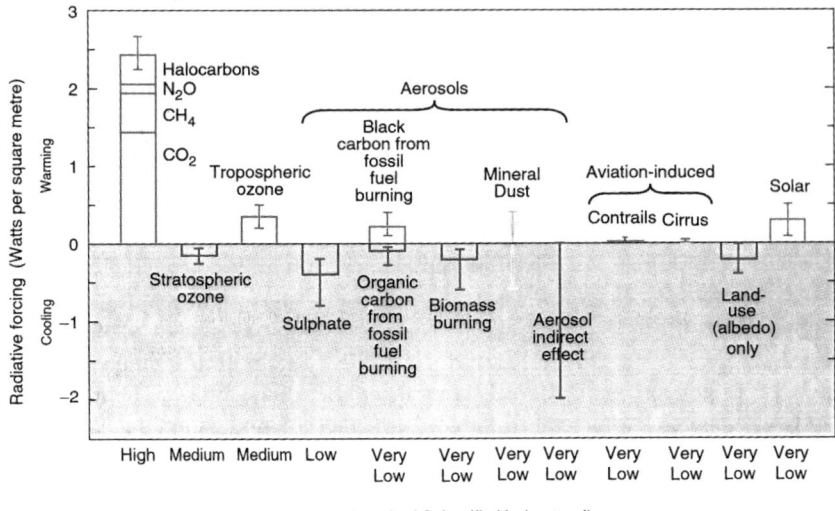

The global mean radiative forcing of the climate system for the year 2000, relative to 1750

Figure 8. The global mean radiative forcing of the climate system for the year 2000, relative to 1750 (from IPCC, 2001).

While there is considerable debate as to the cause of the global warming, the bulk of the scientific community has concluded that the causes of increasing warming trends are dominated by anthropogenic forcing of the global energy balance, with a smaller contribution due to natural variability. Analysis of the global mean radiative forcing of the climate system (Fig. 8) suggests that anthropogenic greenhouse gases (CO_2, CH_4, N_2O, halocarbons, and O_3) are largely responsible for global warming by trapping radiative heat near the Earth's surface (IPCC, 2001).

5.1. Future trends

Emissions of CO_2 due to fossil fuel burning are virtually certain to be the dominant influence in increasing atmospheric CO_2 during the 21st century (Stott et al., 2000; IPCC, 2001). Despite the Kyoto protocol in which countries have pledged to cut back CO_2 emissions, CO_2 emissions continue to rise worldwide. Efforts to reduce CO_2 from the atmosphere have included strategies for tree planting to sequester carbon (Sedjo, 1989; Sampson, 1992). While little is known about how effectively trees will sequester carbon under elevated temperatures and with increasing CO_2 (Karnosky et al., 2001), the process of set-

Ozone Exceedance Days vs. Number of Days Over 90 Degrees F°

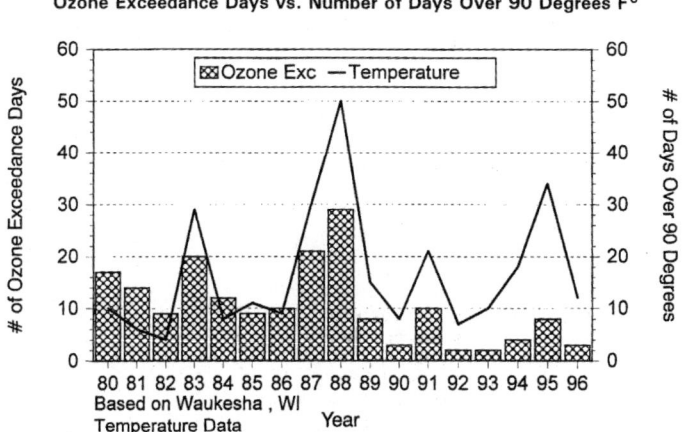

Figure 9. Dependence of ozone exceedance days on temperature from 1980 to 1996 for Waukesha, Wisconsin (from Wisconsin Department of Natural Resources, 1997).

ting up methods to monitor, evaluate, report, verify and certify forestry projects for climate change mitigation is moving forward (Vine et al., 2001) and tree planting is being counted on as one of the key methods to stabilize atmospheric CO_2 concentrations (Swart et al., 2002).

Increases in tropospheric O_3 are believed to have caused a warming effect which is about 15 to 20% of that due to CO_2 and other greenhouse gases (Hauglustaine et al., 1994; Kiehl et al., 1999; Berntsen et al., 2000; Shine, 2001). Tropospheric ozone is also expected to continue to increase through the 21st century (Stevenson et al., 1998; Brasseur et al., 1998; IPCC, 2001). The complicated interrelatedness of increasing global temperature and ozone suggests that tropospheric ozone will increase under global warming for two reasons. First, the reactions in tropospheric O_3 formation (Fig. 6) are enhanced as temperatures increase and there is a well known link of elevated tropospheric O_3 and high temperatures (Fig. 9). Second, as temperatures increase, trees and other plants emit larger amounts of volatile organic compounds (VOC) (Monson et al., 1995) which rapidly react with hydroxyl radical (OH), O_3, and nitrate (NO_x). Such reactions can, among other things, enhance O_3 (Fuentes et al., 2001). Currently, estimated global VOC emissions total 1150 $Tg\,yr^{-1}$, or about an order of magnitude greater than VOC emissions from anthropogenic sources (Guenther et al., 1995; Komenda et al., 2001). Modelled estimates of VOC emission increases in the next century under warming temperatures suggest VOC emissions could double (Constable et al., 1999), although research is needed to more accurately

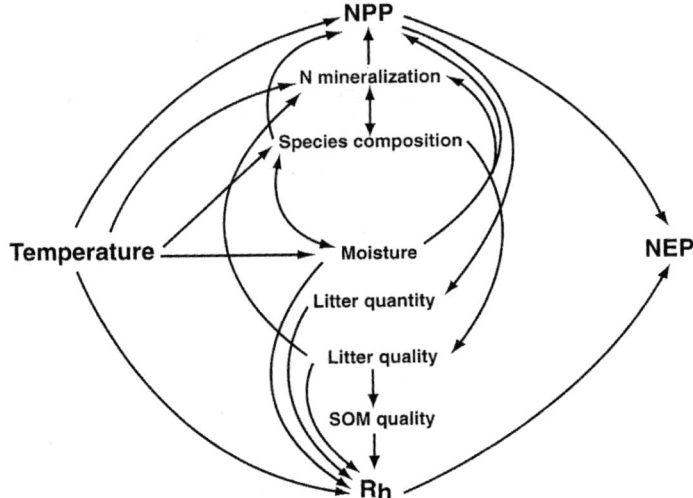

Figure 10. Direct and indirect effects of temperature on net primary production (NPP), heterotrophic respiration (R_h), and net ecosystem production (NEP). Soil organic matter is labeled as SOM (from Shaver et al., 2000).

ascertain how environmental changes will affect VOC emissions (Fuentes et al., 2001).

5.2. Global warming and forests

Temperature affects virtually all chemical and biological processes in plants so it is likely the effects of global warming will be dramatic and complex for forest ecosystems (Melillo et al., 1993; Shaver et al., 2000). For example, the complexity of temperature effects can be seen in an examination of the carbon budget under elevated temperature (Fig. 10) where temperature is seen affecting rates of N mineralization, soil moisture content and precipitation, and measures of growth (NPP) and heterotrophic respiration (R_h). The balance between these two processes determines net ecosystem production (NEP) which is yet difficult to accurately predict (McNulty et al., 1996; Nabuurs et al., 2002) resulting in model projections suggesting that global forests could be carbon sinks or sources in the future (Dixon et al., 1994).

While we know very little about net ecosystem production under global warming, it is generally believed that soil respiration increases exponentially with an increase in soil temperature (Raich and Schlesinger, 1992; Atkin et al., 2000). This is particularly significant as we realize that the 3 m of soil are estimated to contain 2344 Pg of organic carbon (C), which is known

The Global Carbon Cycle

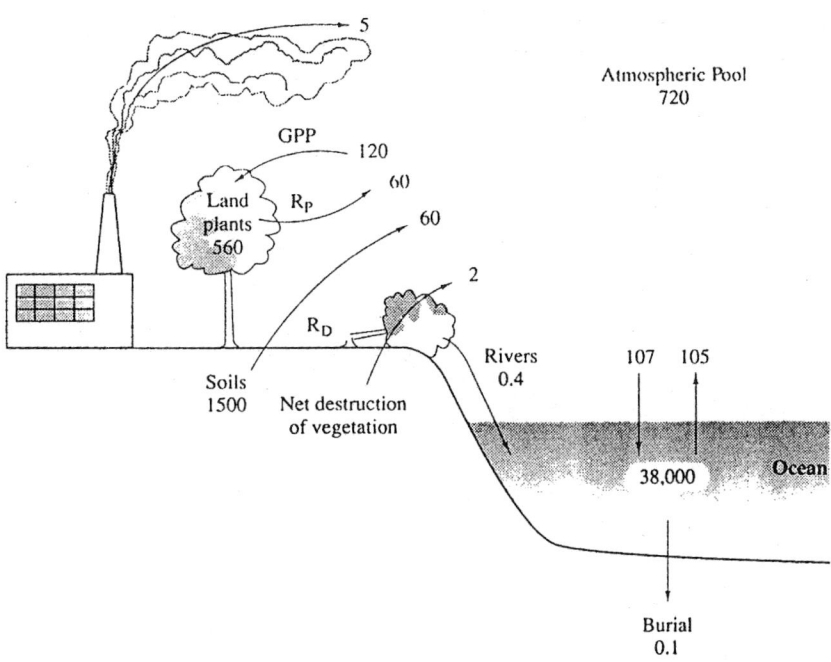

Figure 11. The global carbon cycle. All pools are expressed in units of 10^{15} gC and all fluxes in units of 10^{15} gC/yr (from Schlesinger, 1997).

to interact with the atmosphere (Jobbagy and Jackson, 2000). Considering the importance of forest soils in storing C (Fig. 11), the potential for soil carbon to be respired at unprecedented rates into the atmosphere presents another potential very large source of CO_2 in the atmosphere. Particularly vulnerable to increased soil warming are the forest soils in boreal and arctic regions which represent 20 to 60% of the global soil carbon pool and where low temperatures and permafrost currently limit decomposition (Hobbie et al., 2000). Rates of fine root turnover are also strongly temperature dependent so that there is a great deal of uncertainty as to how root systems will respond to global warming (Pregitzer et al., 2000).

Among the most certain changes predicted to occur under global warming in forest ecosystems are for species composition and ranges to be altered (Pitelka, 1997; Iverson and Prasad, 1998; Bakkenes et al., 2002). Examination of past range changes provides insights into the potential for range changes to occur under the rapid global warming (Davis et al., 1986; Fig. 12). While trees have considerable genetic diversity and capacity to evolve in the face

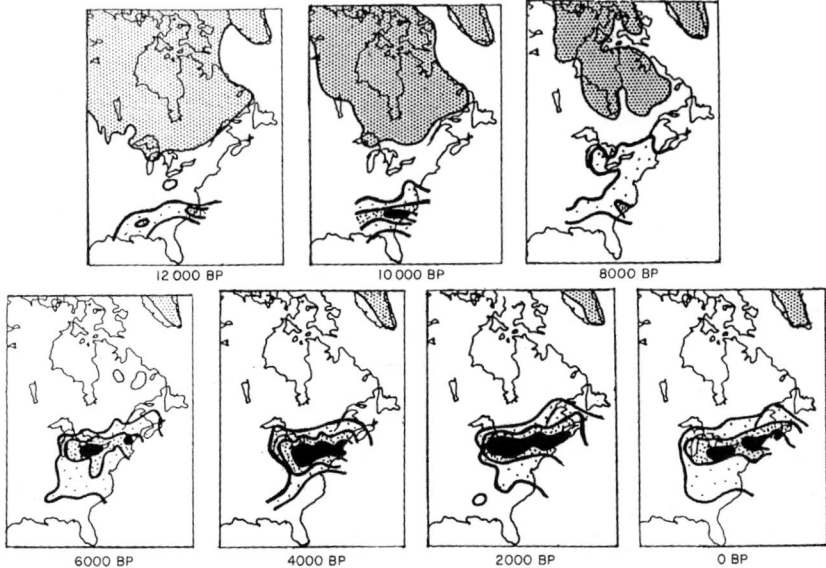

Figure 12.   Isopoll maps of American beech (*Fagus grandifolia*) in the eastern USA from 12000
BP to the present day (isopolls at 1, 5, and 10%) (from Davis et al., 1986).

of changing climate (Bradshaw and McNeilly, 1991), it is likely that major
changes in ranges will take place (Pitelka, 1997). Particularly pronounced
changes will likely occur in boreal regions and boreal treeline species that may
be among the most vulnerable (MacDonald et al., 1993; Makinen et al., 2000;
Lloyd and Fastie, 2002) where trees are currently growing at the cold mar-
gins of the forest. In addition to direct responses to temperature changes,
climate change can affect forests by altering the frequency, intensity, duration and timing of fire, drought, introduced species, insect and pathogen outbreaks, hurricanes, wind storms, ice storms, frost occurrence and landslides
(Dale et al., 2001). The rapid nature of today's climate change challenges trees
adaptive capability to migrate (Davis and Shaw, 2001). This rapid rate of climate change, coupled with land use changes such as habitat fragmentation
by human development that impede gene flow, can be expected to disrupt
the interplay of adaptation and migration, likely affecting forest productivity and threatening the persistence of many species (Davis and Shaw, 2001;
Rehfeldt et al., 2002).

Another major change related to global warming is the noticeable change of
the phenology of bud break and bud set. Earlier dates of average spring bud
break and later dates of fall bud set have been detected resulting in a longer
growing season in forest trees around the world (Menzel and Fabian, 1999;

Menzel, 2000; Penuelas and Filella, 2001; Parmesan and Yohe, 2003). This lengthening of the growing season may have already contributed to increased biomass accumulation (Menzel and Fabian, 1999) and is likely affecting insect and animal phenology and bird migration patterns (Penuelas and Filella, 2001).

Forest insect and disease pests are likely to be also changing as a result of global warming (Cannon, 1998; Harrington et al., 1999; Chakraborty et al., 2000; Volney and Fleming, 2000; Bale et al., 2002). For example, the change in forest tree phenology described above must be met with changes in insect phenology or they will hatch out at a time when there is no foliage at a proper stage for feeding. The effects of asynchrony of insect egg hatch and budbreak was seen over large parts of the northern Great Lakes region in the spring of 2002 as a late spring due to cold temperatures resulted in delayed aspen bud break, after the majority of forest tent caterpillars had hatched. This resulted in a high mortality rate in the otherwise peak cycle forest tent caterpillars (Mattson, personal communication).

The occurrence and abundance of various insects and disease pests are generally predicted to increase under global warming (Chakraborty et al., 2000; Bale et al., 2002). Especially worrisome is the possibility of major forest pests moving northward into temperate forests that were growing in areas with winter temperatures limiting southern forest pests. For example, the pine wood nematode has generally been a major nuisance only in subtropical or southern temperature forests (Suzuki, 1999). This serious pest of pines will likely become a major pest problem in the prime northern pine species such as loblolly pine in the US and Scots pine in Europe. Furthermore, non-indigenous species are likely to be particularly opportunistic under global warming (Cannon, 1998).

The availability of water resources of forests under global warming is likely to become an ever-increasing concern as droughts are predicted to be more common in many parts of the world where forests are already facing common moisture stress problems (Hanson and Weltzin, 2000). This could be a critically important factor, more important than temperature changes, for arid and semi-arid tropical forests that are just marginally alive (Desanker and Justice, 2001; Hulme et al., 2001). See Chapter 3 for more on global warming.

6. Sulfur and nitrogen oxides and acidic deposition

The burning of fossil fuels, particularly coal, has been responsible for deposition of sulfur and nitrogen to forest ecosystems around the world. These pollutants can be transported long distances from tall smokestacks and they can be deposited either in precipitation or in particulate form. The principal forms of impact are acidification, caused by both sulfur and nitrogen, and eutrophication of lakes and streams, caused by nitrogen (Hirst et al., 2000).

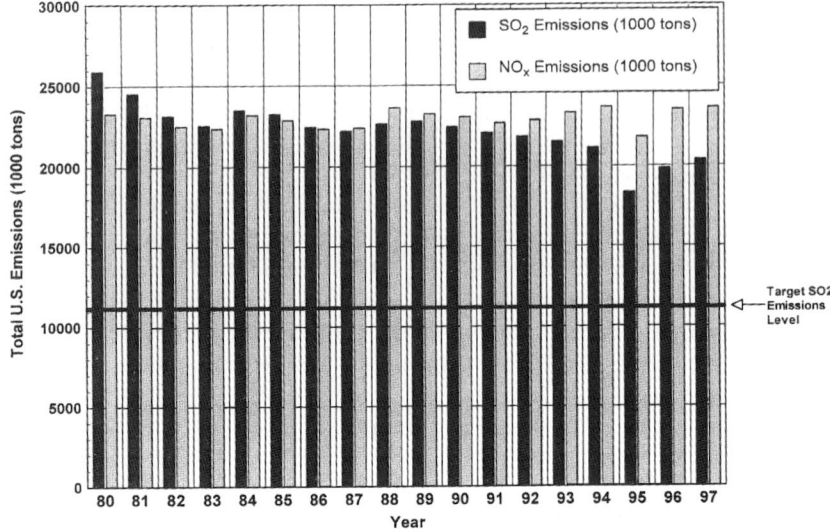

Figure 13. SO$_2$ and NO$_x$ emissions in the United States from 1980 through 1997. The target SO$_2$ emissions were based on the 1980 emission levels (from Lynch et al., 2000).

Forest dieback due primarily to the burning of soft coal, which is particularly high in sulfur content, has occurred over some 2.8 million ha in Europe, primarily in the region of the "Black Triangle" near the common borders of the Czech Republic, Germany, and Poland (Percy, 2002). The International Cooperative Programme on Assessment and Monitoring of Air Pollutant Effects on Forests (ICP Forests) has identified defoliation rates of 39.7% in Poland and 71.9% in the Czech Republic (EC/PHARE, 1999).

While pollution control legislation in the United States (Furiness et al., 1998; Lynch et al., 2000) and Europe (Erisman et al., 1998; Alewell et al., 2000) have produced reductions in sulfur emissions (Figs. 13 and 14) and deposition (Fig. 15), nitrogen emissions have continued to rise (Fowler et al., 1999; Lynch et al., 2000; Galloway and Cowling, 2002; Galloway et al., 2002). Considerable areas of forests in Europe still have nitrogen and sulfur deposition above levels referred to as critical loads (the deposition a natural area can stand without damage) (Hirst et al., 2000). The impacts of acidification on forests are numerous including: soil acidification, leaching of nutrients from foliage and soils, volatilization of ammonia from the soil, mobilization of toxic minerals such as Al from soils, and alteration of fine root turnover, frost hardiness, mycorrhizal fungi associations and foliage retention (Aber et al., 1989; Jeffries and Maron, 1997). Calcium nitration resulting in membrane destabilization has also been associated with decline of spruce trees due to acid rain

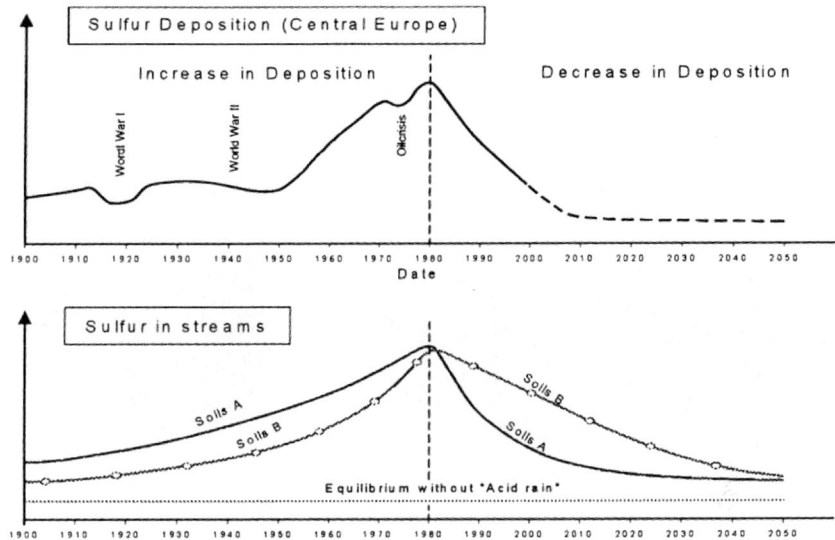

Figure 14.   Trends in sulfur deposition (top) and sulfur in streams (bottom) in Europe from 1900 to 2050 (from Alewell, 2001).

(DeHayes et al., 1999). Forests in which nitrogen deposition is no longer pro-viding a net fertilization effect are referred to as nitrogen saturated. The classic example of this continues to be the forests in the San Bernardino Mountains in the Los Angeles air basin which have received high impacts of nitrogen depo-sition for the past 60 years or more (Bytnerowicz and Fenn, 1996; Bytnerowicz et al. 2002a, 2002b).

Fowler et al. (1999) predicts a six-fold increase in the area of global forest at risk from acidification between 1985 and 2050, with the majority of the increase being from subtropical and tropical forest regions. The majority of these increases will come from developing countries in Asia (Figs. 16 and 17) (Arndt et al., 1997; Lefohn et al., 1999; Streets et al., 2000). Huge increases in both nitrogen and sulfur emissions are predicted for the rapidly industrializing countries of China (Streets and Waldhoff, 2000; Vallack et al., 2001) and India (Parshar et al., 1998). See Chapters 14–18 for more on these pollutants.

7. Other air pollutants

In this short review, we have not attempted to comprehensively describe all air pollutants that impact forests. While the largest acreages worldwide im-pacted by air pollution are those affected by O_3 and acidic deposition, there

SULFATE WET DEPOSITION

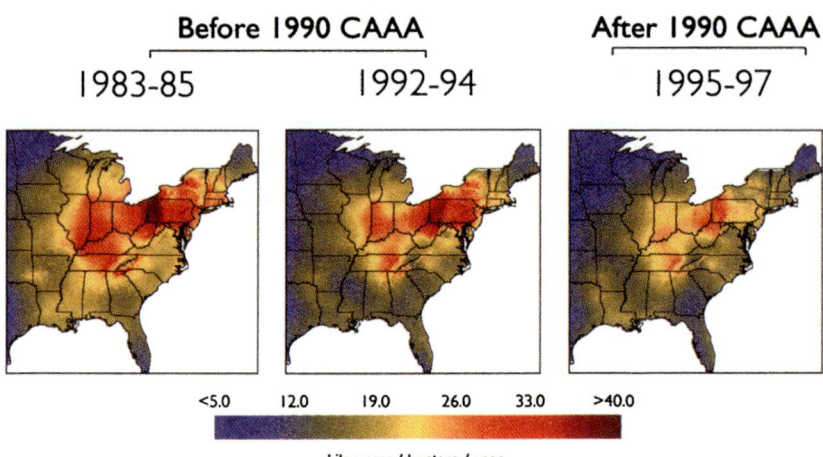

NITRATE WET DEPOSITION

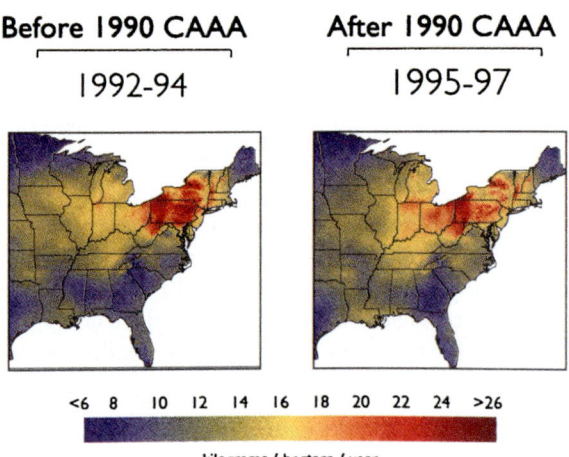

Figure 15.   Recent patterns of wet deposition before and after the implementation of the 1990 Clean Air Act Amendments (CAAA) (from Driscoll et al., 2001).

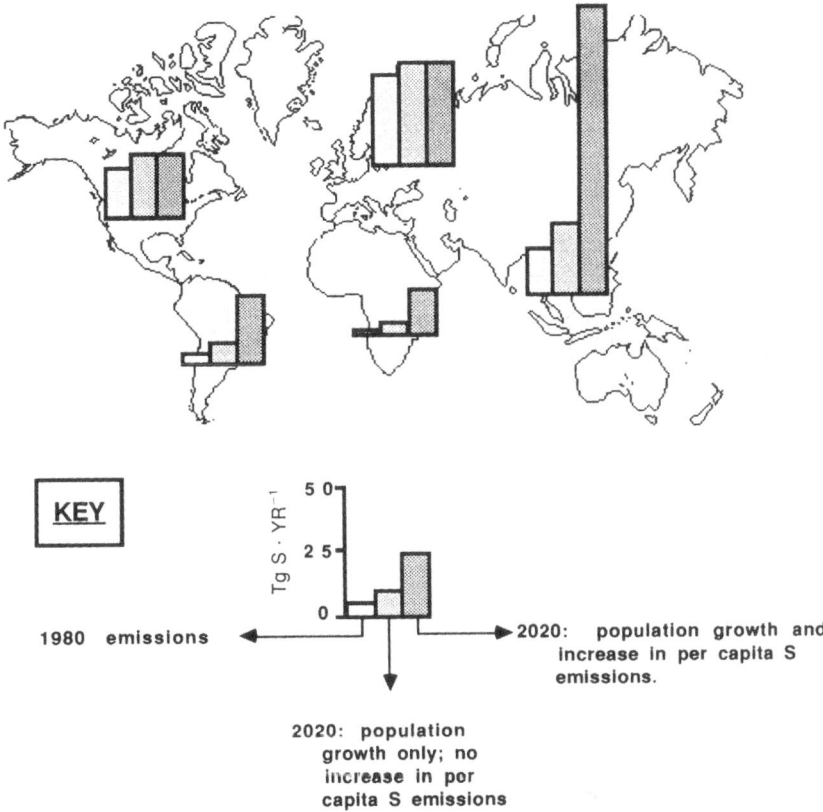

Figure 16. The emission of sulfur(s) to the atmosphere in 1980 compared to 2020 with no change and with a change in per capita energy consumption (from Galloway, 1989).

are large areas of forests worldwide that are impacted by particulate pollution (Kretzschmar, 1994; Edgerton et al., 1999), heavy metals (Mankovska, 1997a, 1997b; Straszewski et al., 2001), gaseous sulfur dioxide (Arndt et al., 1997; Streets et al., 2000; Vallack et al., 2001) and various other photochemical oxidants (Parshar et al., 1998; Streets and Waldhoff, 2000; Vallack et al., 2001; Derwent et al., 2002). With the exception of the regional photochemical oxidant problems, these are generally point-source pollutants around major factories or power plants. While these were very common in the early industrialization periods of western countries, they have largely been cleaned up. However, they are still significant problems in developing countries (Arndt et al., 1997; Vallack et al., 2001; Gupta et al., 2002; Skiba and Davydova-Belitskaya, 2002)

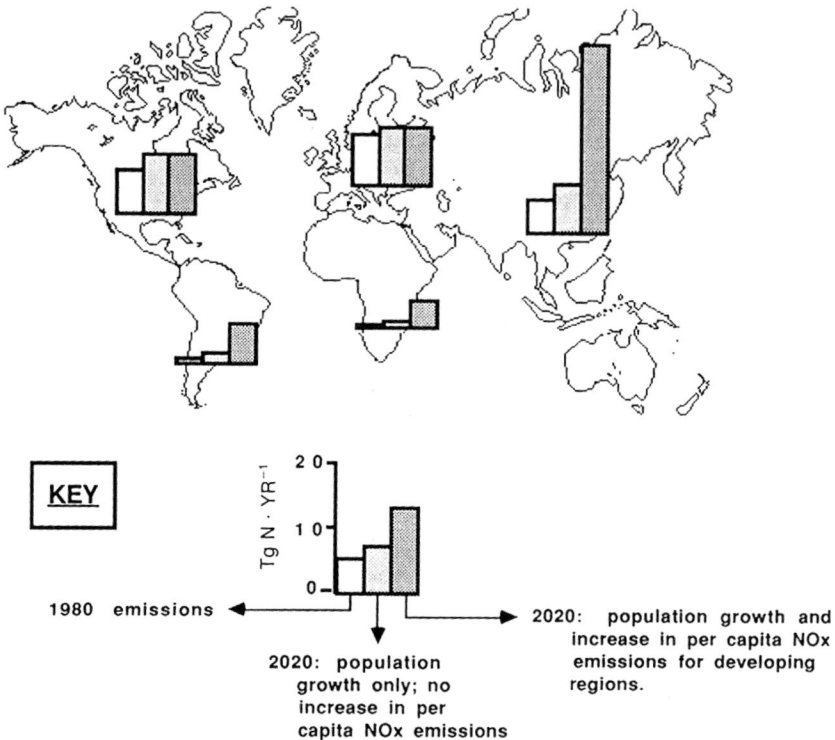

Figure 17. The emission of nitrogen (N) to the atmosphere in 1980 compared to 2020 with no change and with a change in per capita energy consumption (from Galloway, 1989).

and in countries in transition from the communist influence (Mankovska, 1997a, 1997b; Straszewski et al., 2001).

8. Pollutant interactions

Although we have discussed the various air pollutants individually, they seldom occur as individual pollutants (Fig. 6). More often, forests are faced with interacting pollutants which may counteract or exacerbate one another (Krupa and Kickert, 1989; Kickert and Krupa, 1990; Isebrands et al., 2000; Karnosky et al., 2001). For example, both elevated CO_2 and low levels of nitrogen deposition generally have stimulatory effects on forest tree growth and reproduction while excess nitrogen, sulfur oxides, O_3 and other air pollutants generally negatively impact forest ecosystems. Thus, predicting outcomes of multiple pollutant interactions for forest ecosystems is difficult, es-

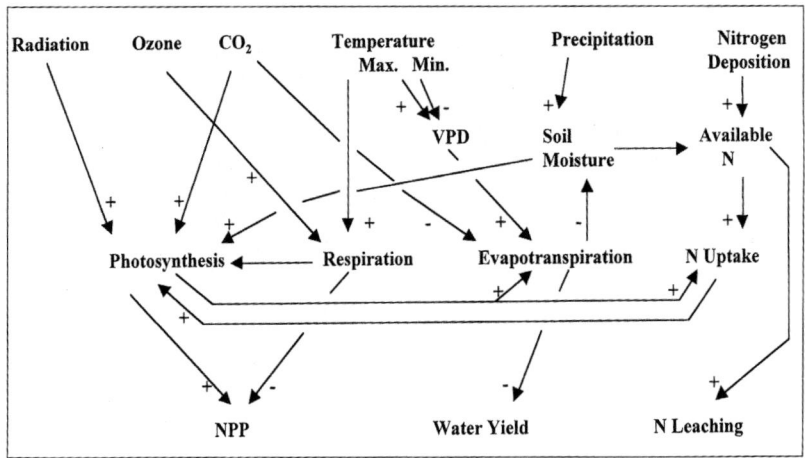

Figure 18. Interactions among environmental factors that are subject to change through human activities, and major processes affecting carbon, water, and nitrogen dynamics in forest ecosystem. (+) Indicates an enhancement, and (−) a suppression, of the receptor process (from Aber et al., 2001).

pecially since these interactions are also affected by other environmental variables such as temperature, relative humidity, and soil moisture (Fig. 18), landuse (Caspersen et al., 2000), and they can also be affected by competitive environment of the forest (Fig. 19). Also, far less research has been done on interacting pollutants than on single pollutants.

In what turned out to be a classic pollutant interaction, many eastern white pine trees in the Ohio River Valley began showing symptoms such as tip-burned and shortened needles, poor needle retention, and stunted growth in the 1960s. The cause was later found to be a synergistic interaction of moderately elevated levels of O_3 and SO_2 (Dochinger et al., 1970; Costonis, 1970).

Probably the most studied interaction with forest trees is that of elevated CO_2 and O_3. Since these two pollutants are increasing in the troposphere at about the same rate (IPCC, 2001), this interaction will impact large areas of future forests. Fowler et al. (1999) estimate nearly 50% of the world's forests will be exposed to O_3 concentrations greater than 60 ppb by the year 2100. Atmospheric CO_2 levels are expected to be doubled over current levels by then (Stott et al., 2000; IPCC, 2001).

This is a complex interaction that is affected by concentrations of the two interacting pollutants, the species and genotypes involved as considerable genetic variation in responses to both of these pollutants occurs, and it can also be affected by other stresses such as competition (McDonald et al., 2002), drought stress or nitrogen additions (Karnosky et al., 1992). Thus, it is not surprising

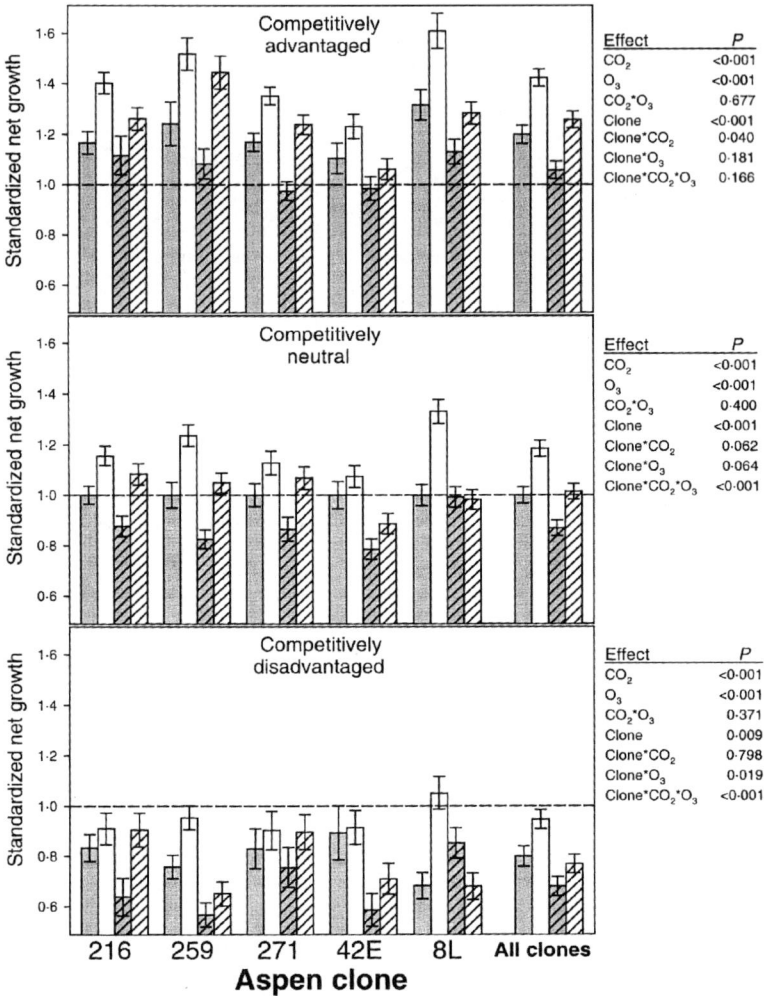

Figure 19. Standardized net growth (SNG) responses averaged during the 1998–2001 period for mixed-clone aspen stands exposed to combinations of ambient and elevated CO_2 and O_3. Bars represent least-squares mean estimates (LS means) ± 1 SE for individual clones, with the average response across clones identified as 'All clones'. Shaded bars, ambient CO_2 treatments; unshaded bars, elevated CO_2 treatments; open bars, ambient O_3 treatments; hatched bars, elevated O_3 treatments. The competition status indices (CSI) for this analysis were means of annual CSI values during the 4-year period, with competitively advantaged (+) and disadvantaged (−) LS means calculated at ± 90 cm values of the CSI covariate. The dashed horizontal lines denote SNG response in competitively 'neutral' (CSI = 0), ambient conditions, for reference. Analysis of covariance (ANCOVA) results for fixed effects of atmospheric treatments, clone and their interactions under competitively advantaged, neutral and disadvantaged conditions are reported next to each panel (from McDonald et al., 2002).

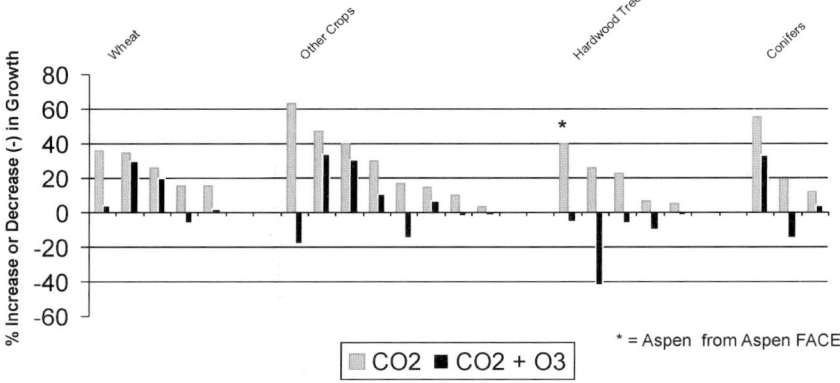

Figure 20. Relative effects of controlled exposure to elevated CO_2 on normalized plant growth under CO_2 alone (striped bars; 500–713 μmol mol^{-1} CO_2) and elevated CO_2 plus ozone (dotted bars) (Modified and expanded from Barnes and Wellburn, 1998). Data presented for wheat (*Triticum aestivum*) taken from Barnes et al. (1995); Rudorff et al. (1996); McKee et al. (1997); Bender et al. (1999) and Hudak et al. (1999); other crops including soybean (*Glycine max*) taken from Heagle et al. (1998) and Miller et al. (1998), tomato (*Lycopersicon esculentum*) taken from Olszyk and Wise (1997) and Hao et al. (2000), rice (*Oryza sativa*) taken from Olszyk and Wise (1997), potato (*Solanum tuberosum*) taken from Donnelly et al. (2001) and Lawson et al. (2001), and corn (*Zea mays*) taken from Rudorff et al. (1996); hardwood trees including hybrid poplars (*Populus hybrids*) taken from Dickson et al. (1998), trembling aspen (*Populus tremuloides*) taken from Volin and Reich (1996), Volin et al. (1998), and Isebrands et al. (2001), and oak (*Quercus petrea*) taken from Broadmeadow and Jackson (2000); and conifers including ponderosa pine taken from David Olszyk (personal communication) and Scots pine (*Pinus sylvestris*) taken from Broadmeadow and Jackson (2000) and Utriainen et al. (2000). Each pair of bars represents one species (Karnosky et al., 2003).

that some conflicting results have been found. For example, with trembling aspen, Volin and Reich (1996) and Volin et al. (1998) suggest that elevated CO_2 ameliorates the effects of O_3 on photosynthesis and growth while Kull et al. (1996), Isebrands et al. (2001), Wustman et al. (2001), McDonald et al. (2002), and Mankovska et al. (2003) suggest that CO_2 does not ameliorate and in some cases it exacerbates the negative effects of O_3. The results of numerous CO_2/O_3 interaction studies that have examined growth or biomass production are shown in Fig. 20. See Chapters 19–23 for more on pollutant interactions.

9. Management of genetic resources for future forests

While there has been a great deal of research on the impacts of air pollution and climate change, there has not yet been much research on the possible consequences of these effects on managed forests (Lindner, 2000). One area particularly important in future managed forests is the genetic makeup of these forests.

Tree breeding and genetic selection has generally involved either plus tree selection followed by progeny testing or provenance testing followed by progeny testing of superior phenotypes. Then, seed orchards have been established and rogued to provide the seed for the next generation. This process has continued with advanced generation selection and breeding in a few commercially important tree species. In all facets of these programs, selection is done based on the conditions prior to selection and for the most part these selections are not done on the basis of predicted pollution and climate scenarios that will be in place during the rotation of the commercial forest. Screening and selection of genotypes suitable for future pollution and climate scenarios is generally thought to be nearly impossible because of the complexity and cost of such programs. Thus, an alternative strategy in which a wider genetic base is maintained in our breeding population is essential for developing future forests (Namkoong, 1991). Maintaining large amounts of genetic diversity will increase the probability that adequate adaptability is maintained to meet rapidly changing environmental conditions (Gregorius, 1986; Müller-Starck, 1989; Koski, 1996; Rehfeldt et al., 1999). Planting a diverse array of species, seed sources, or families, is a hedge against the uncertainty inherent in current projections of warming (Ledig and Kitzmiller, 1992).

Alternative strategies are also needed to insure that gene banks, clone banks, seed zones, seed collection areas and other in-situ conservation strategies are maintained in multiplicative manner such that the changing pollution and/or climate scenarios will not result in the loss of such collections from single vulnerable test sites (Martin, 1996; Hannah et al., 2002). Given the past several decades of "laissez-faire" attitude towards traditional genetic field trials and field conservation efforts, this need to conserve forest genetic resources in multiple amounts may help genetics regain prominence amongst the forestry community.

10. Conclusions and knowledge gaps

10.1. Air monitoring

- The global distribution of air pollution is just being elucidated. An important research need is to develop ways to sort out the relative global versus national contributions to important air pollutants such as NO_x and O_3.
- The actual concentrations of important pollutants, such as NO_x, SO_2, and O_3, in forested areas of the world continue to be largely unknown. Devel-

opment and strategic deployment of relatively inexpensive passive monitors offer many opportunities to address these questions in forested areas.

10.2. Increasing CO_2 and global warming

- CO_2 continues to increase in the atmosphere. The effects of elevated CO_2 on long-term forest productivity, ecosystem stability, and genetic diversity of forests are largely unknown.
- Global warming is now indisputedly taking place at unprecedented rates. The rate of change in species composition and function is largely speculation at best and needs solid research attention. The possible expansion of insect and disease pests under warming climates and less harsh winters has important forest ecosystem implications and is critically important to study. Finally, the usefulness of tree planting and improved forest management for increasing C sequestration in forest ecosystems is largely unknown. Thus, research on carbon budgets and CO_2 mitigation potential of various forestry practices are important research areas to address.

10.3. S and N oxides and acidification

- While great improvements in acidic deposition have been made in several western countries, S and N deposition are increasing in many developing countries and the effects of these problems are not well understood.

10.4. Other air pollutants (particulates, heavy metals, etc.)

- The forests of many of the former Eastern Bloc countries suffered from deposition as particulates and heavy metals, as well as from acidic deposition. The restoration of forest ecosystem degraded by air pollution, particularly in countries in transition from a Communist government, remains a very high priority for much of Eastern Europe.

10.5. Pollutant interactions

- While local large-scale deposition of single pollutants can often cause local damage to forest ecosystems around pollution point sources, most modern air pollution problems are more regional in distribution. Thus, complex air pollution interactions with other air pollutants, with environmental variables such as temperature, relative humidity, light, and soil moisture, and with other stresses such as drought, low fertility, or insect or disease pests often occur. Without exception, pollutant interactions have not been studied to the same detail as have single pollutants. Furthermore, modellers have largely ignored pollutant interactions in their global projections of net ecosystem carbon budgets.

10.6. Managing future forests

- Our future forests will continue to be exposed to unprecedented global environmental changes. Thus, it is critically important that the case for reforestation programs include high genetic diversity. Studies comparing the effects of global change on genetically diverse versus genetically disparate communities are critically needed. Deployment strategies for maximizing forest productivity, while optimizing genetic diversity, are needed for the forest products industry as well as for forests designed to sequester carbon.

Acknowledgements

This research was partially supported by the US Department of Energy, Office of Science (BER) (DE-FG02-95ER62125), the USDA Forest Service Northern Global Change Program, Michigan Technological University, and the USDA Forest Service North Central Research Station. The authors appreciate the valuable comments and suggestions made on the manuscript by Drs. Art Chappelka and Douglas Findley.

References

Aber, J.D., Nadelhoffer, K.J., Steudler, P., Melillo, J.M., 1989. Nitrogen saturation in northern forest ecosystems. Bioscience 39, 378–386.

Aber, J., Neilson, R.P., McNulty, S., Lenihan, J.M., Bachelet, D., Drapek, R.J., 2001. Forest processes and global environmental change: Predicting the effects of individual and multiple stressors. Bioscience 51, 735–751.

Alewell, C., 2001. Predicting reversibility of acidification: The European sulfur story. Water Air Soil Pollut. 130, 1271–1276.

Alewell, C., Manderscheid, B., Gerstberger, P., Matzner, E., 2000. Effects of reduced atmospheric deposition on soil solution chemistry and elemental contents of spruce needles in NE-Bavaria, Germany. J. Plant Nutr. Soil Sci. 163, 509–516.

Arndt, R.L., Carmichael, G.R., Streets, D.G., Bhatti, N., 1997. Sulfur dioxide emissions and sectorial contributions to sulfur deposition in Asia. Atmos. Environ. 31, 1553–1572.

Atkin, O.K., Edwards, E.J., Loveys, B.R., 2000. Response of root respiration to changes in temperature and to relevance to global warming. New Phytol. 147, 141–154.

Aunan, K., Bernstsen, T.K., Seip, H.M., 2000. Surface ozone in China and its possible impact on agricultural crop yields. Ambio 29, 294–301.

Bakkenes, M., Alkemade, J.R.M., Ihle, F., Leemans, R., LaTour, J.B., 2002. Assessing effects of forecasted climate change on the diversity and distribution of European higher plants for 2050. Global Change Biol. 8, 390–407.

Bale, J.S., Masters, G.J., Hodkinson, I.D., Awmack, C., Bezemer, T.M., Brown, V.K., Butterfield, J., Buse, A., Coulson, J.C., Farrar, J., Good, J.E., Harrington, R., Hartley, S., Jones, T.H., Lindroth, R.L., Press, M.C., Symrnioudis, I., Watt, A.D., Whittaker, J.B., 2002. Herbivory in global climate change research: direct effects of rising temperature on insect herbivores. Global Change Biol. 8, 1–16.

Barbo, D.N., Chappelka, A.H., Somers, G.L., Miller-Goodman, M.S., Stolte, K., 1998. Diversity of an early successional plant community as influenced by ozone. New Phytol. 138, 653–662.

Barnes, J.D., Wellburn, A.R., 1998. Air pollutant combinations. In: DeKok, L.J., Stulen, I. (Eds.), Responses of plant metabolism to air pollution and global change. Backhuys Publishers, Leiden, The Netherlands, pp. 147–164.

Barnes, J.D., Ollerenshaw, J.H., Whitfield, C.P., 1995. Effects of elevated CO_2 and/or O_3 on growth, development and physiology of wheat (*Triticum aestivum* L.). Global Change Biol. 1, 129–142.

Bender, J., Herstein, U., Black, C.R., 1999. Growth and yield responses of spring wheat to increasing carbon dioxide, ozone and physiological stresses: A statistical analysis of 'ESPACE–wheat' results. Eur. J. Agron. 10, 185–195.

Benoit, L.F., Skelly, J.M., Moore, L.D., Dochinger, L.S., 1983. The influence of ozone on *Pinus strobus* L. pollen germination. Can. J. Forest Res. 13, 184–187.

Berntsen, T.K., Karlsdóttir, S., Jaffe, D.A., 1999. Influence of Asian emissions on the composition of air reaching the North Western United States. Geophys. Res. Lett. 26, 2171–2174.

Berntsen, T.K., Myhre, G., Stordal, F., Isaksen, I.S.A., 2000. Time evolution of tropospheric ozone and its radiative forcing. J. Geophys. Res. 105, 8915–8930.

Berrang, P.C., Karnosky, D.F., Bennett, J.P., 1989. Natural selection for ozone tolerance in *Populus tremuloides* II. Field verification. Can. J. Forest Res. 19, 519–522.

Berrang, P.C., Karnosky, D.F., Bennett, J.P., 1991. Natural selection for ozone tolerance in *Populus tremuloides*: An evaluation of nationwide trends. Can. J. Forest Res. 21, 1091–1097.

Berrang, P.C., Karnosky, D.F., Mickler, R.A., Bennett, J.P., 1986. Natural selection for ozone tolerance in *Populus tremuloides*. Can. J. Forest Res. 16, 1214–1216

Bradshaw, A.D., McNeilly, T., 1991. Evolutionary response to global climatic change. Ann. Bot. 67, 5–14.

Brasseur, G.P., Kiehl, J.T., Muller, J.-F., Schneider, T., Granier, C., Tie, X., Hauglustaine, D., 1998. Past and future changes in global tropospheric ozone: Impact on radiative forcing. Geophys. Res. Lett. 25, 3807–3810.

Broadmeadow, M.S., Jackson, S.B., 2000. Growth responses of *Quercus petraea, Fraxinus excelsior* and *Pinus sylvestris* to elevated carbon dioxide, ozone, and water supply. New Phytol. 146, 437–451.

Bytnerowicz, A., Fenn, M.E., 1996. Nitrogen deposition in California forests: A review. Environ. Pollut. 92, 127–146.

Bytnerowicz, A., Godzik, B., Fraczek, W., Grodzinska, K., Krywult, M., Badea, O., Barancok, P., Blum, O., Cerny, M., Godzik, S., Mankovska, B., Manning, W., Moravcik, P., Musselman, R., Oszlanyi, J., Postelnicu, D., Szdzuj, J., Varsavova, M., Zota, M., 2002a. Distribution of ozone and other air pollutants in forests of the Carpathian Mountains in central Europe. Environ. Pollut. 116, 3–25.

Bytnerowicz, A., Tausz, M., Alonso, R., Jones, D., Johnson, R., Grulke, N., 2002b. Summer-time distribution of air pollutants in Sequoia National Park, California. Environ. Pollut. 188, 187–203.

Cannon, R.J.C., 1998. The implications of predicted climate change for insect pests in the UK, with emphasis on non-indigenous species. Forestry 72, 237–247.

Caspersen, J.P., Pacala, S.W., Jenkins, J.C., Hurtt, G.C., Moorcroft, P.R., Birdsey, R.A., 2000. Contributions of land-use history to carbon accumulation in US forests. Science 290, 1148–1151.

Chakraborty, S., Tiedemann, A.V., Teng, P.S., 2000. Climate change: Potential impact on plant diseases. Environ. Pollut. 108, 317–326.

Chameides, W., Kasibhatla, P., Yienger, J., Levy II, H., 1994. Continental-scalemetro-ago-plexes, regional ozone pollution, and world food production. Science 264, 74–77.

Chappelka, A.H., Samuelson, L.J., 1998. Ambient ozone effects on forest trees of the eastern United States: A review. New Phytol. 139, 91–108.

Cheung, V.T.F., Wang, T., 2001. Observational study of ozone pollution at a rural site in the Yangtze Delta of China. Atmos. Environ. 35, 4947–4958.

Cobb Jr., F.W., Stark, R.W., 1970. Decline and mortality of smog-injured ponderosa pine. J. For. 68, 147–149.

Coleman, M.D., Dickson, R.E., Isebrands, J.G., Karnosky, D.F., 1995a. Photosynthetic productivity of aspen clones varying in sensitivity to tropospheric ozone. Tree Physiol. 15, 585–592.

Coleman, M.D., Dickson, R.E., Isebrands, J.G., Karnosky, D.F., 1995b. Carbon allocation and partitioning in aspen clones varying in sensitivity to tropospheric ozone. Tree Physiol. 15, 593–604.

Coleman, M.D., Dickson, R.E., Isebrands, J.G., Karnosky, D.F., 1996. Root growth and physiology of potted and field-grown trembling aspen exposed to tropospheric ozone. Tree Physiol. 16, 145–152.

Collins, W.J., Stevenson, D.S., Johnson, C.E., Derwent, D., 2000. The European regional ozone distribution and its link with the global scale for the years 1992 and 2015. Atmos. Environ. 34, 255–267.

Constable, J.V.H., Guentner, A.B., Schimel, D.S., Monson, R.K., 1999. Modelling changes in VOC emissions in response to climate change in the continental United States. Global Change Biol. 5, 791–806.

Costonis, A.C., 1970. Acute foliar injury of eastern white pine induced by sulfur dioxide and ozone. Phytopathology 69, 994–999.

Crutzen, P.J., 1988. Tropospheric ozone: An overview. In: Isaksen, I.S.A. (Ed.), Tropospheric Ozone. Reidel, Norwell, MA, pp. 3–32.

Dale, V.H., Joyce, L.A., McNulty, S., Neilson, R.P., Ayres, M.P., Flannigan, M.D., Hanson, P.J., Irland, I.C., Lugo, A.E., Peterson, C.J., Simberloff, D., Swanson, F.J., Stocks, B.J., Wotton, B.M., 2001. Climate change and forest disturbances. Bioscience 51, 723–734.

Davis, M.B., Woods, K., Webb, S.L., Futyama, R.P., 1986. Dispersal versus climate: Expansion of *Fagus* and *Tsuga* into the Upper Great Lakes region. Vegetatio 67, 93–103.

Davis, M.B., Shaw, R.G., 2001. Range shifts and adaptive responses to Quaternary climate change. Science 292, 673–679.

DeHayes, D.H., Schaberg, P.G., Hawley, G.J., Strimbeck, G.R., 1999. Acid rain impacts on calcium nutrition and forest health. Bioscience 49, 789–800.

DeLucia, E.H., Hamilton, J.G., Naidu, S.L., Thomas, R.B., Andrews, J.A., Finzi, A., Lavine, M., Matamala, R., Mohan, J.E., Hendrey, G.R., Schlesinger, W.H., 1999. Net primary production of a forest ecosystem with experimental CO_2 enrichment. Science 284, 1177–1179.

Derwent, R., Collins, W., Johnson, C., Stevenson, D., 2002. Global ozone concentration and regional air quality. Environ. Sci. Technol. 36, 379–382.

Desanker, P.V., Justice, C.O., 2001. Africa and global climate change: Critical issues and suggestions for further research and integrated assessment modelling. Clim. Res. 17, 93–103.

Dickson, R.E., Coleman, M.D., Riemenschneider, D.E., Isebrands, J.G., Hogan, G.D., Karnosky, D.F., 1998. Growth of five hybrid poplar genotypes exposed to interacting elevated CO_2 and O_3. Can. J. Forest Res. 28, 1706–1716.

Dixon, R.K., Brown, S., Houghton, R.A., Solomon, A.M., Trexler, M.C., Wisniewski, J., 1994. Carbon pools and flux of global forest ecosystems. Science 263, 185–190.

Dochinger, L.S., Bender, F.W., Fox, F.L., Heck, W.W., 1970. Chlorotic dwarf of eastern white pine caused by an ozone and sulphur dioxide interaction. Nature 225, 476.

Dochinger, L.S., Seliskar, C.E., 1970. Air pollution and the chlorotic dwarf disease of eastern white pine. Forest Sci. 16, 46–55.

Donnelly, A., Craigon, J., Black, C.R., Colls, J.J., Landon, G., 2001. Elevated CO_2 increases biomass and tuber yield in potato even at high ozone concentrations. New Phytol. 149, 265–274.

Driscoll, C.T., Lawrence, G.B., Bulger, A.J., Butler, T.J., Cronan, C.S., Egar, C., Lambert, K.F., Likens, G.E., Stoddard, J.L., Weathers, K.C., 2001. Acid Rain Revisited: Advances in Scientific Understanding Since the Passage of the 1970 and 1990 Clean Air Act Amendments. Hubbard Brook Research Foundation. Science Links Publication.

EC/PHARE, 1999. Conservation and sustainable management of forests in Central and Eastern European countries. PHARE Environmental Consortium.

Edgerton, S.A., Bian, X., Doran, J.C., Fast, J.D., Hubbe, J.M., Malone, E.L., Shaw, W.J., Whiteman, C.D., Zhong, S., Arriaga, J.L., Ortiz, E., Ruiz, M., Sosa, G., Vega, E., Limon, T., Guzman, F., Archuleta, J., Bossert, J.E., Elliot, S.M., Lee, J.T., McNair, L.A., Chow, J.C., Watson, J.G., Coulter, R.L., Doskey, P.V., Gaffney, J.S., Marley, N.A., Neff, W., Petty, R., 1999. Particulate air pollution in Mexico City: A collaborative research project. J. Air Waste Manag. Assoc. 49, 1221–1229.

Erisman, J.W., Mennen, M.G., Fowler, D., Flechard, C.R., Spindler, G., Gruner, A., Duyzer, J.H., Ruigrok, W., Wyers, G.P., 1998. Deposition monitoring in Europe. Environ. Monit. Assess. 53, 279–295.

FAO, 2001. State of the World's Forests 2001. The FAO Forestry Department, Rome, Italy.

Fowler, D., Flechard, C., Skiba, U., Coyle, M., Cape, J.N., 1998. The atmospheric budget of oxidized nitrogen and its role in ozone formation and deposition. New Phytol. 139, 11–23.

Fowler, D., Cape, J.N., Coyle, M., Flechard, C., Kuylenstierna, J., Hicks, K., Derwent, D., Johnson, C., Stevenson, D., 1999. The global exposure of forests to air pollutants. Water Air Soil Pollut. 116, 5–32.

Fuentes, J.D., Hayden, B.P., Garstang, M., Lerdau, M., Fitzjarrald, D., Baldocchi, D.D., Monson, R., Lamb, B., Geron, C., 2001. New directions: VOCs and biosphere–atmosphere feedbacks. Atmos. Environ. 35, 189–191.

Furiness, C., Smith, L., Ran, L., Cowling, E., 1998. Comparison of emissions of nitrogen and sulfur oxides to deposition of nitrate and sulfate in the USA by state in 1990. Environ. Pollut. 102 (S1), 313–320.

Galloway, J.N., 1989. Atmospheric acidification projections for the future. Ambio 18, 161–166.

Galloway, J.N., Cowling, E.B., 2002. Reactive nitrogen and the world: 200 years of change. Ambio 31, 64–71.

Galloway, J.N., Cowling, E.B., Seitzinger, S.P., Socolon, R.H., 2002. Reactive nitrogen: Too much of a good thing? Ambio 31, 60–63.

Gielen, B., Ceulemans, R., 2001. The likely impact of rising atmospheric CO_2 on natural and managed *Populus*: A literature review. Environ. Pollut. 115, 335–358.

Gregorius, H.-R., 1986. The importance of genetic multiplicity for tolerance of atmospheric pollution. In: Proceedings of 18th IUFRO World Congress. Ljubljana, Yugoslavia, pp. 195–305.

Guenther, A., Hewitt, C.N., Eickson, D., Fall, R., Geron, C., Graedel, T., Harley, P., Klinger, L., Lerdau, M., McKay, W.A., Pierce, T., Scholes, B., Steinbrecher, R., Tallamraju, R., Taylor, J., Zimmerman, P., 1995. A global model of natural volatile organic compound emissions. J. Geophysiol. Res. 100, 8873–8892.

Gupta, H.K., Gupta, V.B., Rao, C.V.C., Gajghate, D.G., Hasan, M.Z., 2002. Urban ambient air quality and its management strategy for a metropolitan city in India. Environ. Contam. Toxicol. 68, 347–354.

Hall, J.P., Magasi, L., Carlson, L., Stolte, K., Niebla, E., Bauer, M.L., Gonzalez-Vicente, C.E., Hernades-Tejeda, T., 1996. Health of North American forests. North American Forestry Commission Report. Natural Resources Canada, Ottawa.

Hamilton, J.G., DeLucia, E.H., George, K., Naidu, S.L., Finzi, A.C., Schlesinger, W.H., 2002. Forest carbon balance under elevated CO_2. Oecologia 131, 250–260.

Hannah, L., Midgley, G.F., Lovejoy, T., Bond, W.J., Bush, M., Lovett, J.C., Scott, D., Woodward, F.I., 2002. Conservation of biodiversity in a changing climate. Cons. Biol. 16, 264–268.

Hanson, P.J., Weltzin, J.F., 2000. Drought disturbance from climate change: response of United States forests. Sci. Total Environ. 262, 205–220.

Hao, X., Hale, B.A., Ormrod, D.P., Papadopoulos, A.P., 2000. Effects of pre-exposure to ultraviolet-B radiation on responses of ambient and elevated carbon dioxide. Environ. Pollut. 110, 217–224.

Harrington, R., Woiwod, I., Sparks, T., 1999. Climate change and trophic interactions. Trees 14, 146–150.

Hauglustaine, D.A., Granier, C., Brasseur, G.P., Megie, G., 1994. The importance of atmospheric chemistry in the calculation of radiative forcing on the climate system. J. Geophysiol. Res. 99, 1173–1186.

Heagle, A.S., Miller, J.E., Pursley, W.A., 1998. Influence of ozone stress on soybean response to carbon dioxide enrichment: III. Yield and seed quality. Crop Sci. 38, 128–134.

Heck, W.W., Taylor, O.C., Adams, R., Bingham, G., Miller, J., Preston, E., 1982. Assessment of crop loss from ozone. J. Air Pollut. Control Assoc. 32, 353–361.

Hirst, D., Karesen, K., Host, G., Posch, M., 2000. Estimating the exceedance of critical loads in Europe by considering local variability in deposition. Atmos. Environ. 34, 3789–3800.

Hobbie, S.E., Schimel, J.P., Trumbore, S.E., Randerson, J.R., 2000. Controls over carbon storage and turnover in high-latitude soils. Global Change Biol. 6 (S1), 196–210.

Hogsett, W.E., Weber, J.E., Tingey, D., Herstrom, A., Lee, E.H., Laurence, J.A., 1997. Environmental auditing: An approach for characterizing tropospheric ozone risk to forests. Environ. Manag. 21, 105–120.

Horowitz, L.W., Jacob, D.J., 1999. Global impact of fossil fuel combustion on atmospheric NO_x. J. Geophysiol. Res. 104, 23 823–23 850.

Houghton, R.A., 2001. Counting terrestrial sources and sinks of carbon. Clim. Change 48, 525–534.

Hudak, C., Bender, J., Weigel, H.-J., Miller, J., 1999. Interactive effects of elevated CO_2, O_3, and soil water deficient on spring wheat (*Triticum aestivum* L. cv. Nandu). Agronomie 19, 677–687.

Hulme, M., Doherty, R., Ngara, T., New, M., Lister, D., 2001. African climate change: 1900–2100. Clim. Res. 17, 145–168.

IPCC, 2001. Climate change 2001: Impacts, adaptation and vulnerability. Summary for policymakers. Intergovernmental Panel on Climate Change, Geneva.

Isebrands, J.G., Dickson, R.E., Rebbeck, J., Karnosky, D.F., 2000. Interacting effects of multiple stresses on growth and physiological processes in northern forest trees. In: Mickler, R.A., Birdsey, R.A., Hom, J. (Eds.), Responses of Northern U.S. Forests to Environmental Change. Springer-Verlag, New York, pp. 149–180.

Isebrands, J.G., McDonald, E.P., Kruger, E., Hendrey, G., Pregitzer, K., Percy, K., Sober, J., Karnosky, D.F., 2001. Growth responses of *Populus tremuloides* clones to interacting carbon dioxide and tropospheric ozone. Environ. Pollut. 115, 359–371.

Iverson, L.R., Prasad, A.M., 1998. Predicting abundance of 80 tree species following climate change in the eastern United States. Ecol. Monogr. 68, 465–485.

Jacob, D.J., Logan, J.A., Murti, P.P., 1999. Effect of rising Asian emissions on surface ozone in the United States. Geophysiol. Res. Lett. 26, 2175–2178.

Jaffe, D., Anderson, T., Covert, D., Kotchenruther, R., Trost, B., Danielson, J., Simpson, W., Berntsen, T., Karlsdottir, S., Blake, D., Harris, J., Carmichael, G., Uno, I., 1999. Transport of Asian air pollution to North America. Geophysiol Res. Lett. 26, 711–714.

Jeffries, R.L., Maron, J.L., 1997. The embarrassment of riches: Atmospheric deposition of nitrogen and community and ecosystem processes. Trends Ecol. Evolut. 12, 74–77.

Jobbagy, E.G., Jackson, R.B., 2000. The vertical distribution of soil organic carbon and its relation to climate and vegetation. Ecol. Applic. 10, 423–436.

Jonson, J.E., Sundet, J.K., Tarrasón, L., 2001. Model calculations of present and future levels of ozone and ozone precursors with a global and a regional model. Atmos. Environ. 35, 525–537.

Karnosky, D.F., 1977. Evidence for genetic control of response to sulfur dioxide and ozone in *Populus tremuloides* Michx. Can. J. Forest Res. 7, 435–436.

Karnosky, D.F., 1981. Changes in eastern white pine stands related to air pollution stress. Mitteilungen der Forstlichen Bundesversuchsanstalt Wien 137, 41–45.

Karnosky, D.F., Gagnon, Z.E., Reed, D.D., Witter, J.A., 1992. Effects of genotype on the response of *Populus tremuloides* Michx. to ozone and nitrogen deposition. Water Air Soil Pollut. 62, 189–199.

Karnosky, D.F., Gagnon, Z.E., Dickson, R.E., Coleman, M.D., Lee, E.H., Isebrands, J.G., 1996. Changes in growth, leaf abscission, and biomass associated with seasonable tropospheric ozone exposures of *Populus tremuloides* clones and seedlings. Can. J. Forest Res. 16, 23–27.

Karnosky, D.F., Oksanen, E., Dickson, R.E., Isebrands, J.G., 2001. Impacts of interacting greenhouse gases on forest ecosystems. In: Karnosky, D.F., Scarascia-Mugnozza, G., Ceulemans, R., Innes, J. (Eds.), The Impact of Carbon Dioxide and Other Greenhouse Gases on Forest Ecosystems. CABI Publishing, New York, pp. 253–267.

Karnosky, D.F., Percy, K.E., Xiang, B., Callan, B., Noormets, A., Mankovska, B., Hopkin, A., Sober, J., Jones, W., Dickson, R.E., Isebrands, J.G., 2002. Interacting CO_2-tropospheric O_3 and predisposition of aspen (*Populus tremuloides* Michx.) to infection by *Melampsora medusae* rust. Global Change Biol. 8, 329–338.

Karnosky, D.F., Zak, D.R., Pregitzer, K.S., Awmack, C.S., Bockheim, J.G., Dickson, R.E., Hendrey, G.R., Host, G.E., King, J.S., Kopper, B.J., Kruger, E.L., Kubiske, M.E., Lindroth, R.L., Mattson, W.J., McDonald, E.P., Noormets, A., Oksanen, E., Parsons, W.F.J., Percy, K.E., Podila, G.K., Riemenschneider, D.E., Sharma, P., Thakur, R.C., Sober, A., Sober, J., Jones, W.S., Anttonen, S., Vapaavuori, E., Mankovska, B., Heilman, W.E., Isebrands, J.G., 2003. Tropospheric O_3 moderates responses of temperate hardwood forests to elevated CO_2: A synthesis of molecular to ecosystem results from the Aspen FACE project. Funct. Ecol. 17, 289–304.

Keeling, C.M., Whort, T.P., Wahlen, M., Vander Plict, J., 1995. International extremes in the rate of rise of atmospheric carbon dioxide since 1980. Nature 375, 666–670.

Keller, T., 1988. Growth and premature leaf fall in American aspen as bioindications for ozone. Environ. Pollut. 52, 183–192.

Kiehl, J.T., Schneider, T.L., Portmann, R.W., Solomon, S., 1999. Climate forcing due to tropospheric and stratospheric ozone. J. Geophysiol. Res. 104, 31 239–31 254.

Kickert, R.N., Krupa, S.V., 1990. Forest responses to tropospheric ozone and global climate change: An analysis. Environ. Pollut. 68, 29–65.

King, J.S., Pregitzer, K.S., Zak, D.R., Karnosky, D.F., Isebrands, J.G., Dickson, R.E., Hendrey, G.R., Sober, J., 2001. Fine root biomass and fluxes of soil carbon in young stands of paper birch and trembling aspen is affected by elevated CO_2 and tropospheric O_3. Oecologia 128, 237–250.

Komenda, M., Parusel, E., Wedel, A., Koppman, R., 2001. Measurements of biogenic VOC emissions: Sampling, analysis and calibrating. Atmos. Environ. 35, 2069–2080.

Koski, V., 1996. Breeding plans in case of global warming. Euphitica 92, 235–239.

Kretzschmar, J.G., 1994. Particulate matter levels and trends in Mexico City, Sao Paulo, Buenos Aires, and Rio De Janeiro. Atmos. Environ. 28, 3181–3191.

Krupa, S.V., Kickert, R.N., 1989. The greenhouse effect: Impacts of ultraviolet (UV)-B radiation, carbon dioxide (CO_2) and ozone (O_3) on vegetation. Environ. Pollut. 61, 263–392.

Kubiske, M.E., Godbold, D.L., 2001. Growth and function of roots and root systems. In: Karnosky, D., Scarascia-Mugnozza, G., Ceulemans, R., Innes, J. (Eds.), The Impact of Carbon Dioxide and Other Greenhouse Gasses on Forest Ecosystems. CABI Publishing, New York, pp. 325–340.

Kull, O., Sober, A., Coleman, M.D., Dickson, R.E., Isebrands, J.G., Gagnon, Z., Karnosky, D.F., 1996. Photosynthetic response of aspen clones to simultaneous exposures of ozone and CO_2. Can. J. Forest Res. 16, 639–648.

Kurpius, M.R., McKay, M., Goldstein, A.H., 2002. Annual ozone deposition to a Sierra Nevada ponderosa pine plantation. Atmos. Environ. 36, 4503–4515.

Lawson, T., Craigon, J., Black, C.R., Colls, J.J., Tulloch, A.-M., Landon, G., 2001. Effects of elevated carbon dioxide and ozone on the growth and yield of potatoes (*Solanum tuberosum*) grown in open-top chambers. Environ. Pollut. 111, 479–491.

Ledig, F.T., Kitzmiller, J.H., 1992. Genetic strategies for reforestation in the face of global climate change. Forest Ecol. Manag. 50, 153–169.

Lefohn, A.S., Husar, J.D., Husar, R.B., 1999. Estimating historical anthropogenic global sulfur emission patterns for the period 1850–1990. Atmos. Environ. 33, 3435–3444.

Lelieveld, J., Berresheim, H., Borrmann, S., Crutzen, P.J., Dentener, F.J., Fischer, H., Feichter, J., Flatau, P.J., Heland, J., Holzinger, R., Korrmann, R., Lawrence, M.G., Levin, Z., Markowicz, K.M., Mihalopoulos, N., Minikin, A., Ramanathan, V., de Reus, M., Roelofs, G.J., Scheeren, H.A., Sciare, J., Schlager, H., Schultz, M., Slegmund, P., Steil, B., Stephanou, E.G., Steir, P., Traub, M., Warneke, C., Williams, J., Ziereis, H., 2002. Global air pollution crossroads over the Mediterranean. Science 298, 794–799.

Levy II, H., Moxim, W.J., Klonecki, A.A., Kasibhatla, P.S., 1999. Simulated tropospheric NO_x: Its evaluation, global distribution and individual source contributions. J. Geophysiol. Res. 104, 26 279–26 306.

Li, Q., Jacob, D.J., Bey, I., Palmer, P.I., Duncan, B.N., Field, B.D., Martin, R.V., Fiore, A.M., Yantosca, R.M., Parrish, D.D., Simmonds, P.G., Oltmans, S.J., 2003. Transatlantic transport of pollution and its effects on surface ozone in Europe and North America. J. Geophysiol. Res. (In press).

Lin, C.-Y.C., Jacob, D.J., Munger, J.W., Fiore, A.M., 2000. Increasing background ozone in surface air over the United States. Geophysiol. Res. Lett. 27, 3465–3468.

Lin, C.-Y.C., Jacob, D.J., Fiore, A.M., 2001. Trends in exceedances of the ozone air quality standard in the continental United States, 1980–1998. Atmos. Environ. 35, 3217–3228.

Lindner, M., 2000. Developing adaptive forest management strategies to cope with climate change. Tree Physiol. 20, 299–307.

Lindroth, R.L., Kinney, K.K., Platz, C.L., 1993. Responses of deciduous trees to elevated atmospheric CO_2: productivity, phytochemistry and insect performance. Ecology 74, 763–777.

Lindroth, R.L., Roth, S., Kruger, E.L., Volin, J.C., Koss, P.A., 1997. CO_2-mediated changes in aspen chemistry: Effects on gypsy moth performance and susceptibility to virus. Global Change Biol. 3, 279–289.

Liu, S.C., Trainer, M., Fehsenfeld, F.C., Parrish, D.D., Williams, E.J., Fahey, D.W., Hübler, G., Murphy, P.C., 1987. Ozone production in the rural troposphere and the implications for regional and global ozone distributions. J. Geophysiol. Res. 92, 4191–4207.

Lloyd, A.H., Fastie, C.L., 2002. Spatial and temporal variability in the growth and climate response of treeline trees in Alaska. Clim. Change 52, 481–509.

Lynch, J.A., Bowersox, V.C., Grimm, J.W., 2000. Changes in sulfate deposition in eastern USA following implementation of Phase I of Title IV of the Clear Air Act Amendments of 1990. Atmos. Environ. 34, 1665–1680.

MacDonald, G.M., Edwards, T.W.D., Moser, K.A., Pienitz, R., Smol, J.P., 1993. Rapid response of tree-line vegetation and lakes to post climate warming. Nature 361, 243–246.

Makinen, H., Nojd, P., Mielikainen, K., 2000. Climatic signal in annual growth variation of Norway spruce (*Picea abies*) along a transect from central Finland to the Arctic timberline. Can. J. Forest Res. 30, 769–777.

Mankovska, B., 1997a. Deposition of heavy metals in Slovakia—Assessment on the basis of moss and humus analyses. Ekologia 16, 433–442.

Mankovska, B., 1997b. Geochemical mapping of environmental stress by selected elements through foliar analysis. Slovak Geol. Mag. 3, 53–66.

Mankovska, B., Percy, K., Karnosky, D.F., 2003. Impact of greenhouse gases on epicuticular waxes of *Populus tremuloides* Michx: Results from an open-air exposure and a natural O_3 gradient. Ekologia (in press).

Marenco, A., Gouget, H., Nédélec, P., Pagés, J.-P., Karchner, F., 1994. Evidence of a long-term increase in tropospheric ozone from the Pic du Midi data series; Consequences: Positive radiative forcing. J. Geophysiol. Res. 99, 16 617–16 632.

Martin, P.H., 1996. Will forest preserves protect temperate and boreal biodiversity from climate change? Forest Ecol. Manag. 85, 335–341.

McDonald, E.P., Kruger, E.L., Riemenschneider, D.E., Isebrands, J.G., 2002. Competitive status influences tree-growth responses to elevated CO_2 and O_3 in aggrading aspen stands. Funct. Ecol. 16, 792–801.

McKee, I.F., Bullimore, J.F., Long, S.P., 1997. Will elevated CO_2 concentrations protect the yield of wheat from O_3 damage? Plant Cell Environ. 20, 77–84.

McLaughlin, S., Percy, K., 1999. Forest health in North America: Some perspectives on actual and potential roles of climate and air pollution. Water Air Soil Pollut. 116, 151–197.

McNulty, S.G., Vose, J.M., Swank, W.T., 1996. Potential climate change effects on loblolly pine forest productivity and drainage across the southern United States. Ambio 25, 449–453.

Melillo, J.M., McGuire, D., Kicklighter, D.W., Moore III, B., Vorosmarty, C.J., Schloss, A.L., 1993. Global climate change and terrestrial net primary production. Nature 363, 234–240.

Menzel, A., Fabian, P., 1999. Growing season extended in Europe. Nature 397, 659.

Menzel, A., 2000. Trends in phenological phases in Europe between 1951 and 1996. Int. J. Biometeorol. 44, 76–81.

Miller, J.E., Heagle, A.S., Pursley, W.A., 1998. Influence of ozone stress on soybean response to carbon dioxide enrichment: II. Biomass and development. Crop Sci. 38, 122–128.

Miller, P.R., 1973. Oxidant-induced community change in a mixed conifer forest. Adv. Chem. 122, 101–117.

Miller, P.R., 1993. Response of forests to ozone in a changing atmospheric environment. Angew. Bot. 67, 42–46.

Miller, P.R., Tejeda, T.H., 1994. Comparison of ozone exposure characteristics in forested regions near Mexico City and Los Angeles. Atmos. Environ. 28, 141–148.

Miller, P.R., Parmeter, J.R., Taylor, O.C., Cardiff, E.A., 1963. Ozone injury to the foliage of ponderosa pine. Phytopathology 53, 1072–1076.

Miller, P.R., Stolte, K.W., Duriscoe, D.M., Pronos, J., 1996. Evaluating ozone air pollution effects on pines in the western United States. USDA Forest Service Pacific Southwest Research Station, General Technical Report PSW-GTR-155.

Monson, R.K., Lerdau, M.T., Sharkey, T.D., Schimel, D.S., 1995. Biological aspects of constructing volatile organic compound emission inventories. Atmos. Environ. 29, 2989–3002.

Moore, B.D., Cheng, S.H., Sims, D., Seemann, J.R., 1999. The biochemical and molecular basis for photosynthetic acclimation of elevated atmospheric CO_2. Plant Cell Environ. 22, 567–582.

Moxim, W.J., Levy, H., Kasibhatla, P.S., 1996. Simulated global tropospheric PAN: Its transport and impact. J. Geophysiol. Res. 101, 12 621–12 638.

Müller-Starck, G., 1989. Genetic implications of environmental stress in adult forest stands of *Fagus sylvatica* L. In: Scholz, F., Gregorius, H.-R., Rudin, D. (Eds.), Genetic Effects of Air Pollutants in Forest Tree Populations. Springer-Verlag, pp. 127–142.

Nabuurs, G.-J., Pussinen, A., Karjalainen, T., Erhard, M., Kramer, K., 2002. Stemwood volume increment changes in European forests due to climate change—a simulation study with the EFISCEN model. Global Change Biol. 8, 304–316.

Namkoong, G., 1991. Maintaining diversity in breeding for resistance in forest trees. Annu. Rev. Phytopathol. 29, 325–342.

Noormets, A., Sôber, A., Pell, E.J., Dickson, R.E., Podila, G.K., Sôber, J., Isebrands, J.G., Karnosky, D.F., 2001a. Stomatal and nonstomatal control of photosynthesis in trembling aspen (*Populus tremuloides* Mich.) exposed to elevated CO_2 and O_3. Plant Cell Environ. 24, 327–336.

Noormets, A., McDonald, E.P., Kruger, E.L., Isebrands, J.G., Dickson, R.E., Karnosky, D.F., 2001b. The effect of elevated CO_2 and/or O_3 on potential plant level carbon gain in aspen. Trees 15, 262–270.

Norby, R.J., Wullschleger, S.D., Gunderson, C.A., Johnson, D.W., Ceulemans, R., 1999. Tree responses to rising CO_2 in field experiments: implications for the future forest. Plant Cell Environ. 22, 683–714.

Norby, R.J., Todd, D.E., Fults, J., Johnson, D.W., 2001. Allometric determination of tree growth in a CO_2-enriched sweetgum stand. New Phytol. 150, 477–487.

Norby, R.J., Hanson, P.J., O'Neill, E.G., Tschaplinski, T.J., Weltzin, J.F., Hansen, R.A., Cheng, W., Wullschleger, S.D., Gunderson, C.A., Edwards, N.T., Johnson, D.W., 2002. Net primary productivity of a CO_2-enriched deciduous forest and the implications for carbon storage. Ecol. Applic. 12, 1261–1266.

O'Brien, K.L., 2000. Upscaling tropical deforestation: implications for climate change. Clim. Change 44, 311–329.

Oren, R., Ellsworth, D.S., Johnson, K.H., Phillips, N., Ewers, B.E., Maier, C., Schafer, K.V.R., McCarthy, H., Hendrey, G., McNulty, S.G., Katul, G., 2001. Soil fertility limits carbon sequestration by forest ecosystems in a CO_2-enriched atmosphere. Nature 411, 469–472.

Oksanen, E., Sôber, J., Karnosky, D.F., 2001. Interactions of elevated CO_2 and ozone in leaf morphology of aspen (*Populus tremuloides*) and birch (*Betula papyrifera*) in aspen FACE experiment. Environ. Pollut. 115, 437–446.

Olszyk, D.M., Wise, C., 1997. Interactive effects of elevated CO_2 and O_3 on rice and flacca tomato. Agric. Ecosyst. Environ. 66, 1–10.

Parmesan, C., Yohe, G., 2003. A globally coherent fingerprint of climate change impacts across natural systems. Nature 421, 37–42.

Parrish, D.D., Hahn, C.J., Williams, E.J., Norton, R.B., Fehsenfeld, F.C., Singh, H.B., Shetter, J.D., Gandrud, B.W., Ridley, B.A., 1992. Indications of photochemical histories of Pacific air masses from measurements of atmospheric trace species at Point Arena, California. J. Geophysiol. Res. 97, 15 883–15 901.

Parshar, D.C., Kulshrestha, U.C., Sharma, C., 1998. Anthropogenic emissions of NO_x, NH_3 and N_2O in India. Nutr. Cycling Agroecosyst. 52, 255–259.

Penuelas, J., Filella, I., 2001. Responses to a warming world. Science 294, 793–795.

Percy, K.E., 2002. Is air pollution an important factor in forest health? Szaro, R.C., Bytnerowicz, A., Oszlanyi, J. (Eds.), Effects of Air Pollution on Forest Health and Biodiversity in Forests of the Carpathian Mountains. IOS Press, Amsterdam, pp. 23–42.

Percy, K.E., Awmack, C.S., Lindroth, R.L., Kubiske, M.E., Kopper, B.J., Isebrands, J.G., Pregitzer, K.S., Hendrey, G.R., Dickson, R.E., Zak, D.R., Oksanen, E., Sober, J., Harrington, R., Karnosky, D.F., 2002. Altered performance of forest pests under CO_2- and O_3-enriched atmospheres. Nature 420, 403–407.

Peterson, D.L., Arbaugh, M.J., Robinson, L., 1989. Ozone injury and growth trends of ponderosa pine in the Sierra Nevadas. In: Olson, R.K., Lefohn, A.S. (Eds.). In: Effects of Pollution on Western Forests. Transaction Series, Vol. 16. Air and Waste Manag. Association, Pittsburgh, PA, pp. 293–308.

Pitelka, L.F., 1997. Plant migration and climate change. American Scientist 85, 464–473.

Polle, A., Pfirrmann, T., Chakrabarti, S., Rennenberg, H., 1993. The effects of enhanced ozone and enhanced carbon dioxide concentrations on biomass, pigments, and antioxidative enzymes in spruce needles (*Picea abies* L.). Plant Cell Environ. 16, 311–316.

Polle, A., Eiblmeier, M., Sheppard, L., Murray, M., 1997. Responses of antioxidative enzymes to elevated CO_2 in leaves of beech (*Fagus sylvatica* L.) seedlings grown under a range of nutrient regimes. Plant Cell Environ. 20, 1317–1321.

Pregitzer, K.S., King, J.S., Burton, A., Brown, S.E., 2000. Responses of tree fine roots to temperature. New Phytol. 147, 105–115.

Raga, G.B., Raga, A.C., 2000. On the formation of an elevated ozone peak in Mexico City. Atmos. Environ. 34, 4097–4102.

Raich, J.W., Schlesinger, W.H., 1992. The global carbon dioxide flux in soil respiration and its relationship to vegetation and climate. Tellus 44B, 81–99.

Rehfeldt, G.E., Ying, C.C., Spittlehouse, D.L., Hamilton Jr., D.A., 1999. Genetic responses to climate in *Pinus contorta*: Niche breadth, climate change, and reforestation. Ecol. Monogr. 69, 375–407.

Rehfeldt, G.E., Tchebakova, N.M., Parfenova, Y.I., Wykoff, W.R., Kuzmina, N.A., Milyutin, L.I., 2002. Intraspecific responses to climate in *Pinus sylvestris*. Global Change Biol. 8, 912–929.

Romero, H., Ihl, M., Rivera, A., Zalazar, P., Azocar, P., 1999. Rapid urban growth, land-use change and air pollution in Santiago, Chile. Atmos. Environ. 33, 4039–4047.

Rudorff, B.F.T., Mulchi, C.L., Lee, E.H., Rowland, R., Pausch, R., 1996. Effects of enhanced O_3 and CO_2 enrichment on plant characteristics in wheat and corn. Environ. Pollut. 94, 53–60.

Sampson, R.N., 1992. Forestry opportunities in the United States to mitigate the effects of global warming. Water Air Soil Pollut. 64, 157–180.

Schlesinger, W.H., 1997. Biogeochemistry: An Analysis of Global Change. Academic Press, San Diego, CA.

Sedjo, R.A., 1989. Forests to offset the greenhouse effect. J. For. 87, 12–15.

Shaver, G.R., Canadell, J., Chapin III, F.S., Gurevitch, J., Harte, J., Henry, G., Ineson, P., Jonasson, S., Melillo, J., Pitelka, L., Rustad, L., 2000. Global warming and terrestrial ecosystems: A conceptual framework for analysis. Bioscience 50, 871–882.

Shine, K.P., 2001. Atmospheric ozone and climate change. Ozone Sci. Eng. 23, 429–435.

Skiba, Y.N., Davydova-Belitskaya, V., 2002. Air pollution estimates in Guadalajara City. Environ. Mod. Assess. 7, 153–162.

Stevenson, D.S., Johnson, C.E., Collins, W.J., Derwent, R.G., Shine, K.P., Edwards, J.M., 1998. Evolution of tropospheric ozone radiative forcing. Geophysiol. Res. Lett. 25, 3819–3822.

Stohl, A., Trickl, T., 1999. A textbook example of long-range transport: simultaneous observation of ozone maxima of stratospheric and North American origin in the free troposphere over Europe. J. Geophysiol. Res. 104, 30445–30462.

Stott, P.A., Tett, S.F.B., Jones, G.S., Allen, M.R., Mitchell, J.F.B., Jenkins, G.J., 2000. External control of 20th century temperature by natural and anthropogenic forcing. Science 290, 2133–2137.

Straszewski, T., Szdzuj, J., Kubiesa, P., Godzik, S., 2001. Hazard of spruce stands in the Pieniny National Park due to regional and transboundary air pollution. Ecologia 20 (S4), 249–256.

Streets, D.G., Waldhoff, S.T., 2000. Present and future emissions of air pollutants in China: SO_2, NO_x, and CO. Atmos. Environ. 34, 363–374.

Streets, D.G., Tsai, N.Y., Akimoto, H., Oka, K., 2000. Sulfur dioxide emissions in Asia in the period 1985–1997. Atmos. Environ. 34, 4413–4424.

Suzuki, K., 1999. Sustainability of pine forests in relation to pine wilt and decline. In: Proceedings of International Symposium. Tokyo, Japan.

Swart, R., Mitchell, J., Morita, T., Raper, S., 2002. Stabilization scenarios for climate impact assessment. Global Environ. Change 12, 155–165.

Taylor Jr., G.E., Pitelka, L.F., 1991. Genetic diversity of plant populations and the role of air pollution. In: Barker, J.R., Tingey, D.T. (Eds.), Air Pollution Effects on Biodiversity. Van Nostrand Reinhold, New York, pp. 111–130.

Taylor Jr., G.E., 1994. Role of genotype in the response of loblolly pine to tropospheric ozone: Effects at the whole-tree, stand, and regional level. J. Environ. Qual. 23, 63–82.

Teskey, R.O., Dougherty, P., Wiselogel, A.E., 1991. The design and performance of branch chambers suitable for long-term ozone fumigation of foliage in large trees. J. Environ. Qual. 20, 591–595.

Thompson, A.M., 1992. The oxidizing capacity of the Earth's atmosphere: Probable past and future changes. Science 256, 1157–1165.

Tjoelker, M.G., Oleksyn, J., Reich, P.B., 1998. Seedlings of five boreal tree species differ in acclimation of net photosynthesis to elevated CO_2 and temperature. Tree Physiol. 18, 715–726.

Utriainen, J., Janhunen, S., Helmisaari, H.-S., Holopainen, T., 2000. Biomass allocation, needle structural characteristics and nutrient composition in Scots pine seedlings exposed to elevated CO_2 and O_3 concentrations. Trees 14, 475–484.

Vallack, H.W., Cinderby, S., Kuylenstierna, J.C.I., Heaps, C., 2001. Emission inventories for SO_2 and NO_x in developing country regions in 1995 with projected emissions for 2025 according to two scenarios. Water Air Soil Pollut. 130, 217–222.

Vann, D.R., Johnson, A.H., 1995. Design and field operation of an in-situ environmental enclosure for tree branches. Environ. Pollut. 89, 37–46.

Vine, E.L., Sathage, J.A., Makundi, W.R., 2001. An overview of guidelines and issues for the monitoring, evaluation, reporting, verification, and certification of forestry projects for climate change mitigation. Global Environ. Change 11, 203–216.

Volin, J.C., Reich, P.B., 1996. The interaction of elevated CO_2 and O_3 on growth, photosynthesis and respiration of three perennial species grown in low and high nitrogen. Physiol. Plant. 96, 674–684.

Volin, J.C., Reich, P.B., Givnish, T.J., 1998. Elevated carbon dioxide ameliorates the effects of ozone on photosynthesis and growth: species respond similarly regardless of photosynthetic pathway or plant functional group. New Phytol. 138, 315–325.

Volney, W.J.A., Fleming, R.A., 2000. Climate change and impacts of boreal forest insects. Agric. Ecosyst. Environ. 82, 283–294.

Wang, D., Karnosky, D.F., Bormann, F.H., 1986. Effects of ambient ozone on the productivity of *Populus tremuloides* Michx. grown under field conditions. J. Geophysiol. Res. 16, 47–55.

Wang, Y., Jacob, D.J., Logan, J.A., 1998. Global simulation of tropospheric O_3–NO_x–hydrocarbon chemistry 3. Origin of tropospheric ozone and effects of nonmethane hydrocarbons. J. Geophysiol. Res. 103, 10 757–10 767.

Wisconsin Department of Natural Resources. 1997. Wisconsin 1996 Air Quality Report. State of Wisconsin Department of Natural Resources. Bureau of Air Management PUBL-AM-221-97.

Wustman, B.A., Oksanen, E., Karnosky, D.F., Sober, J., Isebrands, J.G., Hendrey, G.R., Pregitzer, K.S., Podila, G.K., 2001. Effects of elevated CO_2 and O_3 on aspen clones varying in O_3 sensitivity: Can CO_2 ameliorate the harmful effects of O_3? Environ. Pollut. 115, 473–481.

Yienger, J.J., Galanter, M., Holloway, T.A., Phadnis, M.J., Guttikunda, S.K., Carmichael, G.R., Moxim, W.J., Levy II, H., 2000. The episodic nature of air pollution transport from Asia to North America. J. Geophysiol. Res. 105, 26 931–26 945.

Air Pollution, Global Change and Forests in the New Millennium
D.F. Karnosky et al., editors

Chapter 2

What is the role of demographic factors in air pollution and forests?

David S. Shriner

*USDA Forest Service, North Central Research Station, 1992 Folwell Avenue,
St. Paul, MN 55108, USA*

David F. Karnosky*

*School of Forest Resources and Environmental Science, Michigan Technological University,
101 U.J. Noblet Forestry Building,1400 Townsend Drive, Houghton, MI 49931, USA
E-mail: karnosky@mtu.edu*

Abstract

The advent of the twenty-first century finds us with many reasons to consider scenarios for global change and air pollution in concert. During the decade of the 1980s, there were large research efforts in Europe and North America focused on the ecological consequences of air pollution and acidic deposition. These efforts, however, largely ignored climate change. During the decade of the 1990s, similar major research efforts were initiated globally with a focus on the consequences of global climate change. At this point, however, there has been limited research focused on understanding the interactions of multiple factors, including air pollution, insects and diseases, and even fewer attempts to incorporate such factors into modeled projections of climate change impacts on forest distribution or productivity. Twenty-first century scenarios for changes in spatial patterns of air quality will not only be driven by natural weather and climate variability, but also by changes in demographics, land use, and economic growth. These factors may result in quite different patterns on a global, regional (continental), or subregional scale from those observed over the past century. Changes in both climate and air pollution are the result of dynamic processes that will influence each other, and that will develop over time. Because of inherent uncertainties in knowledge of the processes affected, our understanding of the magnitude of the responses are equally uncertain. For the community of scientists engaged in research on the relationships among air pollution, climate change, and forest health and productivity, the need of landowners and other decision-makers for science-based adaptive strategies should challenge us, and makes a compelling case for an aggressive research agenda.

---

*Corresponding author.

DOI:10.1016/S1474-8177(03)03002-X

1. Introduction

The words of Kenneth Boulding, the ecological economist, remind us "We have to be prepared to be surprised by the future, but we don't have to be dumbfounded". We are also reminded of the words of former baseball player Yogi Berra, when he once said "If you don't know where you're going, you end up somewhere else". Contemplating the 21st Century from here at its very beginning seems to be much like Yogi described. It is the wild card out there, the thing we might not anticipate, that makes prediction into the future so challenging. In this paper, we examine recent demographics that may help predict this century's trajectories of major air pollutants and climate change.

Our goal should be to break out of our typically reactive mode—to learn to anticipate change and deal with emerging issues in advance, before simple challenge matures into crisis, and before our decision space has shrunk to a consideration of the lesser of evils. We need to devise mechanisms that will enable us to look "beyond the headlights", to see the landscape before us, and to adjust our course to follow the terrain. Lack of foresight has sometimes hampered our collective ability to deal with challenging issues related to air quality, and agencies have sometimes been slow to respond to changes in scientific understanding, or the expectations of the public. Tradition, professional training, and legislative direction have proven to be uncertain guides to the future in a rapidly changing society, and the one thing that seems certain, is that the advent of the twenty first century will bring rapid change.

2. Analysis

As one thinks about global climate change and air pollution and the scenarios for each of them into the twenty-first century, one asks "What do we know about the factors that drive air quality, and what do those factors tell us about the future that complicate our view?" What does the future look like?

There are several ways to view this question, and in this paper we will briefly examine a number of large-scale trends that seem to be redefining our society. Many of the examples will be North American in context, but are relevant to much of the world today, or certainly will be within this century. Largely, they can be categorized into social, demographic, and economic factors; into feedbacks to air quality from those social, demographic, and economic factors through their patterns and distribution on the landscape; and through their interactions with a host of other factors, physical, biological, and social.

We have learned over the past 20 years or so that we can no longer afford the luxury of single factor assessment of issues. This became obvious during the days of atmospheric deposition ("acid rain") assessments (Irving, 1991), and has been reaffirmed repeatedly by the research and assessment activities that have followed (Fox and Mickler, 1996; Mickler et al., 2000; Karnosky et al., 2001). We live in a complex world, and in order for our research to be relevant, we must be able to capture that complexity.

Another thing that seems certain is that over the next century, patterns of social, demographic, and economic activity will shift—globally, nationally, and regionally. Let us consider some of the shifts that have the potential to influence air quality.

2.1. Demographic change

- There will be increased diversity in the American population. By 2050, racial and ethnic minorities will comprise half of all Americans; 86% of immigration is now non-European. Over the next 50 years, 90% of the population growth will come from racial and ethnic minorities.
- America is aging. One hundred years ago, only 4% of the US population was over 65; by 2020, 21% are projected to be over 65, and by 2050, projections are for more than 4 million centenarians (compared to just 70 000 today).
- Shifts in the spatial distribution of population are occurring. In the US, movements out of the North and East, to the South, and West are occurring. Shifts of this nature have been underway for some time, but future projections are for them to continue, or to become more pronounced. People are moving closer to forest resources, bringing NO_x, VOCs and commuter miles with them. On a global scale, nearly 700 000 immigrants relocate to the United States each year, shifting population hemispherically, if not globally. The implication for air pollution, of course, is in the shift of people, cars, jobs, industries, etc, from one place to another, and the resulting changes in precursor emissions, regional atmospheric chemistry, and deposition.
- Shifts are occurring from a manufacturing economy to a service and information technology one. Competition from a global labor market and technological changes making "just in time delivery" a global, not a local concept, and shifting manufacturing emissions globally. It is thus not surprising that total sulfur deposition is expected to rise dramatically in developing countries over the next 50 years (Fig. 1, Fowler et al., 1999).
- A change from rural to urban populations is occurring. Americans are increasingly becoming more urban and suburban, less rural. Political representation will shift more away from rural counties. Between 1969 and 1990, urban areas in the US doubled in spatial extent and now occupy some 3.5% of the US (Dwyer et al., 2003). The expansion of urban areas has not only been in response to population growth; but also to the movement of busi-

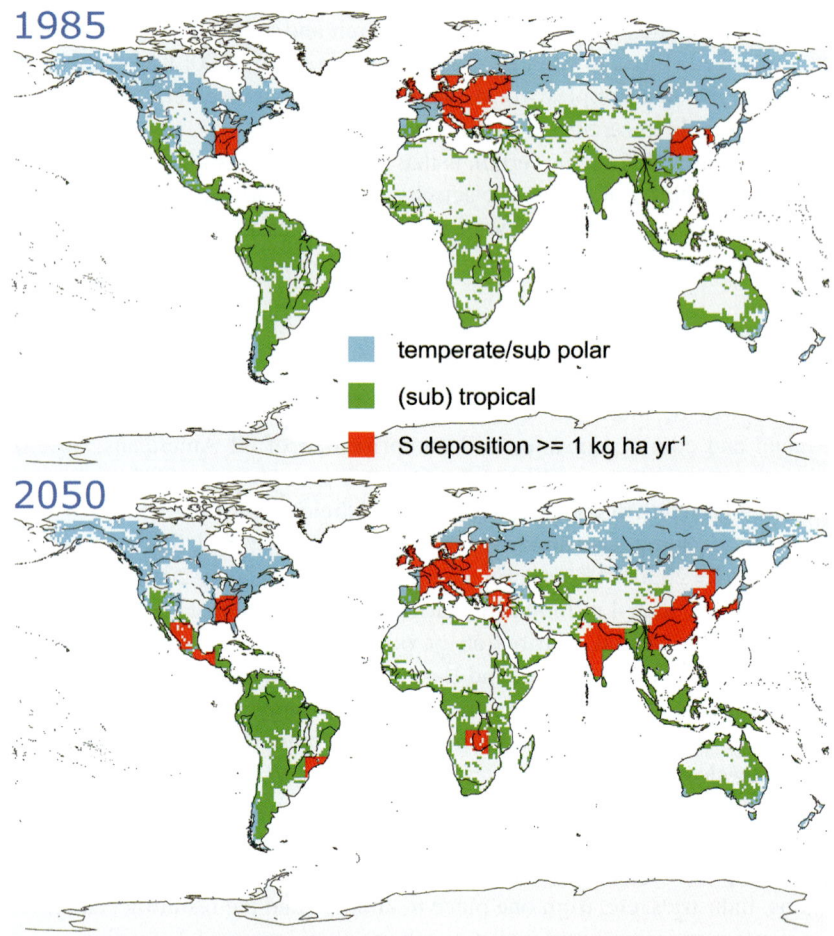

Figure 1. The estimated increase in forest area receiving elevated inputs of sulfur (S) deposition is seen in this map of S deposition in 1985 and the projected S deposition map projected for 2050 (from Fowler et al., 1999).

nesses, industry, and residences from the inner city to the periphery and outlying areas. This movement has been fueled in part, by limitations in living, working and doing business in inner city areas; and in part, by the relative attractiveness of some suburban and ex-urban locations. Many of those who do move to the rural areas will take their urban attitudes, expectations, and incomes with them; they will **reside** in rural America, but will not culturally be **of** rural America, nor economically dependent on it.

2.2. The information revolution

No matter which futurist you consult, virtually all cite the role of information technology in shaping the future. Access to information has been democratized and that trend will only to continue to accelerate. Each of us now has the ability to access data in vast quantities, and to render that data into information for decision-making. Today, there are almost 200 million people on the internet worldwide. This information capacity will increasingly make available virtual businesses, allowing individuals to choose where they live based on their demand for quality of life, and again, potentially shifting the pattern of people on the landscape.

2.3. Material wealth

Wealth and disposable income will continue to increase, on average. But, gaps between wealth and poverty will remain. Disposable wealth, increased leisure time, flexible hours, work at home, educational attainment, and technological sophistication will increase. One futurist predicts that by 2020, 60% of all US consumers will be college educated, work in an information intensive environment, have an income in excess of $50 000 in 1999 dollars, and will have access to high speed, interactive multimedia communication devices in their homes.

2.4. Primary production

American society will continue to see a decreasing reliance on—and identification with—primary production. The shift to a service and information economy will create an even greater gulf between traditional commodity users of forest resources, and the rest of the nation, and may feed resentment on the part of rural, commodity-dependent communities that feel abandoned by the larger society and economy.

2.5. Global integration

There will be greater global integration—manifesting itself in several ways that hold implications for forests and forest health. Markets for material goods, including timber will become increasingly global markets. Markets for eco-tourism and forest recreation will do the same. There will be growing pressures to meet global standards for environmental quality, on the model of the Montreal agreements, and the Kyoto protocol. Global demands on forests worldwide will increase at unprecedented rates as total population goes from 6 bil-

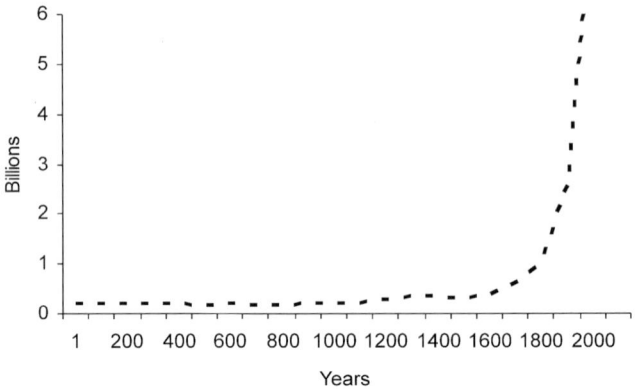

Figure 2.   Worldwide population growth.

lion to 10 or 11 billion (Fig. 2) and environmental pressures continue to limit the extent of "working" forests.

2.6. Environmental quality

The demand for increased environmental quality will continue. The trend of the past 30 years has turned out to not be a passing phenomenon. Current societal demands for open space, clean water, clean air, free-flowing streams, fish and wildlife, endangered species, recreational access to unspoiled places, and a sustainable environmental legacy will remain a dominant feature of the political and policy landscape well into the 21st century. The developing conflicts between developers in natural areas and environmentalist extremists who have recently taken on a radical philosophy of vigilantism towards stopping development (i.e., arson and extreme vandalism) will continue to grow. There is evidence that people are willing to move substantial distances to a preferred retirement location. This is likely to be a logical extension of the increasing mobility of the population. Where people decide to retire can become an important force for change in the distribution of individuals over the landscape. It is not clear to what extent retirement patterns will differ among racial and ethnic groups. However, the retirement opportunities and decisions of individuals in these groups are likely to have important implications for shifts in the location of the US population in the years ahead.

2.7. Air pollution and climate change research

The established pattern of chronic under-investment in air pollution and climate change research will likely continue. Despite greater wealth of individuals, we can anticipate increased competition for limited federal dollars, over the next 30 years or more. The reason for this apparent contradiction lies in our aging population. As baby-boomers retire over the next few years, the ratio of workers to pensioners will change, stretching the ability of the government to meet social needs, and intensifying the competition for federal dollars.

So what do these changes mean for air quality? Changing lifestyles, population patterns, and age structures may create patterns of regional air quality which differ significantly from the present over the course of the next century. Regulatory measures have resulted in up to 15% decreases in recent years in one-hour peak ozone (O_3) concentrations in the areas previously with the highest number of exceedances. However, projections are that the spatial extent of areas designated as non-attainment under the new 8-hour standard will increase, perhaps already due to some of the kinds of changes we have discussed. Ozone is potentially the most damaging of the major air pollutants in terms of negative impacts to forest growth and species composition (Fig. 3). In the eastern US, 20% of the forested land area is in counties that exceeded the National Ambient Air Quality Standard for O_3 in 1989. Use of the standard as the measure of impacted area probably underestimates the actual impact, however, since many sensitive plant species are known to experience growth reductions at levels as low as 50–60 ppb.

The forests of North America are vast and diverse, representing over 750 million hectares in the US and Canada. Projections for the effects of Global Climate Change on forest health include both increased growth and range of some forests, but also an increased frequency of declining health in others in response to increased biotic and abiotic stresses associated with climate warming (Houghton et al., 2001; IPCC, 2001). It is this more complex future environment, where changes in temperature and precipitation not only have direct consequences for forest growth, but indirect consequences on the atmospheric environment, and on insect and pathogen populations, that we are challenged to understand.

Collectively, the physiological effects of air pollutants are anticipated to either predispose forest trees to other stresses, or to amplify the negative effects of other stresses. McLaughlin and Percy (1999) recently provided insight for the eastern US, on the spatial consistency between the patterns of the most frequent occurrence of major disease problems as documented by forest surveys, and the patterns of the highest levels of ozone and acidic deposition across the region. While the data are inadequate to establish a statistical relationship between forest health and air quality, the circumstantial evidence is, none the less intriguing.

From a physiological perspective, the effects of increasing CO_2 and increasing temperature on net primary production, the effects of temperature on the frequency and severity of drought, and the effects of increasing nitrogen on

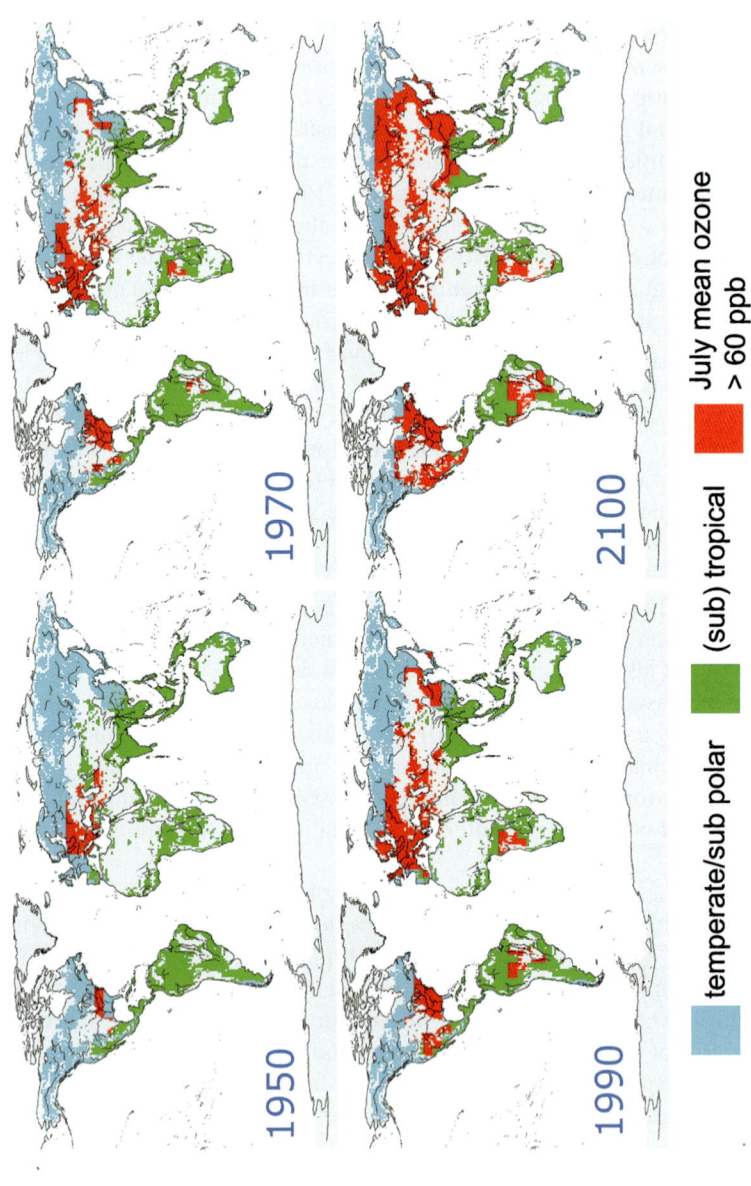

Figure 3. The increase in forest area receiving elevated ozone (July mean ozone > 60 ppb) from 1950 to 1990 and a projection for the year 2100 (from Fowler et al., 1999).

nutrient uptake and allocation to shoots and leaching from soils appear to be the most compelling components of climate change that could significantly influence forest sensitivity to biotic and abiotic stresses.

A wide variety of controlled studies now indicate that increasing CO_2 will lead to significant enhancement in leaf photosynthetic rates and increased net primary production of forest trees (Norby et al., 1999). Although very limited data are available for more mature forest trees (Karnosky, 2003), results to date suggest that positive growth responses are rather robust, in terms of absence of limitations by variations in supply of other resources. Warming temperatures, on the other hand, affects many essential forest processes and in many different ways. Most notably, the carbohydrate economy may be improved by extending the length of the growing season, but adversely impacted by the increased respiratory costs of temperature driven metabolic processes. In addition, warming temperatures will increase evapotranspiration, thus magnifying potential limitations of water supply on the amount and allocation of growth.

There is limited information available about the response of forest trees to air pollutant interactions, and several authors have suggested that increasing CO_2 might mitigate ozone-caused growth reductions (Volin and Reich, 1996; Volin et al., 1998). However, recent results with aspen bring this model into question (Isebrands et al., 2001; McDonald et al., 2002).

We believe at the policy level, current global models used for projecting the distribution and condition of forests in response to global change do consider increases in CO_2 and water availability, and increased temperature on forest growth. At present, these models do not consider the potential implications of regional air pollutants on forest growth processes, or on the predisposition of forest trees to other forms of biotic and abiotic stress (Percy et al., 2002).

3. Future pollutant scenarios

The current temporal trends of acidifying pollutant emissions show a marked contrast between the total emissions, which are increasing globally, but decreasing in Europe and North America. These trends will, over the next decade, appreciably reduce the areas of forest in Europe and North America currently receiving inputs of acidity in excess of 2 keq H^+ ha^{-1} annually. By 2050, more than half of the global forest subject to acidifying inputs in excess of that level will be in the tropics and subtropics—again, a significant shift in spatial pattern (Fig. 1) that appears to be closely linked to demographics as can be seen in the close relationship between the population density (Fig. 4) and pollutant deposition (Galloway et al., 2002; Galloway and Cowling, 2002).

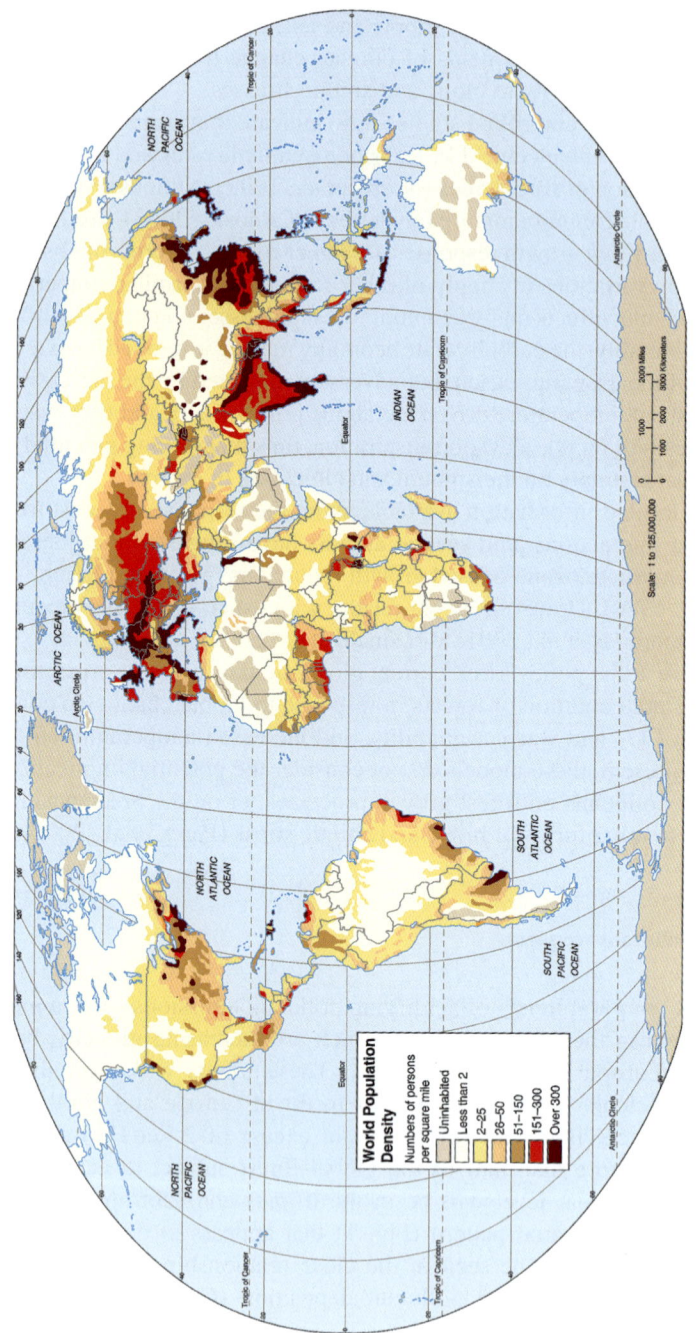

Figure 4. Worldwide population density (from US Department of Agriculture).

The photochemical oxidants, and O_3 in particular, have been shown to be major pollutants for forests globally, and to have increased rapidly throughout the last half century. Model forecasts of the global exposure of forest to O_3 (Fowler et al., 1999; Derwent et al., 2002) indicate continued increases in exceedance of thresholds for physiological effects, and probably also growth decreases (Fig. 3). The timescale of the projections is sufficiently long that important interactions between the effects of climate, CO_2 and O_3 on the responses of forests to O_3 are likely (Kickert and Krupa, 1990). The scale of the interactions is also likely to be of the same order as the direct effect of the O_3, so that significant positive or negative feedback is likely. The steady increase in deposition of nitrogen compounds with time is also likely to interact with climate change and CO_2 (Norby, 1998). Thus, the net effect on forest composition, productivity, and sensitivity to environmental stresses becomes very difficult to predict. There is little doubt that given the scope for environmental changes in the chemical and physical climates to which forests will be exposed over the next century that current models could simulate dramatic changes to the composition of the forested landscape. It is much more likely that the actual effects which occur will be unforeseen, and that they may be either positive or negative, and only extensive monitoring and research programs can provide the evidence sufficiently early to avoid significant social and economic consequences.

4. Research needs

1. Innovative and integrated approaches are needed to study the effects of CO_2, temperature, precipitation, O_3, N and S deposition on forests, and the potential interactions of these factors with forest insect pests and disease organisms.
2. An increased basic understanding of current forest disturbances is needed to determine how disturbances will interact with air pollution and climate change to shape future forests.
3. Prediction of climate and change in climate at scales of resolution relevant to ecosystems on the landscape will require major advances in modelling and some experimental ground truthing.
4. Understanding of the relationships between climate/weather patterns, trace gases, and the interactions of these factors with biological species at multiple scales is needed.
5. Integration of climate change, air pollution impacts, and land use change into major ecological models are needed to increase our predictive capabilities for forest growth, biodiversity, and ecosystem functions.

6. Improved integration of socioeconomic considerations into the development of future scenarios of climate response and air pollution impacts is needed to better understand the demographic implications of global change.

5. Summary

In summary, 21st century scenarios for changes in spatial patterns of air quality will not only be driven by natural weather and climate variability, but also by changes in demographics, land use, and economic growth. These factors may result in quite different patterns on a global, regional, or continental scale from those observed over the past century. Alterations in both climate and air pollution are the result of dynamic processes that will influence each other, and that will develop over time. Because of inherent uncertainties in our knowledge of the processes affected, our understanding of the magnitude of the response is equally uncertain. For the community of scientists engaged in research on the relationships among air pollution, climate change, and forest health and productivity, the needs of landowners and other decision makers for science-based adaptive strategies should challenge us all. Furthermore, a compelling case for an aggressive research agenda regarding research on the impacts of air pollution and climate change can easily be made.

Acknowledgements

This research was partially supported by the US Department of Energy, Office of Science (BER) (DE-FG02-95ER62125), the USDA Forest Service Northern Global Change Program, Michigan Technological University, and the USDA Forest Service North Central Research Station. The authors thank reviewers Art Chappelka and Scott Enebak for their helpful comments and suggestions on the manuscript.

References

Derwent, R., Collins, W., Johnson, C., Stevenson, D., 2002. Global ozone concentrations and regional air quality. Environ. Sci. Technol. 36, 379–382.
Dwyer, J.F., Nowak, D.J., Noble, M.H., 2003. Sustaining urban forests. J. Arboric. 29, 49–55.
Fowler, D., Cape, J.N., Coyle, M., Flechard, C., Kuylenstierna, J., Hicks, K., Derwent, D., Johnson, C., Stevenson, D., 1999. The global exposure of forests to air pollutants. Water Air Soil Pollut. 116, 5–32.
Fox, S., Mickler, R.A., 1996. Impact of Air Pollutants on Southern Pine Forests. Springer-Verlag, New York.

Galloway, J.N., Cowling, E.B., 2002. Reactive nitrogen and the world: 200 years of change. Ambio 31, 64–71.

Galloway, J.N., Cowling, E.B., Seitzinger, S.P., Socolon, R.H., 2002. Reactive nitrogen: Too much of a good thing? Ambio 31, 60–63.

Houghton, J.T., Ding, Y., Griggs, D.J., Noguer, M., van der Linden, P.J., Xiaosu, D., 2001. Climate Change 2001: The Scientific Basis. Cambridge Univ. Press, Cambridge, MA.

IPCC, 2001. Climate change 2001: Impacts, adaptation and vulnerability. Summary for policymakers. Intergovernmental Panel on Climate Change, Geneva.

Irving, P.M. 1991. Acid deposition: State of science and technology. The National Acid Precipitation Assessment Program. Washington, DC.

Isebrands, J.G., McDonald, E.P., Kruger, E., Hendrey, G., Pregitzer, K., Percy, K., Sober, J., Karnosky, D.F., 2001. Growth responses of *Populus tremuloides* clones to interacting carbon dioxide and tropospheric ozone. Environ. Pollut. 115, 359–372.

Karnosky, D.F., 2003. Impacts of elevated atmospheric CO_2 on forest trees and forest ecosystems: Knowledge gaps. Environ. Internat. 29, 161–169.

Karnosky, D.F., Scarascia-Mugnozza, G., Ceulemans, R., Innes, J., 2001. The Impact of Carbon Dioxide and Other Greenhouse Gases on Forest Ecosystems. CABI Press.

Kickert, R.N., Krupa, S.V., 1990. Forest responses to tropospheric ozone and global climate change: An analysis. Environ. Pollut. 68, 29–65.

McDonald, E.P., Kruger, E.L., Riemenschneider, D.E., Isebrands, J.G., 2002. Competitive status influences tree-growth responses to elevated CO_2 and O_3 in aggrading aspen stands. Funct. Ecol. 16, 792–801.

McLaughlin, S., Percy, K.E., 1999. Forest Health in North America: Some perspectives on actual and potential roles of climate and air pollution. Water Air Soil Pollut. 116, 151–197.

Mickler, R.A., Birdsey, R.A., Hom, J., 2000. Responses of Northern U.S. Forests to Environmental Change. Springer-Verlag, New York.

Norby, R.J., 1998. Nitrogen deposition: A component of global change analyses. New Phytol. 139, 189–200.

Norby, R.J., Wullschleger, S.D., Gunderson, C.A., Johnson, D.W., Ceulemans, R., 1999. Tree responses to rising CO_2 in field experiments: implications for the future forest. Plant Cell Environ. 22, 683–714.

Percy, K.E., Awmack, C.S., Lindroth, R.L., Kubiske, M.E., Kopper, B.J., Isebrands, J.G., Pregitzer, K.S., Hendrey, G.R., Dickson, R.E., Zak, D.R., Oksanen, E., Sober, J., Harrington, R., Karnosky, D.F., 2002. Altered performance of forest pests under CO_2- and O_3-enriched atmospheres. Nature 420, 403–407.

Volin, J.C., Reich, P.B., 1996. The interaction of elevated CO_2 and O_3 on growth, photosynthesis and respiration of three perennial species grown in low and high nitrogen. Physiol. Plant. 96, 674–684.

Volin, J.C., Reich, P.B., Givnish, T.J., 1998. Elevated carbon dioxide ameliorates the effects of ozone on photosynthesis and growth: species respond similarly regardless of photosynthetic pathway or plant functional group. New Phytol. 138, 315–325.

Air Pollution, Global Change and Forests in the New Millennium
D.F. Karnosky et al., editors
© 2003 Elsevier Ltd. All rights reserved.

57

Chapter 3

Changing atmospheric carbon dioxide: A threat or benefit?

D.F. Karnosky*, P. Sharma, R.C. Thakur, M. Kinouchi, J. King

*School of Forest Resources and Environmental Science, Michigan Technological University,
101 U.J. Noblet Forestry Building,1400 Townsend Drive, Houghton, MI 49931, USA
E-mail: karnosky@mtu.edu*

M.E. Kubiske

*USDA Forest Service, North Central Research Station, Forestry Sciences Laboratory,
5985 Highway K, Rhinelander, WI 54501, USA*

R.A. Birdsey

USDA Forest Service, 11 Campus Boulevard, Suite 200, Newtown Square, PA 19073, USA

Abstract

Atmospheric carbon dioxide (CO_2) concentrations are rapidly increasing, having risen by about 100 ppm over the last century. Atmospheric CO_2 is the basic photosynthetic building block of plants and is respired to generate the plant's energy. Atmospheric CO_2 in enhanced conditions is like an "all-you-can-eat buffet" for trees. But are there hidden threats from rising atmospheric CO_2? Are the world's forests sinks or sources of CO_2? What will they be in the future? Can we slow the rise in atmospheric CO_2 with more intensive forestry efforts? How effective can intensive forestry practices be in sequestering carbon? Simultaneous with increasing CO_2, other greenhouse gases such as NO_x and O_3 are also rapidly increasing in the atmosphere. Will these gases be beneficial or detrimental to trees under elevated CO_2? This chapter will address these questions and will use our current understanding of forest function to gain insight into how future forests will be affected by increasing greenhouse gases. Finally, we address the potential of afforestation, reforestation, agroforestry and forest management for increasing carbon sequestration.

1. Introduction

Atmospheric carbon dioxide (CO_2) currently limits the growth potential and growth rates of forest trees. However, it is rising at the rate of about $1\frac{1}{2}$–2% per

---

*Corresponding author.

DOI:10.1016/S1474-8177(03)03003-1

year and is expected to double preindustrial-period concentrations in the 21st century (Keeling et al., 1995). Does this mean that the extra atmospheric CO_2 will be beneficial or detrimental to forest trees? For example, will the rapidly rising temperature associated with elevated atmospheric CO_2 and other greenhouse gases cause massive diebacks, declines and changes in species ranges? These questions will be addressed in this paper and we will examine the potential for large scale afforestation and reforestation efforts to slow the rise in atmospheric CO_2.

2. Growth responses

Thousands of studies have been conducted to examine the impacts of elevated atmospheric CO_2 on plant photosynthesis, growth, and physiology (see reviews by Bazzaz and Fajer, 1992; Ceulemans and Mousseau, 1994; Mooney and Koch, 1994; Lee and Jarvis, 1995; Curtis and Wang, 1998; Ceulemans et al., 1999; Medlyn et al., 1999; Norby et al., 1999; Luo et al., 1999; Körner, 2000; Karnosky et al., 2001; Medlyn et al., 2001). Yet, no consensus has been developed as to whether long-term forest productivity will increase, decrease, or remain the same under elevated CO_2. While photosynthetic enhancement averages 60% under elevated CO_2 and growth enhancement averages about 27% (Norby et al., 1999), most studies have been conducted for relatively short time periods and they have been conducted in artificial conditions where plants are not exposed to other stresses or to competition (Körner, 2000; Karnosky, 2003).

2.1. Long-term studies

Long-term (we define "long-term" as 3 years or longer) studies of forest tree growth under elevated CO_2 are highly variable (0–97% enhancement) in their results suggesting that species (Norby et al., 1992; Rey and Jarvis, 1997; Tissue et al., 1997; Jach et al., 2000; Pokorný et al., 2001; Ceulemans et al., 2002; Dijkstra et al., 2002; Percy et al., 2002), genotype (Isebrands et al., 2001), soil fertility (Johnson et al., 1998; Oren et al., 2001; Sigurdsson et al., 2001; Maroco et al., 2002), and competition (McDonald et al., 2002) can affect responses. Studies where growth enhancement was found generally attributed a substantial portion of this effect to increased leaf area under elevated CO_2 in young trees (Taylor et al., 2001; Maroco et al., 2002) and enhanced photosynthetic capacity (Tissue et al., 1997; Noormets et al., 2001). While a general long-term growth trend is to have a large initial increase in growth followed by little or no increase after that

(Centritto et al. 1999a, 1999b), some studies have reported continued increases in growth enhancement throughout their studies (Pokorný et al., 2001; Dijkstra et al., 2002; Percy et al., 2002).

Whether or not growth increases will continue in closed-canopy conditions where leaf area index is not likely to be affected by elevated CO_2 and where CO_2 enhancement of photosynthesis is limited by shade is still unknown. However, Norby et al. (2002) has provided the first evidence of sustained increase in forest productivity in a closed-canopy forest. Certainly, the long-term nature of forest ecosystem responses to elevated atmospheric CO_2 is still a critical research need and one that is complicated by the coupled nature of carbon and nitrogen cycles and how these feedback to forest growth and productivity through litter fall, litter decomposition, and biogeochemical processes such as mineralization and nitrification needs critical investigation (Woodward, 2002).

It is important to note that increased photosynthesis or growth is not equivalent to increased C sequestration. The size of the C pool is not necessarily correlated with growth but rather "residence time", or how long the accumulated C actually remains in the ecosystem. Because no studies have followed forest stand development from generation to maturity, there is no evidence that residence time increases with increased exposure to CO_2, and in fact the opposite may be true.

2.2. Co-occurring stresses

Trees and forests are being exposed simultaneously to increasing atmospheric CO_2 and other stresses such as air pollution, excess nitrogen deposition or nitrogen deficiency, drought, and increasing temperatures. How will these co-occurring stresses affect responses to elevated CO_2?

2.3. Nitrogen

Globally, forests are facing steadily increasing levels of nitrogen additions related to deposition from the burning of fossil fuels (Galloway, 1989; Norby, 1998). Since most forest soils are nitrogen poor, the initial response of forest trees to increasing nitrogen levels has generally been to accelerate growth (Lloyd, 1999; Schraml et al., 2002). Studies of elevated atmospheric CO_2 and increased nitrogen deposition have generally seen greater CO_2 enhancement under higher nitrogen levels (Kerstiens et al., 1995; Murray et al., 2000a). For forests near or at "nitrogen saturation", as are many of the European forests whose soils have been contaminated by excessive nitrogen deposition for the past 100 years, it is less likely that CO_2 enhancement will occur. To verify this hypothesis, there is a need to conduct elevated CO_2 studies in those forests to

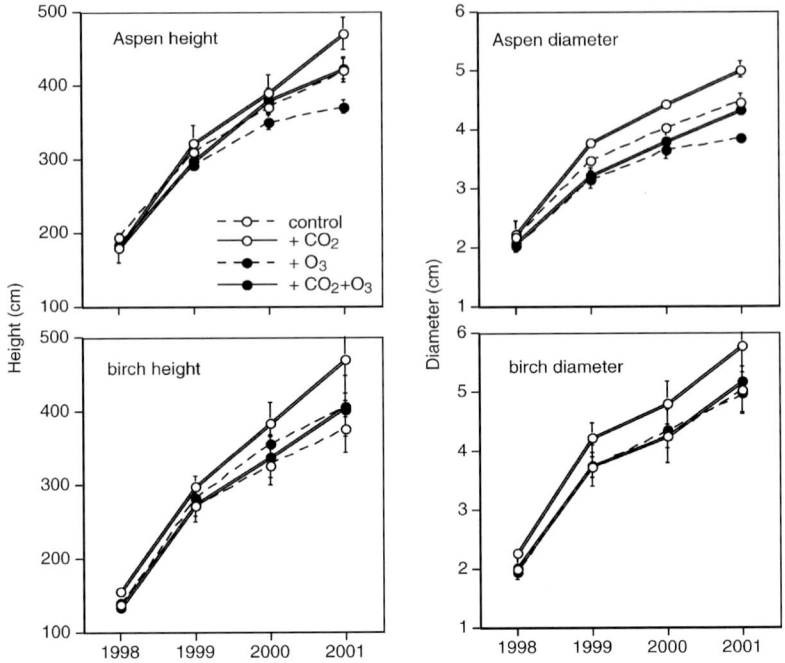

Figure 1. Impacts of interacting elevated atmospheric CO_2 and O_3 on height and diameter growth for trembling aspen (*Populus tremuloides*) and paper birch (*Betula papyrifera*) (Aspen data from Percy et al., 2002).

more accurately predict forest ecosystem responses to co-occurring elevated CO_2 and high nitrogen.

2.4. Ozone

Tropospheric ozone (O_3) is increasing worldwide at about the same rate as CO_2 (Fowler et al., 1999; IPCC, 2001). Currently, about 25% of the world's forests are exposed to damaging levels of O_3 and nearly 50% of the world's forests are expected to be exposed to damaging O_3 levels by the year 2100 (Fowler et al., 1999). Since CO_2 generally enhances growth and O_3 usually suppresses growth, it is not surprising that co-occurring $CO_2 + O_3$ generally cancel out one another (Isebrands et al., 2001; Percy et al., 2002; McDonald et al., 2002). In the largest and longest running $CO_2 + O_3$ interaction study, the growth enhancement by elevated atmospheric CO_2 is completely lost when O_3 is present at concentrations already occurring over much of the eastern United States (Isebrands et al., 2001; Percy et al., 2002; McDonald et al., 2002; Fig. 1).

2.5. Drought

As the climate changes, it is likely that the frequency and severity of droughts will increase (Cubasch et al., 2001). Since water availability largely controls tree growth (Maurer et al., 1999), there is a considerable interest developing in investigating elevated CO_2/drought interactions (Wullschleger et al., 2002). Will growth enhancement by elevated CO_2 be negated under water-limiting environments? Or, will the decreased stomatal conductance and improved water use efficiency under elevated atmospheric CO_2 mean that growth stimulation in water-limited environments is greater than in water-rich environments?

Various elevated CO_2/drought interaction studies suggest that the interaction depends on severity and frequency of drought (Centritto et al. 1999a, 1999b), and tree species (Tschaplinski et al., 1995; Samuelson and Seiler, 1993; Catovsky and Bazzaz, 1999). Competitive interactions among species under elevated CO_2 can be strongly influenced by water availability (Groninger et al., 1995; Catovsky and Bazzaz, 1999) suggesting that forest community structure may be altered in water-limited environments for forests exposed to elevated CO_2.

2.6. High temperature events

As global warming continues, there will be more frequent occurrence of high temperature stress events (Mearns et al., 1984). Whether these will have more or less effect on forest tree growth under increasing CO_2 is not known. While both increasing temperature and CO_2 are predicted to increase light-saturated photosynthetic rates and apparent quantum yields (Lewis et al., 1999), the positive effect of warmer temperatures on the CO_2-induced stimulation of photosynthesis and growth is not always observed (Olszyk et al., 1998; Tjoelker et al. 1998a, 1998b; Bruhn et al., 2000; Kellomaki and Wang, 2001). Regarding high temperature stress in combination with elevated atmospheric CO_2, some studies have shown that elevated CO_2 decreases the adverse effects of high temperature (Kriedmann et al., 1976; Faria et al. 1996, 1999; Wayne et al., 1998). Others have found an increased susceptibility to heat stress when trees are grown under elevated CO_2 (Bassow et al., 1994). Clearly, this interaction is critically important for adaptation of tree species to the rapidly changing temperature regimes and needs further examination.

3. Global warming: effects on forest ecosystems

As CO_2 and other greenhouse gases (tropospheric O_3, methane, nitrogen oxides, etc.) are increasing in the atmosphere, they are trapping radiant energy,

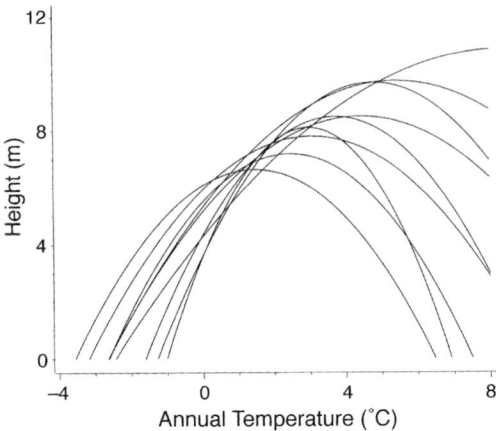

Figure 2.   Response functions of mean annual temperature as a predictor of height for nine populations for *Pinus contorta latifolia* (from Rehfeldt et al., 1999).

causing global warming (IPCC, 2001; Houghton et al., 2001). In the past century, mean global surface temperatures have increased 0.6 °C (Houghton et al., 2001). The recent warming trend is indisputable (Mann et al., 1998) and the link to anthropogenic causes is clear (IPCC, 2001; Houghton et al., 2001). The global climate is predicted to become increasingly warmer through this century with mean temperatures expected to increase from 1–5 °C (Stott et al., 2000; Stott and Kettleborough, 2002).

3.1. Species richness and range

Rising temperatures and elevated atmospheric CO_2 both tend to increase forest growth and productivity (Hasenauer et al., 1999; Caspersen et al., 2001) as do the related changes in seasonal phenology that are linked to global warming (Menzel and Fabian, 1999; Peñuelas and Filella, 2001; Parmesan and Yohe, 2003). Paradoxically, these same changes will first likely increase forest productivity but in the long term may cause major forest dieback and decline resulting in species shifts, migrations and/or extinctions (Davis and Shaw, 2001; Rehfeldt et al. 1999, 2002). Trees are adapted to their environment and populations have optimal temperatures for growth (Rehfeldt et al., 1999; Fig. 2). This large genetic variability in response has allowed tree species to adapt to change. This variability, in combination with long-distance seed and pollen dispersal has facilitated previous migrations, during both glaciation and post glaciation periods (Davis and Shaw, 2001). Currently, however, the unparalleled rate of the current global warming, along

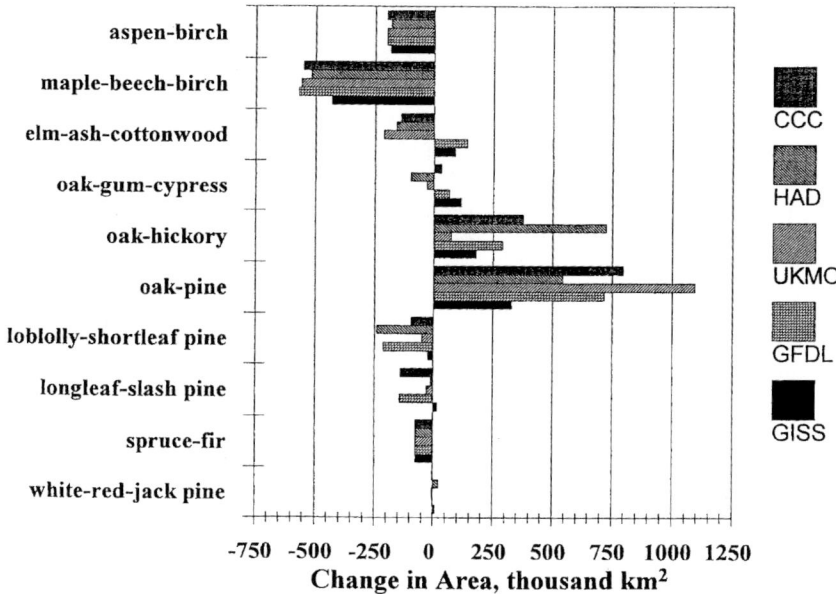

Figure 3. Potential change in area occupied by forest types in the Eastern US under five climate change scenarios (from: Iverson and Prasad, 2001). Note that the analysis has a firm boundary at the Canadian border; thus, the summary statistics do not include potentially compensating changes farther north.

with the widespread fragmentation of habitats due to anthropogenic disturbances has ecologists concerned that many species will not be able to adapt or migrate fast enough in the next century; thus, resulting in major changes in species richness and forest community types (Price et al., 1999; Currie, 2001; Iverson and Prasad, 2001; Fig. 3).

3.2. Soil

The response of root and microbial respiration to global warming is critically important in determining the response of vegetation to global climatic change (Atkin et al., 2000) and in determining if northern latitude forests will be sources or sinks of carbon (C) in the future (Rustad and Fernandez, 1998; Melillo et al., 2002). Thus, understanding how belowground systems respond to warming temperatures will be critical.

Soil warming will largely impact two main systems: the tree roots, particularly fine roots, and the microbial communities involved in litter decomposition. Root respiration is very sensitive to temperature as fine root turnover and root growth both increase with increasing temperature. Root respiration provides the energy and C skeletons necessary for ion uptake and the synthesis and maintenance of root biomass (Atkin et al., 2000). It is also a major source of CO_2 loss in plants, with 8–52% of the CO_2 fixed in photosynthesis being released back into the atmosphere by root respiration (Lambers et al., 1996). Increased soil respiration has been demonstrated to occur under elevated CO_2 (King et al., 2001a; Schlesinger and Lichter, 2001) and under elevated temperatures (Peterjohn et al., 1994; Atkin et al., 2000; Pregitzer et al., 2000). Increased frequency of hot days under global warming is likely to substantially increase soil respiration (Atkin et al., 2000).

Fine root production and turnover rates are likely tightly linked to the impact of new C from the canopy during the growing season (Pregitzer et al., 2000) so they are closely linked to whole-canopy assimilation. Both fine root growth and turnover have been shown to increase under elevated CO_2 (Tingey et al., 2000) and temperatures (King et al., 1999). However, no perceivable change in allocation of carbon to roots versus shoots has yet been detected (van Noordwijk et al., 1998).

In addition to plant respiration, litter decomposition plays a central role in regulating the balance between C assimilation and release to the atmosphere (Rustad and Fernandez, 1998). It is well known that litter decomposition is increased dramatically under increasing temperatures (Rustad and Fernandez, 1998; McHale et al., 1998; MacDonald et al., 1999; Melillo et al., 2002). However, it is less clear what effects elevated CO_2 have on litter decomposition (Norby et al., 2000). Predicted decreases under elevated CO_2 are largely predicated on changes in foliar chemical composition, which can be substantial (King et al., 2001b; Lindroth et al., 2001) or on altered microbial populations (Larson et al., 2002; Phillips et al., 2002) or their function (Zak et al. 2000a, 2000b).

3.3. Global warming and pests

Changes in atmospheric CO_2, other greenhouse gases, and related global warming, are likely to have major impacts on the occurrence, distribution, and abundance of both insect (Kopper et al., 2002; Percy et al., 2002; Kopper and Lindroth, 2003) and disease pests (Karnosky et al., 2002) that could have immense implications for the growth, productivity and biogeochemistry of future forests. The reader is referred to Chapter 1 for more information on this subject.

3.4. Net effects on carbon storage

These changes in species composition, soil biogeochemistry, and pests will surely have some effect on C storage in forest ecosystems. The net effect will

Table 1. Global carbon stocks in vegetation and the top mass of soils (from FAO, 2001)

Biome	Area (10^6 km^2)	Global C Stocks (Gt C)		
		Vegetation	Soils	Total
Tropical forests	17.6	212	216	428
Temperature forests	10.4	59	100	159
Boreal forests	13.7	88	471	559
Tropical savannas	22.5	66	264	330
Temperate grasslands	12.5	9	295	304
Deserts/semideserts	45.5	8	191	199
Tundra	9.5	6	121	127
Wetlands	3.5	15	225	240
Croplands	16.0	3	128	131
Total	151.2	466	2011	2477

depend on factors such as the degree and timing of warming, changes in precipitation and moisture availability, threshold responses, ecosystem resilience, and interactions with co-occurring stresses (Scheffer et al., 2001). Model simulations have generated hypotheses about future change in carbon storage. An emerging hypothesis from multiple models is that the early effect of a modest temperature increase is higher ecosystem productivity and carbon storage, but without a corresponding precipitation increase, ecosystems could rapidly shift to drought-induced dieback and release of carbon back to the atmosphere (Aber et al., 2001). It should be noted here that decomposition, not productivity, regulates C storage. For example, boreal forests have historically had relatively high C storage even though they have low productivity because they have low decomposition rates. In contrast, C storage in the tropical soils is limited even though productivity is high as decomposition is very high. Thus, temperature and moisture drive decomposition rates.

4. Afforestation and reforestation for carbon sequestration

Forest vegetation represents a major pool in the global cycle (Schroeder et al., 1997). Forest vegetation alone contains 350–460 Gt C (Dixon et al., 1994; FAO, 2001). Forest soils store another 2000 Gt C (FAO, 2001, Table 1). The exact size of these C pools remains an active research question as does the effect of elevated greenhouse gases and the associated global warming on these pools (Ciais et al., 1995; Fan et al., 1998; Field and Fung, 1999; Fung, 2000; Kaiser, 2000; Field, 2001; Pacala et al., 2001; Brown, 2002). We know that harvesting has a major impact on C storage of individual forest tracts (Pypker

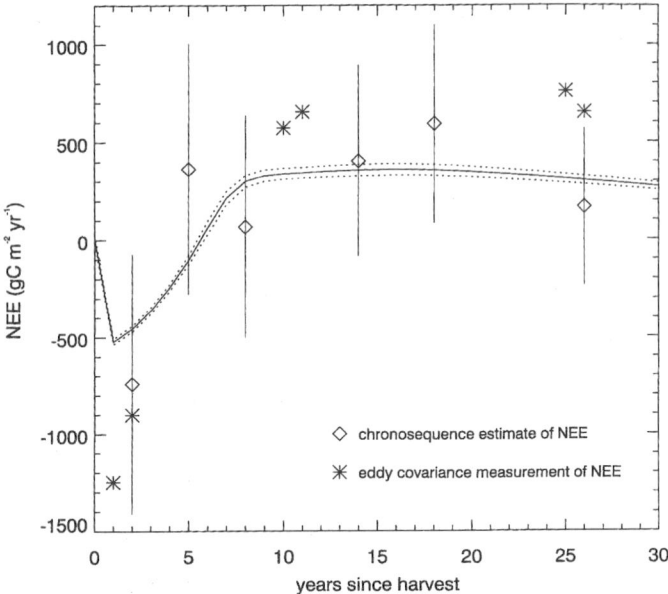

Figure 4. The effects of harvesting on C budgets is shown here in this modeled mean (solid line) and interannual standard deviation (dotted line) for simulated net ecosystem carbon exchange following harvest in a slash pine plantation at a Florida site. Eddy flux measurements of 10 EE from actual harvest site data are also shown (from Thornton et al., 2002).

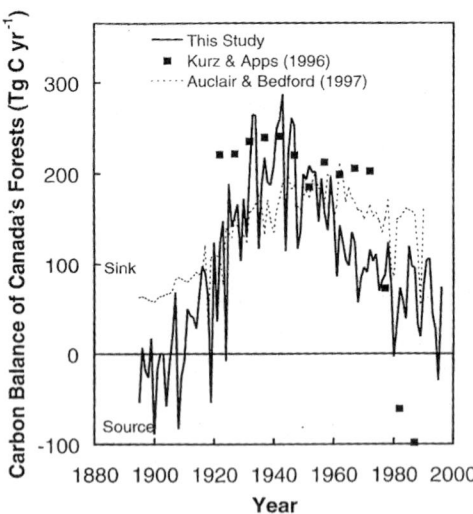

Figure 5. Comparison of C balance of Canada's forests as estimated by Chen et al. (2000) (this study), Kurz and Apps (1996) and Auclair and Bedford (1997) (from Chen et al., 2000).

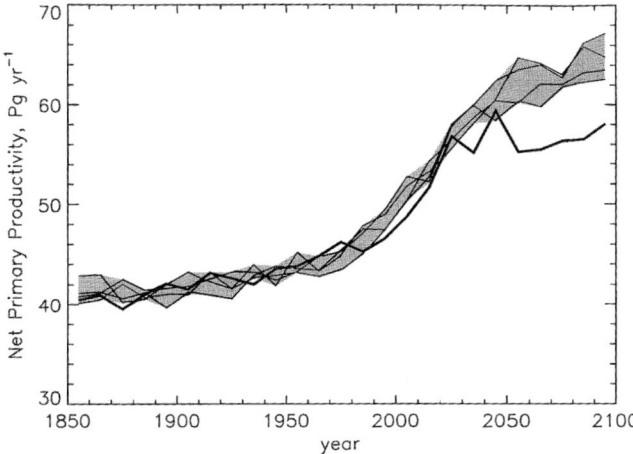

Figure 6. Global annual net primary productivity predicted by various dynamic global vegetation models (from White et al., 1999).

and Fredeen, 2002; Thornton et al., 2002; Fig. 4) and that larger forest regions are highly dynamic in their carbon balance (Kurz and Apps, 1999; Chen et al., 2000; Fig. 5) and particularly sensitive to disturbances such as insects, diseases and fire (Kurz and Apps, 1999; Goodale et al., 2002).

It is a demanding challenge to predict how forests will change in C sink or source strength under a rapidly changing climate and in the face of anthropogenic disturbances (Schroeder et al., 1997; Caspersen et al., 2001). Generally, forest productivity is predicted to increase under the changing climate for the next 50 years before stabilizing. Increases in photorespiration and maintenance respiration may cause a leveling off of photosynthesis stimulation and increasing soil respiration may limit productivity (White et al., 1999; Cox et al., 2000; Fig. 6). However, it should be noted that the C sink strength of forests is highly variable depending on forest types, forest age, moisture availability, local climate, and soil carrying capacity (Lloyd, 1999; Fig. 7).

Clearly, better-managed and fully-stocked forests and forests established with genetically superior selections offer considerable opportunities for increasing the C sink strength of the world's forests. For example, currently over 80% of the estimated C sink in northern hemisphere forests occurs in one-third of the forest (Goodale et al., 2002). It has been estimated that another 757–1144 Mt C/yr could be sequestered in the world's forests by afforestation, reforestation, improved forest management, and increased agroforestry (Kaiser, 2000; Table 2). This increased C sequestration potential under more intensive forestry practices is being closely examined as a method to decrease the buildup of CO_2 in the atmosphere (Schroeder, 1991;

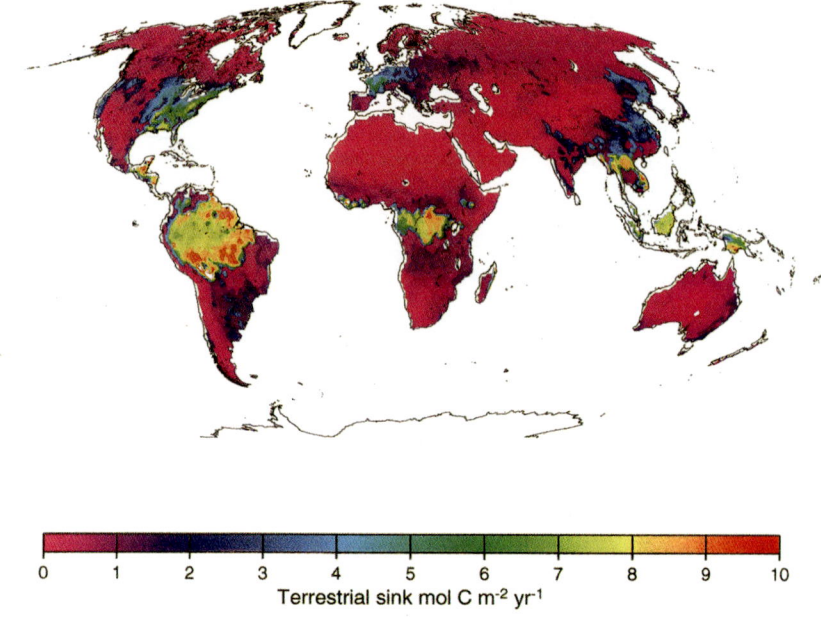

$$0 \quad 1 \quad 2 \quad 3 \quad 4 \quad 5 \quad 6 \quad 7 \quad 8 \quad 9 \quad 10$$

Terrestrial sink mol C m^{-2} yr^{-1}

Figure 7. The modeled global terrestrial CO_2 sink (from Lloyd, 1999).

Nilsson and Schopfhauser, 1995; Boman and Turnbull, 1997; Winjum and Schroeder, 1997; Fig. 8).

4.1. Traditional forests

Increasing forest productivity via intensive forestry offers considerable possibilities to sequester C including:

- using more productive trees,
- using more intensive forest management,
- increasing the acreage in afforestation and reforestation,
- substituting wood for other materials (such as plastic or metal), and
- substituting wood for other fuels (i.e., coal or oil) (Nilsson and Schopfhauser, 1995; Marland and Schlamadinger, 1997; Winjum et al., 1998; Johnsen et al., 2001; Liski et al., 2001; Table 3).

All forest biomes have undergone major changes in distribution since the last ice age (18 000 years ago) (FAO, 2001). The resulting forests now occupying these lands are not necessarily the optimal trees for forest productivity. Simply changing the species composition in a reforestation program can have a large

Table 2. Effects of various activities on carbon storage (from Kaiser, 2000)

	Carbon (10^{-6} tonn C yr^{-1})
Deforestation	up to 1788
Newly planted and regrowing forests	197 to 584
Better management of:	
• croplands	125
• grazing lands	240
• forests	170
Changes in land use	
• agroforestry	390
• cropland to grassland	38
Other	42

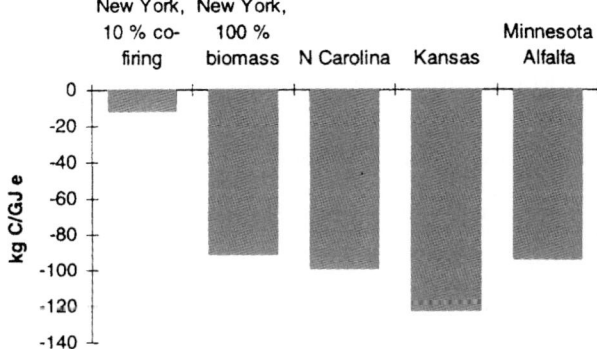

Figure 8. Reduced carbon dioxide emissions compared with a cool-based system (Kg C GJ^{-1} of electricity) (from Boman and Turnbull, 1997).

impact on "C" sequestration potential as can be seen for a comparison of three northern conifers in northern Michigan (Fig. 9).

The C sequestration potential of various afforestation/reforestation activities also depends on the site and forest management practices and forest type. Typical sequestration rates are 0.8–2.4 tons of C per year in boreal forests, 0.7–7.5 tons in temperate forests, and 3.2–10 tons in the tropical forests (Brown et al., 1996). Expansion of the world's plantations (Table 4) could increase C sequestration by some 38 Gt through the year 2050 (Brown et al., 1996; Tables 5 and 6).

Carbon sequestration policy

Under the Kyoto Protocol (Murray et al., 2000b) national emissions or uptake of C by forests are accounted on an annual basis and expressed

Table 3. There are a number of ways in which forestry research can improve the capacity of forest trees to sequester carbon

---

Near Term

Genetics

- Species changes on an existing site (optimizing species)
- Selection of the best seed sources, hybrids, or clones available today

Management

- Fertilization
- Weed control
- Irrigation
- Pest control
- Planting density

Future

Genetics and Biotechnology

- Clones selected for future climate (higher temperature, higher CO_2, etc.)
- Improve stress tolerance through genetic engineering (Metting et al., 2001)
- Genetic engineering of altered carbon allocation to stems and roots (Metting et al., 2001)
- Genetic engineering of decreased height growth, optimized photoperiod response, altered branching patterns, pest resistance or improved world chemistry (Tuskan, personal communication)
- Genetic engineering of decreased volatile organic compound production (Tuskan, personal communication)
- Genetic engineering of compounds to decrease rates of decomposition

Soils

- Selecting and/or engineering mycorrhizae to optimize tree growth rates
- Engineering "carbon-holding" soils

---

as tons of CO_2 released or sequestered (FAO, 2001). While underdeveloped countries can get C credits for decreasing forest destruction, the 39 developed countries ("Annex I") can achieve part of their greenhouse gas emission targets by enhancing sinks through CO_2 absorption in terrestrial ecosystems through land-use change and forestry (Murray et al., 2000b; Schulze et al., 2002). Thus, the Kyoto Protocol and other national policies have generated worldwide interest in forestry practices that will result in increasing C sequestration (Cannell, 1999; Sohngen and Sedjo, 2000; Lindner, 2000; Harmon, 2001; Scholes and Noble, 2001; Fang et al., 2001). Furthermore, they have generated a rudimentary "carbon market" that is developing in financial and technological transfers to support forestry projects that sequester C or protect C stocks (Smith et al., 2000; McCarl and Schneider, 2001; Plantinga and Mauldin, 2001). Some countries have already initiated massive

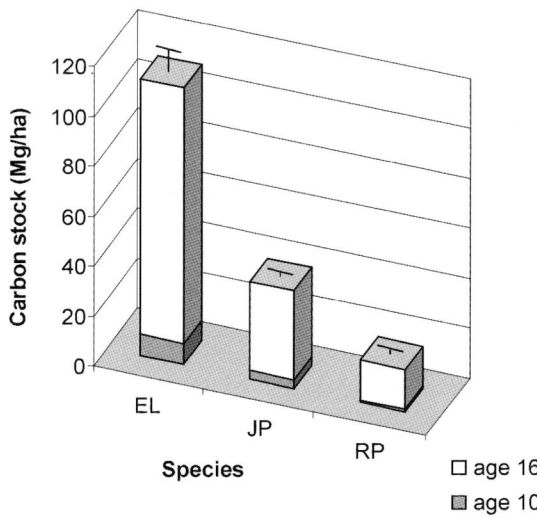

Figure 9. Comparison of the relative carbon storage potential at ages 10 and 16 of three northern temperate conifers (E.L. = European larch [*Larix decidua* Mill.], JP = jack pine [*Pinus banksiana* Lamb.], and RP = red pine [*Pinus resinosa* Ait.]) (from Kinouchi et al., 2003).

Table 4. The world's forests (from FAO, 2001)

Region	Forest (10^6 ha)	% of all forests	Natural forest (10^6 ha)	Forest plantations
Africa	650	17	642	8
Asia	548	14	432	116
Europe	1039	27	1007	32
North America	549	14	532	18
Oceania	198	5	194	3
South America	886	23	875	10
	3869	100	3682	187

tree planting programs including Canada (van Kooten et al., 1999) and China (Fang et al., 2001). For example, in a program aimed at sequestering C, restoring deforested landscapes, and stopping desertification, China has committed $12.7 billion to planting hundreds of millions of trees over the next decade (Xu et al., 2001).

4.2. Urban forests

Urban forests cover large and expanding areas of developed and developing countries around the world and offer considerable opportunities for in-

Table 5. Potential contribution of afforestation/reforestation and agroforestry activities to global carbon sequestration, 1995–2050 (from Brown et al., 1996)

Activity	Gt
Tropical afforestation/reforestation	16.7
Temperate afforestation/reforestation	11.8
Tropical agroforestry	6.5
Boreal afforestation/reforestation	2.3
Temperate agroforestry	0.7
	38.0

Table 6. Rotation periods and corresponding mean annual increments (MAIs) (from Nilsson and Schopfhauser, 1995)

Region/species	Rotation period (yr)	MAI ($m^3 ha^{-1} yr^{-1}$)
New Zealand/coniferous	25	25
Australia/coniferous	30	25
Australia/deciduous	30	22
South Africa/coniferous	25	16
Tropical Africa/deciduous	30	17
Tropical Africa/coniferous	30	15
Tropical Latin America/coniferous	15	18
Tropical Latin American/deciduous	20	25
Temperate Latin America/coniferous + deciduous	30	22
Tropical Asia/coniferous + deciduous	20	16
US south/coniferous	35	15
US temperate/coniferous	50	10
Europe:		
Nordic countries	60	5
EC-9	40	8
Central	60	6
Southern	20	10
Eastern	60	6
Former Soviet Union	80	3
Canada	60	4
China	80	2.3
Temperate Asia	40	12.0

creasing C sequestration with tree planting. Approximately 3.5% of the US is currently classified as urban with the area in metropolitan areas having increased threefold between 1950 and 1990 (Dwyer et al., 2003). It has been estimated that urban trees currently store 700 million tons of C, with an annual sequestration rate of 22.8 million tons (Nowak and Crane, 2002, Ta-

Table 7. Estimated carbon storage, gross and net annual sequestration, number of trees, and percent tree cover for ten US cities (from Nowak and Crane, 2002)

City	Storage (t C)	Annual sequestration		No. Trees ($\times 10^3$)
		Gross (t C/yr)	Net (t C/yr)	
New York, NY	1 225 200	38 400	20 800	5 212
Atlanta, GA	1 220 200	42 100	32 200	9 415
Sacramento, CA	1 107 300	20 200	na	1 733
Chicago, IL	854 800	40 100	na	4 128
Baltimore, MD	528 700	14 800	10 800	2 835
Philadelphia, PA	481 000	14 600	10 700	2 113
Boston, MA	289 800	9 500	6 900	1 183
Syracuse, NY	148 300	4 700	3 500	891
Oakland, CA	145 800	na	na	1 588
Jersey City, NJ	19 300	800	600	136

ble 7). In addition, the air-conditioning and space-heating savings reduce a significant number of CO_2 emissions (as much as 0.29 t/ha) (Heisler, 1974; McPherson, 1998).

Tree planting remains a viable option for most urban areas around the world as mortality has generally far outnumbered replanting in most cities over the past century (Gerhold et al., 2001). Planting can be done along streets, median strips, parking lots, parks, gardens, cemeteries, and in some little thought of places such as rooftops and under power lines with utility-compatible trees (Karnosky and Karnosky, 1985; Johnson and Gerhold, 2001; Fig. 10). In addition to sequestering C, urban trees help ameliorate the microclimate, provide wildlife habitat, increase property values, impact human moods, absorb air pollutants, conserve water, reduce soil erosion, and decrease noise pollution (Karnosky, 1984; Berrang and Karnosky, 1983).

With the rapid rate of urbanization that is occurring worldwide, there is clearly a need for expanded research into the role of urban forests and urban tree planting in C sequestration. The little research to date shows considerable potential for increasing C sequestration in urban forests.

4.3. Can forests be long-term carbon sinks?

Because of widespread interest in using forest C sequestration to supplement emissions reduction as a strategy to reduce atmospheric CO_2, the question of sink permanence is often raised. As the previous discussion has clearly shown, there is uncertainty about maintaining C storage in ecosystems affected by multiple stresses. But uncertainty about forest growth and health is not new. Even in the absence of climate change forests have some risk of damage and loss of C from disturbances such as harvesting, insects, diseases, and weather.

Figure 10. Urban trees have the potential to sequester carbon (see Table 7). Here, an urban tree in State College, Pennsylvania, is being excavated to examine total carbon storage of this street tree (photos by Henry Gerhold) (from Johnson, 2002).

In the 1980s, forests in the Northeastern US lost approximately 44.6 million tons/yr of C due to harvesting (some remains sequestered in solid wood products and landfills), and 11.2 million tons/yr of C due to mortality from all causes (some remains sequestered in coarse woody debris) (Birdsey, 1992). Together these annual losses comprised 2.3 percent of the live biomass. Despite atmospheric changes such as increased CO_2 and ozone concentrations, and significant amounts of sulfur and nitrogen deposition, Northeastern forests remain healthy overall and have sustained sufficient growth to allow significant accumulation of C. An important factor in maintaining healthy forests is the combined federal, state, and landowner effort to protect forests from fire, insects and diseases.

Historical data from the Northeastern US show that afforestation can successfully sequester C for more than 100 years. Large areas of forests in the Northeastern US are regrowing on agricultural land that was abandoned in the

1800s. According to recent forest inventory data, northeastern forests are currently accumulating C in live biomass at a rate of about 1 t ha^{-1} yr^{-1} (Birdsey and Heath, 1995). Johnson and Strimbeck (1995) measured the change in biomass and soil C over a 33-year period for 23 stands of aggrading sugar maple in Vermont, ranging in age (at initial measurement in 1957–1959) from 47 to 97 years, and found that only 2 stands lost biomass while 21 gained biomass. The average gain in biomass for all stands was 1.8 t ha^{-1} yr^{-1}. There was no detectable change in soil C over this period.

Some news reports have concluded that because field experiments now show that the effects of increasing atmospheric CO_2 are less than previously thought, the potential use of forests as C sinks is less feasible (e.g., Ferber, 2001). However, most analyses that attempt to determine the potential of forests for enhancing C sequestration do not include the effects of increased CO_2 in the estimates to begin with (e.g., Brown et al., 1996). The effectiveness of C sequestration strategies will depend on the continued management and protection of forest resource productivity and health provided by the nation's land managers.

5. Conclusions and knowledge gaps

The process of photosynthesis, which results in assimilation of C into plants, is currently limited by CO_2 availability in most tree species. Thus, increasing atmospheric CO_2 generally increases photosynthesis and growth in the short-term for trees growing individually or in small numbers in chambers. Less certain is how forest trees will respond in forest stands, under competition, and exposed to other stresses as well as co-occurring greenhouse gases such as O_3 which have a tendency to diminish or completely eliminate growth enhancement under elevated CO_2. Thus, long-term growth and productivity studies under realistic forest conditions and with co-occurring natural stresses are needed to reduce the uncertainty in this issue.

The world's forests vary dramatically in terms of being C sources or sinks such that the largest sinks are highly productive tropical forests. Also, strong sinks are the temperate forests of North America, Europe and Asia. Boreal forests are weak sinks, or in some cases sources, but they may contain large soil C pools. Global warming will likely affect the sink strength of all the world's forests but especially vulnerable to change are the temperate and boreal forests which may undergo large-scale changes in species richness and range. The rates of change are yet largely uncertain and use of genetically diverse planting stock will slow the rate of change and increase the adaptability of regeneration stock.

Intensive forestry efforts, including afforestation, reforestation, agroforestry, and improved forest management intensity offer opportunities to slow the rise in atmospheric CO_2. These efforts can also be used to displace a significant amount of fossil fuels further slowing the rise of atmospheric CO_2. In addition, these "C forestry" practices can have many positive environmental impacts, especially when practiced on degraded lands. Although climate change and interactions with multiple stresses are likely to increase the risk of losing C storage in forests, there is yet no evidence that the current impact of climate change and other stresses on C storage has been significant. Using C sequestration to slow the rise in atmospheric CO_2 is very likely to remain a viable part of a comprehensive strategy to manage greenhouse gases for at least the next several decades. Certainly, this remains a very important forest research topic as no one has studied "C" budgets of forest stands from establishment through harvest.

Acknowledgements

This research was partially supported by the US Department of Energy, Office of Science (DE-FG02-95ER62125), the USDA Forest Service Northern Global Change Program, Michigan Technological University, and the USDA Forest Service North Central Research Station. The authors thank reviewers Art Chappelka and Erick Schilling for their helpful comments and suggestions on the manuscript.

References

Aber, J., Neilson, R.P., McNulty, S., Lenihan, J.M., Bachelet, D., Drapek, R.J., 2001. Forest processes and global environmental change: predicting the effects of individual and multiple stressors. Bioscience 51 (9), 735–751.

Atkin, O.K., Edwards, E.J., Loveys, B.R., 2000. Response of root respiration to changes in temperature and its relevance to global warming. New Phytol. 147, 141–154.

Auclair, A.N.D., Bedford, J.A., 1997. Century trends in the volume balance of boreal forests: Implications for global CO_2 balance. In: Oechel, W.C., et al. (Eds.), Global Change and Arctic Terrestrial Ecosystems. Ecol. Stud. 124, 452–472.

Bassow, S.L., McConnaughay, K.D.M., Bazzaz, F., 1994. The response of temperature for tree seedlings grown in elevated CO_2 to extreme temperature events. Ecol. Applic. 4, 593–603.

Bazzaz, F.A., Fajer, E.D., 1992. Plant life in a CO_2-rich world. Scientific Amer. 266, 68–74.

Berrang, P., Karnosky, D.F., 1983. Street Trees for Metropolitan New York. New York Botanical Garden.

Birdsey, R.A., 1992. Carbon storage and accumulation in United States forest ecosystems. Gen. Tech. Rep. WO-59, US Department of Agriculture, Forest Service, Washington, DC.

Birdsey, R.A., Heath, L.S., 1995. Carbon changes in US forests. In: Joyce L.A. (Ed.), Productivity of America's Forests and Climate Change General Technical Report RM-271. US Department of Agriculture, Forest Service, Ft. Collins, CO, pp. 56–70.

Boman, U.R., Turnbull, J.H., 1997. Integrated biomass energy systems and emissions of carbon dioxide. Biomass Bioenerg. 13 (6), 333–343.

Brown, S., 2002. Measuring carbon in forests: current status and future challenges. Environ. Pollut. 116, 363–372.

Brown, S., Sathaye, J., Cannel, M., Kauppi, P., 1996. Management of forests for mitigation of greenhouse gases. In: Watson, R.T., Zinyowera, M.C., Moss, R.H. (Eds.), Climate Change 1995. Impacts, Adaptations, and Mitigation of Climate Change: Scientific-technical Analyses. IPCC Assessment Report, Working Group II. Cambridge Univ. Press, Cambridge, UK, pp. 773–797.

Bruhn, D., Leverenz, J.W., Saxe, H., 2000. Effects of tree size and temperature on relative growth rate and its components of *Fagus sylvatica* seedlings exposed to two partial pressures of atmospheric [CO_2]. New Phytol. 146, 415–425.

Cannell, M.G.R., 1999. Growing trees to sequester carbon in the UK: Answers to some common questions. Forestry 72 (3), 237–247.

Caspersen, J.P., Pacala, S.W., Jenkins, J.C., Hartt, G.C., Moorcroft, P.R., Birdsey, R.A., 2001. Contributions of land-use history to carbon accumulation in US forests. Science 290 (5494), 1148–1151.

Catovsky, S., Bazzaz, F.A., 1999. Elevated CO_2 influences the responses of two birch species to soil moisture: implications for forest community structure. Global Change Biol. 5, 507–518.

Centritto, M., Lee, H.S.J., Jarvis, P.G., 1999a. Interactive effects of elevated [CO_2] and drought on cherry (*Prunus avium*) seedlings. New Phytol. 141, 129–140.

Centritto, M., Magnani, F., Lee, H.S.J., Jarvis, P.G., 1999b. Interactive effects of elevated [CO_2] and drought on cherry (*Prunus avium*) seedlings. II. Photosynthetic capacity and water relations. New Phytol. 141, 141–153.

Ceulemans, R., Mousseau, M., 1994. Effects of elevated atmospheric CO_2 on woody plants. New Phytol. 127, 425–446.

Ceulemans, R., Janssens, I.A., Jach, M.E., 1999. Effects of CO_2 enrichment on trees and forests: Lessons to be learned in view of future ecosystem studies. Ann. Bot. 84, 577–590.

Ceulemans, R., Jach, M.E., Van De Velde, R., Lin, J.X., Stevens, M., 2002. Elevated atmospheric CO_2 alters wood production, wood quality and wood strength of Scots pine (*Pinus sylvestris* L) after three years of enrichment. Global Change Biol. 8, 153–162.

Chen, J., Chen, W., Liu, J., Cihlar, J., 2000. Annual carbon balance of Canada's forests during 1895–1996. Global Biogeochem. Cycles 14 (3), 839–849.

Ciais, P., Tans, P.P., Trolier, M., White, J.W.C., Francey, R.J., 1995. A large northern hemisphere terrestrial CO_2 sink indicated by the $^{13}C/^{12}C$ ratio of atmospheric CO_2. Science 269, 1098–1102.

Cox, P.M., Betts, R.A., Jones, C.D., Spall, S.A., Totterdell, I.J., 2000. Acceleration of global warming due to carbon-cycle feedbacks in a coupled climate model. Nature 408, 184–187.

Cubasch, U., Meehl, G.A., Boer, G.J., Stouffer, R.J., Dix, M., Nada, A., Senior, C.A., Ruper, S., Yap, K.S., et al., 2001. Projections of future climate change. In: Houghton, J.T., Ding, Y., Griggs, D.J., Noguer, M., Van der Linden, P.J., Dai, S., Maskell, K., Johnson, C.A. (Eds.), Climate Change 2001: The Scientific Basis. Cambridge Univ. Press, Cambridge, UK, pp. 525–582.

Currie, D.J., 2001. Projected effects of climate change on patterns of vertebrate and tree species richness in the conterminous United States. Ecosystems 4, 216–225.

Curtis, P.S., Wang, X., 1998. A meta-analysis of elevated CO_2 effects on woody plant mass, form, and physiology. Oecologia 113, 299–313.

Davis, M.B., Shaw, R.G., 2001. Range shifts and adaptive responses to quaternary climate change. Science 292, 673–679.

Dijkstra, P., Hymus, G., Colavito, D., Vieglais, D.A., Cundari, C.M., Johnson, D.P., Hungate, B.A., Hinkle, C.R., Drake, B.G., 2002. Elevated atmospheric CO_2 stimulates aboveground biomass in a fire-regenerated scrub-oak ecosystem. Global Change Biol. 8, 90–103.

Dixon, R.K., Brown, S., Houghton, R.A., Solomon, A.M., Trexler, M.C., Wisniewski, J., 1994. Carbon pools and flux of global forest ecosystems. Science 263, 185–190.

Dwyer, J.F., Nowak, D.J., Noble, M.H., 2003. Sustaining urban forests. J. Arboric. 29, 49–55.

Fan, S., Gloor, M., Mahlman, J., Pacala, S., Sarmiento, J., Takahashi, T., Tans, P., 1998. A large terrestrial carbon sink in North America implied by atmospheric and oceanic carbon dioxide data and models. Science 282, 442–446.

Fang, J., Chen, A., Peng, C., Zhao, S., Ci, L., 2001. Changes in forest biomass carbon storage in China between 1949 and 1998. Science 292, 2320–2322.

FAO, 2001. State of the World's Forests 2001. The FAO Forestry Department, Rome, Italy.

Ferber, D., 2001. Superweeds, and a sinking feeling on carbon sinks. Science 293, 142.

Fowler, D., Cape, J.N., Coyle, M., Flechard, C., Kuylenstierna, J., Hicks, K., Derwent, D., Johnson, C., Stevenson, D., 1999. The global exposure of forests to air pollutants. Water Air Soil Pollut. 116, 5–32.

Faria, T., Wilkins, D., Besford, B., Vaz, M., Pereira, J.S., Chaves, M.M., 1996. Growth at elevated CO_2 leads to down-regulation of photosynthesis and altered response to high temperature in *Quercus suber* L. seedlings. J. Exp. Bot. 47, 1755–1761.

Faria, T., Vaz, M., Schwanz, P., Polle, A., Pereira, J.S., Chaves, M.M., 1999. Responses of photosynthetic and defence systems to high temperature stress in *Quercus suber* L. seedlings grown under elevated CO_2. Plant Biol. 1, 365–371.

Field, C.B., 2001. Plant physiology of the "missing" carbon sink. Plant Physiol. 125, 25–28.

Field, C.B., Fung, I.Y., 1999. The not-so-big US carbon sink. Science 285, 544–545.

Fung, I., 2000. Variable carbon sinks. Clim. Change 290, 1313.

Galloway, J.N., 1989. Atmospheric acidification projections for the future. Ambio 18, 161–166.

Gerhold, H.D., Lacasse, N.L., Wandell, W.N., 2001. Landscape Tree Factsheets, 3rd Edition. The Pennsylvania State University, College of Agricultural Science.

Goodale, C.L., Apps, M.J., Birdsey, R.A., Field, C.B., Heath, L.S., Houghton, R.A., Jenkins, J.C., Kohlmaier, G.H., Kurz, W., Liu, S., Nabuurs, G.-J., Nilsson, S., Shvidenko, A.Z., 2002. Forest carbon sinks in the northern hemisphere. Ecol. Applic. 12 (3), 891–899.

Groninger, J.W., Seiler, J.R., Zedaker, S.M., Berrang, P.C., 1995. Effects of elevated CO_2, water stress, and nitrogen level on competitive interactions of simulated loblolly pine and sweetgum stands. Can. J. Forest Res. 25, 1077–1083.

Harmon, M.E., 2001. Carbon sequestration in forests. Addressing the scale question. J. For. 99 (4), 24–29.

Hasenauer, H., Nemani, R.R., Schadauer, K., Running, S.W., 1999. Forest growth response to changing climate between 1961 and 1990 in Austria. Forest Ecol. Manag. 122, 209–219.

Heisler, G.M., 1974. Trees and human comfort in urban areas. J. For. 72, 466–469.

Houghton, J.T., Ding, Y., Griggs, D.J., Noguer, M., van der Linden, P.J., Xiaosu, D., 2001. Climate change 2001: The scientific basis. Cambridge Univ. Press, Cambridge, UK.

IPCC, 2001. Climate change 2001: Impacts, adaptation and vulnerability. Summary for policymakers. Intergovernmental Panel on Climate Change, Geneva.

Isebrands, J.G., McDonald, E.P., Kruger, E., Hendrey, G., Pregitzer, K., Percy, K., Sober, J., Karnosky, D.F., 2001. Growth responses of *Populus tremuloides* clones to interacting carbon dioxide and tropospheric ozone. Environ. Pollut. 115, 359–371.

Iverson, L.R., Prasad, A.M., 2001. Potential changes in trees species richness and forest community types following climate change. Ecosystems 4, 186–199.

Jach, M.E., Laureysens, I., Ceulemans, R., 2000. Above- and below-ground production of young Scots pine (*Pinus sylvestris* L.) trees after three years of growth in the field under elevated CO_2. Ann. Bot. 85, 789–798.

Johnson, D.W., Thomas, R.B., Griffin, K.L., Tissue, D.T., Ball, J.T., Strain, B.R., Walker, R.F., 1998. Effects of carbon dioxide and nitrogen on growth and nitrogen uptake in ponderosa and loblolly pine. J. Environ. Qual. 27, 414–425.

Johnson, A.D., 2002. Carbon storage in roots of urban tree cultivars. Ph.D. Thesis. Pennsylvania State University.

Johnson, A.D., Gerhold, H.D., 2001. Carbon storage by utility-compatible trees. J. Arboric. 27 (2), 57–68.

Johnson, A.H., Strimbeck, G.R., 1995. Thirty-three year changes in above- and below-ground biomass in northern hardwood stands in Vermont. In: Hom, J. (Ed.), Proceedings, 1995 Meeting of the Northern Global Change Program. USDA Forest Service, Northeastern Forest Experiment Station, Radnor, PA, pp. 169–174. Gen. Tech. Rep. NE-214.

Johnsen, K.H., Wear, D., Oren, R., Teskey, R.O., Sanchez, F., Will, R., Butnor, J., Markewitz, D., Ritcher, D., Rials, T., Allen, H.L., Seiler, J., Ellsworth, D., Maier, C., Katul, G., Dougherty, P.M., 2001. Meeting global policy commitments: carbon sequestration and southern pine forests. J. For. 99 (4), 14–21.

Karnosky, D.F., 2003. Impacts of elevated CO_2 on forest trees and forest ecosystems: Knowledge gaps. Environ. Internat. 29, 161–169.

Karnosky, D.F., 1984. Urban forestry comes of age. Orion Nature Quart. 3 (4), 46–53.

Karnosky, D.F. and Karnosky, S.L., 1985. Improving the quality of urban life with plants. New York Botanical Garden Institute of Urban Horticulture Pub. No. 2.

Karnosky, D.F., Scarascia-Mugnozza, G., Ceulemans, R., Innes, J. (Eds.), 2001. The Impact of Carbon Dioxide and Other Greenhouse Gases on Forest Ecosystems. CABI Publishing, New York.

Karnosky, D.F., Percy, K.E., Xiang, B., Callan, B., Noormets, A., Mankovska, B., Hopkin, A., Sober, J., Jones, W., Dickson, R.E., Isebrands, J.G., 2002. Interacting elevated CO_2 and tropospheric O_3 and predisposes aspen (*Populus tremuloides* Michx.) to infection by rust (*Melampsora medusae* f.sp. *tremuloidae*). Global Change Biol. 8, 329–338.

Kaiser, J., 2000. Panel estimates possible carbon 'sinks'. Science 288, 942–943.

Keeling, C.M., Whort, T.P., Wahlen, M., Van der Plict, J., 1995. International extremes in the rate of rise of atmospheric carbon dioxide since 1980. Nature 375, 666–670.

Kellomaki, S., Wang, K.-Y., 2001. Growth and resource use of birch seedlings under elevated carbon dioxide and temperature. Ann. Bot. 87, 669–682.

Kerstiens, G., Toenend, J., Heath, J., Mansfield, T.A., 1995. Effects of water and nutrient availability on physiological responses of woody species to elevated CO_2. Forestry 68, 303–315.

King, J.S., Pregitzer, K.S., Zak, D.R., 1999. Clonal variation in above- and below-ground growth responses of *Populus tremuloides* Michaux: Influence of soil warming and nutrient availability. Plant Soil 217, 130–199.

King, J.S., Pregitzer, K.S., Zak, D.R., Sober, J., Isebrands, J.G., Dickson, R.E., Hendrey, G.R., Karnosky, D.F., 2001a. Fine root biomass and fluxes of soil carbon in young stands of paper birch and trembling aspen as affected by elevated atmospheric CO_2 and tropospheric O_3. Oecologia 128, 237–250.

King, J.S., Pregitzer, K.S., Zak, D.R., 2001b. Correlation between the chemistry of foliage and litter in sugar maple (*Acer saccharum* Marsh.) as influenced by elevated CO_2 and varying N availability, and its decomposition. Oikos 94, 403–416.

Kinouchi, M., Jones, W.S., Megown, K.A., King, J.S., Karnosky, D.F., 2003. The relative ability of northern conifers to sequester carbon in the Upper Peninsula of Michigan. Northern J. Appl. For. (submitted).

Kopper, B.J., Lindroth, R.L., Nordheim, E.V., 2002. CO_2 and O_3 effects on paper birch (Betulaceae: *Betula papyrifera* Marsh.) phytochemistry and white-marked tussock moth (Lymantriidae: *Orgyia leucostigma J.E. Sm.*) performance. Environ. Entomol. 30 (6), 1119–1126.

Kopper, B.J., Lindroth, R.L., 2003. Effects of elevated carbon dioxide and ozone on the genotypic response of aspen phytochemistry and the performance of an herbivore. Oecologia 134 (1), 95–103.

Körner, C., 2000. Biosphere responses to CO_2 enrichment. Ecol. Applic. 10 (6), 1590–1619.

Kriedmann, P.E., Sward, R.Y., Downton, W.J.S., 1976. Vine response to carbon dioxide enrichment during heat therapy. Austr. J. Plant Physiol. 3, 605–618.

Kurz, W.A., Apps, M.J., 1996. Retrospective assessment of carbon flows in Canadian boreal forests. In: Apps, M.J., Price, D.T. (Eds.), Forest Ecosystems, Forest Management and the Global Carbon Cycle. In: NATO ASI Ser. I, Vol. 40, pp. 173–182.

Kurz, W.A., Apps, M.J., 1999. A 70-year retrospective analysis of carbon fluxes in the Canadian forest sector. Ecol. Applic. 9 (2), 526–547.

Lambers, H., Atkin, O.K., Scheurwater, I., 1996. Respiratory patterns in roots in relation to their functioning. In: Waisel, Y., Eshel, A., Kafkaki, U. (Eds.), Plant Roots. The Hidden Half. Marcel Dekker, New York, pp. 323–362.

Larson, J.L., Zak, D.R., Sinsabaugh, R.L., 2002. Microbial activity beneath temperate trees growing under elevated CO_2 and O_3. Soil Sci. Soc. Amer. J. 66, 1848–1856.

Lee, H.S.J., Jarvis, P.G., 1995. Trees differ from crops and from each other in their responses to increases in CO_2 concentration. J. Biogeogr. 22, 323–330.

Lewis, J.D., Olszyk, D., Tingey, D.T., 1999. Seasonal patterns of photosynthetic light response in Douglas-fir seedlings subjected to elevated atmospheric CO_2 and temperature. Tree Physiol. 19, 243–252.

Lindner, M., 2000. Developing adaptive forest management strategies to cope with climate change. Tree Physiol. 20, 299–307.

Lindroth, R.L., Kopper, B.J., Parsons, W.F.J., Bockheim, J.G., Sober, J., Hendrey, G.R., Pregitzer, K.S., Isebrands, J.G., Karnosky, D.F., 2001. Effects of elevated carbon dioxide and ozone on foliar chemical composition and dynamics in trembling aspen (*Populus tremuloides*) and paper birch (*Betula papyrifera*). Environ. Pollut. 115, 395–404.

Liski, J., Pussinen, A., Pingoud, K., Mäkipää, R., Karjalainen, T., 2001. Which rotation length is favourable to carbon sequestration? Can J. Forest Res. 31, 2004–2013.

Lloyd, J., 1999. The CO_2 dependence of photosynthesis, plant growth responses to elevated CO_2 concentrations and their interaction with soil nutrient status, II. Temperate and boreal forest productivity and the combined effects of increasing CO_2 concentrations and increased nitrogen deposition at a global scale. Funct. Ecol. 13, 439–459.

Luo, Y., Reynolds, J., Wang, Y., Wolfe, D., 1999. A search for predictive understanding of plant responses to elevated [CO_2]. Global Change Biol. 5, 143–156.

MacDonald, N.W., Randlett, D.L., Zak, D.R., 1999. Soil warming and carbon loss from a Lakes States spodosol. Soil Sci. Soc. Amer. J. 63, 211–218.

Mann, M.E., Bradley, R.S., Hughes, M.K., 1998. Global-scale temperature patterns and climate forcing over the past six centuries. Nature 392, 779–787.

Marland, G., Schlamadinger, B., 1997. Forests for carbon sequestration or fossil fuel substitution? A sensitivity analysis. Biomass Bioenerg. 13 (6), 389–397.

Maroco, J.P., Breia, E., Faria, T., Pereira, J.S., Chaves, M.M., 2002. Effects of long-term exposure to elevated CO_2 and N fertilization on the development of photosynthetic capacity and biomass accumulation in *Quercus suber* L. Plant Cell Environ. 25, 105–113.

Maurer, S., Egli, P., Spinnler, D., Körner, C.H., 1999. Carbon and water fluxes in Beech-Spruce model ecosystems in response to long-term exposure to atmospheric CO_2 enrichment and increased nitrogen deposition. Funct. Ecol. 13, 748–755.

McCarl, B.A., Schneider, U.W., 2001. Greenhouse gas mitigation in US agriculture and forestry. Science 294, 2481–2482.

McDonald, E.P., Kruger, E.L., Riemenschneider, D.E., Isebrands, J.G., 2002. Competitive status influences tree-growth responses to elevated CO_2 and O_3 in aggrading aspen stands. Funct. Ecol. 16 (6), 792–801.

McHale, P.J., Mitchell, M.J., Bowles, F.P., 1998. Soil warming in a northern hardwood forest: trace gas fluxes and leaf litter decomposition. Can. J. Forest Res. 28, 1365–1372.

McPherson, E.G., 1998. Atmospheric carbon dioxide reduction by Sacramento's urban forest. J. Arboric. 24, 215–223.

Mearns, L.O., Katz, R.W., Schneider, S.H., 1984. Extreme high temperature events: changes in their probabilities with changes in the mean temperature. J. Climatol. Appl. Meteorol. 23, 1601–1613.

Medlyn, B.E., Badeck, F.-W., De Pury, D.G.G., Barton, C.V.M., Broadmeadow, M., Ceulemans, R., De Angelis, P., Forstreuter, M., Jach, M.E., Kellomaki, S., Laitat, E., Marek, M., Philippot, S., Rey, A., Strassemeyer, J., Laitinen, K., Liozon, R., Portier, B., Roberntz, P., Wang, K., Jarvis, P.G., 1999. Effects of elevated $[CO_2]$ on photosynthesis in European forest species: a meta-analysis of model parameters. Plant Cell Environ. 22, 1475–1495.

Medlyn, B.E., Barton, C.V.M., Broadmeadow, M.S.J., Ceulemans, R., De Angelis, P., Forstreuter, M., Freeman, M., Jackson, S.B., Kellomaki, S., Laitat, E., Rey, A., Roberntz, P., Sigurdsson, B.D., Strassemeyer, J., Wang, K., Curtis, P.S., Jarvis, P.G., 2001. Stomatal conductance of forest species after long-term exposure to elevated CO_2 concentration: a synthesis. New Phytol. 149, 247–264.

Melillo, J.M., Steudler, P.A., Aber, J.D., Newkirk, K., Lux, H., Bowles, F.P., Catricala, C., Magill, A., Ahrens, T., Morrisseau, S., 2002. Soil warming and carbon-cycle feedbacks to the climate system. Science 298, 2173–2176.

Menzel, A., Fabian, P., 1999. Growing season extended in Europe. Nature 397, 659.

Metting, F.B., Smith, J.L., Amthor, J.S., Izaurralde, R.C., 2001. Science needs and new technology for increasing soil carbon sequestration. Clim. Change 51, 11–34.

Mooney, H.A., Koch, G.W., 1994. The impact of rising CO_2 concentrations on the terrestrial biosphere. Ambio 23, 74–76.

Murray, M.B., Smith, R.I., Friend, A., Jarvis, P.G., 2000a. Effect of elevated CO_2 and varying nutrient supply application rates on physiology and biomass accumulation of Sitka spruce (*Picea sitchensis*). Tree Physiol. 20, 421–434.

Murray, B.C., Prisley, S.P., Birdsey, R.A., Sampson, R.N., 2000b. Carbon sinks in the Kyoto Protocol. J. For. 98 (9), 6–11.

Nilsson, S., Schopfhauser, W., 1995. The carbon-sequestration potential of a global afforestation program. Clim. Change 30, 267–293.

Noormets, A., Sôber, A., Pell, E.J., Dickson, R.E., Podila, G.K., Sôber, J., Isebrands, J.G., Karnosky, D.F., 2001. Stomatal and nonstomatal control of photosynthesis in trembling aspen (*Populus tremuloides* Mich.) exposed to elevated CO_2 and O_3. Plant Cell Environ. 24, 327–336.

Norby, R.J., 1998. Nitrogen deposition: A component of global change analyses. New Phytol. 139 (1), 189–200.

Norby, R.J., Cotrufo, M.F., Ineson, P., O'Neill, E.G., Canadell, J.G., 2000. Elevated CO_2, litter chemistry, and decomposition: A synthesis. Oecologia 127 (2), 153–165.

Norby, R.J., Gunderson, C.A., Wullschleger, S.D., O'Neill, E.G., McCracken, M.K., 1992. Productivity and compensatory responses of yellow-poplar trees in elevated CO_2. Nature 357, 322–324.

Norby, R.J., Wullschleger, S.D., Gunderson, C.A., Johnson, D.W., Ceulemans, R., 1999. Tree responses to rising CO_2 in field experiments: implications for the future. Plant Cell Environ. 22, 683–714.

Norby, R.J., Hanson, P.J., O'Neill, E.G., Tschaplinski, T.J., Weltzin, J.F., Hansen, R.A., Cheng, W., Wullschleger, S.D., Gunderson, C.A., Edwards, N.T., Johnson, D.W., 2002. Net primary productivity of a CO_2-enriched deciduous forest and the implications for carbon storage. Ecol. Applic. 12 (5), 1261–1266.

Nowak, D.J., Crane, D.E., 2002. Carbon storage and sequestration by urban trees in the USA. Environ. Pollut. 116, 381–389.

Olszyk, D., Wise, C., VanEss, E., Tingey, D., 1998. Elevated temperature but not elevated CO_2 affects long-term patterns of stem diameter and height of Douglas-fir seedlings. Can. J. Forest Res. 28, 1046–1054.

Oren, R., Ellsworth, D.S., Johnson, K.H., Phillips, N., Ewers, B.E., Maier, C., Schafer, K.V.R., McCarthy, H., Hendrey, G., McNulty, S.G., Katul, G., 2001. Soil fertility limits carbon sequestration by forest ecosystems in a CO_2-enriched atmosphere. Nature 411, 469–472.

Pacala, S.W., Hurtt, G.C., Baker, D., Peylin, P., Houghton, R.A., Birdsey, R.A., Heath, L., Sundquist, E.T., Stallard, R.F., Ciais, P., Moorcroft, P., Caspersen, J.P., Shevliakova, E., Moore, B., Kohlmaier, G., Holland, E., Gloor, M., Harmon, M.E., Fan, S.-M., Sarmiento, J.L., Goodale, C.L., Schimel, D., Field, C.B., 2001. Consistent land- and atmosphere-based US carbon sink estimates. Science 292, 2316–2320.

Parmesan, C., Yohe, G., 2003. A globally coherent fingerprint of climate change impacts across natural systems. Nature 421, 37–42.

Peñuelas, J., Filella, I., 2001. Responses to a warming world. Science 294, 793–795.

Percy, K.E., Awmack, C.S., Lindroth, R.L., Kubiske, M.E., Kopper, B.J., Isebrands, J.G., Pregitzer, K.S., Hendrey, G.R., Dickson, R.E., Zak, D.R., Oksanen, E., Sober, J., Harrington, R., Karnosky, D.F., 2002. Altered performance of forest pests under CO_2- and O_3-enriched atmospheres. Nature 420, 403–407.

Peterjohn, W.T., Melillo, J.M., Steudler, P.A., Newkirk, K.M., Bowles, F.P., Aber, J.D., 1994. Responses of trace gas fluxes and N availability to experimentally elevated soil temperatures. Ecol. Applic. 4 (3), 617–625.

Phillips, R.L., Zak, D.R., Holmes, W.E., White, D.C., 2002. Microbial community composition and function beneath temperate trees exposed to elevated atmospheric carbon dioxide and ozone. Oecologia 131 (2), 236–244.

Plantinga, A.J., Mauldin, T., 2001. A method for estimating the cost of CO_2 mitigation through afforestation. Clim. Change 49, 21–40.

Pokorný, R., Šalanská, P., Janouš, D., 2001. Growth and transpiration of Norway spruce trees under atmosphere with elevated CO_2 concentration. Ekologia 20 (1), 14–28.

Pregitzer, K.S., King, J.S., O'Neill, E.G., Burton, A.J., Brown, S.E., 2000. Responses of tree fine roots to temperature. New Phytol. 147, 105–115.

Price, D.T., Peng, C.H., Apps, M.J., Halliwell, D.H., 1999. Simulating effects of climate change on boreal ecosystem carbon pools in central Canada. J. Biogeogr. 26, 1237–1248.

Pypker, T.G., Fredeen, A.L., 2002. Ecosystem CO_2 flux over two growing seasons for a sub-Boreal clearcut 5 and 6 years after harvest. Agric. For. Meteorol. 114, 15–30.

Rehfeldt, G.E., Ying, C.C., Spittlehouse, D.L., Hamilton Jr., D.A., 1999. Genetic responses to climate change in *Pinus contorta*: niche breadth, climate change, and reforestation. Ecol. Monogr. 69 (3), 375–407.

Rehfeldt, G.E., Tchebakova, N.M., Parfenova, Y.I., Wykoff, W.R., Kuzmina, N.A., Milyutin, L.I., 2002. Intraspecific responses to climate in *Pinus sylvestris*. Global Change Biol. 8, 912–929.

Rey, A., Jarvis, P.G., 1997. Growth response of young birch trees (*Betula pendula* Roth.) after four and a half years of CO_2 exposure. Ann. Bot. 80, 809–816.

Rustad, L.E., Fernandez, I.J., 1998. Soil warming: Consequences for foliar litter decay in a spruce-fir forest in Maine, USA. Soil Sci. Soc. Amer. J. 62, 1072–1080.

Samuelson, L.J., Seiler, J.R., 1993. Interactive role of elevated CO_2, nutrient limitations, and water stress in the growth responses of red spruce seedlings. Forest Sci. 39 (2), 348–358.

Scheffer, M., Carpenter, S., Foley, J., Folke, C., Walker, B., 2001. Catastrophic shifts in ecosystems. Nature 413, 591–596.

Schlesinger, W.H., Lichter, J., 2001. Limited carbon storage in soil and litter of experimental forest plots under increased atmospheric CO_2. Nature 411 (6836), 466–469.

Schraml, C., Herschbach, C., Eiblmeier, M., Rennenberg, H., 2002. Consequences of elevated CO_2, augmented nitrogen-deposition and soil type on the soluble nitrogen and sulphur in the phloem of beech (*Fagus sylvatica*) and spruce (*Picea abies*) in a competitive situation. Physiol. Plant. 115, 258–266.

Scholes, R.J., Noble, I.R., 2001. Storing carbon on land. Science 294, 1012–1013.

Schroeder, P., 1991. Can intensive management increase carbon storage in forests? Environ. Manag. 15 (4), 475–481.

Schroeder, P., Brown, S., Mo, J., Birdsey, R., Cieszewski, C., 1997. Biomass estimation for temperate broadleaf forests of the United States using inventory data. Forest Sci. 43 (3), 424–434.

Schulze, E.-D., Valentini, R., Sanz, M.-J., 2002. The long way from Kyoto to Marrakesh: Implications of the Kyoto Protocol negotiations for global ecology. Global Change Biol. 8, 505–518.

Sigurdsson, B.D., Thorgeirsson, H., Linder, S., 2001. Growth and dry-matter partitioning of young *Populus trichocarpa* in response to carbon dioxide concentration and mineral nutrient availability. Tree Physiol. 21, 941–950.

Smith, J., Mulongoy, K., Persson, R., Sayer, J., 2000. Harnessing carbon markets for tropical forest conservation: towards a more realistic assessment. Environ. Conserv. 27 (3), 300–311.

Sohngen, B., Sedjo, R., 2000. Potential carbon flux from timber harvest and management in the context of a global timber market. Clim. Change 44, 151–172.

Stott, P.A., Kettleborough, J.A., 2002. Origins and estimates of uncertainty in predictions of twenty-first century temperature rise. Nature 416, 723–726.

Stott, P.A., Tett, S.F.B., Jones, G.S., Allen, M.R., Mitchell, J.F.B., Jenkins, G.J., 2000. External control of 20th century temperature by natural and anthropogenic forcings. Science 290, 2133–2137.

Taylor, G., Ceulemans, R., Ferris, R., Gardner, S.D.L., Shao, B.Y., 2001. Increased leaf area expansion of hybrid poplar in elevated CO_2. From controlled environments to open-top chambers and to FACE. Environ. Pollut. 115, 463–472.

Thornton, P.E., Law, B.E., Gholz, H.L., Clark, K.L., Falge, E., Ellsworth, D.S., Goldstein, A.H., Monson, R.K., Hollinger, D., Falk, M., Chen, J., Sparks, J.P., 2002. Modeling and measuring the effects of disturbance history and climate on carbon and water budgets in evergreen needleleaf forests. Agric. For. Meteor. 113, 185–222.

Tingey, D.T., Phillips, D.L., Johnson, M.G., 2000. Elevated CO_2 and conifer roots: effects on growth, life span and turnover. New Phytol. 147, 87–103.

Tissue, D.T., Thomas, R.B., Strain, B.R., 1997. Atmospheric CO_2 enrichment increases growth and photosynthesis of *Pinus taeda*: a 4 year experiment in the field. Plant Cell Environ. 20, 1123–1134.

Tjoelker, M.G., Oleksyn, J., Reich, P.B., 1998a. Seedlings of five boreal trees species differ in acclimation of net photosynthesis to elevated CO_2 and temperature. Tree Physiol. 18, 715–726.

Tjoelker, M.G., Oleksyn, J., Reich, P.B., 1998b. Temperature and ontogeny mediate growth response to elevated CO_2 in seedlings of five boreal tree species. New Phytol. 140, 197–210.

Tschaplinski, T.J., Stewart, D.B., Hanson, P.J., Norby, R.J., 1995. Interactions between drought and elevated CO_2 on growth and gas exchange of seedlings of three deciduous tree species. New Phytol. 129, 63–71.

van Kooten, G.C., Krcmar-Nozic, E., Stennes, B., van Gorkom, R., 1999. Economics of fossil fuel substitution and wood product sinks when trees are planted to sequester carbon on agricultural lands in western Canada. Can. J. For. Res. 29, 1669–1678.

van Noordwijk, M., Martikainen, P., Bottner, P., Cuevas, E., Rouland, C., Dhillion, S.S., 1998. Global change and root function. Global Change Biol. 4, 759–772.

Wayne, P.M., Reekie, E.G., Bazzaz, F.A., 1998. Elevated CO_2 ameliorates birch response to high temperature and frost stress: implications for modeling climate-induced geographic range shifts. Oecologia 114, 335–342.

White, A., Cannell, M.G.R., Friend, A.D., 1999. Climate change impacts on ecosystems and the terrestrial carbon sink: a new assessment. Global Environ. Change 9, S21–S30.

Winjum, J.K., Schroeder, P.E., 1997. Forest plantations of the world: their extent, ecological attributes, and carbon storage. Agric. Forest Meteorol. 84, 153–167.

Winjum, J.K., Brown, S., Schlamadinger, B., 1998. Forest harvests and wood products: sources and sinks of atmospheric carbon dioxide. Forest Sci. 44, 272–284.

Woodward, F.I., 2002. Potential impacts of global elevated CO_2 concentrations on plants. Current Opinion Plant Biol. 5, 207–211.

Wullschleger, S.D., Tschaplinski, T.J., Norby, R.J., 2002. Plant water relations at elevated CO_2— implications for water-limited environments. Plant Cell Environ. 25, 319–331.

Xu, D., Zhang, X.-Q., Shi, Z., 2001. Mitigation potential for carbon sequestration through forestry activities in southern and eastern China. Mitig. and Adapt. Strat. for Global Change 6, 213–232.

Zak, D.R., Pregitzer, K.S., Curtis, P.S., Holmes, W.E., 2000a. Atmospheric CO_2 and the composition and function of soil microbial communities. Ecol. Applic. 10, 47–59.

Zak, D.R., Pregitzer, K.S., King, J.S., Holmes, W.E., 2000b. Elevated atmospheric CO_2, fine roots and the response of soil microorganisms: a review and hypothesis. New Phytol. 147, 201–222.

Air Pollution, Global Change and Forests in the New Millennium
D.F. Karnosky et al., editors
© 2003 Elsevier Ltd. All rights reserved.

Chapter 4

Tropospheric ozone: A continuing threat to global forests?

K.E. Percy*

*Natural Resources Canada, Canadian Forest Service-Atlantic Forestry Centre,
P.O. Box 4000, Fredericton, New Brunswick E3B 5P7, Canada*

A.H. Legge

Biosphere Solutions, 1601 11th Avenue Northwest, Calgary, Alberta T2N 1H1, Canada

S.V. Krupa

*Department of Plant Pathology, University of Minnesota, 495 Borlaug Hall,
1991 Buford Circle, St. Paul, MN 55108, USA*

Abstract

Ozone (O_3) has a critical role in tropospheric chemistry. It absorbs radiation in the infrared and ultraviolet regions and is very reactive and biologically toxic at appropriate levels of exposure. At the earth's surface, O_3 is subject to long-range transport and is the most pervasive air pollutant affecting the world's forests today. The existence of O_3 has been known since 1840 and smog-induced foliar injury on plants was first identified in the 1950s. Levels were $\sim$ 10–15 ppb during the second half of the 1800s, compared with 30–40 ppb measured as the global background today. By 2100, fully 50% (17 million km^2) of world forests are predicted to be exposed to O_3 at concentrations > 60 ppb. Ozone induces a variety of symptoms and pattern of injury that are dependant upon species, genotype, leaf position on the plant, leaf age, exposure dynamics, and meteorological factors or growth conditions. It is absolutely essential to have knowledge on species sensitivities, O_3 profiles and toxicity concentrations for the species under investigation before diagnosis can be confirmed. Ozone is generally detrimental to tree growth and ecosystem productivity, often through induced changes in patterns of carbon allocation or pre-disposition to insects and disease. The development of ozone exposure–forest response relationships that are scientifically defensible and applicable in air quality regulation has been difficult due to serious limitations encountered in scaling-up experimental data. In terms of air quality regulations, North America and Europe have adopted different approaches toward ambient ozone standard setting, with Europe opting for an approach that protects vegetation. The US and Canada, in their individual countries, implement separate or identical standards to protect both human health and the environment.

---

*Corresponding author.

DOI:10.1016/S1474-8177(03)03004-3

1. Introduction

The earth's atmosphere is highly dynamic and very complex (Finlayson-Pitts and Pitts, 2000). In the troposphere, temperature decreases with increasing altitude due to the heating effect near the earth's surface, where incoming solar radiation is absorbed. The troposphere is separated from the stratosphere above (there, the temperature increases with altitude) by the tropopause. These characteristics result in the trapping of air pollutants within the troposphere and their chemical transformation, destruction, scavenging or deposition onto vegetation and other surfaces (Finlayson-Pitts and Pitts, 2000). Ozone (O_3) is the most pervasive, and one of the most damaging air pollutants to forests (Sandermann et al., 1997) worldwide. It is also somewhat unique in its presence both within the troposphere (0–15 km, $\sim 10\%$ of the total column) and the stratosphere (15–50 km, $\sim 80\%$ of the total column). In the stratosphere, O_3 (so-called "good ozone") is essential for life on earth, as it absorbs biologically damaging UV-C radiation < 290 nm (Krupa, 2000). However, in the troposphere, O_3 (so-called "bad ozone") is extremely phytotoxic to plants at appropriate exposures (US EPA, 1996).

1.1. Historical aspects of surface-level ozone

The existence of tropospheric O_3 has been known since 1840. The German scientist Schönbein noted the presence of an electrical odor during thunderstorms and proposed the name ozone for it (Schönbein, 1840). By the mid 1850s, routine O_3 measurements were being conducted at more than 300 locations in Europe (Houzeau, 1858). However, it was not until the late 1940s that O_3 was confirmed as a major constituent of smog then impinging on the Los Angeles region. Unlike London during the 1950s, the Los Angeles smog was found to contain strongly oxidizing, eye-watering, and plant-killing pollutants (Finlayson-Pitts and Pitts, 2000).

Following the identification of smog-induced foliar injury on plants by Middleton et al. (1950), Haagen-Smit and co-workers conducted a classic series of laboratory experiments that confirmed that the injury reported under ambient conditions could be reproduced by concurrently exposing plants under sunlight to polluted air containing nitrogen dioxide (NO_2) and alkenes (hydrocarbons), primary pollutant precursors to catalytic O_3 formation (Haagen-Smit et al., 1952). Later, Richards et al. (1958) demonstrated that O_3 was the smog constituent that caused foliar injury to California grapes, and Heggestad and Middleton (1959) reported extensive crop damage in the eastern United States. Today, O_3 is a human health concern and is damaging vegetation in much of the industrialized world, and increasingly, in the rapidly industrializing world as well (Mauzerall and Wang, 2001).

1.2. Source and formation of surface-level ozone

Ozone has a critical role in tropospheric chemistry. It absorbs radiation in the infrared and ultraviolet regions and at the same time is very reactive and biologically toxic at appropriate levels of exposure (Finlayson-Pitts and Pitts, 2000). In its former role, O_3 is a major contributor to the global greenhouse effect. Tropospheric O_3 ($+0.40$ $W m^{-2}$) is estimated to be the third largest contributor to global mean radiative forcing, after carbon dioxide (CO_2; $+1.56$ $W m^{-2}$) and methane (CH_4; $+0.47$ $W m^{-2}$) (Ramaswamy et al., 2001).

There are natural and anthropogenic sources for surface level O_3 (Guenther et al., 2000; Trainer et al., 2000). Natural sources include: (1) lightning activity during thunderstorms; (2) downward intrusions of naturally produced O_3 from the stratosphere; and (3) biological processes that emit two key precursor pollutant groups, volatile organic compounds (VOC) and nitrogen oxides (NO_x). In urban areas, natural sources of VOCs and NO_x are less important than manmade sources. With the significant expansion of city centers and the resultant rise of urban conurbations, many former rural forests are now exposed to the photochemical products of urban VOC and NO_x emissions.

Sillman (1999), Kley et al. (1999) and Jenkin and Clemitshaw (2000) have provided detailed summaries of O_3 formation at the surface. The general mechanisms for catalytic O_3 formation from the oxidation of VOCs and NO_x by sunlight are well characterized. Oxides of nitrogen are emitted into the atmosphere mainly as NO (nitric oxide) from the combustion of fossil fuels. Nitric oxide is converted rapidly (reaction (1)) during daytime to NO_2 (nitrogen dioxide) via reaction with O_3 already present at the surface.

$$NO + O_3 \rightarrow NO_2 + O_2 \qquad (1)$$

Nitrogen dioxide is then converted back to NO by photolysis (reaction (2)). Overall, this reaction generates no net flux in chemistry and a photo-stationary state is reached where concentrations of NO and NO_2 are related to O_3:

$$NO_2 + h\nu(\text{wavelength} < 420 \text{ nm}) \rightarrow O(^3P) \qquad (2)$$

$$O(^3P) + O_2(+M) \rightarrow O_3(+M) \qquad (3)$$

However, other daytime processes (Jenkin and Clemitshaw, 2000) convert NO_x and those are the result of the photo oxidation of carbon monoxide (CO) and VOCs. The intermediate compounds produced during those processes generate highly reactive free radicals, including the hydro-peroxy radical (HO_2) and organo-peroxy radicals (RO_2). These radicals also convert NO to NO_2 (reactions (4) and (5)), but in so doing, do not consume O_3. Hence, the photolysis of NO_2 (reaction (2)) followed by reaction (3) results in a net source

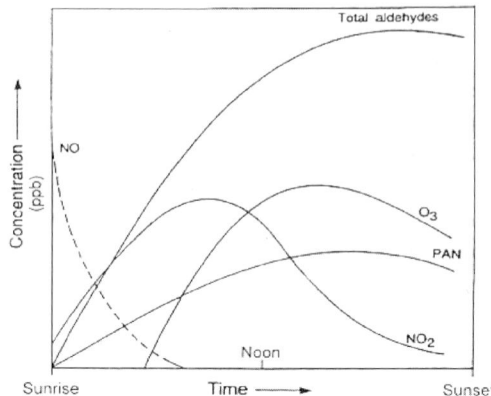

Figure 1. Idealized diurnal pattern of reactants (NO_2, and total aldehydes or hydrocarbons) and products (i.e., O_3 and PAN) in photochemical air pollution (from Krupa, 1997).

of O_3.

$$HO_2 + NO \rightarrow OH + NO_2 \qquad (4)$$

$$RO_2 + NO \rightarrow RO + NO_2 \qquad (5)$$

Although other reactions occur during the daytime, under certain conditions, night-time conversion of NO_x to secondary aerosols (including nitric acid, HNO_3) at some geographic locations and their subsequent transport and deposition are of considerable importance (Takemoto et al., 2001).

Krupa (1997) provided an idealized daily pattern of urban O_3 and its precursors to highlight the predominance of daytime O_3 formation. Nitric oxide emissions and VOCs are high during the early morning due to heavy traffic (Fig. 1). These emissions scavenge some of the O_3 present (reaction (1)) leading to the production of NO_2. The NO concentrations reach a peak later in the day when the atmosphere is able to oxidize NO without completely consuming O_3. This process is maintained due to the increased presence of hydrocarbons (aldehydes) in the atmosphere, leading to a much greater NO_2: NO ratio and a peak in O_3 concentrations in mid to late afternoon. Subsequently during the day, NO_2 conversion decreases (lower sunlight) and new emissions of NO deplete the existing O_3. The cycle may then repeat itself the following day.

1.3. *Transport of surface ozone*

There is clear evidence for O_3 formation and/or transport at regional (NAR-STO, 2000), continental (Chameides et al., 1994) and inter-continental scales

(Derwent et al., 2002)). The presence of high O_3 concentrations is governed by the occurrences of stagnant air masses on one to several days at different spatial scales. Stagnant conditions are characterized by sufficient sunlight, high air temperature, low wind speeds and abundant precursor emissions at the local and regional scales. In addition, O_3 concentrations at the surface are governed by thermally driven meteorological phenomena: (a) up-slope, down-slope air flow in mountainous areas (e.g., Fortress Mountain, Alberta, Canada (Legge and Krupa, 1990)), (b) land–sea or land–lake circulation (e.g., Lake Michigan, (LADCO, 1995)), (c) nocturnal jets or eddy circulation (e.g., Mexico City, (Jáuregui, 2002; Bravo and Torres, 2000)), and (d) convergence zones of two air masses flowing in the opposite directions (e.g., NE US (see Solomon et al., 2000)).

Examples of long-range transport of O_3 and related pollutants include SE Canada, NE US, along the Mexico–US border (Guinnup and Collom, 1997; Roberts et al., 1996) and much of Europe (Borrell et al., 1997). Long-range transport is also important in western North America, sometimes exceeding 500 km (Blumenthal et al., 1997). Similarly, transport from Western Europe has been suggested to be partially responsible for some of the elevated O_3 concentrations in Norway (Schjoldager, 1981). Many geographically specific studies in North America and Europe show that O_3 concentrations > 100 ppb are observed when NO_x–VOC and/or O_3-laden air is transported into or downwind from an urban area, or derived from recirculation of polluted air during more than one dial cycle (Beck and Grennfelt, 1994; Blumenthal et al., 1997; Reid et al., 1996; Thuillier, 1997). It is important to note that peak concentrations of O_3 and other oxidants usually occur when a relatively high regional background level is augmented by additional O_3 by plume transport from urban areas and even from rural point sources rich in NO_x (Blumenthal et al., 1997). Normally hydrocarbons are the limiting factor in the photochemical O_3 production in the city centers, and NO_x is the limiting factor away from urban centers (Sillman, 1999). Therefore, NO_x-rich point source plumes in the rural areas downwind from urban centers are very important.

Ozone concentrations aloft are higher than the levels measured at the surface (Blumenthal et al., 1997; Hidy, 1994; Pisano et al., 1997; Reid, 1994). Evidence shows that O_3 can be carried over aloft from one day to the next and be brought to the ground when the night-time atmospheric inversion breaks up the next day (Beck and Grennfelt, 1994; Blumenthal et al., 1997; McKendry et al., 1997; Niccum et al., 1995; Roussel et al., 1996). Estimates from some of the studies indicate that carry over can account for up to 100 ppb and for nearly 50–100% of the surface O_3 in rural areas.

2. Ozone distribution at the surface and trends

2.1. Global ozone distribution

Tropospheric O_3 is the most pervasive air pollutant affecting global forests. Levels were $\sim$ 10–15 ppb more than a century ago, compared with 30–40 ppb measured as the global background today (Finlayson-Pitts and Pitts, 2000). Taking a concentration of 60 ppb as likely to be phytotoxic to sensitive forest vegetation, Fowler et al. (1999) have used a global model (STOCHEM) to simulate exposure of forests for the years 1860, 1950, 1970, 1990 and 2100. Model results show no exposure to $O_3 > 60$ ppb in 1860 and only 6% exposed by 1950 (Fig. 2). Of those forests, 75% were in temperate latitudes (parts of south-western/eastern North America, Mexico, central/eastern Europe) and 25% in the tropics.

The area of global forests exposed to $O_3 > 60$ ppb expanded threefold between 1950 and 1970 (Fig. 2). Most of Europe, eastern and parts of western North America, southern Scandinavia, parts of central/eastern Asia, South America and Africa were affected, with greatest increases (2.9–12% of forests) in tropical and subtropical forests. By 1990, some 24% of global forests were exposed to $O_3 > 60$ ppb. Fowler et al. (1999) predict that by 2100, 50% (17 million km^2) of the world's forests will be exposed to $O_3 > 60$ ppb. Large increases in the affected forest areas are calculated for both temperate–subpolar and tropical–subtropical forests.

Maximum daily surface O_3 concentrations are lowest over remote marine areas (20–40 ppb O_3) and remote tropical forests (20–40 ppb O_3). However, rural areas are exposed to daily maximums between 50–120 ppb O_3. In comparison, highest daily maximum concentrations of 100–400 ppb O_3 occur over urban–suburban areas (Krupa et al., 2001). For most of the year, a country (UK) may be a net sink for O_3, with production only exceeding losses in the photochemically active months (Coyle et al., 2003).

2.2. Trends in surface ozone concentrations

Three types of trends in surface O_3 concentrations are apparent: (1) a huge increase in the extent of O_3 and the forest areas at risk; (2) a decrease in maximum 1-hour O_3 concentrations, at least in the northern hemisphere countries having O_3 precursor control programs in place; and (3) an increase in background O_3 concentrations over much of the world. In the United States, trends in surface O_3 are based upon both 1-hour and 8-hour data. Over the past 20 years (379 sites; 1982–2001), national ambient O_3 levels decreased by 18% based on 1-hour data and by 11% based on 8-hour data (US EPA, 2002). For the period 1981–2001, the downward trend ($> 10\%$) in 1-hour maximum O_3

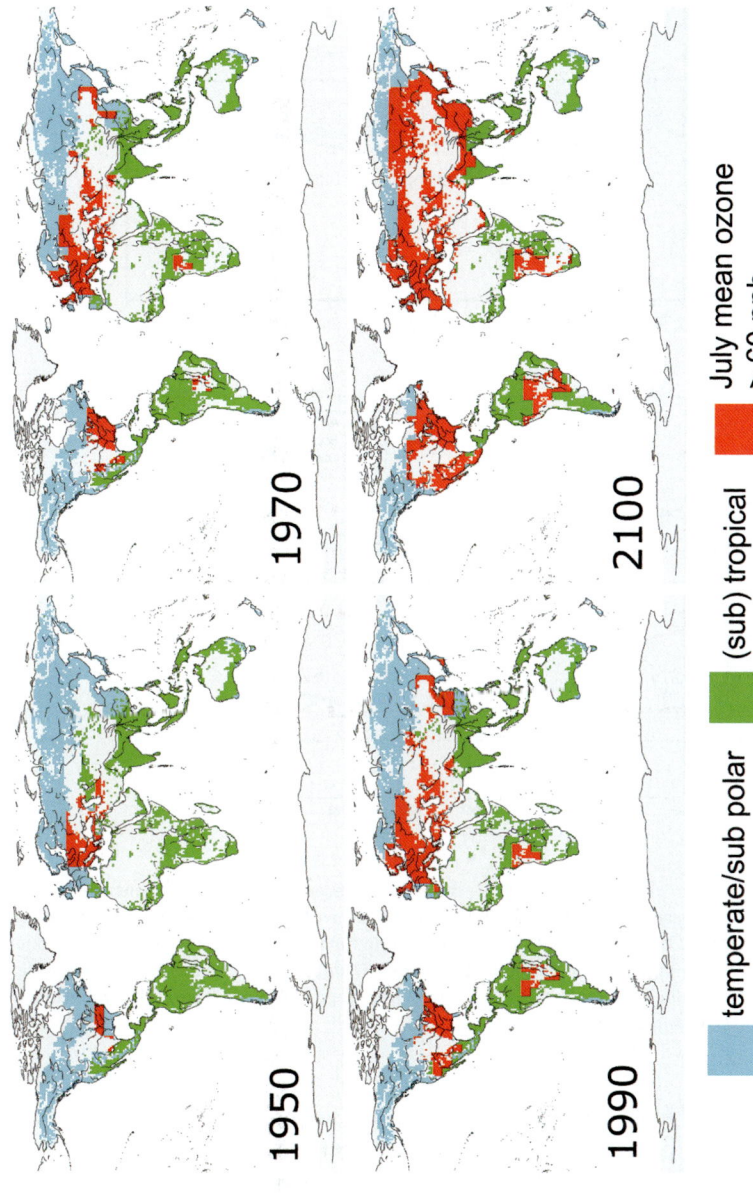

Figure 2. Global forest area where July peak surface O_3 concentration exceeds 60 ppb (from Fowler et al., 1999).

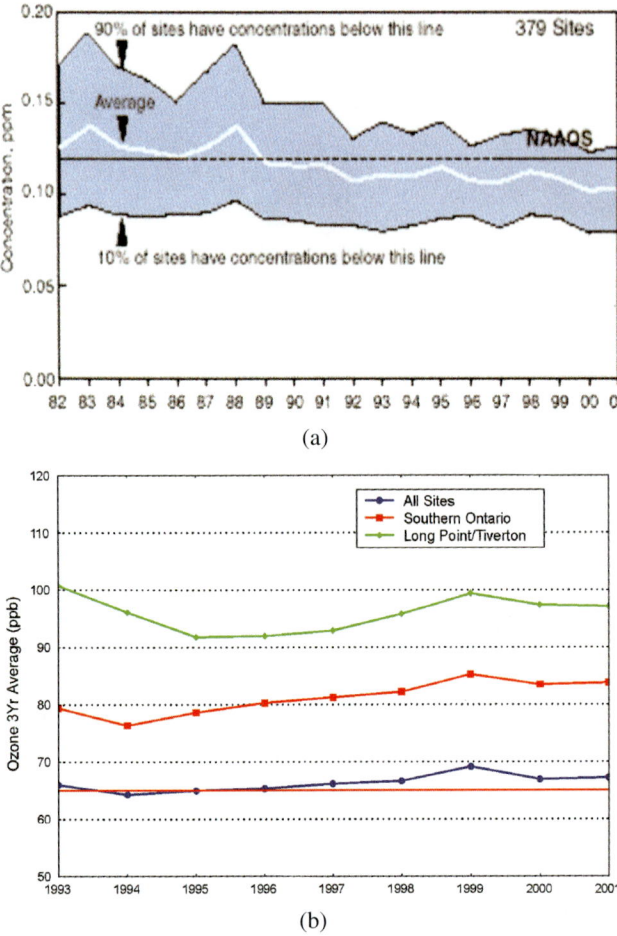

Figure 3. Comparison of surface-level O_3 trends and concentrations in North America and Europe. (a) Twenty-year trend in annual 4th highest 8-hour average O_3 at 379 ground locations in the United States (US EPA, 2002); (b) Ten-year trend in running 3-year annual 4th highest 8-hour average O_3 at 126 ground locations in Canada (from Dann, 2001); (c) Number of exceedances of the threshold value of 150 µg m^{-3} (75 ppb) O_3 in Europe during 2000 (from Hjellbrekke and Solberg, 2002). (*Continued on next page*)

levels occurred in every geographic area (Fig. 3(a)). Nearly all regions experienced improvement in 8-hour O_3 levels between 1981–2001 except the north central region, which showed little change. The west and northeast exhibited the most substantial reductions as measured by both 1-hour and 8-hour data (US EPA, 2002).

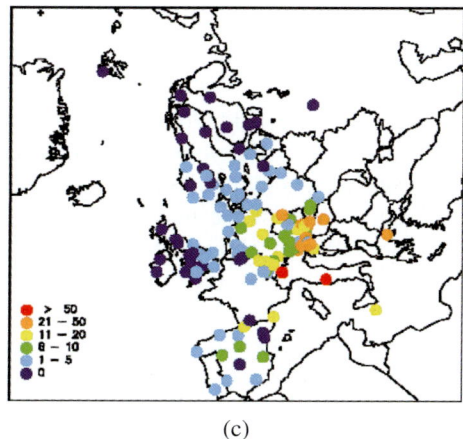

(c)

Figure 3. (*Continued*)

Across the US, highest 1-hour O_3 concentrations are typically found at suburban sites and concentrations there have decreased by 20% since 1982. However, national improvements at rural monitoring locations have been slower. One-hour O_3 levels in 2001 were 11% below those in 1982 but < 1% lower than levels in 1991 (US EPA, 2002). Since 1996, 1-hour rural O_3 concentrations have been greater than the corresponding values at urban sites. Between 1992 and 2001, the average 8-hour O_3 level in 33 national parks actually increased nearly 4%.

These trends mirror the patterns in other northern countries and regions. In Canada, trends in surface O_3 are tracked using the 3-year running average of 4th highest 8-hour daily maximum O_3 (ppb) concentration. National ambient O_3 levels averaged across all locations showed a decrease from 70 ppb in 1991 to 62 ppb in 1993 (Dann, 2001). Since 1993, the trend, except for the years 1996 and 2000, has shown increasing levels (70 ppb) through 2001 (Fig. 3(b)). Parallel trends have been observed for locations in southern Ontario. Levels decreased from 88 ppb in 1991 to 74 ppb in 1993, followed by an increase to 88 ppb in 1999 and 87 ppb in 2001.

Systematic O_3 monitoring throughout Europe began in the late 1980s and the corresponding trends are less certain than in North America (Matyssek and Innes, 1999). There are no clear trends in the alpine regions, with episodes occurring downwind of industrialized areas in northern and central regions. Nevertheless, O_3 episodes occur over the entire continent every summer. During 2000, surface-level O_3 concentrations were monitored at 124 ground locations in 26 countries (Hjellbrekke and Solberg, 2002). The 1-hour critical level for O_3 formulated by the Economic Commission for Europe (ECE) for the protection of vegetation, 150 µg m^{-3} ($\sim$ 75 ppb), was exceeded in 2000 at 95 (77%)

of European ground-level monitoring sites (Fig. 3(c)). Exceedance was considerable in the central parts of Europe. Highest hourly mean values were reported for two Italian locations (126, 128 ppb) and > 100 ppb O_3 were recorded at locations in Germany, Austria, Switzerland, Poland, Slovenia, Sweden and Denmark during a continental-scale episode around June 20. The lowest maximum hourly concentration measured was 47 ppb (Spitsbergen, Norway).

In the most recent years, some of the highest surface O_3 concentrations recorded have been in the Valley of Mexico (Mexico city urban metro-plex), representing a unique situation that deserves a brief explanation. At night and during early morning (especially during the dry season), the down slope winds from surrounding mountains and the converging circulation induced by the heat island combine to restrict the lateral dispersion of pollutants. The vertical dilution of urban gases is further restricted by the frequent (more than 70%) surface inversions observed during that period. Once the stable layer is heated from below by abundant insolation, turbulent mixing dilutes the pollutants in the vertical plane, while the regional winds, that descend to the ground (usually with a northerly component), transport pollutants to the southern suburbs, where usually the highest levels of O_3 (average 1-hour concentrations of 0.3–0.5 ppm) are observed (Jáuregui, 2002; Bravo and Torres, 2000). The Mexican O_3 Air Quality Standard (MOAQS) is a 1-hour mean of 0.11 ppm, not to be exceeded more than once per year. According to Bravo and Torres (2000), as a gross average, the MOAQS is exceeded during more than 4–5 hours per day on at least 300 days per year.

Emberson et al. (2001) have summarized some historical O_3 data for developing countries and regions. In India, maximum hourly concentrations up to 166 ppb O_3 have been reported in Delhi, with northern and western parts of the country exposed to higher concentrations. In Pakistan, surface O_3 levels increased away from Lahore, with 6-hour weekly means in peri-urban areas reaching 72 ppb during 1993–1994. In Taiwan, daily mean O_3 concentrations > 120 ppb frequently occurred between 1994 and 1997. Similarly hourly mean O_3 concentrations > 100 ppb have been recorded in Cairo and Alexandria, Egypt.

2.3. Natural surface-level ozone

Knowledge of O_3 dose and flux-uptake dynamics is essential to fully understand cause–effect relationships. Also, policy makers are required to establish and regularly review ambient O_3 (and other air pollutants) air quality standards and critical levels or loads to protect human health and vegetation (welfare). As noted previously, there is a natural background O_3, produced primarily from precursor pollutants originating from non-anthropogenic processes (Finlayson-Pitts and Pitts, 2000; Trainer et al., 2000). The levels of such background O_3 concentrations influence the maxima observed at any geographic location (Derwent et al., 2002). Thus, among the key questions being asked is the nature of the background O_3 concentrations and their significance in more remote forested areas.

To address the atmospheric portion of this issue, three different approaches have been used in determining the so-called background ambient O_3 concentrations: (a) examining historical measurement records from the late 1800s (pre-industrial times); (b) measuring current ambient O_3 concentrations at pristine geographic locations; and (c) application of photochemical models. In the context of approach (b), it is highly questionable whether there are any truly pristine locations in the world today that are not being influenced by human activity, and frequently criteria used to select such sites are not well described (see also Section 2.2).

Typical annual variations in tropospheric O_3 at Moncalieri, Italy from 1868 to 1893, Montsouris, France from 1876 to 1886 and Zagreb and Croatia in 1900 show peak values around 10 ppb (Marenco et al., 1994; Sandroni et al., 1992). In contrast, recent O_3 concentrations in the most unpolluted parts of Europe averaged between 20 and 45 ppb (Volz and Kley, 1988; Janach, 1989). A typical average tropospheric O_3 concentration of 30–40 ppb is found essentially everywhere in the world today (Finlayson-Pitts and Pitts, 2000; Derwent et al., 2002). Data from relatively remote European sites indicate a 1–2% annual increase in average O_3 concentrations during the period 1960–1990 (Janach, 1989). Such increases were attributed to greater emissions of NO_x associated with fossil fuel combustion (Volz and Kley, 1988; Janach, 1989).

Our future understanding of the issue of background O_3 concentrations is expected to improve considerably with the rapidly increasing use of passive samplers in ecological effects research (Krupa and Legge, 2000; Bytnerowicz et al., 2001). Passive samplers are inexpensive, relatively easy to use if deployed properly in the field, and do not require electrical power to operate. It is most interesting to note that some 150 years ago measurements of ambient O_3 levels were made through the use of passive samplers (Houzeau, 1858). Although passive samplers can be used over wide geographic areas, unattended for prolonged periods (1 week to a month), they provide average or total O_3 concentrations for the sampling duration. Mazzali et al. (2002) used krieging and iterative procedures based on the functional dependence of O_3 on the elevation above sea level and time of day to map vegetation exposures in a complex terrain of 80×40 km area in the southern side of the European Alps. Tuovinen (2002) achieved a rather simple model formulation to predict AOT40 (Accumulated dose Over a Threshold of 40 ppb, the European definition of exposure, Fuhrer and Achermann, 1999), by approximating the frequency distributions of hourly O_3 concentrations by their Gaussian probability. He concluded that it is possible to obtain reasonable AOT40 values even in the absence of continuous measurement data.

However, in many cases hourly ambient O_3 concentrations do not exhibit a frequency distribution that is Gaussian (Nosal et al., 2000). Furthermore, weekly or bi-weekly means or seasonal concentration summation (e.g., AOT40) methods cannot capture the dynamic changes of the atmosphere and the plant biology (Krupa and Kickert, 1997). To address that issue, Krupa and Nosal (2001) used a stochastic, Weibull probability model to predict hourly ambient O_3 concentrations from single weekly passive sampler data (weekly averages) to simulate the corresponding continuous measurements. More recently, Krupa et al. (2003) have developed a multi-variate statistical model for achieving the same purpose, but with the inclusion of meteorological variables (global radiation, air temperature, relative humidity and wind speed, variables that also influence plant O_3 uptake through stomata). If such efforts are coupled to multi-point plant response measurements, meaningful cause–effect relationships can be derived regarding the nature of the so-called background O_3 concentrations and their significance in more remote forested areas.

3. Effects of ozone on forest trees

Ozone is generally detrimental to tree growth and forest productivity. Numerous studies have been conducted on the impacts of O_3 on plant growth and biomass accumulation (see reviews by Chappelka and Samuelson, 1998; Matyssek and Innes, 1999). However, many of those studies are confounded by the artificial conditions imposed by exposure chambers and limited by the available space in such chambers to allow only single trees or small numbers of largely young, immature trees or saplings. Therefore, it is unlikely that such investigations can be used to assess risk at the landscape level.

3.1. Ozone symptomatology and species sensitivity

The symptomatology of O_3 injury on plants was first described in the 1950s (Richards et al., 1958; Heggestad and Middleton, 1959). At the present time, there is a large literature base on the development and recognition of O_3-induced foliar injury symptoms in forest trees (ICP Forests, http://www.ozone.wsl.ch). A comprehensive description, along with images of visible injury and a listing of species sensitivity, has also been provided by Krupa et al. (1998). There are a large number of tree species that are relatively sensitive to O_3 and some of these are listed in Table 1. After long-term, chronic exposure to relatively low concentrations, common symptoms of O_3 injury in broadleaf plants include changes in pigmentation or bronzing, chlorosis and premature senescence of leaves or flowers. After short-term, acute exposure to relatively high concentrations, symptoms may include bleaching on the upper or lower leaf surface, flecking and stippling on the upper surface or bifacial necrosis (Krupa et al., 1998). On conifers, chronic exposure results in flecking and mottling of needles and often, premature loss of needles, and clear bands of chlorotic tissue (banding) and reddish-brown tip burn are common following acute exposure.

Several factors should be considered when attempting to determine O_3 injury. Ozone causes a variety of symptoms on many plants and the pattern of injury can depend on species, genotype, foliar position on the plant, leaf age, exposure dynamics and meteorological factors or growth conditions. It is essential to have knowledge about species sensitivities, O_3 profiles and toxicity concentrations for the species in question. Other stressors may also induce visible symptoms that mimic those of O_3. The use of native or transplanted indicator plants having a range of sensitivity and for which O_3 symptoms are clearly defined is extremely useful in diagnosis, especially if indicator plants are co-located with surface-level passive O_3 sampling or continuous monitoring systems. Another approach is the use of clonal field plantations containing genotypes of varying O_3 sensitivity such as trembling aspen (*Populus tremuloides* Michx.) co-located with continuous ground-level O_3 monitoring in which biomass growth, morbidity and mortality are measured over a multi-season period (Karnosky et al., this volume).

3.2. Biochemical action of ozone

The first interface between surface-level O_3 and the forest is the tree canopy. The main pathway of entry of O_3 into the leaf is through the stomata, although there is evidence for limited trans-cuticular movement. As O_3 is extremely soluble in water (although it is pH dependent), leaf surface dissolution may also be important, particularly in the late summer when a thin film of water (dew) may be present on the upper leaf surface. Once inside the leaf interior, O_3 dissolves in the fluids of the sub-stomatal cavity. However, the concentration of ozone within the leaf is essentially zero. It is the free radicals that are considered to be the initiators of cellular response. Enzymatic response within the leaf is extremely rapid, and can occur within minutes (Noormets et al., 2000). The movement and biochemical action of O_3 in the plant is well described and three major metabolic events induced by O_3 have been identified: (1) increased turnover of antioxidant systems; (2) production of symptoms similar to wounding, especially ethylene production; and (3) decline in photosynthesis (Heath, 1999).

These events have led Heath (1999) to present the following hypothesis for the induction of injury: (1) rapid entry of O_3 overwhelms the plant's antiox-

Table 1. List of some tree species that are relatively sensitive to O_3. Modified from Krupa et al. (1998) and ICP-Forests (http://www.ozone.wsl.ch)

Common name	Latin name
Ailanthus	*Ailanthus altissima* (Mill.) Swingle
Alder	*Alnus* spp., *Alnus viridis* DC.
European Alder	*Alnus glutinosa* (L.) Gaertner
Aleppo Pine	*Pinus halepensis* Miller
Ash	*Fraxinus* spp., *Fraxinus angustifolia* Vahl subsp. *angustifolia*
Aspen	*Populus tremula* L., *Populus tremuloides* Michx.
Austrian Pine	*Pinus nigra* Arnold
Beech	*Fagus sylvatica* L.
Birch	*Betula pendula* Roth
Black Poplar	*Populus nigra* L.
Cluster Pine	*Pinus pinaster* Aiton
Black Locust	*Robinia pseudoacacia* L.
Cornicabra	*Pistacia terebinthus* L.
Dogwood	*Cornus sanguinea* L.
Eastern White Pine	*Pinus strobus* L.
Field Maple	*Acer campestre* L.
Grey Alder	*Alnus incana* (L.) Moench
Green Ash	*Fraxinus pennsylvanica* Marshall
Hawthorn	*Crataegus* spp.
Hornbeam	*Carpinus betulus* L.
Italian Maple	*Acer granatense* Boiss.
Italian Stone Pine	*Pinus pinea* L.
Larch	*Larix* spp.
Mastic	*Pistacia lentiscus* L.
Narrow-leaved Ash	*Fraxinus angustifolia* Vahl subsp. *angustifolia*
Norway Maple	*Acer platanoides* L.
Norway Spruce	*Picea abies* (L.) Karst.
Plane-trees	*Platanus* spp.
Ponderosa Pine	*Pinus ponderosa* Laws.
Poplar	*Populus* spp.
Rowan, Rowan Tree, European Mountain-Ash	*Sorbus aucuparia* L.
Wild Black Cherry	*Prunus serotina* Ehrh.
Sassafras	*Sassafras albidum* (Nutt.) Nees.
Small-leaved Lime	*Tilia cordata* Miller
Smooth-leaved Elm	*Ulmus minor* Miller
Snowberry	*Symphoricarpos alba* (L.) S.F. Blake
Sweet Cherry, Wild Cherry	*Prunus avium* L.
Sweetgum	*Liquidambar styraciflua* L.
Sycamore Maple	*Acer pseudoplatanus* L.
Yellow Poplar	*Liriodendron tulipifera* L.
English Walnut	*Juglans regia* L.
White Mulberry	*Morus alba* L.
White Willow	*Salix alba* L.
Willow	*Salix* spp.
Wych Elm	*Ulmus glabra* Miller

idant response; (2) the membrane is altered such that permeability, transport and triggering mechanisms are no longer correct for the state of the cell, and the rate of ion movement and sensitivities to messages becomes either too slow, too fast or too reactive to maintain homeostasis; (3) stress-induced ethylene is transformed into a toxic product; (4) a decline in m-RNA (messenger-Ribonucleic acid) leads to a reduction in the level of Rubisco (Ribulose biphosphate carboxylase), a lowered rate of CO_2 fixation and lowered productivity; (5) the stomata close and limit CO_2 fixation; and (6) transport of sugar occurs from the source to the sinks. Carbon costs to the plant may include increased antioxidant and ethylene production as well as translocation of sugars out of source cells to sinks.

3.3. Evolution of ozone tolerance or resistance

In many industrialized regions, surface-level O_3 concentrations are high enough to cause subtle shifts in composition of natural (and semi-natural) plant communities. As such, O_3 is thought to constitute a novel evolutionary challenge for natural and managed ecosystems. In fact, there is a growing consensus that O_3 levels in many industrialized and industrializing regions are high enough to drive selection for resistant individuals within wild plant communities (Barnes et al., 1999). In the case of trembling aspen, the most widely distributed tree species in North America, there is new evidence from managed stands that rapid selection may occur over a very few years near major urban areas (Karnosky et al., 2003). That work also provides important new field evidence for the previously described (McDonald et al., 2002) competitive interaction among genotypes in the presence of elevated O_3.

The role of competitive status within a stand is a strong cofactor determining tree-growth response to O_3. It is very important, however, to remember that, depending upon genotype, negative O_3 effects in aspen may be stronger with competitive advantage or competitive disadvantage (McDonald et al., 2002). What is of concern is that in some tree species, sensitive individuals exhibit a higher genetic multiplicity and diversity. The decline of pollution sensitive tress, therefore, may result in a partial genetic depletion of forest stands through the loss of frequent alleles with potential adaptive significance to altered stress regimes of the future (Longauer et al., 2001).

In sensitive genotypes, acute O_3 injury, usually observed as the development of foliar lesions, resembles the hypersensitive response of plants to pathogens (Sandermann, 1996). An oxidative burst occurs as the initial reaction to both O_3 exposure and pathogen infection and similar signal molecules have been implicated in induction of the hypersensitive response and O_3 injury (Schraudner et al., 1998). In O_3-tolerant genotypes, either the oxidative burst is suppressed or oxidative damage is highly localized, thereby restricting the extent of foliar lesions (Koch et al., 2000). The plant antioxidant system, which scavenges naturally occurring reactive oxygen compounds, could function as a primary mechanism to alleviate the oxidative burden resulting from O_3 exposure. The ascorbate (vitamin C)–glutathione (a tripeptide) cycle has been the most intensively studied and generally there is a positive correlation of O_3 tolerance with levels of antioxidants and antioxidant enzyme activities (Conklin and Last, 1995). However, transgenic plants that have been engineered to overproduce ascorbate-glutathione antioxidant enzymes have provided mixed results regarding O_3 tolerance, depending upon the species, the cellular compartment in which the enzyme is expressed and the isozyme chosen (Mullineaux and Creissen, 1999). Numerous other antioxidant compounds are found in plants and their role in O_3 tolerance requires attention. In addition to signaling processes and biochemical protective mechanisms, plants may express differential tolerance depending upon the rate of influx of O_3 into the leaf interior. Uptake may be affected by stomatal density or by guard cell response to oxidative conditions. It appears that O_3 does not directly affect stomatal closure (Torsethaugen et al., 1999; Karnosky et al., 2003), but acts indirectly, perhaps by influencing CO_2 "fixation" or altering hormone levels. In any case, the role of stomates in influencing O_3 tolerance has not been fully characterized.

4. Case studies of forest ecosystem response to ozone

4.1. North America

McLaughlin and Percy (1999) provided a retrospective analysis of the four most prominent case studies on air pollution and forest responses in North America. They concluded that changes in depth and vigor of root systems, shifts in pool sizes and allocation patterns of carbon, and changes in supply rates of nitrogen and calcium, caused by O_3 and acidic deposition (singly or combined), represent important shifts in ecological function in diverse forest types across a large geographic area in North America. The authors predicted that the influence of those process-level changes on future health of North American forests could be greatly increased by changes in climate. Chappelka and Samuelson (1998) reported that O_3 repeatedly induces foliar injury on a number of tree species throughout much of the eastern US. However, as symptom expression is influenced by a number of endogenous and exogenous factors, it has proven difficult to confirm cause and effect at the routine biomonitoring level.

Apparent growth declines in unmanaged southeastern pines were reported in the early 1980s by Sheffield and Cost (1987). Process research conducted through the Southern Commercial Forest Research Cooperative has been summarized in Fox and Mickler (1996). Those authors have compiled a summary of forest characteristics, biotic and abiotic stresses along with the results from a large number of integrated field and experimental research projects. Studies at ambient levels have provided evidence that moderate O_3 levels can increase water stress and reduce growth in larger trees. McLaughlin and Downing (1995, 1996) evaluated seasonal growth patterns of mature loblolly pine (*Pinus taeda* L.) trees growing in eastern Tennessee for O_3 effects. Although levels of O_3, rainfall and temperature varied widely over the period, they identified significant influences of O_3 on stem growth patterns using regression analysis. Effects of ozone exposures interacted with soil moisture and high air temperatures to reduce short-term rates of stem expansion (McLaughlin and Downing, 1995). Observed O_3 responses were rapid, occurring within 1–3 days of exposure to average O_3 concentrations at > 40 ppb.

Since the mid 1950s, much of the mixed conifer forests in Southern California has been exposed to some of the highest concentrations of O_3 in North America. Long-term exposure to high levels of O_3 and oxidants in the San Bernardino forest has produced the classic example of hierarchical forest response to O_3 (Miller and McBride, 1999). Effects from foliar level to succession stages have been documented. Average concentrations of 50–60 ppb O_3 were sufficient to cause foliar injury, early needle loss, decreased nutrient availability, reduced carbohydrate production, lower vigor, decreased height/diameter growth and increased susceptibility to bark beetles (Miller et al., 1989). The concurrence of drought, long-term reduction in precipitation and high O_3 (Arbaugh et al., 1999) has contributed to a period of growth decline in ponderosa (*Pinus ponderosa* Laws.) and Jeffrey (*Pinus jeffreyii* Grev. and Balf.) pines and bigcone Douglas fir (*Pseudotsuga macrocarpa* (Vasey) Mayr.). Decreases in radial basal area growth rates during 1950–1975 were 25–45% (ponderosa and Jeffrey pines) and 28% (bigcone Douglas fir). Other factors such as high nitrogen deposition (Bytnerowicz et al., 1999) and stand developmental changes due to fire suppression have also contributed. Older trees are the most vulnerable and additional stress imposed by high O_3 may render these trees more vulnerable to bark beetle attack (Arbaugh et al., 1999). It is interesting that, under diminishing annual average O_3 concentrations, Miller et al. (1989) have reported an improvement (1974–1988) in the foliar injury index. This recovery is not expected to prevail indefinitely due to changing precipitation patterns.

4.2. Valley of Mexico studies

Early reports of O_3 damage to forests in Mexico date from the mid 1970s when O_3-induced chlorotic mottle and premature needle senescence was observed

Table 2. A comparison of surface-level, O_3-related characteristics of Mexico City and the Los Angeles area. From Krupa (1997), modified originally from Miller (1993)

Characteristic	Mexico City	Los Angeles
Latitude	25°19′N	00°34′N
Elevation	2250 m MSL	104 m MSL
Topography	Mountain valley	Ocean front, inland mountain range
Rainfall period	Mainly summer	Mainly winter
Periods of high concentrations	Summer, between rains and even higher in winter	Summer
UV-A (320–400 nm)[a]	Relatively high	Lower than Mexico City
Vegetation	No respite from O_3	Respite during winter from O_3
Impact on natural ecosystems (area) (sensitive, major tree species)	Ajusco, Desierto de los Leones (approximately >80 km downwind from Mexico City) (*Pinus hartwegii*)	San Bernardino National Forest (approximately >120 km downwind from Los Angeles) (*Pinus ponderosa*)

[a]The peak wavelength for the photolysis of NO_2 is ~ 398 nm (UV-A band), a requirement for photochemical O_3 production.

on Hartweg pine (*Pinus hartwegii* Lindl.), Chihauha pine (*Pinus leiophylla* (Engelm) Shaw.), and Montezuma pine (*Pinus montezumae* var. *lindleyi*) near Mexico city (Krupa and de Bauer, 1976). In the early 1980s, a sudden decline of sacred fir (*Abies religiosa* (H.B.K.) Schdl. and Cham.) was observed in a national park situated southwest of Mexico City (Desierto de los Leones). This was thought to be due to O_3 (deBauer et al., 1985). Further investigation revealed an ongoing reduction in ring width since the early 1970s. The overall problem, including visible foliar injury, appears to be the product of complex interactions between chronic O_3 exposures and other growth-regulating factors.

Table 2 provides a comparison of surface-level, O_3-related characteristics of Mexico City (an area with some of the highest concentrations in the world at the present time) and the Los Angeles area (the area where O_3-induced vegetation injury was first reported in the 1950s). The main difference between the two cities is that, although high O_3 concentrations persist throughout the year in Mexico City, the winter season is a period of respite for O_3 levels in Los Angeles. In the Valley of Mexico, nine different native pine species represent 43% of the total forested land cover. Based on foliar injury, several of these species exhibit moderately high to very high sensitivity to O_3 (Miller et al., 2002). Of concern are the reductions in their normal growth rates, an issue that requires a detailed study.

4.3. European forest health monitoring

One-third of Europe is covered in forest. Despite concerns (Ferretti et al., 1999), annual crown condition surveys remain the main tool in the overview of European forest condition. In the most recent (2001) assessment, 22.4% of 132 000 trees assessed were classified as moderately or severely defoliated (UN-ECE and EC, 2002). In 2000, exceedance of the European critical level for O_3 was considerable over large parts of central and Eastern Europe (Hjell-brekke and Solberg, 2002). Although O_3 may adversely affect tree growth, unequivocal evidence for O_3-induced foliar injury has only been found at few locations. Research suggests that risks exist for European forests, but that those risks need to be validated at the stand level (Matyssek and Innes, 1999).

The impact of O_3 on European forests has been reviewed in Skarby et al. (1998). Evidence for O_3 stress was at that time inferred and was derived in the main from controlled fumigation experiments using young saplings. In the late 1990s, no relationship existed between needle loss in the field and O_3 levels. In comparison with the effects of extended drought or nutrient stress, the effects of O_3 were stated to be much milder with respect to tree growth. Therefore, Skarby et al. (1998) concluded that a link between the occurrence of O_3 and forest damage had not at that time been unequivocally established.

Although accelerated growth trends (attributed to excess nitrogen) have been found in parts of northern (EFI, 2002), most of central, and some parts of southern Europe, decreasing trends have been found where exposure to pollutants (including O_3) or exceptional climatic conditions prevailed. In 2001, visible injury assessments were carried out on 53 intensive monitoring plots located in 9 countries. Ozone injury was reported on the main tree species growing at 18 of those plots. In central Europe, visible ozone damage on European beech (*Fagus sylvatica* L.) was reported on 27% of the plots (UN-ECE and EC, 2003). However, by 2002, the precise role of O_3 in those declining growth trends remains to be elucidated.

5. Ozone air quality standards in North America and critical levels in Europe

Ambient air quality standards in North America are based on the best available scientific knowledge and understanding which is balanced by social, economic and political considerations at the time they are set. These air quality standards provide policy makers and regulators with a measure of air quality for air contaminants for the purpose of compliance with legislation (IUAPPA, 1995). Ambient air quality standards do not assume the existence of a concentration threshold for receptor response and, therefore, target values are often substituted for the regulatory purposes. In Europe, the United Nations Economic Commission for Europe (UN-ECE) has taken an approach to managing ambient air quality that is quite different from North America, with the concept of critical levels being applied. Critical levels are defined as concentrations of pollutants in the atmosphere above which direct adverse effects on receptors, such as plants, ecosystems or materials, may occur according to present knowledge (UN-ECE, 1988; Tema Nord, 1994). Critical levels are set based solely on the best available scientific knowledge and understanding and acknowledgement that a threshold concentration for receptor response exists. They are not used directly for air quality compliance purposes but rather as an integral part of emission control/abatement strategies (Bull, 1991). That being said, the critical levels concept implicitly assumes that all adverse effects should be prevented regardless of the economic costs of reducing pollutant emissions. A more in-depth discussion of these matters is provided in Legge et al. (1995) and Ashmore (2002).

5.1. North American standards for ozone

In the United States, the Clean Air Act requires the Environmental Protection Agency (US EPA) to propose and promulgate the form and level of National Ambient Air Quality Standards (NAAQS) for selected ubiquitous pollutants that are known to endanger public health and welfare and to issue primary and secondary standards for them. A primary standard is defined as one of attainment and maintenance that allows an adequate margin of safety for sensitive population groups to protect the public health. A secondary standard is one of attainment and maintenance which protects the public welfare from any known or anticipated adverse effects associated with the presence of the pollutant in ambient air (US EPA, 1996). Welfare effects relate to impacts on vegetation, crops, ecosystems, visibility, climate and man-made material to name a few. Primary and secondary NAAQS can be different or they can be the same. The previous primary O_3 standard which was set in 1979, is currently being phased out and is the maximum hourly O_3 concentration of 0.12 ppm (120 ppb) not be exceeded more than 1 day per calendar year. The US EPA has replaced the 1979 primary O_3 NAAQS with a new primary O_3 NAAQS set at 0.08 ppm (80 ppb) calculated as the 3-year average of the annual fourth highest daily maximum 8-hour O_3 concentrations measured at each monitor within an area (Federal Register, 1997).

Before setting a new NAAQS, the US EPA conducts an open and rigorous scientific assessment process. First, a criteria document for the pollutant in question is prepared by the US EPA's National Center for Environmental Assessment (NCEA) that summarizes the current state of the science in the peer-reviewed scientific literature as it relates to public health and welfare. Based on the science in the criteria document, the US EPA's Office of Air Quality Planning and Standards (OAQPS) develops a 'staff paper' that includes recommendations to either retain and/or revise the current NAAQSs. Both documents undergo rigorous scientific peer review by the US EPA's Science Advisory Board's Clean Air Scientific Advisory Committee (CASAC), as well as comment from the public, industry and other interest groups. Based on all of the above, the US EPA Administrator then decides whether or not to maintain or revise the current NAAQSs.

In 1998, Canada began development of a new ambient O_3 air quality standard under the Canada-wide Accord on Environmental Harmonization under the leadership of the Canadian Council of Ministers of the Environment (CCME). The fundamental concepts accepted as prerequisites for a new standard were as follows: (1) a numerical target value; (2) a time line by which the target is to be achieved; and (3) a reporting protocol. Canada adopted a Canada-Wide Standard (CWS) for ozone in 2000 (CCME, 2000). The CWS for O_3 has been set at 65 ppb with an 8-hour averaging time with achievement based upon the fourth highest annual measurement, averaged over 3 consecutive years. This target is to be achieved by 2010. Community-oriented monitoring sites in census metropolitan areas with greater than 100 000 people will be included, but rural and source-specific sites will be excluded for CWS achievement determination.

5.2. European critical levels for ozone

Critical levels for vegetation for ground-level ozone in Europe have evolved through a series of UN-ECE workshops: Bad Harzburg, Federal Republic of Germany, 1988 (UN-ECE, 1988); Egham, United Kingdom, 1992 (Ashmore and Wilson, 1992); Bern, Switzerland, 1993 (Fuhrer and Achermann, 1994); Kuopio, Finland, 1996 (Kärenlampi and Skärby, 1996); Gerzensee, Switzerland, 1999 (Fuhrer and Achermann, 1999); and Göteborg, Sweden, 2002 (Karlsson et al., 2003). Based on the available scientific literature at the time, the UN-ECE (1988) provisionally defined the critical levels for short-term O_3 exposures as 75 ppb (150 $\mu g\, m^{-3}$) for a 1-hour mean and 30 ppb (60 $\mu g\, m^{-3}$) for an 8-hour mean and for longer-term O_3 exposures as 25 ppb (50 $\mu g\, m^{-3}$) for a 7-hour daily mean (0900–1600) averaged over the vegetation growing period.

As scientific knowledge and understanding advanced and the workshop process proceeded, however, a new two-stage approach to critical levels was proposed by the UN-ECE called Level I and Level II (Fuhrer and Achermann, 1994). The recommendation was made for critical levels for O_3 for forests, agricultural crops and semi-natural vegetation to be based on the accumulated mean hourly exposure of vegetation to ozone over an O_3 concentration threshold of 40 ppb expressed as ppbh or ppmh and was called AOT40. The first-stage or Level I approach was not to consider any biotic or abiotic factors that might modify the response of vegetation to ozone. Level I was to be used to assess maximum potential risk to vegetation from ozone exposure. The second-stage or Level II approach, however, was to incorporate the modifying influence of biotic factors, as well as abiotic factors, on the responses of vegetation to ozone. Level II was to be used to more realistically quantify impacts of ozone exposure and their economic consequences to vegetation.

The provisional Level I AOT40 value for forests has been set at 10 000 ppbh (10 ppmh) and is calculated for daylight hours with global clear-sky radiation greater than 50 W m^{-2} during a six-month period (April–September). The long-term provisional Level I AOT40 value for semi-natural vegetation and agricultural crops has been set at 3000 ppbh (3 ppmh) calculated for daylight hours with global clear-sky radiation greater than 50 W m^{-2} for the appropriate 3-month growing season (Kärenlampi and Skärby, 1996). No Level II values have been established at this time as the underlying science for Level II is a work in progress.

The UN-ECE reported in Göteborg, Sweden (Karlsson et al., 2003) that one of the main difficulties in establishing O_3 critical levels at Level II is to determine the impact of ambient ozone on mature trees under field conditions. Data from Scandinavia suggest that a lower AOT40 value is required to protect the most sensitive tree species growing in northern Europe, under long-day, low vapor pressure deficit (VPD) conditions and short growing seasons. It was concluded that a stepwise procedure to enable a Level II estimate of ozone critical levels for forests is needed. The first step was the development of ozone uptake–biomass response relationships from experimental data using young trees. The following three approaches were suggested as possibilities: (1) modification of AOT with response factors; (2) MPOC (Maximum Permissible Ozone Concentration); and (3) flux-based concept. The final consensus from the Göteborg workshop indicated that the effective dose to forest trees based upon stomatal uptake should be implemented for future ozone critical levels for forests and that the AOT approach should be retained as a provisional measure.

North America and Europe have clearly taken different approaches in ambient ozone standard setting, with the latter opting for an approach to specifically protect vegetation. European research is now focused on incorporating the ozone uptake metric into the next generation of Level II, flux-based critical levels for ozone. It is clear that this process will serve as a useful guide for an equivalent standard-setting approach in North America (Mauzerall and Wang, 2001; Karlsson et al., 2003).

6. Uncertainties in current scientific understanding on ozone and forests

6.1. Uncertainties due to experimental methodologies

Much of our knowledge of tree species responses to O_3 is largely derived from experimental exposure in continuous stir-tank reactors, growth chambers and open-top chambers (Manning and Krupa, 1992; Chappelka and Samuelson, 1998), although a few studies relate to ambient field observations, e.g., the San Bernardino Mountains of Southern California (Miller and McBride, 1999). However, some recent field studies include chamber-less, free air O_3 enrichment (Karnosky et al., 2003). For instance, the exposure of tree communities to O_3 in Free Air Carbon Dioxide Enrichment (FACE) facilities has created an opportunity to study multi-trophic, long-term (5–10 years) responses of stand-level forest ecosystems under inter-annual climate variability and relatively unaltered experimental microclimate. At Aspen FACE (Wisconsin, USA), O_3 effects are being studied in terms of carbon sequestration, physiological processes, growth and productivity, competitive interactions and stand dynamics, and interactions with pests and ecosystem processes, such as foliar decomposition, mineral weathering, and nutrient cycling. Overall during 1998–2002, the response to O_3 exposure at growing season, daytime hourly average concentrations between 46–56 ppb (max. AOT40 25 000 ppbh; max CWS 96 ppb) has been remarkably consistent from leaf to ecosystem levels in pure aspen (five genotypes) and aspen/birch stands (Karnosky et al., 2003). Feedbacks to growth include a large reduction in both height and diameter growth at the population level (Percy et al., 2002). When averaged across five clones ranging from tolerant to highly sensitive to O_3, height growth began to diverge following 3 years of fumigation and was 12% reduced from growth in controls (Fig. 4(a)). Diameter growth, however, diverged between treatments almost immediately after year 2 of fumigation and was 13% reduced from that in the controls at year 4 (Fig. 4(b)). Physiological and genetic responses are cascading through the ecosystem, leading to a large reduction in stand net primary productivity and biomass (Karnosky et al., 2003).

Feedbacks to plant growth have also been shown in free-air O_3 exposure due to alteration of chemical plant defenses leading to a "bottom up" driven increased performance of some important herbivores (Percy et al., 2002). Trembling aspen normally accumulates high concentrations of phenolic glycosides (PG) that have important roles as protective agents against pests. Significantly, PG concentrations decreased following aspen exposure to O_3 leading to a large increase in forest tent caterpillar female pupal mass (Fig. 5), a surrogate for insect performance in the most important defoliator of deciduous forest trees in North America. It is important to emphasize that only through studies such as the Aspen free-air, field, stand-level, multitrophic studies can longer-term

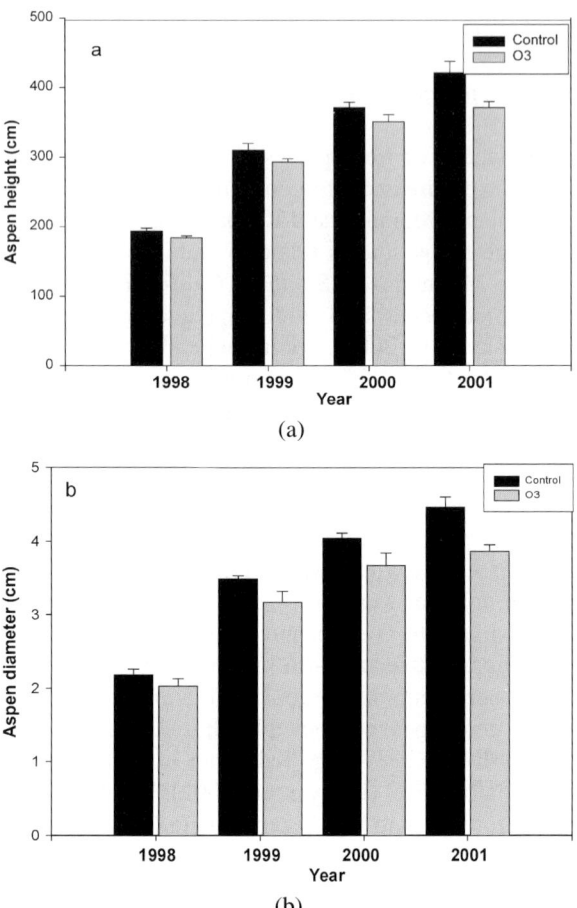

Figure 4. Effect of 4 years of free-air exposure at Aspen FACE on growth of trembling aspen (*Populus tremuloides* Michx.) averaged across five clones varying in sensitivity to O_3. (a) Trend in trembling aspen height growth; (b) trend in trembling aspen diameter growth. Data are means and 1 SE averaged across three FACE rings. Modified from Percy et al. (2002).

changes in forest function be studied and, importantly, forest-dose response relationships better defined.

Independent of the various experimental methods used, there are a number of uncertainties associated with our current understanding. In most investigations, saplings rather than mature tree responses were examined. Although efforts are being made to scale the results to mature trees (Samuelson and Kelly, 2001; Kolb and Matyssek, 2001), there are other issues: (1) Use of inadequate number of O_3 exposure treatments to fully define the response surface (Khuri

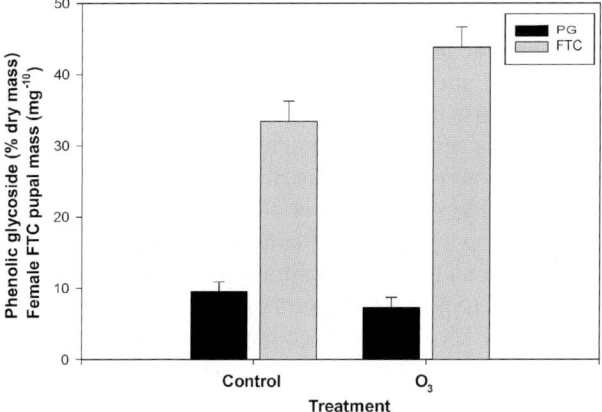

Figure 5. The effect of O_3 exposure at Aspen FACE on trembling aspen (*Populus tremuloides* Michx.) foliar susceptibility to herbivorous insects. Change in chemical leaf defence as manifested by concentrations of the important defensive metabolites, phenolic glycosides (PS); consequential change in performance of the important lepidopteran, leaf-chewing insect, the forest tent caterpillar (FTC) as assessed by female pupal mass. Modified from Percy et al. (2002).

and Cornell, 1996; Myers, 1971), (2) Use of O_3 treatments that do not simulate the stochasticity of the ambient fluxes and thus, realism, (3) Use of uni- or bi-variate systems and lack of emphasis on multi-variate systems, and (4) Use of monocultures, thus eliminating interspecies competition. Individuals growing in multi-species assemblages may respond differently to O_3 than individuals of the same species growing alone. For example, response of ponderosa pine to O_3 was greater when growing in the presence of a grass competitor than in its absence (Andersen and Grulke, 2001).

6.2. Definition of ozone exposure–forest response relationships

Kickert and Krupa (1991) provided a comprehensive review of the literature on modeling plant response to tropospheric O_3. In that context, cause–effect relationships have been established through the use of empirical or statistical and mechanistic or process models. Each approach has its advantages and disadvantages. Much of the effort in the US during the last decade has been directed to statistical models that have relied heavily on correlations. However, correlation does not necessarily mean causality (Snedcor and Cochran, 1978) and should have an underlying meaning. Furthermore, many such empirical models have not performed equally well every time. They are single-point models using season end biomass as the dependent variable. Such efforts cannot account for the dynamics of the atmosphere and the corresponding plant physiological phenology-dependent processes of avoidance, compensation or stress repair and the consequent stochastic cause–effect relationships (Krupa and Kickert, 1997). The associated uncertainty can be addressed by the use of multi-point models that use response measurements as a time series.

Another source of uncertainty is the definition of the O_3 concentrations. Using average values for non-normally distributed populations of data such as the ambient O_3 concentrations is inappropriate (Krupa and Kickert, 1997). More importantly, average values cannot explain the temporal or spatial dynamics in the O_3 concentrations and their flux. Similarly summation methods rely on a single threshold value (e.g., SUM06, sum of all hourly concentrations equal to or above 60 ppb, US EPA (1996)). In ecological research a single threshold is inappropriate, as opposed to a range of such values (Woodwell, 1975).

Until now, numerous studies have relied on the use of measured air concentrations of O_3 at some height in establishing cause–effect relationships (US EPA, 1996). In reality, it is the actual O_3 concentration at the canopy level and absorbed by the plant that results in an effect. In recent years, a great deal emphasis has been directed by scientists within the European Community to adapt an ozone flux-based approach to protect vegetation (see Karlsson et al., 2003). However, concerted efforts in that direction have not taken place in North America.

At the field monitoring scale, large and well-coordinated programs have attempted to relate forest condition/health at national or supra-national scales to O_3 exposure. Despite a large incidence of foliar injury (27% of plots) in a given year due to O_3, cause–effect linkage is rarely achieved (UN-ECE and EC, 2003). It is clear, however, that in some cases air pollutants (including O_3) have been successfully linked with changes in forest condition/health. Retrospective analysis (Percy, 2002) indicates that if spatial/temporal scales of the stressor (including O_3) are considered, appropriate indicators are measured, ecosystem function is investigated and if there is continuity of investigation such as described in Section 4.1 and summarized in Table 3, then the role of O_3 in relation to other man-made and natural stresses can be elucidated.

One important factor that must be remembered when cause–effect relationships are studied or when dose–response functions are being developed is the fact that plants are seldom exposed to one air pollutant singly, but usually a number of pollutants together. Although a simultaneous occurrence of air pollutants like SO_2, NO, NO_2, O_3 or NH_3 at phytotoxic levels is unusual (Fangmeier et al., 2002), co-exposure of forests to pollutant mixtures does occur even if the patterns of exposure differ spatially and temporally by pollutant. Although our knowledge of the effects of pollutant mixtures is far from satisfactory owing to experimental constraints imposed in running the required multifactorial experiments, the combined effect of O_3 and SO_2 has been shown to be either synergistic or antagonistic (Fangmeier et al., 2002) and the interac-

Table 3.   Retrospective analysis of degree of success in documenting the role of air pollution as an important factor in forest health. Modified from Percy (2002)

Successful when:	Unsuccessful when:
Network-level monitoring is succeeded by process-oriented research across spatial and temporal scales of stressors	Systematic monitoring is disconnected from process-oriented research
Appropriate indicators of ecosystem function are investigated at an intensity/extent appropriate to stressors and ecosystem	Endpoints measured are inappropriate or unresponsive to stressors
Systematic monitoring is not stratified on stressor distribution patterns	System protocols are not developed for single agents and integrated into multi-variant systems
Investigations in essential processes/cycles are integrated with investigations in ecosystem resilience through pests, genetics, succession, etc.	Hierarchical nature of forest response is not recognized
There is continuity of investigation as in long-term San Bernardino Mountain Forest-Oxidant Case Study	Dominant role of air pollutants in predisposition to other stressors is not recognized

tion of these two important globally distributed pollutants must be considered, particularly given the predicted increase in geographic co-occurrence of sulfur and O_3 during this century. Area of the world's forests at risk from S is predicted to increase 114% by 2050, with the largest proportional increases, as for O_3, being in tropical and subtropical forests (Fowler et al., 1999).

7. Conclusions

Ozone is a natural constituent of the surface layer in which the world's forests grow. However, man-made emissions of the O_3 precursors NO_x and VOCs have led to a large increase in average surface-level O_3 in the northern hemisphere during the past 100 years. In many regions of the world, O_3 concentrations are now damaging to vegetation, materials and human health. Between 1990 and 2100, the percentage of world forests exposed to damaging O_3 levels is expected to increase from 24% to fully 49%, or 17 million km^2 (Fowler et al., 1999).

The underlying processes responsible for the formation and deposition of O_3 are reasonably well understood. Despite some remaining uncertainties, the rate of O_3 deposition onto a given forest area can be calculated once ambient concentrations are known and information on key environmental variables such as wind velocity, air temperature, solar radiation and degree of leaf surface wetness is available. Indeed, recent progress in scientific understanding has led to the development of O_3 mass budgets at the country level (Coyle et al., 2003). Yet, true risk assessment remains problematic due to scientific uncertainty around the magnitude of O_3 flux into the plant. Indeed, standards and critical levels currently used by regulators remain based upon some index of O_3 exposure alone.

New approaches employing passive O_3 samplers are providing valuable data on the patterns of exposure in forested regions. Case studies have documented O_3 damage to health and function across diverse forest types and over a wide geographic range. New evidence from free-air experiments is pointing to the multi-trophic nature of forest ecosystem response to longer-term, lower levels of O_3 and to the important role of O_3 in predisposition to pests and environmental change. When assessing risk to forests from O_3, one must remember that North American ambient air quality standards that are set for compliance regulation purposes, do not in fact assume the existence of a concentration threshold for receptor response. In contrast, European critical levels are set based solely upon the best available scientific knowledge and understanding and the presumption that a threshold concentration for receptor response exists. The future development of new flux-based critical levels in Europe and biologically based dose–response functions in North America will allow policy makers for the first time to more accurately predict O_3 risk to the world's forests in the future.

References

Andersen, C.P., Grulke, N.E., 2001. Complexities in understanding ecosystem response to ozone stress. Human Ecol. Risk Assess. 7, 1169–1182.

Arbaugh, M.J., Peterson, D.L., Miller, P.R., 1999. Air pollution effects on growth of ponderosa pine, jeffrey pine and bigcone Douglas-fir. In: Miller, P.R., McBride, J.R. (Eds.), Oxidant Air Pollution Impacts in the Montane Forests of Southern California. A Case Study of the San Bernardino Mountains. Springer-Verlag, New York, pp. 179–207.

Ashmore, M.R., 2002. Air quality guidelines and their role in pollution control policy. In: Bell, J.N.B., Treshow, M. (Eds.), Air Pollution and Plant Life, 2nd Edition. Wiley, Chichester, England, pp. 417–429.

Ashmore, M.R., Wilson, R.B., 1992. Critical Levels of Air Pollutants for Europe. Background papers prepared for the United Nations Economic Commission for Europe Workshop on Critical Levels, Egham, United Kingdom, March 23–26, 1992. Air Quality Division, Department of the Environment, London, United Kingdom.

Barnes, J.D., Bender, J., Lyons, T., Borland, A., 1999. Natural and man-made selection for air pollution resistance. J. Exp. Bot. 50, 1423–1435.

Beck, J.P., Grennfelt, P., 1994. Estimate of ozone production and destruction over northwestern Europe. Atmos. Environ. 28, 129–140.

Blumenthal, D.L., Lurmann, F., Kumar, N., Dye, T., Ray, S., Korc, M., Londergan, R., Moore, G., 1997. Transport and mixing phenomena related to ozone exceedances in the Northeast US. Report STI-996133–1710, Sonoma Technology, Santa Rosa, CA.

Borrell, P., Builtjes, P.J.H., Hov, Ø., Grennfelt, P. (Eds.), 1997. Photo-oxidants, Acidification and Tools: Policy Applications of EUROTRAC Results. In: Transport and Chemical Transformation of Pollutants in the Troposphere, Vol. 10. Springer-Verlag, Berlin, Heidelberg.

Bravo, A.H., Torres, J.R., 2000. The usefulness of air quality monitoring and air quality impact studies before the introduction of reformulated gasolines in developing countries. Mexico City: a real case study. Atmos. Environ. 34, 499–506.

Bull, K.R., 1991. The critical loads/levels approach to gaseous pollutant emission control. Environ. Pollut. 69, 105–123.

Bytnerowicz, A., Krupa, S., Cox, R. (Eds.), 2001. Proceedings of the International Symposium on Passive Sampling of Gaseous Air Pollutants in Ecological Effects Research. In: The Scientific World, Vol. 1, pp. 461–462 (Preface).

Bytnerowicz, A., Fenn, M.E., Miller, P.R., Arbaugh, M.J., 1999. Wet and dry pollutant deposition to the mixed conifer forest. In: Miller, P.R., McBride, J.R. (Eds.), Oxidant Air Pollution Impacts in the Montane Forests of Southern California. A Case Study of the San Bernardino Mountains. Springer-Verlag, New York, pp. 235–269.

Chameides, W.L., Kasibhatla, P.S., Yienger, J., Levi II, H., 1994. Growth of continental-scale metro-agro-plexes, regional ozone pollution, and world food production. Science 264, 74–77.

CCME (Canadian Council of Ministers of the Environment), 2000. Canada-Wide Standards for Particulate Matter (PM) and Ozone. Available at http://www.ccme.ca/assets/pdf/pmozone_standard_e.pdf.

Chappelka, A.H., Samuelson, L.J., 1998. Ambient ozone effects on forest trees of the eastern United States: a review. New Phytol. 139, 91–108.

Conklin, P.L., Last, R.L., 1995. Differential accumulation of antioxidant mRNAs in *Arabidopsis thaliana* exposed to ozone. Plant Physiol. 109, 203–212.

Coyle, M., Smith, R., Fowler, D., 2003. An ozone budget for the UK: using measurements from the national ozone monitoring network; measured and modelled metereological data, and a 'big-leaf' resistance analogy model of dry deposition. Environ. Pollut. 123, 115–123.

Dann, T., 2001 Trends and levels of CWS Pollutants, Presentation to an AWMA Speciality Conference on Canada Wide Standards, Toronto, March 7, 2001.

deBauer, M.L., Tejeda, T.H., Manning, W.J., 1985. Ozone causes needle injury and tree decline in *Pinus hartwegii* at high altitudes in the mountains around Mexico City. J. Air Pollut. Control. Assoc. 35, 838.

Derwent, R., Collins, W., Johnson, C., Stevenson, D., 2002. Global ozone concentrations and regional air quality. Environ. Sci. Technol. 36, 379–382.

Emberson, L.D., Ashmore, M.R., Murray, F., Kuylenstierna, J.C.I., Percy, K.E., Izuta, T., Zheng, Y., Shimizu, H., Sheu, H., Liu, C.P., Agrawal, M., Wahid, M., Abdel-latif, N.M., van Tienhoven, M.N., de Bauer, L.I., Domingos, M., 2001. Impacts of air pollutants on vegetation in developing countries. Water Air Soil Pollut. 130, 107–118.

EFI (European Forest Institute), 2002. Nitrogen deposition appears to be the main cause of increased forest growth in Europe, http://www.efi.fi/news/2002/recognition.html.

Fangmeier, A., Bender, J., Weigel, H.-J., Jäger, H.-J., 2002. Effects of pollutant mixtures. In: Bell, J.N.B., Treshow, M. (Eds.), Air Pollution and Plant Life, 2nd Edition. Wiley, Chichester, England, pp. 251–272.

Federal Register, 1997. National Ambient Air Quality Standards for Ozone, Rules and Regulations, Environmental Protection Agency, 40 CFR Part 50, Final Rule. 62(138), July 18, 1997, p. 38856.

Ferretti, M., Bussotti, F., Cenni, E., Cozzi, A., 1999. Implementation of quality assurance procedures in the Italian programs of forest condition monitoring. Water Air Soil Pollut. 116, 371–376.

Finlayson-Pitts, B.J., Pitts Jr., J.N., 2000. Chemistry of the Upper and Lower Atmosphere. Academic Press, San Diego, CA.

Fowler, D., Cape, J.N., Coyle, M., Flechard, C., Kuylenstierna, J., Hicks, K., Derwent, D., Johnson, C., Stevenson, D., 1999. The global exposure of forests to air pollutants. Water Air Soil Pollut. 116, 5–32.

Fuhrer, J., Achermann, B. (Eds.), 1994. Critical Levels for Ozone – A UN-ECE Workshop Report. UN-ECE Convention on Long-Range Transboundary Air Pollution. Bern, Switzerland, November 1–4, 1993. Federal Research Station for Agricultural Chemistry and Environmental Hygiene, Liebefeld-Bern, Switzerland.

Fuhrer, J., Achermann, B. (Eds.), 1999. Critical Levels for Ozone – Level II, UN-ECE Workshop Report. UN-ECE Convention of Long-Range Transboundary Air Pollution. Gerzensee, Switzerland, April 11–15, 1999. Swiss Agency for the Environment, Forests and Landscape, Berne, Switzerland.

Fox, S., Mickler, R.A. (Eds.), 1996. Impact of Air Pollutants on Southern Pine Forests. Springer-Verlag, New York.

Guenther, A., Geron, C., Pierce, T., Lamb, B., Harley, P., Fall, R., 2000. Natural emissions of nonmethane volatile organic compounds, carbon monoxide, and oxides of nitrogen from North America. Atmos. Environ. 34, 2205–2230.

Guinnup, D., Collom, R., 1997. Final Report: Summary and Integration of Results. Air Quality Analysis Workgroup, Ozone Transport Assessment Group, Ozone Transport Commission. Office of Air Quality Planning and Standards, US EPA, Research Triangle Park, NC.

Haagen-Smit, A.J., Brunelle, M.F., Haagen-Smit, J.W., 1952. Ozone cracking in the Los Angeles area. Rubber Chem. Technol. 32, 1134–1142.

Heath, R.L., 1999. Biochemical processes in an ecosystem: How should they be measured? Water Air Soil Pollut. 116, 279–298.

Heggestad, H.E., Middleton, J.T., 1959. Ozone in high concentrations as cause of tobacco leaf injury. Science 129, 208–210.

Hidy, G.M., 1994. Atmospheric Sulfur and Nitrogen Oxides. Academic Press, San Diego, CA.

Hjellbrekke, A.-G., Solberg, S., 2002. Ozone measurements 2000, EMEP/CCC Report 5/2002, Kjeller, Norway.

Houzeau, A., 1858. Preuve de la présence dans l'atmosphère d'un nouveau principe gazeux, l'oxygène naissant. C. R. Acad. Sci. Paris 46, 89–91.

IUAPPA (International Union of Air Pollution Prevention and Environmental Protection Associations), 1995. Clean Air Around the World—National Approaches to Air Pollution Control, 3rd Edition. IUAPPA, Brighton, England.

Janach, W.E., 1989. Surface ozone: trend details, seasonal variations, and interpretation. J. Geophys. Res. 94, 18289–18295.

Jáuregui, E., 2002. The climate of the Mexico City air basin: its effects on the formation and transport of pollutants. In: Fenn, M.E., de Bauer, L.I., Hernández-Tejeda, T. (Eds.), Urban Air Pollution: Resources at Risk in the Mexico City Air Basin. Springer-Verlag, New York, pp. 86–117.

Jenkin, M.E., Clemitshaw, K.C., 2000. Ozone and other secondary photochemical pollutants: chemical processes governing their formation in the planetary boundary layer. Atmos. Environ. 34, 2499–2527.

Kärenlampi, L., Skärby, L., 1996. Critical Levels for Ozone in Europe: Testing and Finalizing the Concepts, UN-ECE Workshop Report. UN-ECE Convention on Long-Range Transboundary Air Pollution. Kuopio, Finland, April 15–17,1996. Department of Ecology and Environmental Science, University of Kuopio, Kuopio, Finland.

Karlsson, P.E., Selldén, G., Pleijel, H., (Eds.) 2003. Establishing Ozone Critical Levels II. UN-ECE Workshop Report. IVL Report B 1523. IVL Swedish Environmental Research Institute, Gothenburg, Sweden.

Karnosky, D.F., Pregitzer, K.S., Hendrey, G.R., Percy, K.E., Zak, D.R., Lindroth, R.L., Mattson, W.J., Kubiske, M., Podila, G.K., Noormets, A., McDonald, E., Kruger, E.L., King, J., Mankovska, B., Sober, A., Awmack, C.S., Callan, B., Hopkin, A., Xiang, B., Hom, J., Sober, J., Host, G., Riemenschneider, D.E., Zasada, J., Dickson, R.E., Isebrands, J.G., 2003. Impacts of interacting CO_2 and O_3 on trembling aspen: results from the aspen FACE experiment. Funct. Ecol. 17, 289–304.

Khuri, A.I., Cornell, J.A., 1996. Response Surfaces: Designs and Analyses, 2nd Edition. Marcel Dekker, New York.

Kickert, R.N., Krupa, S.V., 1991. Modeling plant response to tropospheric ozone: A critical review. Environ. Pollut. 70, 271–383.

Kley, D., Kleinmann, M., Sandermann, H., Krupa, S., 1999. Photochemical oxidants: state of the science. Environ. Pollut. 100, 19–42.

Koch, J.R., Creelman, R.A., Eshita, S.M., Seskar, M., Mullet, J.E., Davis, K.R., 2000. Ozone sensitivity in hybrid poplar correlates with insensitivity to both salicylic acid and jasmonic acid. The role of programmed cell death in lesion formation. Plant Physiol. 123, 487–496.

Kolb, T.E., Matyssek, R., 2001. Limitations and perspectives about scaling ozone impacts in trees. Environ. Pollut. 115, 373–393.

Krupa, S.V., 1997. Air Pollution, People and Plants. APS Press, St. Paul, MN.

Krupa, S.V., 2000. Commentary: Ultraviolet (UV)-B radiation, ozone and plant biology. Environ. Pollut. 110, 193–194.

Krupa, S.V., de Bauer, L.I., 1976. La Ciudad dana los Pinos del Ajusco. Panagfa 4, 5–7.

Krupa, S.V., Kickert, R.N., 1997. Ambient ozone (O_3) and adverse crop response. Environ. Rev. 5, 55–77.

Krupa, S.V., Legge, A.H., 2000. Passive sampling of ambient, gaseous air pollutants: an assessment from an ecological perspective. Environ. Pollut. 107, 31.

Krupa, S., Nosal, M., 2001. Relationships between passive sampler and continuous ozone (O_3) measurement data in ecological effects research. The Scientific World 1, 593–601.

Krupa, S., Nosal, M., Peterson, D.L., 2001. Use of passive ozone (O_3) samplers in vegetation effects assessment. Environ. Pollut. 112, 303–309.

Krupa, S.V., Tonneijck, A.E.G., Manning, W.J., 1998. Ozone. In: Flagler, R.B. (Ed.), Recognition of Air Pollution Injury to Vegetation: A Pictorial Atlas, 2nd Edition. Air and Waste Management Association, Pittsburg, PA, pp. 1–28.

Krupa, S., Nosal, M., Ferdinand, J.A., Stevenson, R.E., Skelly, J.M., 2003. A multi-variate statistical model integrating passive sampler and meteorology data to predict the frequency distributions of hourly ambient ozone (O_3) concentrations. Environ. Pollut. 124, 173–178.

LADCO, 1995. Lake Michigan Ozone Study: Lake Michigan Ozone Control Program, Vol. II-Technical Data Base. Lake Michigan Air Directors Consortium, Des Plaines, Il.

Legge, A.H., Krupa, S.V. (Eds.), 1990. Acidic Deposition: Sulphur and Nitrogen Oxides. Lewis Publishers, Chelsea, MI.

Legge, A.H., Grunhäge, L., Nosal, M., Jäger, H.-J., Krupa, S.V., 1995. Ambient ozone and adverse crop response: An evaluation of North American and European data as they relate to exposure indices and critical levels. Angew. Bot. 69, 192–205.

Longauer, R., Gomory, D., Paule, L., Karnosky, D.F., Mankovska, B., Muller-Starck, G., Percy, K., Szaro, R., 2001. Selection effects of air pollution on gene pools of Norway spruce, European Silver fir and European beech. Environ. Pollut. 115, 405–411.

Marenco, A., Gouget, H., Nédélec, P., Pagés, J.P., Karcher, F., 1994. Evidence of a long-term increase in tropospheric ozone from Pic du Midi data series: Consequences: Positive radiative forcing. J. Geophys. Res. Atmos. 99, 16617–16632.

Manning, W.J., Krupa, S.V., 1992. Experimental methodology for studying the effects of ozone on crops and trees. In: Lefohn, A.S. (Ed.), Surface Level Ozone Exposures and their Effects on Vegetation. Lewis Publishers, Chelsea, MI, pp. 93–156.

Matyssek, R., Innes, J.L., 1999. Ozone: a risk factor for trees and forests in Europe? Water Air Soil Pollut. 116, 199–226.

Mauzerall, D.L., Wang, X., 2001. Protecting agricultural crops from the effects of tropospheric ozone exposure: reconciling science and standard setting in the United States, Europe and Asia. Annu. Rev. Energy Environ. 26, 237–268.

Mazzali, C., Angelino, E., Gerosa, G., Ballarin-Denti, A., 2002. Ozone risk assessment and mapping in the Alps based on data from passive samplers. Proceedings of the International Symposium on Passive Sampling of Gaseous Air Pollutants in Ecological Effects Research. The Scientific World 1, 1023–1035.

McDonald, E.P., Kruger, E.L., Riemenschneider, D.E., Isebrands, J.G., 2002. Competitive status influences tree-growth response to elevated CO_2 and O_3 in aggrading aspen stands. Funct. Ecol. 16, 792–801.

McKendry, I.G., Steyn, D.G., Lundgren, J., Hoff, R.M., Strapp, W., Anlauf, K.G., Froude, F., Martin, J.B., Banta, R.M., Oliver, L.D., 1997. Elevated ozone layers and vertical downmixing over the Lower Fraser Valley, British Columbia. Atmos. Environ. 31, 2135–2146.

McLaughlin, S.B., Percy, K.E., 1999. Forest health in North America: some perspectives on potential roles of climate and air pollution. Water Air Soil Pollut. 116, 151–197.

McLaughlin, S.B., Downing, D.J., 1995. Interactive effects of ambient ozone and climate measured on growth of mature forest trees. Nature 374, 252–254.

McLaughlin, S.B., Downing, D.J., 1996. Interactive effects of ambient ozone and climate measured on growth of mature loblolly pine trees. Can. J. Forest Res. 26, 670–681.

Middleton, J.T., Kendrick, J.B., Schwalm, H.W., 1950. Injury to herbaceous plants by smog or air pollution. Plant Disease Rep. 34, 245–252.

Miller, P.R., 1993. Response of forests to ozone in a changing atmospheric environment. J. Appl. Bot. (Angew. Bot.) 67, 42–46.

Miller, P.R., McBride, J.R. (Eds.), 1999. Oxidant Air Pollution Impacts in the Montane Forests of Southern California. A Case Study of the San Bernardino Mountains. Springer-Verlag, New York.

Miller, P.R., McBride, J.R., Schiling, S.L., Gomez, A.P., 1989. Trend in ozone damage to conifer forests between 1974 and 1988 in the San Bernardino Mountains in southern California. In: Olson, R.K., Lefohn, A.S. (Eds.), Transactions Air and Waste Management Association symposium, Effects of Air Pollution on Western Forests. Air & Waste Management Association, Pittsburgh, PA, pp. 309–323.

Miller, P.R., de Bauer, L.I., Hernández-Tejeda, T., 2002. Oxidant exposure and effects on pines in forests in the Mexico City and Los Angeles, California air basins. In: Fenn, M.E., de Bauer, L.I., Hernández-Tejeda, T. (Eds.), Urban Air Pollution and Forests: Resources at Risk in the Mexico City Air Basin. In: Ecol. Study Series. Springer-Verlag, New York, pp. 225–242.

Mullineaux, P., Creissen, G., 1999. Manipulating oxidative stress responses using transgenic plants: successes and dangers. In: Altman, A., Ziv, M., Izhar, S., Plant, S. (Eds.), Biotechnology and In Vitro Biology in the 21st Century. Kluwer Academic Publishers, Amsterdam, pp. 525–532.

Myers, R.H., 1971. Response Surface Methodology. Allyn and Bacon, Boston.

NARSTO (North American Research Strategy for Tropospheric Ozone), 2000. The Narsto Ozone Assessment—Critical Reviews. Atmos. Environ. 34, 1853–2332.

Niccum, E.M., Lehrman, D.E., Knuth, W.R., 1995. The influence of meteorology on the air quality in the San Luis Obispo County–Southwestern San Joaquin Valley Region for 3–6 August 1990. Special Issue for the Regional Photochemical Measurement and Modeling Studies Specialty Conference. J. Appl. Meteorol. 34, 1834–1847.

Nosal, M., Legge, A.H., Krupa, S.V., 2000. Application of a stochastic, Weibull probability generator for replacing missing data on ambient concentrations of gaseous pollutatants. Environ. Pollut. 108, 439–446.

Noormets, A., Podila, G.K., Karnosky, D.F., 2000. Rapid response of antioxidant enzymes to O_3-induced oxidative stress in *Populus tremuloides* clones varying in O_3 tolerance. For. Genet. 7, 339–342.

Percy, K.E., 2002. Is air pollution an important factor in forest health? In: Szaro, R.C., Bytnerowicz, A., Oszlanyi, J. (Eds.), Effects of Air Pollution on Forest Health and Biodiversity in Forests of the Carpathian Mountains. IOS Press, Amsterdam, pp. 23–42.

Percy, K.E., Awmack, C.S., Lindroth, R.L., Kubiske, M.E., Kopper, B.J., Isebrands, J.G., Pregitzer, K.S., Hendrey, G.R., Dickson, R.E., Zak, D.R., Oksanen, E., Sober, J., Harrington, R., Karnosky, D.F., 2002. Altered performance of forest pests under CO_2- and O_3-enriched atmospheres. Nature 420, 403–407.

Pisano, J.T., McKendry, I., Steyn, D.G., Hastie, D.R., 1997. Vertical nitrogen dioxide and ozone concentrations measured from a tethered balloon in the Lower Fraser Valley. Atmos. Environ. 31, 2071–2078.

Ramaswamy, V., Boucher, O., Haigh, J., Hauglustaine, D., Haywood, J., Myhre, G., Nakajima, T., Shi, G.Y., Solomon, S., 2001. Radiative forcing of climate change. In: Houghton, J.T., Ding, Y., Griggs, D.J., Noguer, M., van der Linden, P.J., Dai, X., Maskell, K., Johnson, C.A. (Eds.), Climate Change 2001: The Scientific Basis. Cambridge Univ. Press, New York, pp. 350–416.

Reid, N.W., 1994. The Southern Ontario Study (SONTOS): Data Report, Summer Field Campaign 1993 (13 July to 27 August 1993). Ministry of the Environment and Energy, Toronto, Ontario, Canada.

Reid, N.W., Niki, H., Hastie, D., Shepson, P., Roussel, P., Melo, O., Mackay, G., Drummond, J., Schiff, H., Poissant, L., Moroz, W., 1996. The Southern Ontario Study (SONTOS): Overview and case studies for 1992. Atmos. Environ. 30, 2125–2132.

Richards, B.L., Middleton, J.T., Hewitt, W.B., 1958. Air pollution with relation to agronomic crops. V. Oxidant stipple of grape. Agron. J. 50, 559–561.

Roberts, P.T., Coe, D.L., Dye, T.S., Ray, S.E., Arthur, M., 1996. Summary of Measurements Obtained during the 1996 Paso Del Norte Ozone Study, Final Report. Report STI-996191–1603-FR, Sonoma Technology, Santa Rosa, CA.

Roussel, P.B., Lin, X., Camacho, F., Laszlo, S., Taylor, R., Melo, O.T., Shepson, P.B., Hastie, D.R., Niki, H., 1996. Observations of ozone and precursor levels at two sites around Toronto, Ontario, during SONTOS 92. Atmos. Environ. 30, 2145–2155.

Samuelson, L., Kelly, J.M., 2001. Scaling ozone effects from seedlings to forest trees. New Phytol. 149, 21–41.

Sandermann, H., 1996. Ozone and plant health. Annu. Rev. Phytopathol. 34, 347–366.

Sandermann, H., Wellburn, A.R., Heath, R.L. (Eds.), 1997. Forest Decline and Ozone: A Comparison of Controlled Chamber and Field Experiments. Springer-Verlag, New York.

Sandroni, S., Anfossi, D., Viarengo, S., 1992. Surface ozone levels at the end of the nineteenth century in South America. J. Geophys. Res. Atmos. 97, 2535–2539.

Schjoldager, J., 1981. Ambient ozone measurements in Norway, 1975–1979. JAPCA 31, 1187–1190.

Schraudner, M., Moeder, W.V., Camp, W., Inzé, D., Langebartels, C., Sandermann, H., 1998. Ozone-induced oxidative burst in the ozone biomonitor plant tobacco W3. Plant J. 16, 235–245.

Schönbein, C.F., 1840. Recherches sur la nature de l'odeur qui se manifeste dans certaines actions chimiques. C. R. Acad. Sci. Paris 10, 706–710.

Sheffield, R.M., Cost, N.D., 1987. Behind the decline: why are natural pine stands in the Southeast growing slower? J. Forests 85, 29–33.

Sillman, S., 1999. The relation between ozone, NO_x and hydrocarbons in urban and polluted rural environments. Atmos. Environ. 33, 1821–1846.

Skarby, L., Ro-Poulson, H., Wellburn, F.A.M., Sheppard, L.J., 1998. Impacts of ozone on forests: a European perspective. New Phytol. 139, 109–122.

Snedcor, G.W., Cochran, W.G., 1978. Statistical Methods. Iowa State Univ. Press, Ames, IA, USA.

Solomon, P., Cowling, E., Hidy, G., Furiness, C., 2000. Special Issue: the NARSTO Ozone Assessment—Critical Reviews. Comparison of Scientific Findings from Major Ozone Field Studies in North America and Europe. Environ. Pollut. 34, 1885–1920.

Takemoto, B.K., Bytnerowicz, A., Fenn, M.E., 2001. Current and future effects of ozone and atmospheric nitrogen deposition on California's mixed conifer forests. Forest Ecol. Manag. 144, 159–173.

Tema Nord, 1994. Critical Levels for Tropospheric Ozone—Concepts and Criteria Tested for Nordic Conditions. L. Skärby (Coordinator). Nordic Council of Ministers, Copenhagen, Denmark.

Thuillier, R.H., 1997. Summary Description of Meteorological Conditions During SJ-VAQS/AUSPEX, Final Report. Research and Development, Report 005–97.7, Pacific Gas and Electric Company, San Ramon, CA.

Torsethaugen, G., Pell, E.J., Assmann, S.M., 1999. Ozone inhibits guard cell K^+ channels implicated in stomatal opening. Proc. Nat. Acad. Sci. USA 96, 13577–13582.

Trainer, M., Parrish, D.D., Goldan, P.D., Roberts, J., Fehsenfeld, F.C., 2000. Review of observation-based analysis of the regional factors influencing ozone concentrations. Atmos. Environ. 34, 2045–2061.

Tuovinen, J.-P., 2002. Assessing vegetation exposure to ozone: is it possible to estimate AOT40 by passive sampling? Environ. Pollut. 119, 203–214.

UN-ECE, 1988. ECE Critical Levels Workshop, Final Draft Report. Bad Harzburg, Federal Republic of Germany, March 14–18, 1988.

UN-ECE and EC 2002. The Condition of Forests in Europe: 2002 Executive Report, Federal Research Centre for Forestry and Forest Products (BFH), Germany.

UN-ECE and EC, 2003. 2003 Executive Report: The Condition of Forests in Europe. Federal Research Centre for Forestry and Forest Products (BFH), Germany (in press).

US EPA (United States Environmental Protection Agency), 1996. Air Quality Criteria for Ozone and Other Photochemical Oxidants, Vol. II. US EPA, National Center for Environmental Assessment, Research Triangle Park, NC, EPA-600/P-93/00bF. United States Environmental Protection Agency, Washington, D.C.

US EPA (United States Environmental Protection Agency), 2002. Ground-level ozone: nature and sources of the pollutant. Available at http://www.epa.gov/airtrends/ozone.html. United States Environmental Protection Agency, Washington, D.C.

Volz, A., Kley, D., 1988. Evaluation of the Montsouris series of ozone measurements made in the nineteenth century. Nature 332, 240–242.

Woodwell, G.M., 1975. The threshold problem in ecosystems. In: Levin, S.A. (Ed.), Ecosystem Analysis and Prediction. Society for Industrial and Applied Mathematics, Philadelphia, PA, pp. 9–21.

Air Pollution, Global Change and Forests in the New Millennium
D.F. Karnosky et al., editors
© 2003 Elsevier Ltd. All rights reserved.

119

Chapter 5

Regional scale risk assessment of ozone and forests

G. Gerosa*

*DMF, Department of Mathematics and Physics, Università Catolica del Sacro Cuore,
via Musei 2, 25121 Brescia, Italy*

A. Ballarin-Denti

*Department of Mathematics and Physics, Catholic University of Brescia,
via Musei 41, 25121 Brescia, Italy*

Abstract

The ozone exposure risk for vegetation in Lombardy (Northern Italy) has been assessed by the AOT40 exposure index, based on data taken from the existing local monitoring networks covering 5 growing seasons (1994 to 1998). One-square kilometer exposure maps were obtained by using geostatistic techniques (ordinary kriging) followed by an altitude detrendization of measurement's temporal series to account for the domain's large topographic heterogeneity. Risk areas (Level I maps) were identified using a GIS and overlaying the ozone-critical-level exceedance maps on the distribution maps of forests and sensitive species.

The critical ozone exposure level of 10 000 ppb h, adopted by UN/ECE protocols, is exceeded over the whole Lombardy Territory over the 6-month growing season. The highest risk areas are the northwest pre-alpine and alpine belt, directly impacted by the photo-oxidant plume generated by the Milan urban area. Difficulties met in creating a proper Level II risk assessment for forests in mountain areas have been bypassed by comparing ozone exposures with summer climate features. Soil water availability was assumed not to be a significant modifying factor in the mountains of this region because of the frequent summer rains, whereas the opposite held true for wind ventilation which is generally weak. Field surveys have reported foliar injuries attributable to ozone in different species of forest trees and shrubs, which provide further evidence of potentially phytotoxic ozone levels.

---

* Corresponding author.

DOI:10.1016/S1474-8177(03)03005-5

1. Introduction

High concentrations of tropospheric ozone and photo-oxidants represent a major environmental concern in most European countries, particularly in the Alpine and Mediterranean regions (Sandroni et al., 1994; Staffelbach et al., 1997), because of their possible negative effects on agricultural and forest ecosystems. The ozone contribution to biological damages and productivity loss, although dependent on species-specific genetic features and environmental conditions, is a function of absorbed dose and, therefore, of the true physiological exposure to this pollutant.

An assessment of photo-oxidants impact on vegetation at regional scale might be based on exposure values obtained from ambient ozone concentrations, aimed at estimating absorbed doses and at predicting plant organisms' responses based on their specific diversity. Although different dose-injury relationships have been reported from studies performed on different grass and tree species in open-top chambers (Fuhrer et al. 1997), it is often difficult to assess damages at territorial scale due to the uncertainty in the determination of pollutant doses realistically absorbed by the vegetation in open field conditions.

The actual uptake at leaf level is in fact influenced by several meteorological factors, such as temperature, precipitation, humidity, evapotranspiration rate and soil water content, all able to affect the stomatal opening and, therefore, the true absorbed dose. The difficulty to collect reliable data on all the necessary parameters at adequate territorial scale and the need to obtain their estimates from model computations make ozone risk assessments for vegetation often limited to the determination of exposures as a first-step approximation.

The consequent land mapping of exposures indexes leads to the so-called Level I assessment as defined in the UN/ECE protocols framework.

2. Methods

In order to identify the risk areas connected to photo-oxidative stress in the forest domains of the Lombardy region in Italy (Fig. 1), Level I maps have been generated on the basis of the AOT40 index—the long-term cumulative exposure index adopted by many ozone risk assessment protocols in Europe. The selected time base covers the vegetative seasons of five years from 1994 to 1998.

The AOT40 (Accumulated exposure Over a Threshold of 40 ppb) is defined as the sum of the differences between the ozone hourly concentrations and the offset threshold of 40 ppb, calculated for all daylight hours (global radiation $\geqslant 50$ W/m^2) of the whole vegetative season which is conventionally set

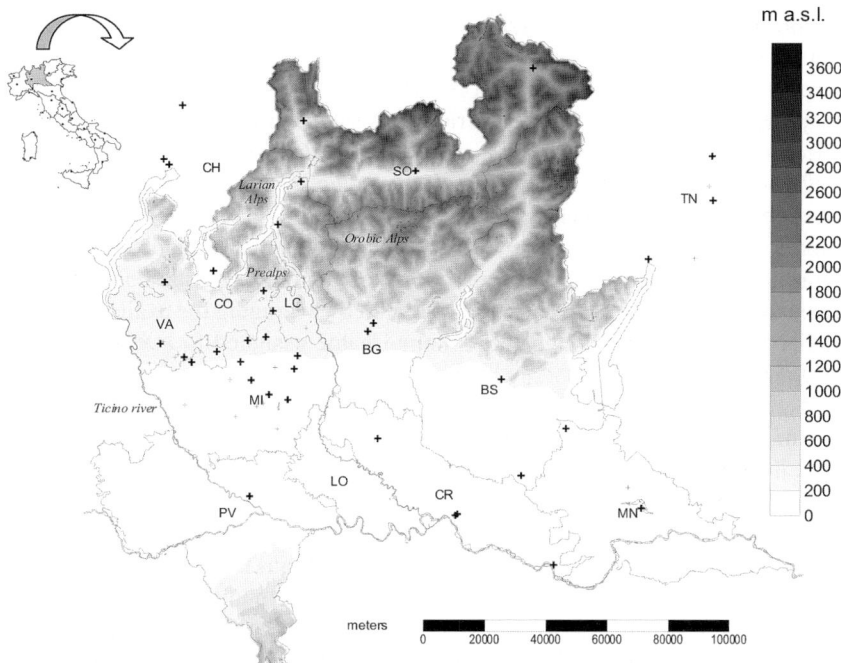

Figure 1. Tropospheric ozone survey network: dislocation of monitoring stations. Stations selected for risk assessment procedure are indicated with bold crosses. Province labels: BG = Berg-
amo, BS = Brescia, CO = Como, CR = Cremona, LC = Lecco, LO = Lodi, MI – Milano, MN =
Mantova, PV = Pavia, SO = Sondrio, VA = Varese, TN = Trento, CH = Switzerland (Canton
Ticino).

between April and September for forest vegetation (Kärenlampi and Skärby,
1996).

$$\text{AOT40} := \sum_{[O_3]_i > 40 \text{ ppb}} \left([O_3]_i - 40\right). \tag{1}$$

According to several investigations performed on biomass production decrease
in forest species (Fuhrer et al., 1997), a critical level of 10 000 ppb h has been
adopted by recent UN-ECE protocols and chosen on the assumption that, above
it, the appearance of injury symptoms might be expected in the most sensitive
species.

2.1. Data collection and mapping procedure

The following procedure has been adopted in the present study for mapping
ozone exposure levels: (0) selection of monitoring stations to be accounted for;

(1) raw data collection and database organization; (2) calculation of the AOT40 within the reference period; (3) compensation for missing data; (4) spatial interpolation; (5) generation of final AOT40 maps as average of the five AOT40 maps of the five-year reference period; (6) intersection of AOT40 maps with specific plant receptors maps (vegetation spatial distribution).

Ozone concentration and solar radiation data in the five-year period were obtained as hourly means from the measurements recorded by the regional monitoring stations network and then structured in a proper database. To compensate missing data of solar radiation by some monitoring stations, the mean radiative season for each year was calculated by averaging the available global radiation measurements and the obtained value assigned to stations lacking sufficient data.

The monitoring stations used for AOT40 calculation have been selected according to criteria of (i) sampling efficiency (data capturing $\geqslant 70\%$), (ii) location in rural areas, and (iii) territorial significance. Stations classified as rural, semi-rural or sub-urban (Class A, D and B according to Italian regulations) have been chosen and selected in the listed order. Stations classified as urban (Class C) have been excluded to prevent bias linked to higher ozone-destruction rates by nitrogen oxide (NO) emitted from industrial and transport combustion processes in urban areas. This criterion has been violated only in two cases in order not to leave a whole province domain for each year unrepresented. Lastly, when two or more stations were closer than 5 km, only the station with the highest AOT40f has been chosen to prevent overrepresentation of the area. The configuration of the regional monitoring network (Fig. 1) had not been constant during the reference time domain as far as data sampling efficiency and space location of monitoring stations were concerned. Consequently, the number of stations selected for each year has been varying from 13 to 22, to which 7 more stations were added out of the geographic domain (taken from the monitoring networks of Switzerland and Italian Trento region) in order to improve the system boundary conditions (Table 1).

AOT40 index for crops was calculated over the trimester April 1–June 30, thus modifying the time span suggested by the Kuopio UN-ECE protocol (Kärenlampi and Skärby, 1996). According to this protocol, the critical exposure level for crops has been determined by choosing summer wheat (*Triticum aestivum*) as indicator species. Since wheat is harvested in Italy—differently from what happens in central Europe—before the end of June, we decided to choose accordingly a reference period more representative of the real ozone exposure for wheat.

AOT40 is a cumulative index; this means that any lack of hourly concentration values may produce an underestimate of the overall exposure. The problem has been tackled, under the simplifying assumption of the equi-distribution of the missing data into the different daily hours, by correcting the monthly

Table 1. Ozone monitoring stations. A cross in the last 5 columns indicates the stations which have been selected for the risk assessment procedure

Name	Province or State	Class[a]	Typology	Altitude m a.s.l.	X UTM[b]	Y UTM[b]	'94	'95	'96	'97	'98
Bergamo (S. Giorgio)	BG	C	Urban	249	1 551 800	5 059 540	+	+	+		
Bergamo (Goisis)	BG	A	Rural	249	1 553 660	5 062 285				+	+
Brescia (Broletto)	BS	B	Urban park	149	1 595 495	5 043 695	+	+	+		+
Gambara	BS	D	Semi-rural	51	1 601 980	5 011 530	+	+	+		+
Colico	LC	D	Semi-rural	218	1 529 740	5 109 475				+	
Nibionno	LC	D	Semi-rural	310	1 520 640	5 066 450				+	
Varenna	LC	D	Semi-rural	220	1 522 115	5 095 155	+	+	+	+	
Erba	CO	A	Rural	323	1 517 445	5 073 135		+	+	+	+
Cremona (Libertà)	CR	C	Urban	45	1 581 200	4 998 750	+	+			+
Cremona (Cavour)	CR	B	Sub-urban	45	1 580 520	4 998 300				+	
Crema	CR	B	Sub-urban	79	1 555 170	5 023 910				+	+
Casalmaggiore	CR	B	Semi-rural	23	1 612 575	4 981 835				+	
Mantova (Tè)	MN	B	Sub-urban	18	1 641 120	5 000 670	+	+	+	+	
Castiglione D. Stiviere	MN	B	Semi-rural	116	1 616 488	5 027 282	+	+	+	+	
Agrate Brianza	MI	B	Sub-urban	162	1 527 650	5 047 000					+
Pioltello	MI	D	Sub-urban	123	1 525 610	5 036 750	+	+			
Legnano	MI	B	Sub-urban	199	1 493 695	5 049 200	+	+	+		
Carate Brianza	MI	B	Semi-rural	256	1 518 250	5 057 650	+				+
Meda	MI	B	Sub-urban	221	1 512 230	5 056 500			+	+	
Limbiate	MI	B	Sub-urban	186	1 509 850	5 049 350		+	+	+	
Vimercate	MI	B	Sub-urban	194	1 528 750	5 051 350		+	+	+	
Cormano	MI	B	Sub-urban	146	1 513 450	5 043 250		+			
Milano (Lambro Park)	MI	D	Urban park	122	1 519 350	5 038 500					+
Pavia	PV	B	Urban park	77	1 512 960	5 004 610	+	+	+	+	+
Sondrio	SO	B	Sub-urban	307	1 567 210	5 113 100	+		+	+	
Bormio	SO	B	Semi-rural	1225	1 605 380	5 147 330			+	+	+
Chiavenna	SO	B	Semi-rural	333	1 530 480	5 129 790				+	+
Varese (Vidoletti)	VA	B	Sub-urban	382	1 484 800	5 075 965	+		+	+	+
Saronno	VA	B	Sub-urban	212	1 501 900	5 052 710	+	+	+	+	+
Gallarate	VA	A	Rural	238	1 483 415	5 055 385				+	+
Castellanza	VA	A	Rural	217	1 491 235	5 050 853				+	+
Mendrisio	CH-TI	D	Sub-urban	350	1 496 700	5 077 750	+	+	+	+	+
Bodio	CH-TI	B	Sub-urban	320	1 490 400	5 134 850	+	+	+	+	+
Brione Sopra Minusio	CH-TI	D	Semi-rural	480	1 483 050	5 113 200	+	+	+	+	+
Monte Cimetta	CH-TI	A	Rural, remote	1650	1 481 300	5 115 050	+	+		+	
Trento (Park)	TN	A	Rural	203	1 664 490	5 103 275	+	+	+	+	+
Grumo	TN	D	Semi-rural	228	1 664 160	5 118 126	+			+	+
Riva Del Garda	TN	D	Sub-urban	73	1 643 220	5 083 750	+	+	+	+	+

[a]Letters indicates the station typology according to Italian regulations: A = rural, not directly interested by urban emission sources; B = urban or sub-urban; C = traffic area; D = semi-rural or peripheric sub-urban.

[b]Geographic coordinates are referred to the Mercatore Universal Transvers projection.

"raw" AOT40 by a factor representing the reciprocal of the sampling efficiency for each month

$$AOT40 = \sum_m \frac{AOT40_{raw_m} \cdot N_{hours_m}}{N_{ValidHours_m}} \tag{2}$$

where N_{hours} is the total number of hours in the considered period, $N_{ValidHours}$ is the number of hours with valid measurements and the subscript m refers to the mth month.

Maps of spatial distribution for ozone exposure over the whole region were obtained by interpolating single-point AOT40 data on a 1×1 km grid.

2.2. Ozone elevation dependency and geostatistics

Tropospheric ozone concentrations strongly depend on elevation which is in addition a function of precursor levels and solar radiation density. Orographic roughness and clustering of emission sources on valley bottoms lead, in mountain areas, to ozone concentration gradients (and therefore of AOT40 values) even at quite short distances. As a consequence, also the shape of daily cycles of ozone production and depletion is dependent on the station's geographic features and specifically on elevation.

Two different interpolation techniques (Fowler et al., 1995; Loibl et al., 1994), which account for elevation and structure of production-depletion cycles, have been critically evaluated and then applied with some modifications in the present study. Both techniques trace a "basic" trend surface of ozone distribution over the domain. Local variations, due to deviations of measured data from model data, are then interpolated by means of geostatistic techniques such as ordinary kriging (Goovaerts, 1997) or *inverse weighted distance.* The final distribution is obtained by re-assembling the trend surface with the interpolated deviations.

The Loibl technique is based on a model function which takes into account time and elevation dependence of ozone hourly mean concentrations in mountain regions. Its algorithm requires a heavy computational load since more than 4000 hourly maps have to be calculated by *kriging* and then grouped together. Moreover, the *kriging* procedure is by itself only partially automatic since it relies on human expertise in order to choose the correct semi-variogram model, on which the accuracy of kriging depends.

In the present study the Loibl et al. (1994) algorithm was modified with proper parameters to better fit the structure of local ozone data and to calculate over the whole domain the "expected" AOT40 over a six-month growing season as a function of the mean relative altitude of each grid square. The relative altitude was defined as the difference between the mean altitude of the considered grid square and the lowest mean altitude in a 5 km neighborhood. Dif-

ferences between the expected and single-point measured AOT40s were calculated and these residuals krigged over the whole domain after the best fitting of
the semi-variogram model. The residuals map was added with the "expected"
AOT40 map to obtain the elevation-dependent AOT40 estimate map.

The second technique we considered, originally developed by Fowler et
al. (1995) and adopted by the UK Photochemical Oxidants Review Group
(1997), is based on the assumption that the structure of the ozone production-
depletion cycles and the elevation dependence can be described by one indicator only, the Ratio between the AOT40 measured during daylight hours
and that ($AOT40_{mix}$) calculated during the maximal remixing hours (12:00–
18:00).

$$\text{Ratio} = \frac{\text{AOT40}}{\text{AOT40}_{mix}} \qquad (3)$$

The assumption originates from the observation that, during the hours of maximal atmosphere remixing, ozone concentrations measured at different elevations are almost similar, in all nearby stations, substantial differences being
observed only in the evening and morning hours. Stations located in the plain
or near emission sources show Ratio values near 1, whereas remote and more
elevated stations, subject to a lower nighttime ozone depletion, have higher
values, ranging between 1 and 2.

By means of an analytical relationship between the Ratio and the elevation
parameter, Ratio values can therefore be predicted for any point of the territory
whose elevation is known. Ratio maps are then obtained by applying this relationship, generally linear, to the terrain digital elevation model of the reference
grid. The Ratio parameter, therefore, reflects the basic structural features of
ozone distribution. The $AOT40_{mix}$ instead reflects the local characteristics and
can be spatially interpolated with ordinary geostatistical techniques.

Finally, AOT40 maps of the whole daylight hours are obtained from
$AOT40_{mix}$ and Ratio maps by reversing equation (3)

$$\text{AOT40}(x, y) = \text{AOT40}_{mix}(x, y) \cdot \text{Ratio}(x, y) \qquad (4)$$

and applying equation (4) to every element (x, y) of the grid.

The reliability of this method in Lombardy land conditions is grounded on
the elevated thickness of the boundary mixing layer during summertime, which
extends to more than 2000 m a.g.l. due to strong thermal atmospheric turbulence. Moreover, $AOT40_{mix}$ is relatively independent from the elevation, thus
satisfying the stationary condition requested for the application of the ordinary
kriging and allowing the use of this powerful geostatistical technique for interpolation. Differently from the inverse distance weighting method adopted
by the UK-PORG investigators, the ordinary kriging approach is also able to
produce estimate variance maps allowing researchers to evaluate the estimate's

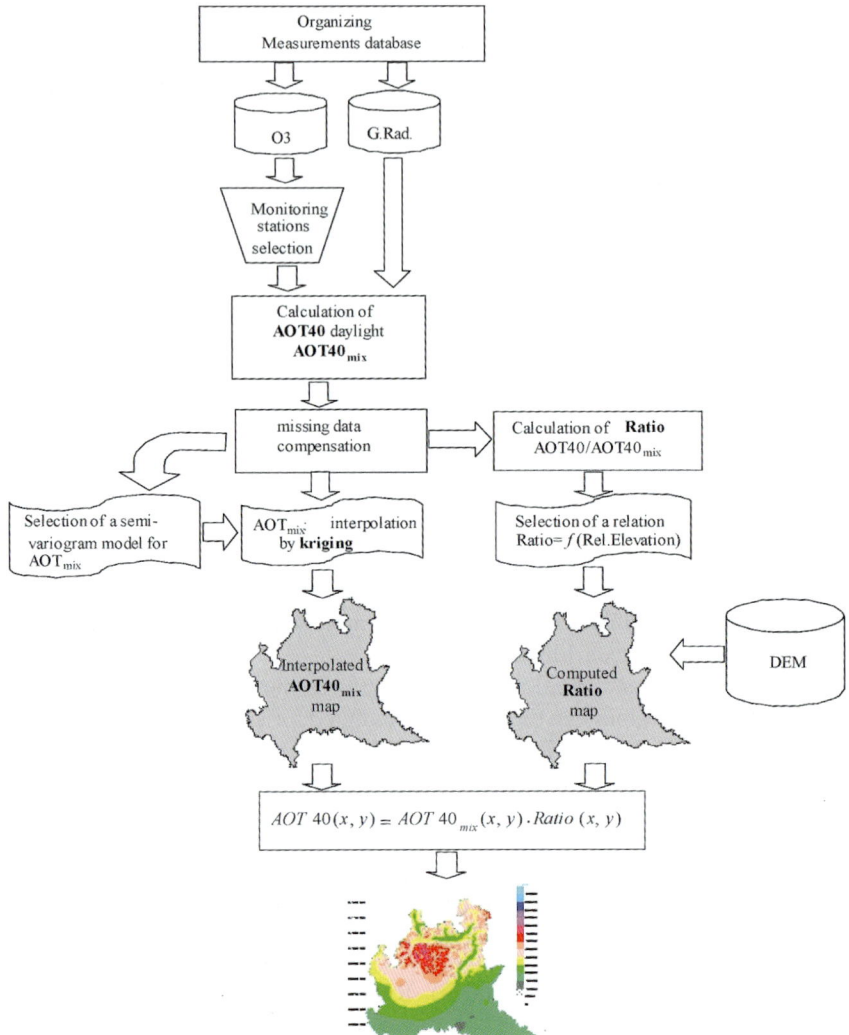

Figure 2. Production of AOT40 maps (outline).

own accuracy. The comprehensive technique scheme is shown in Fig. 2, while details of the semi-variogram functions used in kriging procedure are reported in Table 2.

In order to improve the consistency with the orographic characteristics of the Lombardy territory and to allow an acceptable exposure representation also in higher elevation areas, we decided to modify this technique by replacing

Table 2. Parameters of the semi-variogram functions used in the kriging procedure

AOT40$_{mix}$	Year				
	1998	1997	1996	1995	1994
Semi-variogram model	Linear[a]	Linear	Linear	Linear	Exponential[a]
Range	70.102	87.294	81.343	66.213	16.544
Sill	7.40E+07	1.37E+08	9.59E+07	1.56E+08	8.14E+07
Nugget	0	3.90E+06	0	0	0
RSS	5.37E+16	1.77E+16	4.82E+16	1.87E+17	6.00E+16
Precision (last iteration fit)	1.08E−07	1.11E−10	8.78E−08	1.71E−16	1.73E−07

[a]The function of the linear semi-variogram model is

$$\gamma(h) = \begin{cases} C_0 + C[h], & h < A, \\ C[A], & h \geqslant A \end{cases}$$

and that of the exponential model is $\gamma(h) = C[1 - e^{-h}]$. The parameters C, C_0, A and $C[A]$ represent respectively the scale, the nugget, the range and the sill, while h indicates the spatial lag.

the relative altitude in place of the absolute altitude of the monitoring stations. Consequently, the ratio–altitude function has been changed in terms of Ratio–Relative Altitude (Ratio $= 1.166626 + 0.000682 \cdot$ [Relative Altitude]; $R = 0.88$; $n = 115$; $F = 392$, Significance $p < 0.0001$).

2.3. GIS and geostatistics tools

Finally, to smooth yearly variations of summer ozone levels, five consecutive yearly maps have been averaged according to recommendations put forward by UN/ECE Task Force for mapping (UBA, 1996; Posch et al., 1998).

Ozone risk areas (Level I mapping) have been spotted by means of a GIS (Esra ARC/INFO) and by intertwining exposure exceedances maps with receptors distribution maps obtained from the Territorial Information System of the Lombardy Region at synthesis scale (1 : 250 000).

The procedure of land use classes aggregation adopted to obtain receptors maps as reported in Table 3 and relative notes. CORINE land cover classes are also reported for comparison. Different broadleaf forest managements were not distinguished (high trunk wood and copse). Poplar plantations were aggregated to broadleaf forests.

Receptors maps were calculated as cover percentage for every 1 × 1 km grid element. The intertwining procedure to obtain Level I maps considered valid only grid elements in which the receptor coverage was at least 1 ha.

Table 3. Receptor classes used in the mapping procedure and their relation with the original land cover classes from the Lombardy Region SIT 250 source. The CORINE Land Cover Classes system is also reported for comparison

Lombardy region SIT 250 land cover classes	CORINE land cover classes	Aggregated land cover classes (receptors)
Water	Closed Urban	Areas not included
Barren area	Open Urban	in the analysis
Quarry	Industrial area	
Industrial and commercial building	Road	
Residential building	Airport	
Big facility and infrastructure	Quarry	
Open space	Waste disposal	
Green Area	Building site	
	Sport area	
	Beach, barren area, etc.	
	Icefield	
	Swamp	
	River	
	Water basin	
	Urban green area	
Sowing-field	Sowing-field	Crops
Rice-field	Rice-field	
Wooden agrarian plantation (orchard, vineyard, etc.)	Vineyard	
	Orchard	
	Olive-grove	
	Stable meadow	
	Annual crop and permanent crop	
	Complex crop system	
	Crop with natural areas	
Pasture lawn	Natural pasture area	Semi-naturals
Uncultivated	Moor and shrubs	
	Trees and shrubs area in evolution	
	Rare vegetation area	
	Fire area	
Broadleaves wood/forest. Copse	Broadleaves forest	Broadleaves forests
Broadleaves wood/forest.	Mixed forest	
High trunk wood		
Wood plantation (poplar)		
Resinous wood/forest	Conifers forest	Conifers forests

Notes:

– Different broadleaves forest managements were not distinguished (High trunk wood and copse);
– Poplar plantations were aggregated to broadleaves forest;
– The Green Area class has been included in the "area non-included in the analysis" because this class interest quali urban parks, cemeteries, little gardens and green areas with infrastructures;
– Uncultivated area class has been included in the "Semi-natural receptors" because it includes a lot of mountain areas covered by shrubs, bushes, rocks vegetation, etc.).

The GSTAT (Pebesma and Wesseling, 1998, http://www.geog.uu.nl/gstat/) geostatistical tool has been employed in this study. More general tools for different exposure indices computations and for grid choice and manipulation have also been developed (source codes are freely available upon request to the authors).

Surface wind maps were drawn by ordinary kriging starting from hourly measurements obtained by the meteorological monitoring network of the Lombardy Region. Soil water availability was assessed by the Regional Agrometeorological Service (ERSAL) using a simple soil water balance model based on climatic data of the last 30 years (Mariani, 1997; Maracchi et al., 1992), taking into account and properly kriging soil available water capacity, soil depth and covers obtained from the ERSAL pedologic database based on taxonomic units distribution in the region.

Finally, a Digital Elevation Model with 1×1 km grid resolution was generated for the whole domain using data obtained by the Territorial Information System (SIT) of the Lombardy Region and then employed into the mapping procedure.

3. Results and discussion

3.1. Level I risk assessment

The maps obtained by using the two approaches presented in the previous section (UK-PORG and Loibl) provided quite consistent results in terms of proportion of surface exposed to different ozone levels (Table 4) and the exposure spatial patterns turned out to be very similar. By comparison with the UK-PORG "modified" approach, the "modified" Loibl method gave slightly

Table 4. Regional cover (surface percentage) for each forest exposure class (AOT40f computed with three different methods: see references in the text)

Exposure classes AOT40f (ppm.h)	0–5	5–10	10–15	15–20	20–25	25–30	30–35	35–40	40–45	45–50	50–55	55–60	60–65	65–70	70–75
Methods[a]															
UK-PORG (Fowler et al., 1995)	–	–	0.7	33.5	15.9	14.8	20.1	9.5	3.9	1.4	0.3	–	–	–	–
Loibl et al., 1994, modified	–	0.1	1.1	32.8	16.4	12.5	15.3	8.1	5.3	4.2	2.8	1.0	0.2	–	–
UK-PORG modified	–	–	0.1	34.2	18.4	16.6	17.0	7.7	3.4	1.7	0.7	0.2	–	–	–

[a]See references in the text.

higher exposure estimates in the more elevated sites and lower estimates in the valley bottoms. Estimated and measured exposures have been compared selecting the used data set or an independent validation dataset with scattered available measurements (Gerosa et al., 1999; Vecchi and Valli, 1999). In both cases the UK-PORG modified technique showed the lowest residual square's sum (RSS); therefore maps based on this procedure were used for the following risk analysis. The estimate accuracy was quite satisfactory: the kriging's relative standard deviation resulted less than 15% in the most risky areas, with values under 5% in the most densely monitored zones.

The whole region proves to exceed the critical level established for forests (10 000 ppb h), and the most risky areas appear to be located in the northwestern alpine and pre-alpine belt, where exceedances up to 6 times above the critical level have been recorded (Fig. 3). This typical pattern reflects the space distribution of the broad photo-oxidant plume generated by NO_x and VOCs emission within the urban and industrial area located around Milan. In fact, summer breezes push the locally produced ozone from the plain over the mountains, where its lower destruction rate leads to an increase in the background levels.

The distribution of surface areas affected by different ozone exposure classes exhibits a bi-modal feature (Table 4). The largest exposure class lays between 10 000 and 15 000 ppb h (the lower in terms of dose values) and covers 34% of the regional territory; the second most relevant class (between the 30 000 and 35 000 ppb h) covers about 17% of the whole territory. If we consider areas with neither the highest nor the lowest exposure, a major portion of territory (about 60%) shows quite high exposure values ranging between 20 000 and 40 000 ppb h. Fig. 4 shows the percentage of forest areas affected by the different exposure classes.

If we examine forest species distribution (Fig. 5), the most critical conifer forests are located around the northwestern Orobic Alps on both the northern and southern sides. Also, Scotch pine forests in the natural parks located immediately north of Milan are subject to strong stress pressures.

Broadleaved forests (Fig. 5(a)) are spread over the most ozone-risky subalpine belt which stretches from Western Orobic Alps to Larian Alps and Pre-Alps. The distribution area of beech (*Fagus sylvatica*), the forest species for which the critical AOT40f level was established, perfectly matches the core of this area. The diffusion area of black cherry (*Prunus serotina*), alloctone species particularly sensitive to ozone and thus potentially suitable to be used as a bioindicator, covers the Ticino Valleys and the upper western plain. Poplar (*Populus* spp.) cultivation areas in the plain are subject only to a slight exceedance of the critical level, with the exception of those located in the northern belt of Pavia, Lodi and Cremona provinces and the southern side of the Milan province.

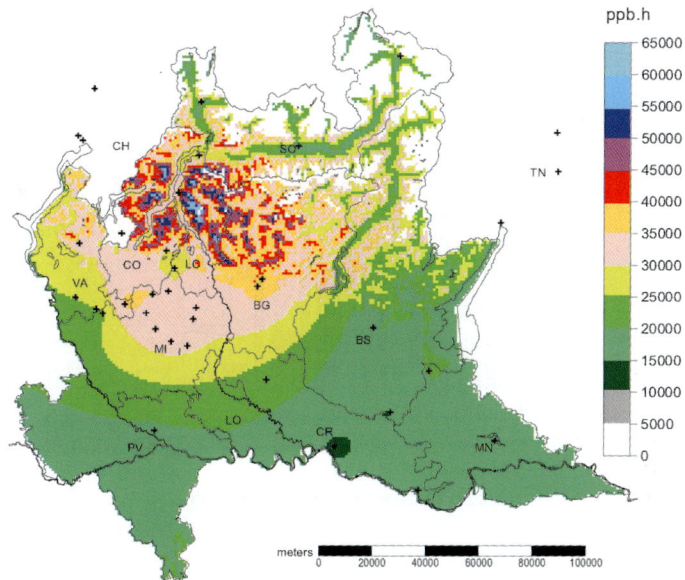

(a) UK-PORG "modified" mapping techniques

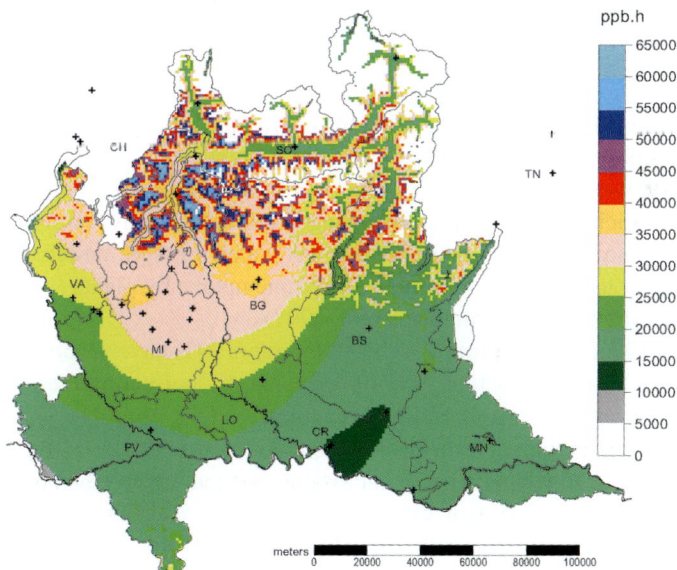

(b) Loibl "modified" mapping techniques

Figure 3. AOT40f in Lombardy. Exposures are expressed in ppmh (April 1st–September 30th, 1994–1998). Crosses indicate ozone monitoring stations. Elevated zones (> 2000 m a.s.l.) are blanked.

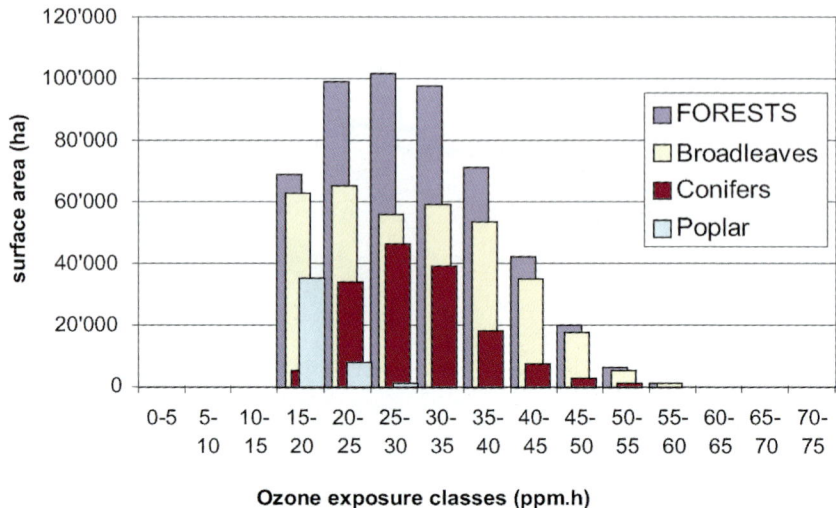

Figure 4. Forests involved in different ozone exposure classes.

3.2. Toward a Level II assessment

Ozone exposure indices offer only a rough approximation of the actual physically absorbed dose which determines the overall plant physiological response. This "real" dose depends on the stomatal conductance that in turn is influenced by a set of meteoclimatic factors able to trigger the complex physiological processes which regulate stomatal opening. Among these environmental physical factors, soil water potential, leaf-to-air water pressure deficit and wind intensity play a major role in influencing ozone uptake by plants (Davidson et al., 1992; Fuhrer, 1995; Grünhage et al., 1997).

Fuhrer (1996a, 1996b), and Posch and Fuhrer (1999) have proposed an easy approach to assess the actual AOT40 for wheat by correcting the calculated AOT40 with multiplicative empirical functions (values from 0 to 1) related to the soil water content, vapor pressure deficit (Emberson et al., 1998).

This approach, developed for ozone risk assessment on crops, is much more difficult to extend to forests because no similar empirical functions have been reported so far for forest populations. We tried, therefore, a semi-quantitative approach to evaluate the weight of these modifying factors on forests in our study case. Ozone exposures maps have been compared with the monthly wind speed and soil water availability distributions of a dry, moist and wet year, obtained from a 30 years historical series respectively as the 10th, 50th, and 90th percentile.

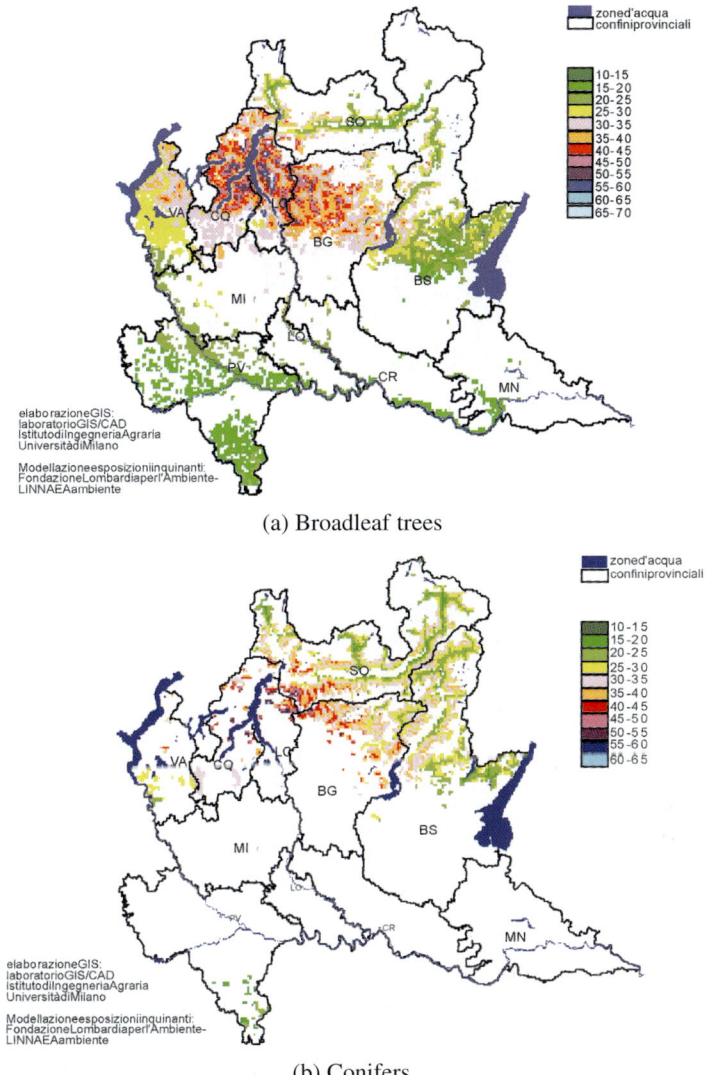

(a) Broadleaf trees

(b) Conifers

Figure 5. Exposures for broadleaf and conifer forests.

The vapor pressure deficit parameter has not been taken into account because of an excess in missing or unreliable data on atmospheric relative humidity, but it is worth mentioning that the summer season in this region is often characterized by elevated humidity.

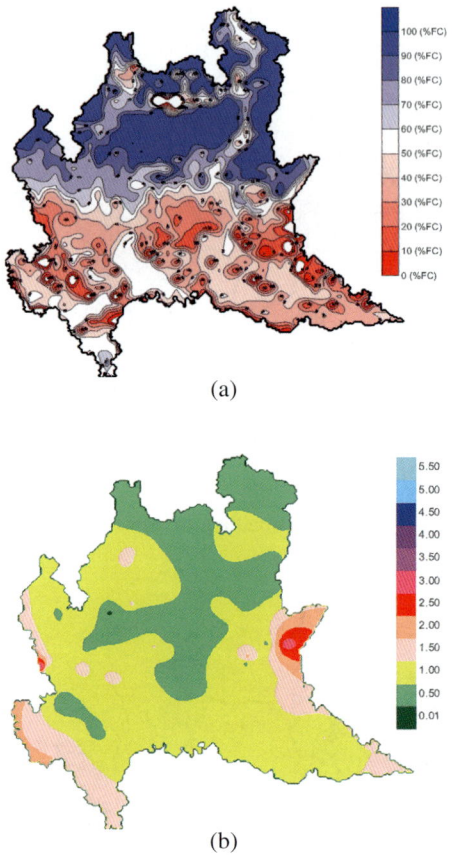

Figure 6. (a) Soil water content (expressed as % field capacity) in the vegetative semester of the dry year (Q10, $n = 30$ years); (b) Daily mean ventilation (vegetative semester 1994–1998).

During the summer the local insubric climate determines frequent midday rainfalls on the northern hillsides (Maracchi et al., 1992) and hot-and-humid conditions on the plains in the south. As a consequence, the weight of soil water content as the modifying factor for forests—which are located mainly on the hillsides (Fig. 6(a))—may be reduced.

On the other hand, the generally low wind speed seems to play a major role as modifying factor for forests' exposure to ozone (Fig. 6(b)). Because turbulence is reduced in weak ventilation conditions and the ambient ozone concentration surrounding the plants is lowered by dry deposition and/or gas phase reactions.

A significant reduction of the effective AOT40 for forests could be reasonably expected, even though its amount is at present unpredictable because wind measurements are not readily extrapolated in mountain areas—where most forests are located—and only modeling approaches (e.g., MINERVE, Geai, 1987) may help in future developments to reach this goal.

3.3. Survey on observed biological symptoms

Given the high AOT40 levels recorded in the region, both the reduced biological productivity and visible leaf injury symptoms might be expected.

As a qualitative and preliminary check, we searched for data on whether symptomatic ozone leaf responses were reported or not for sensitive species in areas at risk based on the AOT40 exposure maps. We collected and critically reviewed all available ozone leaf injury observations reported independently by different authors for forest and crop species over the region in the same or different years.

Mignanego et al. (1992), Schenone et al. (1995) and the Belgiovine et al. (1999) reported the onset and development of leaf necrosis on tobacco sensitive cultivar Bel-W3 observed in several ozone biomonitoring campaigns; symptoms' extension and AOT40 levels reported in these studies fit well with the estimated AOT40 for plain zones. The same outcome fits the OTC experiments on crops (*Phaseolus vulgaris, Triticum aestivum, Trifolium repens*) performed by the same investigators (Schenone et al., 1992; Fumagalli et al., 1997, Fumagalli et al., 1999)in two sites located in the regional plain aimed to assess the quantitative relation between ozone exposure and yield loss.

Remarkable effects both at biochemical and visible level were reported in *Fagus sylvatica, Picea abies* and *Larix decidua* by Ballarin-Denti et al. (1995, 1998) and Rabotti and Ballarin-Denti (1998) at two forest intensive monitoring plots located in a mountain region at the northern edge of the area subject to the photo-oxidant plume of Milan.

Bussotti et al. (2000) and Cozzi and Ferretti (1999) reported visible leaf symptoms clearly attributable to ozone exposure for different forest native species in a recent survey performed in several sites in the Valtellina Valley in the northern part of the region. They validated the recorded symptoms by comparison with those observed at the Swiss OTC facility of Lattecaldo (Skelly et al., 1998, 1999)—located very close to the Switzerland-Lombardy border, only 50 km north of Milan—and by microscope histology analysis (Gravano, *personal communication*). AOT40 levels at which visible symptoms were observed in non-filtered OTCs well agree with the lower limit of the AOT40s we estimated for the sites where injuries on symptomatic species were reported in field conditions.

Cozzi et al. (1998), Bussotti (*personal communication*) and Gerosa carried out in 1998–2000 other on-field surveys in three broadleaf forest sites located in the highest-risk area (Moggio, Pian dei Resinelli, Forest nursery of Curno) and reported symptoms on a variety of trees, shrubs, vines and herbs species, namely: *Acer pseudoplatanus, Ailanthus altissima, Carpinus betulus, Corylus avellana, Fagus sylvatica, Fraxinus excelsior, Laburnum anagyroides, Ostrya carpinifolia, Parthenocyssus quinquefolia, Populus nigra, Prunus avium, Rosa spp., Rubus idaeus, Rubus ulmifolius, Salix alba, Salix glabra, Sambucus ebulus, Sorbus aucuparia, Sorbus aria, Tilia platyphyllos, Ulmus glabra, Viburnum lantana, Vitis vinifera; Alchemilla vulgaris, Aquilegia einseleana, Atropa belladonna, Centaurea nigrescens, Euphorbia amygdaloides, Mycelis ruralis, Rumex alpinum, Thalictrum minus, Trifolium pratense, Valeriana montana, Veronica urticifolia.*

On the contrary, they found only few and weak symptoms in other sparse observations carried out in the lowest-risk zones at the end of August in 1998 and 1999.

4. Conclusions

In this study the UN/ECE methodology for assessing ozone risks to forests was applied in a regional context to describe the small-scale distribution of ecosystems at risk. Due to the presence of large emission sources the Lombardy region is one of the most exposed areas in Europe. Some information is available from European-scale mapping exercises, but small-scale information is lacking. Small-scale ozone risk assessment in regions with complex topography needs accurate interpolation methodologies able to account for ozone elevation dependency. Two different techniques have been modified, simplified and made suitable for an appropriate use at regional scale. Both algorithms gave satisfactory results but the mapping algorithm originally developed by Fowler et al. (1995) and modified by introducing the relative height concept was considered to be more reliable. The present study represents the first attempt to draw ozone risk maps in Italy.

Level I risk maps are useful to locate the areas suffering potential ozone risk, but nevertheless are intrinsically limited because these maps are not suitable to predict the real scale of effects on ecosystems. To overcome this limitation a level II evaluation is requested.

Level II ozone risk evaluation is more difficult to achieve for mountain forests than for crops which are cultivated in the plain. To bypass this obstacle, at least in a semi-quantitative way, a contextual analysis of climate features like ventilation, summer rain, potential evapotranspiration, pluviometric deficit or more detailed soil water balance should be done to better assess Level I calculated exposures. In any case, Level II maps, able to assess the real scale of *effects*, rely on the availability of detailed territorial databases and short-time-resolution meteorological data or their modeled substitutes which are not always easy to obtain at reduced scale.

Maps produced in the present regional case study have proved that ozone exposures in Lombardy are very high both for forests and crops and possibly able to cause visible injury symptoms. Such exposure levels can determine deep modifications in the plant metabolism which can in turn influence both the productivity and stability of natural ecosystems.

Survey campaigns have confirmed the on-field presence of evident foliar injury symptoms in forest populations of the mountainous target of the photo-oxidant plume generated in the lower urban and industrial plain.

Acknowledgements

This project was supported by the Lombardy Region, Department of the Environment, and by the Lombardy Foundation for the Environment as a part of the Regional Air Quality Recovery Plan (PRQA) of the Lombardy Region (2000). The authors are grateful to the Institute of Agricultural Engineering of the University of Milan for the GIS support; to Dr. L. Mariani and M. Russo of the Regional Agrometeorological Service (ERSAL) for soil water balance data and calculations; to Dr. A. Cozzi and M. Ferretti (Linnaea Ambiente of Florence) for their contributions; to Dr. F. Spinazzi for the programming support; to Dr. F. Bussotti, Dr. E. Gravano, Dr. A. Gubertini and the Regional Forest Agency (ARF) for the symptoms validation in the field.

References

Ballarin-Denti, A., Rabotti, G., Tagliaferri, A., Rapella, A., 1995. Novel decline symptoms in an alpine forest system and biochemical indicators of air pollution stress. Life Chem. Rep. 13, 11–119.

Ballarin-Denti, A., Cocucci, S.M., Di-Girolamo, F., 1998. Environmental Pollution and forest stress: A multidisciplinary study on alpine forest ecosystems. Chemosphere 36 (4–5), 1049–1054.

Belgiovine, N., Bergonzi, S., Fumagalli, I., Sormani, L., Mignanego, L., Mietto, S., Ballarin-Denti, A., Brambilla, E., Mazzali, C., 1999. Progetto biennale di Biomonitoraggio della qualità dell'Aria in Provincia di Milano, Final Project Report, Provincia di Milano, Assessorato all'Ambiente, Corso di Porta Vittoria 27, Milano, Italy, p. 70.

Bussotti, F., Mazzali, C., Cozzi, A., Ferretti, M., Gravano, E., Gerosa, G., Ballarin-Denti, A., 2000. Ozone levels and symptoms on vegetation in an alpine valley (North Italy). In: Air Pollution, Global Change and Forests in the New Millennium, 19th International Meeting for Specialists in Air Pollution Effects on Forest Ecosystems. May 28–31, 2000. Michigan Technological University, Houghton, MI, USA.

Cozzi, A., Cenni, E., Ferretti, M., 1998. Condizione degli alberi nel 1997 nelle Unità Territoriali di Riferimento del lago Maggiore, Oltrepò Pavese, Valchiavenna-Lago di Como, Valtellina, Serio-Brembo, Val Camonica, Lago d'Idro-Chiese e Lago di Garda-Mincio (Livello I). Final Project Report, Vol 1. Regione Lombardia, Azienda Regionale delle Foreste, Milano, Italy.

Cozzi, A., Ferretti, M., 1999. Indagine sui sintomi fogliari visibili attribuibili ad Ozono sulla vegetazione spontanea in Valtellina. Rapporto LINNAEA Ambiente, Firenze.

Davidson, S.R., Ashmore, M.R., Garretty, C., 1992. Effects of ozone and water deficit on the growth and physiology of *Fagus sylvatica*. Forest Ecol. Manag. 51, 187–193.

Emberson, L.D., Ashmore, M.R., Cambridge, H.L., 1998. Development of Methodologies for Mapping Level II Critical Levels of Ozone. DETR Report n. EPG. 1/3/82, Imperial College Centre for Environmental Technology, London.

Fowler, D., Smith, R.I., Coyle, M., Weston, K.J., Davies, T.D., Ashmore, M.R., Brown, M., 1995. Quantifying the fine scale (1 km × 1 km) exposure and effects of ozone. Part 1. Methodology and application for effects on forests. Water Air Soil Pollut. 85 (3), 1479–1484.

Fuhrer, J., 1995. Critical level for ozone to protect agricultural crops: interaction with water availability. Water Air Soil Pollut. 85, 1355–1360.

Fuhrer, J., 1996a. Key elements in ozone risk analysis. In: Knoflacher, M., Schneider, J., Soja, G. (Eds.), 1996 Exceedance of Critical Loads and Levels, CLRTAP Workshop Report. Vienna, Austria, pp. 1–17.

Fuhrer, J., 1996b. The critical level for effects of ozone on crops and the transfer to mapping. In: Kärenlampi, L., Skärby, L. (Eds.), 1996—Critical Levels for Ozone in Europe: Testing and Finalizing the Concepts, UN-ECE workshop report. Kuopio, Finland, pp. 27–43.

Fuhrer, J., Skärby, L., Ashmore, M.R., 1997. Critical levels for ozone effects on vegetation in Europe. Environ. Pollut. 97, 91–106.

Fumagalli, I., Mignanego, L., Ambrogi, R., 1999. Effetti dell'inquinamento atmosferico sui vegetali: 10 anni di ricerche condotte dall'ENEL. Acqua Aria 10, 109–118.

Fumagalli, I., Mignanego, L., Violini, G., 1997. Effects of tropospheric ozone on white clover plants exposed in open-top chambers or protected by the antioxidant ethylenediurea (EDU). Agronomie 17, 271–281.

Geai, P., 1987. Methode d'interpolation et de reconstitution tridimensionalle d'un champ de vent: le code d'analyse objectie MINERVE. Rep. ARD-AID: E34-E11, EDF, Chatou, France.

Gerosa, G., Spinazzi, F., Ballarin-Denti, A., 1999. Tropospheric ozone in alpine forest sites: Air quality monitoring and statistical data analysis. Water Air Soil Pollut. 116, 345–350.

Goovaerts, P., 1997. Geostatistics for Natural Resources Evaluation. Oxford Univ. Press, New York.

Grünhage, L., Jäger, H.J., Haenel, H.D., Hanewald, K., Krupa, S., 1997. PLATIN (PLant-ATmosphere INteraction) II: Co-occurrence of high ambient ozone concentrations and factor limiting plant absorbed dose. Environ. Pollut. 98, 51–60.

Kärenlampi, L., Skärby, L. (Eds.), 1996. Critical Levels for Ozone in Europe: Testing and Finalizing the Concepts (UN-ECE Workshop Report). Department of Ecology and Environmental Science, University of Kuopio, Kuopio, Finland.

Loibl, W., Winiwarter, W., Kopcsa, A., Züger, J., Baumann, R., 1994. Estimating the spatial distribution of ozone concentrations in complex terrain using a function of elevation and day time and kriging techniques. Atmos. Environ. 28, 2557–2566.

Maracchi, G., Bindi, M., Conese, C., Mariani, L., 1992. Guida Agrometeorologica della Lombardia. ERSAL, Milano, Italy. p. 108.

Mariani, L. (Ed.), 1997. Caratterizzazione Agroclimatica dei Consorzi di Bonifica Lombardi. SIB-ITeR Project Final Report. ERSAL, Milano Due Palazzo Canova, Segrate (MI), Italy, p. 40.

Mignanego, L., Biondi, G., Schenone, G., 1992. Ozone biomonitoring in northern Italy. Environ. Monit. Assess. 21, 141–159.

Pebesma, E.J., Wesseling, C.G., 1998. Gstat, a program for geostatistical modelling, prediction and simulation. Comput. Geosci. 24 (1), 17–31.

Posch, M., Hetteling, J.-P., de Smet, P.A.M., Downing, R.J. (Eds.), 1998. Calculation and mapping of critical thresholds in Europe: Cordination Center for Effects Status report 1997. RIVM Rep. 259101007. Bilthoven, Netherlands.

Posch, M., Fuhrer, J., 1999. Mapping Level II exceedance of ozone critical levels for crops on a European scale: the use of correction factors. In: Fuhrer, Achermann (Eds.), Critical levels for ozone–Level II. Environmental Documentation N. 115. Swiss Agency for Environment, Forest and Landscape. Bern, Switzerland, pp. 49–53.

Rabotti, G., Ballarin-Denti, A., 1998. Biochemical responses to abiotic stress in beech (*Fagus sylvatica* L.) leaves. Chemosphere 36 (4–5), 871–875.

Sandroni, S., Bacci, P., Boffa, G., Pellegrini, U., Ventura, A., 1994. Tropospheric ozone in the pre-alpine and Alpine Regions. Sci. Total Environ. 156, 169–182.

Schenone, G., Botteschi, G., Fumagalli, I., Montinaro, F., 1992. Effects of ambient air pollution in open-top chambers on bean (*Phaseolus vulgaris* L.). I. effects on growth and yield. New Phytol. 122, 689–697.

Schenone, G., Fumagalli, I., Mignanego, L., Violini, G., 1995. Effects of ambient ozone on bean (*Phaseolus vulgaris* L.): results of an experiment with the antioxidant EDU in the Po plain (Italy) in the 1993 season. Responses of Plants to Air Pollution. Biologic and Economic Aspects. Agricoltura Mediterranea, 104–108.

Skelly, J.M., Innes, J.L., Savage, J.E., Snyder, K.R., Vander-Heyden, D., Zhang, J., Sanz, M.J., 1999. Observation and confirmation of foliar ozone symptoms of native plant species of Switzerland and southern Spain. Water Air Soil Pollut. 116, 227–234.

Skelly, J.M., Innes, J.L., Snyder, K.R., Savage, J.E., Hug, C., Landolt, W., Bleuler, P., 1998. Investigations of ozone induced injury in forests of southern Switzerland: Field surveys and open-top chamber experiments. Chemosphere 36 (4–5), 994–1000.

Staffelbach, T., Neftel, A., Blattner, A., Gut, A., Fahrni, M., Stähelin, J., Prévôt, A., Hering, A., Lehning, M., Neininger, B., Bäumle, M., Kok, G.L., Dommen, J., Hutterli, M., Anklin, M., 1997. Photochemical oxidant formation over Southern Switzerland. I. Results from summer 1994. J. Geophys. Res. 102, 23345–23362.

UBA, 1996. Manual on Methodologies and Criteria for Mapping Critical Levels/Loads and geographical areas where they are exceeded. UN/ECE Convention on Long-Range Transboundari Air Pollution. Federal Environmental Agency (Umweltbundesamt), Texte 71/96, Berlino.

UK-PORG, 1997. Ozone in the United Kingdom. Fourth report of the Photochemical Oxidants Review Group. United Kingdom. ISBN 1-870393-30-9.

Vecchi, R., Valli, G., 1999. Ozone assessment in the southern part of the Alps. Atmos. Environ. 33, 97–109.

Air Pollution, Global Change and Forests in the New Millennium
D.F. Karnosky et al., editors
© 2003 Elsevier Ltd. All rights reserved.

Chapter 6

Limitations and perspectives about scaling ozone impacts in trees

T.E. Kolb*

School of Forestry, Northern Arizona University, Box 15018, Flagstaff, AZ 86011-5018, USA

R. Matyssek

Lehrstuhl für Forstbotanik, Technische Universität München, Am Hochanger 13, D-85354 Freising, Germany

Abstract

We review the need for scaling effects of ozone (O_3) from juvenile to mature forest trees, identify the knowledge presently available, and discuss limitations in scaling efforts. Recent findings on O_3/soil nutrient and O_3/CO_2 interactions from controlled experiments suggest consistent scaling patterns for physiological responses of individual leaves to whole-plant growth, carbon allocation, and water use efficiency of juvenile trees. These findings on juvenile trees are used to develop hypotheses that are relevant to scaling O_3 effects to mature trees, and these hypotheses are examined with respect to existing research on differences in response to O_3 between juvenile and mature trees. Scaling patterns of leaf-level physiological response to O_3 have not been consistent in previous comparisons between juvenile and mature trees. We review and synthesize current understanding of factors that may cause such inconsistent scaling patterns, including tree-size related changes in environment, stomatal conductance, O_3 uptake and exposure, carbon allocation to defense, repair, and compensation mechanisms, and leaf production phenology. These factors should be considered in efforts to scale O_3 responses during tree ontogeny. Free-air O_3 fumigation experiments of forest canopies allow direct assessments of O_3 impacts on physiological processes of mature trees, and provide the opportunity to test current hypotheses about ontogenetic variation in O_3 sensitivity by comparing O_3 responses across tree-internal scales and ontogeny.

*Corresponding author.

DOI:10.1016/S1474-8177(03)03006-7

1. Introduction

Tropospheric ozone (O_3) is a likely contributing factor to tree decline in some North American and European forests (Taylor et al., 1994; Sandermann et al., 1997; Chappelka and Samuelson, 1998; Skärby et al., 1998; Matyssek and Innes, 1999; McLaughlin and Percy, 1999). For logistical reasons, most data on tree response to O_3 is from controlled exposures of juvenile trees in either indoor or outdoor chambers often using potted plants that do not experience the actual rigors of natural site conditions. Thus, there is growing concern about whether O_3 response data from juvenile trees can be extrapolated to mature trees (e.g., Samuelson and Edwards, 1993; Grulke and Miller, 1994; Samuelson and Kelly, 1996; Fredericksen et al., 1996b; Kolb et al., 1997; Matyssek and Innes, 1999; Baumgarten et al., 2000; Wieser, 2002). Although O_3 response data for mature forest trees are becoming available from free-air O_3 exposure experiments (Tjoelker et al., 1995; Matyssek and Innes, 1999; Karnosky et al., 2001), such experiments can concentrate only on a limited number of tree species. Therefore, information on changes in O_3 impacts during tree ontogeny is still needed.

In this paper, we address two aspects of scaling O_3 impacts from juvenile to mature trees. First, we examine results from juvenile trees for consistency in O_3 responses within the individual plant. Based on these results, we formulate hypotheses that may provide insight on differences in O_3 response between juvenile and mature forest trees. Second, we address studies that have compared O_3 response between juvenile and mature forest trees to highlight factors that may lead to changes in O_3 sensitivity during forest maturation. Throughout the paper, we offer perspectives about existing gaps in knowledge that currently limit scaling of O_3 responses from juvenile to old trees.

1.1. Consistency patterns of O_3 impacts within trees

Studies conducted on young trees under phytotron or chamber conditions have been the prevailing means to provide insights into physiological mechanisms of O_3 impact (Matyssek et al., 1995b; Matyssek et al., 1997). However, such studies have included a broad spectrum of plant responses at the biochemical and molecular levels (e.g., Kangasjärvi et al., 1994; Sandermann, 1996; Langebartels et al., 1997; Sandermann et al., 1997). Such highly resolved response mechanisms may not apply directly to larger trees or forest stands because of temporal variability, tree ontogeny, and multi-factorial environmental influences that characterize forest sites (Baldocchi, 1993). Therefore, incorporation of such complexity into quantitative scaling concepts that reach the stand level is problematic (Jarvis, 1993). Nevertheless, knowledge about underlying mechanisms is indispensably valuable for understanding ecophysiological tree responses to stresses such as O_3 (Waring, 1993). Thus, ecophysiological processes and responses may be regarded as 'scaling units' that integrate tree performance across functional levels and ontogenetic stages (Reynolds et al., 1993; Matyssek et al., 1995b).

The action of O_3 must be viewed in the context of other environmental impacts that may influence sensitivity to stress (Mooney et al., 1991; Ellenberg, 1996). From the many factors that can affect O_3 impact to trees highlighted in recent reviews (e.g., Kolb et al., 1997; Skärby et al., 1998; Matyssek and Innes, 1999), we use mineral nutrition and carbon dioxide (CO_2) as examples of how resource availability can modify O_3 sensitivity in trees (Polle et al., 2000). We address controlled experiments with young *Betula pendula* and *Fagus sylvatica* as examples of the current state of knowledge of consistency patterns in O_3 response within young trees. These species were chosen because they represent a contrast in shoot growth phenology (indeterminate versus determinate), and both are sensitive to O_3. The functional basis of scaling from leaf-level to whole-tree responses is demonstrated by ecophysiological relationships which are examined for plant-internal consistency and relevance to mature trees.

1.2. Nutrition as a modifier of O_3 impact

1.2.1. Metabolic and leaf responses

In an outdoor chamber experiment, young *Betula pendula* cuttings were grown throughout the growing season either under O_3 free-air (control) or chronic day/night O_3 exposure of 90/40 nll^{-1}. Both exposure regimes were supplied with low or high nutrition by irrigation with a balanced fertilizer. Both O_3 exposure and high nutrition increased the $\delta^{13}C$ content in tree cellulose (Fig. 1(A); Matyssek et al., 1992; Saurer et al., 1995). The decrease in discrimination against ^{13}C during photosynthesis that occurred in response to high fertilization was associated with a lower concentration of CO_2 in the mesophyll intercellular space (C_i), as predicted by Farquhar et al. (1989a), and lower stomatal conductance (G_W), indicating increased photosynthetic water-use efficiency (WUE; Schulze and Hall, 1982).

O_3 exposure also increased $\delta^{13}C$, but in contrast to the effect of nutrition, C_i either increased or did not change in response to O_3. This result was surprising as an increase in $\delta^{13}C$ is often associated with a decrease in C_i (Farquhar et al., 1989a). As an explanation, the CO_2-binding activity of the PEP-carboxylase enzyme (PEPC) increased in response to O_3 exposure (Luethy-Krause et al., 1990; Saurer et al., 1995; Gerant et al., 1996; Landolt et al., 1997), and RUBISCO activity was inhibited by O_3. PEPC can increase tissue $\delta^{13}C$ as its ^{13}C discrimination is negligible relative to that of the RUBISCO enzyme (Farquhar et al., 1989b). PEPC is part of an anaplerotic pathway that

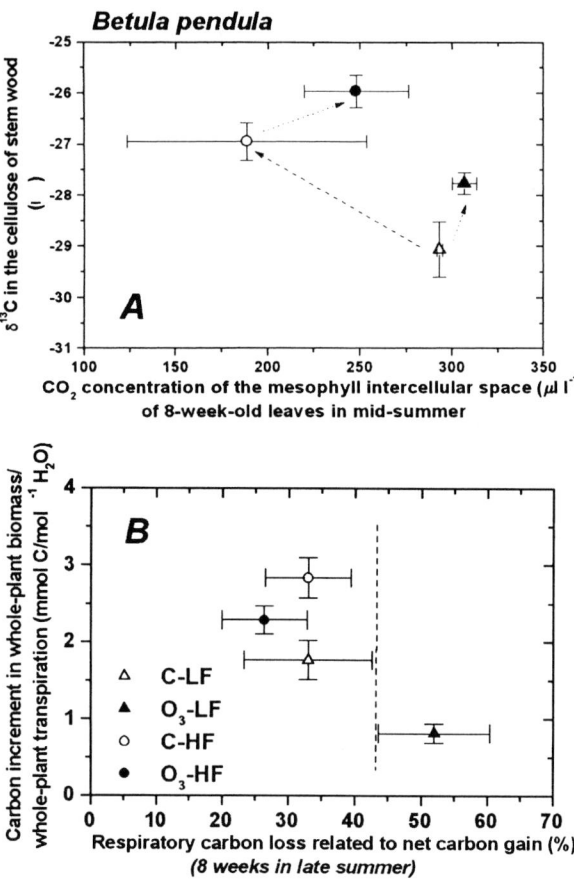

Figure 1. (A) Relationship between the $^{13}C/^{12}C$ isotope ratio in the cellulose of stem wood and the CO_2 concentration of the mesophyll intercellular space of 8-week-old leaves of young birch (*Betula pendula*) plants grown under four combinations of O_3 exposure and nutrient supply (data adapted from Saurer et al., 1995). (B) Relationship between carbon incorporated into whole-plant biomass and transpiration, and the proportion of respiration in net carbon gain of whole-plant foliage for a period of 8 weeks in late summer (data adapted from Maurer and Matyssek, 1997). Abbreviations: C-LF = O_3-free air/low-fertilized; O_3-LF = O_3-exposed/low-fertilized; C-HF = O_3-free air/high-fertilized; O_3-HF = O_3-exposed/high-fertilized (data as means ± SD).

feeds the oxalacetate pool of the citric acid cycle under conditions of high demand for substrate and energy (Wiskich and Dry, 1985). Such a demand is created by O_3 in leaves because of increased respiratory costs of detoxification and repair processes (Rennenberg et al., 1996; Maurer et al., 1997; Polle et al., 2000). Low WUE and reduced stomatal limitation of CO_2 uptake (Farquhar et al., 1989b) in plants exposed to O_3 resulted from distinct declines in carboxylation efficiency associated with extensive collapse of mesophyll cells (Matyssek et al., 1991; Gunthardt-Goerg et al., 1997). Exposure to O_3 decreased G_W only at low fertilization, but did not prevent a decline in WUE (Maurer et al., 1997). O_3 uptake by leaves was similar between levels of fertilization, as O_3 caused G_W of low-fertilized plants to decrease towards levels at high fertilization.

1.2.2. *Whole-plant response*

Betula pendula displays indeterminate shoot growth, thus formation of new leaves can continue throughout most of the growing season as long as resource availability is high. Given this feature, new leaf production may compensate for the premature loss of older, injured leaves. On the other hand, low nutrient availability may restrict new leaf production, which increases the need for maintaining injured, older leaves (Laurence et al., 1994; Fredericksen et al., 1995). This contrasting growth pattern characterized the experimental plants of *Betula pendula*, as the proportion of O_3-injured leaves was high at low fertilization (Maurer and Matyssek, 1997). Respiratory costs of foliage as a proportion of whole-plant carbon balance were higher in low-fertilized plants because of the high number of O_3-injured leaves (Fig. 1(B)), as these plants relied on the photosynthesis of injured foliage that had been formed early in the season. However, O_3 inhibited branch formation and leaf size at high fertilization (Matyssek et al., 1992). Because stomata responded to chronic O_3 exposure by partial closure only at low fertilization (Maurer et al., 1997), daily and seasonal canopy transpiration per unit of foliage area decreased only in low-fertilized plants. It appears that the 'opportunity costs' (Stitt and Schulze, 1994) were more favorable for high-fertilized plants to form new leaves rather than maintain older injured leaves as nitrogen is often reallocated from O_3-injured leaves to new leaves which stimulates their photosynthetic performance (Beyers et al., 1992; Greitner et al., 1994; Pell et al., 1994; Temple and Riechers, 1995).

The impact of O_3 on carbon allocation between root and shoot of *Betula* depended on fertilization level. O_3 decreased the root/shoot biomass ratio (R/S) more at low fertilization than at high fertilization (Maurer and Matyssek, 1997). Moreover, O_3 did not affect specific root length or carbon allocation among root diameter classes, but increased the leaf mass per area ratio (LMA), regardless of fertilization level. The dramatic decrease in root growth caused by O_3 at low fertilization was related to a high carbon demand for maintaining foliage (Polle et al., 2000) and to disruption of assimilate translocation from injured foliage (Spence et al., 1990; Matyssek et al., 1992; Gunthardt-Goerg et al., 1993). In fact, respiratory costs were high in low-fertilized plants under O_3 exposure, contributing to the low WUE of whole-plant production

(Fig. 1(B)). The effect of O_3 on WUE of single leaves (Fig. 1(A)) was consistent with the effect on the whole plant (Fig. 1(B)).

1.2.3. Synopsis

The processes linked together in Fig. 2 form a mechanistic basis for scaling O_3 effects through the tree, and may represent an ecologically meaningful scheme because they reflect the low nutritional status present in many forests. After entering the leaf through stomata, O_3 increases respiration (defense, repair) and stimulates PEPC activity which increases tissue ^{13}C content. Leaf growth decreases, LMA and stomatal density increase, and transpiration decreases because of partial stomatal closure. In parallel, P_N decreases through biochemical signals arising from the primary impact of O_3 on the mesophyll apoplast that damage chloroplasts (Sandermann et al., 1997). Tissue collapse also decreases P_N and WUE, disrupts both translocation within the leaves and phloem structure and function, and decreases carbon allocation to stem and root growth (Friend and Tomlinson, 1992; Kelly et al., 1993; Smeulders et al., 1995). Even though R/S stays low, plant growth decreases because of decreased leaf size, P_N, and branching, and the occurrence of premature leaf loss. Plants grown

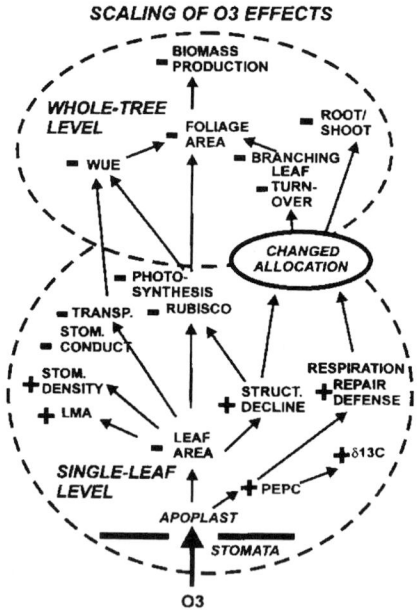

Figure 2. Scaling scheme of ozone effects on plant internal processes in young birch (*Betula pendula*) plants.

at high nutrient availability may respond to O_3 exposure at the whole-plant level by having lower respiratory costs, a smaller decline in WUE and R/S, but an increase in compensatory leaf turn-over that may partially counteract O_3 impacts on growth.

1.3. O_3 impact modified by 'CO$_2$ fertilization'

In contrast to *Betula pendula*, *Fagus sylvatica* displays determinate growth, and completes foliage production during a few weeks in spring. Under favorable circumstances, a second flush (i.e., the lammas shoot) may follow in early summer, adding new foliage to the crown (Roloff, 1985; Ceulemans and Mousseau, 1994; Gruber, 1998; Saxe et al., 1998). In another phytotron experiment, three-year-old *Fagus sylvatica* plants were exposed throughout the growing season either to a regime that mimicked the climate and ambient fluctuating O_3 levels from a rural field site, or to a regime with twice-ambient O_3 levels (Grams et al., 1999; Häberle, personal communication). Both O_3 regimes were combined with contrasting exposure to CO_2 (ambient levels up to 400 µl l^{-1} versus 700 µl l^{-1}), a 'pollutant' that represents another kind of nutrient which may counteract carbon limitation in plant growth (Ceulemans et al., 1999). Mineral nutrients and water were non-limiting throughout the experiment. To prevent transitional responses to CO_2, the plants were acclimated to the CO_2 treatments during one year prior to the O_3 treatments.

1.3.1. Leaf responses

Fagus sylvatica displayed a leaf response to high O_3 and ambient CO_2 levels commonly observed under controlled chamber conditions: P_N declined relative to the control (ambient O_3 and CO_2) treatment (Lippert et al., 1996; Zeuthen et al., 1997; Dixon et al., 1998) due to decreases in both the light and dark reactions of photosynthesis (Grams et al., 1999). In contrast, elevated CO_2 increased P_N compared with the control and ameliorated the negative effects of high O_3 on P_N (cf. Ceulemans and Mousseau, 1994; Manes et al., 1998; Volin et al., 1998). Photosynthetic parameters were not down regulated by elevated CO_2 (Mousseau et al., 1996; Heath and Kerstiens, 1997), although foliar chlorophyll and N levels decreased (Curtis and Wang, 1998). While elevated CO_2 can decrease G_W and O_3 uptake (Saxe et al., 1998), such a reduction in G_W occurred in the *Fagus* plants only during early summer, and G_W did not differ among the four treatments throughout the remainder of the growing season (Grams et al., 1999). As root density in the soil increased during midsummer under elevated CO_2, G_W was apparently not affected by constraints on root performance as reported elsewhere (Arp, 1991; Kerstiens et al., 1995; Saxe et al., 1998). As a consequence, the ameliorating effect of elevated CO_2

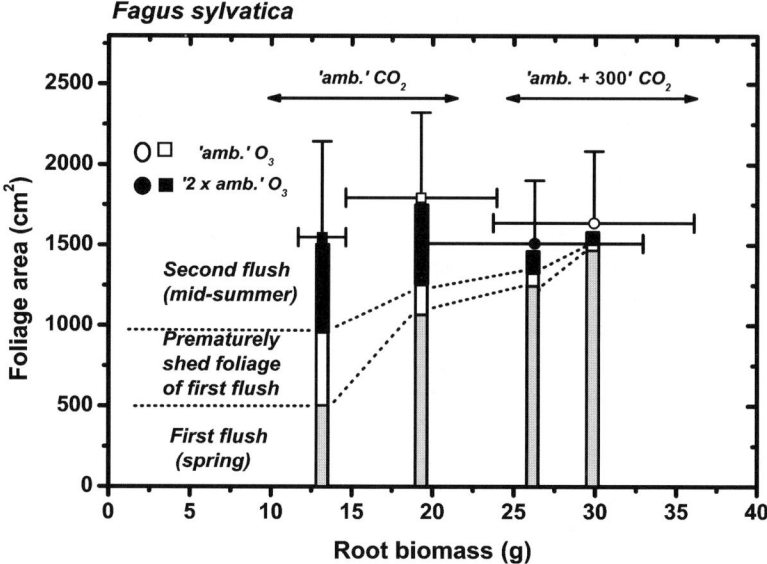

Figure 3. Relationship between root biomass and foliage area of young beech (*Fagus sylvatica*) growing under four combinations of ambient or twice-ambient ozone and CO_2 exposure (data as means ± SD). Also depicted are the proportions of first-, second-flush, and prematurely shed foliage (data from Häberle, personal communication).

on O_3 stress likely resulted from compensatory responses at the biochemical level (Anegg and Langebartels, unpublished data).

1.3.2. Whole-plant response and synopsis

Whole-plant biomass production of *Fagus* consistently reflected photosynthetic performance in each treatment: biomass was lowest at high O_3 and ambient CO_2, and highest at ambient O_3 and elevated CO_2 (El Kohen et al., 1993; Rey and Jarvis, 1997). Elevated CO_2 ameliorated the negative effects of high O_3 on biomass (Volin et al., 1998). R/S was lowest under the high O_3-ambient CO_2 regime (Matyssek and Innes, 1999).

Decreased root growth caused by high O_3 in *Fagus sylvatica* with its determinate shoot growth resulted in an R/S response similar to that of *Betula* when its indeterminate growth was restricted to the early part of the growing season by low nutrition (see above). In both cases, the need for maintaining foliage under O_3 impact apparently reduced carbon allocation to roots (Mooney and Winner, 1991). In *Fagus*, lammas shoot formation in mid-summer was a strong sink for carbon under conditions of ambient CO_2 and high O_3, and helped to compensate for the decline in P_N and loss of first-flush foliage (Häberle, personal communication; Fig. 3). Root growth was favored at high CO_2 when demand for repair, detoxification, and compensatory foliage growth was low in the absence of severe O_3 impact (Fig. 3). These studies suggest that resistance to O_3 is promoted by tree capability to create carbon sinks that facilitate compensatory responses such as high rates of leaf turnover in *Betula* or lammas shoot formation in *Fagus*. Creation of these sinks can strongly depend on nutrient availability (e.g., soil nutrition and CO_2). Long-term persistence of O_3 tolerance through elevated CO_2 (Polle et al., 2000) however, is questionable. In general, the current state of knowledge allows for scaling of ecophysiological processes and responses to O_3 across functional levels within juvenile trees.

1.4. Hypotheses for scaling air pollutant impacts to mature trees and stand conditions

Given the lack of knowledge about mature forest trees, we use findings from multi-factor experiments on juvenile trees to form hypotheses that may be relevant to scaling O_3 responses across tree ontogeny and to stand conditions (Table 1). If O_3 and low nutrition constrain shoot growth in trees of indeterminate growth habit, carbon may be allocated to maintain foliage formed early in the season. The shoot growth habit of such plants resembles that of trees with determinate growth, although the change in growth habit is caused by O_3 stress and resource limitation. Many determinate species have a high capacity to store carbohydrates that can support compensatory mechanisms leading to O_3 tolerance in older leaves (Laurence et al., 1994). Therefore, we hypothesize that trees with determinate growth, including most mature trees of indeterminate species (Kozlowski and Pallardy, 1997), respond to O_3 stress by decreasing carbon allocation to non-green organs, and to sinks associated with defense or tolerance to other stresses, more than trees with indeterminate growth because of allocation to maintain older leaves if surplus carbon is available (Hypothesis 1).

Replacement of O_3-damaged foliage leads to shifts in carbon allocation from root to shoot growth in trees with determinate growth by recurrent flushing or the formation of lammas shoots. These extra shoot flushes can be initiated by high availability of limiting resources such as mineral nutrients. However, if supply of both mineral nutrients and CO_2 are high, carbon reserves are available for allocation to O_3 detoxification and repair of damaged tissues in older leaves; under this condition, the second flush is not required, and extra internal carbon is allocated to root growth (Hypothesis 2).

O_3 often increases leaf respiration and decreases P_N, whereas high CO_2 often increases P_N with effects on respiration being unclear (Ceulemans et al., 1999; Drake et al., 1999). High respiration costs and low P_N may constrain growth of old trees more than juvenile trees provided that respiration really

Table 1.   Hypotheses derived from multi-factor studies relevant to scaling O_3 impacts from juvenile to mature forest trees (see text for details)

Hypothesis 1: Carbon allocation to stem and roots and sinks associated with defense or tolerance to other stresses is reduced by chronic O_3 stress more in trees with determinate growth than trees with indeterminate growth.

Hypothesis 2: In trees with determinate growth, recurrent flushing or formation of lammas shoots will increase as a compensatory response to chronic O_3 stress if soil mineral nutrition is high; however, high atmospheric CO_2 and high mineral nutrition will increase carbon available for O_3 detoxification and tissue repair, and ameliorate O_3 damage to older leaves of the first flush, thus favoring carbon allocation to roots versus lammas shoots.

Hypothesis 3: Supply of labile carbon available for defense, compensation, and repair responses to O_3 stress will be lower in mature than juvenile trees because of higher respiratory costs for maintaining living tissues and lower photosynthetic rate of mature trees.

Hypothesis 4: Stomatal aperture limits O_3 uptake into leaves more for mature than juvenile trees because of greater resistance to water transport in larger trees.

Hypothesis 5: Scaling O_3 effects from juvenile to mature trees must consider modification of tree competitive ability as mediated through structural and functional changes in resource uptake by canopy and roots.

Hypothesis 6: Scaling O_3 impacts from juvenile to mature trees must consider influences of O_3 on mycorrhizae, and secondary effects of mycorrhizae on stress tolerance and pest resistance.

Hypothesis 7: Light limitations on photosynthesis of shaded leaves increases O_3 sensitivity because of decreased carbon available for defense and repair.

---

limits growth of old trees (Waring and Schlesinger, 1985; Ryan and Waring, 1992; Ryan et al., 1997; Hunt et al., 1999). Therefore, we hypothesize that supply of labile carbon available for defense, compensation, and repair responses to O_3 stress will be lower in mature than juvenile trees (Hypothesis 3).

If water limitations constrain the growth of mature trees more than juveniles because of increasing resistance to water transport with increasing tree size (Waring and Silvester, 1993; Yoder et al., 1994; Ryan et al., 1997) or other factors (Becker et al., 2000), then reductions in G_W can limit O_3 uptake and foliar damage more for mature trees than for juveniles (Hypothesis 4; Reich, 1987; Kolb et al., 1997).

We also hypothesize that changes in carbon allocation by O_3 stress leading to decreased canopy leaf area and root growth and related changes in crown and root architecture modifies tree competitive ability (Hypothesis 5; Kuppers, 1994; Bazzaz, 1997). Further, influences of O_3 on carbon allocation to mycorrhizae (Andersen and Scagel, 1997), which have often been overlooked, and their secondary effects on stress tolerance and pest resistance in trees (Ericsson et al., 1996; Gehring et al., 1997; Langebartels et al., 1997), may play a key role for scaling O_3 effects to the stand level (Hypothesis 6).

Shaded portions of the canopy occur in many mature forest trees but are typically absent in juvenile trees grown in the open or under controlled chamber conditions. Photosynthate available for repair and defense against O_3 stress may be limiting in shade leaves (Tjoelker et al., 1993; Matyssek et al., 1995a; Matyssek et al., 1995b; Tjoelker et al., 1995; Fredericksen et al., 1996a), and shade leaf morphology may promote O_3 injury (Bennett et al., 1992). Therefore, we hypothesize that light limitation increases O_3 sensitivity of leaves and this limitation must be considered in attempts to scale O_3 impacts across tree ontogeny (Hypothesis 7).

2. O_3 impacts on juvenile and mature trees

2.1. Effects of O_3 on leaf physiology of juvenile and mature trees

Effects of O_3 on some aspects of leaf chemistry and carbon relations appear to be qualitatively similar for juvenile and mature trees. For example, changes in N and carbohydrate concentrations in response to O_3 stress occurred in the same direction in leaves of seedlings exposed to O_3 in indoor or outdoor open-top chambers, and mature trees exposed with whole-tree or branch chambers for *Pseudotsuga menziesii* (Smeulders et al., 1995) and *Quercus rubra* (Samuelson et al., 1996). Similarly, levels of ascorbate, an anti-oxidant compound, did not vary among 8-, 15-, and 36-year-old trees of *Picea abies* growing in the same forest (Wieser, 2002). These few studies suggest similar metabolic responses to O_3 injury in leaves of juvenile and mature trees, and no fundamental changes in foliar defensive mechanisms against oxidant injury during tree maturation.

Another issue that is relevant to scaling leaf-level impacts of O_3 from juvenile to mature trees is the amount of O_3 exposure that causes the onset of leaf injury. A recent study of *Fagus sylvatica* (Baumgarten et al., 2000) showed that visible symptoms of O_3 damage (leaf discoloration) on seedlings exposed to O_3 in an indoor chamber started at a higher cumulative exposure (AOT40) than for mature trees exposed to ambient concentrations in the forest (Fig. 4). However, when exposure was expressed as cumulative O_3 uptake into the leaf mesophyll (calculated from ambient O_3 concentration and stomatal conductance), damage symptoms started at a similar cumulative O_3 uptake for seedlings and mature trees (Fig. 4). Similar results were reported for *Picea abies* for the relationship between cumulative O_3 uptake and reduction of P_N by O_3 exposure (Wieser, 2002). In contrast, visible foliar damage symptoms to sun leaves of *Prunus serotina* seedlings growing in an open field started at lower cumulative O_3 uptake into the leaf compared with sun leaves of mature trees in a nearby forest (Fig. 5), suggesting lower capacity for defense or repair for mature trees

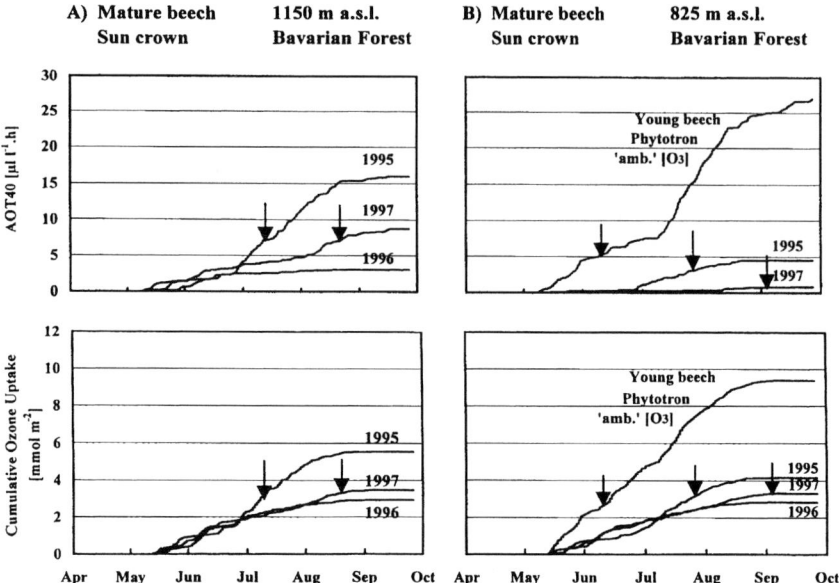

Figure 4. AOT40 (sum of external O_3 exposure > 40 $nl l^{-1}$; cf. Fuhrer and Achermann, 1999) and cumulative O_3 uptake into the leaf mesophyll for mature beech (*Fagus sylvatica*) trees in the Bavarian forest at two sites (1150 m a.s.l., 825 m a.s.l.) and over three years (1995, 1996, 1997), and for beech seedlings exposed to O_3 in a phytotron experiment. The arrow indicates the onset of visible leaf discoloration. O_3 uptake is based on two-sided leaf area. (After data in Baumgarten et al., 2000).

at low rates of uptake. However, damage symptoms on seedling leaves eventually exceeded symptoms on mature trees because of higher cumulative O_3 uptake resulting from greater G_W (Fredericksen et al., 1996b). These results for *Prunus serotina* are consistent with Hypothesis 3 (Table 1), whereas the results for *Fagus sylvatica* and *Picea abies* are not.

Lower capacity for defense or repair of O_3 damage for mature trees may be a consequence of a decline in P_N during tree aging (Fredericksen et al., 1996c; Ryan et al., 1997; Hubbard et al., 1999; Niinemets et al., 1999; Bond, 2000; Kolb and Stone, 2000). Sunlit leaves of old trees often have lower P_N than sunlit leaves of young trees because of reduced G_W likely caused by greater frictional resistance to water flow through xylem of large trees (Becker et al., 2000; Bond and Ryan, 2000; Mencuccini and Magnani, 2000). The tissues most responsible for frictional resistance to water flow in xylem are not understood well, but current evidence suggests that intercellular connections in the xylem (e.g., pits, perforation plates), branch junctions, and bud scars are important (Zimmermann, 1983; Tyree and Ewers, 1991; Rust and Huttl, 1999).

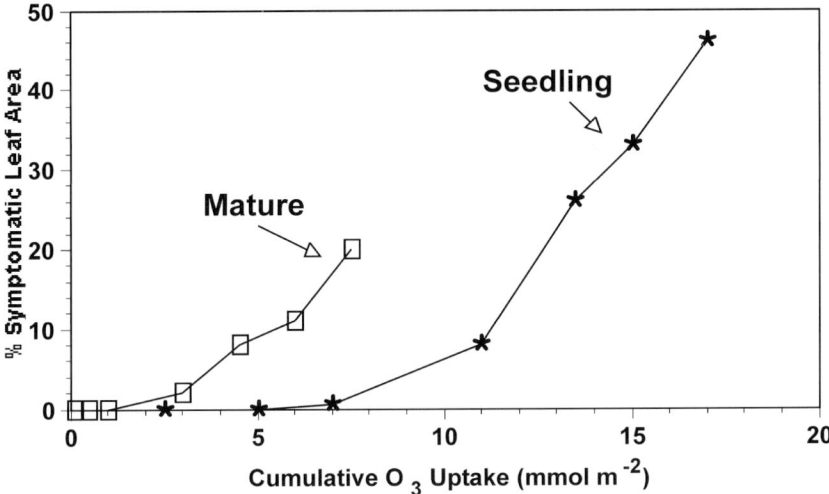

Figure 5. Average percent of leaf area with visible symptoms of O_3 injury (dark adaxial stipple) on shoots of unshaded seedlings and the upper canopy of mature forest trees of *Prunus serotina* during 1994 in central Pennsylvania versus cumulative O_3 uptake. The seedlings were growing in an open field, and the mature trees were growing in an adjacent, natural forest. Cumulative O_3 uptake was calculated as the product of stomatal conductance to O_3 and O_3 concentrations using hourly concentrations > 60 ppb. N = all leaves on 3 shoots for each of 5 seedlings, and all leaves on 5 shoots for each of 4 mature trees (after data in Fredericksen et al., 1996b).

These sources of resistance should be more important in large trees because of their greater size and complexity. Also, large trees may close stomates earlier in the day than small trees because of lower leaf water potential (Kolb and Stone, 2000), greater stomatal sensitivity to vapor pressure deficit, and lower soil-to-leaf hydraulic conductance (Hubbard et al., 1999; Ryan et al., 2000). These factors suggest less O_3 uptake into the leaf mesophyll of mature versus juvenile trees, consistent with Hypothesis 4 described earlier (Table 1).

Several studies have quantitatively compared effects of O_3 on leaf condition of juvenile and mature trees (Rebbeck et al., 1992; Rebbeck et al., 1993; Samuelson and Edwards, 1993; Grulke and Miller, 1994; Hanson et al., 1994; Samuelson, 1994a; Fredericksen et al., 1995; Fredericksen et al., 1996b; Samuelson and Kelly, 1996; Momen et al., 1996; Momen et al., 1997). O_3 exposure had greater negative effects on carbon balance of sun leaves of juvenile versus mature trees in all studies, except those on *Quercus rubra* (Fig. 6). For *Pinus ponderosa*, O_3 exposure decreased P_N similarly for juvenile and mature trees (Fig. 6), but stimulated foliar respiration rate more in juveniles (Momen et al., 1996), indicating greater negative impacts of O_3 on leaf carbon balance of juveniles. Effects of O_3 on leaf respiration rate of *Sequoiadendron giganteum* juvenile and mature trees (Grulke and Miller, 1994) were similar to those

Leaf Response to O_3: Mature vs. Juvenile Trees, Sun Leaves

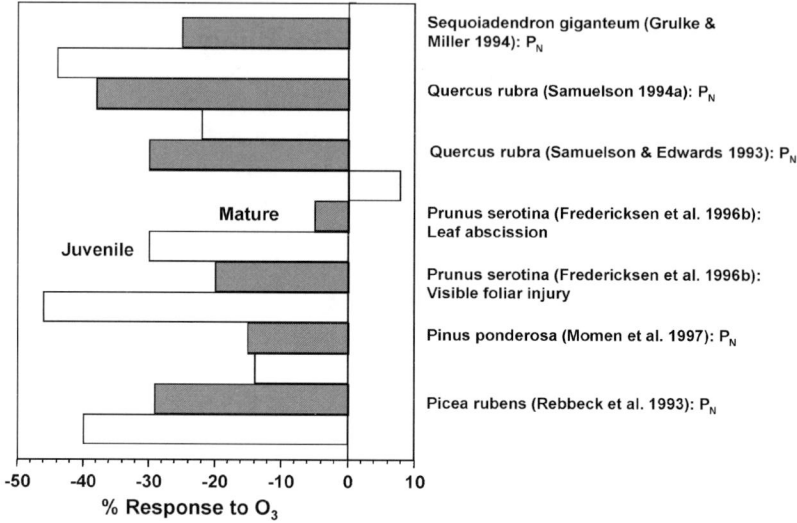

Figure 6. Percent response to O_3 for sun leaves of juvenile and mature trees of several tree species. Response variables are listed for each study (P_N = net photosynthetic rate). For studies using controlled exposures (Rebbeck et al., 1993; Grulke and Miller, 1994; Samuelson and Edwards, 1993; Samuelson, 1994a; Momen et al., 1997), the percent response was calculated for the highest level of exposure compared with the lowest level. For field studies where trees were exposed only to ambient O_3 concentrations (Fredericksen et al., 1996b), the percent response is the mean amount of leaf abscission or visible foliar O_3 injury for each tree size.

reported for *Pinus ponderosa*. Greater negative impacts of O_3 on P_N of juvenile compared with mature trees is suggested for *Prunus serotina* based on greater visual symptoms of O_3 injury in juveniles, and the strong negative relationship between these symptoms and P_N for this species (Samuelson, 1994b; Fredericksen et al., 1996b). In contrast to the findings for the other species, O_3 exposure reduced P_N of mature *Quercus rubra* trees more than for juveniles (Fig. 6). The mechanisms that underlie this result for *Quercus rubra* are not clear, but may include stimulation of G_W, and thus O_3 uptake, in mature trees by strong carbohydrate sinks such as seed development (Maier and Teskey, 1992; Luxmoore et al., 1995), a reduction of G_W and O_3 uptake in the container-grown seedling used in the study because of restrictions on root growth, and frequent watering of trees in the experiment that may have minimized differences in water relations between the juvenile and mature trees. In summary, available evidence suggests that O_3 has less effect on the carbon balance of sun leaves in mature trees than sun leaves in juveniles, with *Quercus rubra* being a notable exception to this trend. Low O_3 uptake by leaves of mature trees resulting from low G_W (Hypothesis 4, Table 1) may override

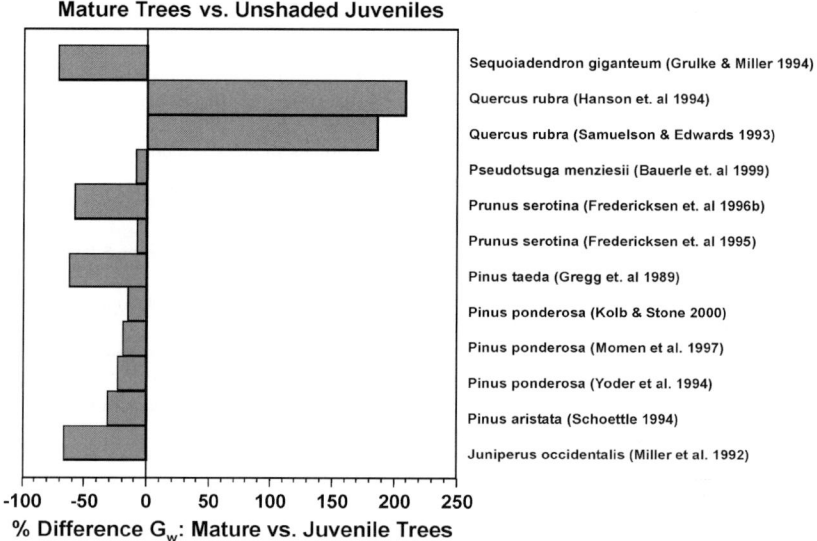

Mature Trees vs. Unshaded Juveniles

Sequoiadendron giganteum (Grulke & Miller 1994)

Quercus rubra (Hanson et. al 1994)

Quercus rubra (Samuelson & Edwards 1993)

Pseudotsuga menziesii (Bauerle et. al 1999)

Prunus serotina (Fredericksen et. al 1996b)

Prunus serotina (Fredericksen et. al 1995)

Pinus taeda (Gregg et. al 1989)

Pinus ponderosa (Kolb & Stone 2000)

Pinus ponderosa (Momen et al. 1997)

Pinus ponderosa (Yoder et al. 1994)

Pinus aristata (Schoettle 1994)

Juniperus occidentalis (Miller et al. 1992)

-100   -50   0   50   100   150   200   250

% Difference G_w: Mature vs. Juvenile Trees

Figure 7. Percent difference in stomatal conductance (G_W) between sunlit leaves of mature trees and unshaded juveniles for several tree species. Negative values indicate lower G_W of mature trees versus juveniles, and positive values indicate greater G_W of mature trees versus juveniles. The differences were calculated based on seasonal or daily maximum values where possible; otherwise, seasonal mean values were used.

possible shortages of carbon available for defense, compensation, and repair responses (Hypothesis 3, Table 1), and thus limit leaf-level damage.

The difference in response of sun leaves to O_3 exposure between juvenile and mature trees (Fig. 6) can be partly explained by differences in G_W. Several studies have shown lower G_W, and thus lower O_3 uptake assuming the same O_3 concentration at the leaf surface, for mature trees than juveniles (Fig. 7). Again, *Quercus rubra* is an exception to this trend; G_W was much greater for mature trees than container-grown juveniles (Fig. 7). Regardless of the direction in the difference in G_W between juvenile and mature trees, the tree size with the greatest G_W (Fig. 7) also had the largest negative response to O_3 exposure (Fig. 6). Interestingly, this linkage between leaf-level physiological response to O_3 and G_W is the same as that reported for comparisons among species by Reich (1987). These findings suggest a consistent scaling pattern of O_3 impacts on the carbon balance of sun leaves during tree ontogeny.

2.2. Factors influencing differences in G_W between juvenile and mature trees

Given the linkages among leaf physiological response to O_3 and G_W, understanding of how environmental factors influence differences in G_W between

Overstory Mature Trees vs. Understory, Shaded Juveniles

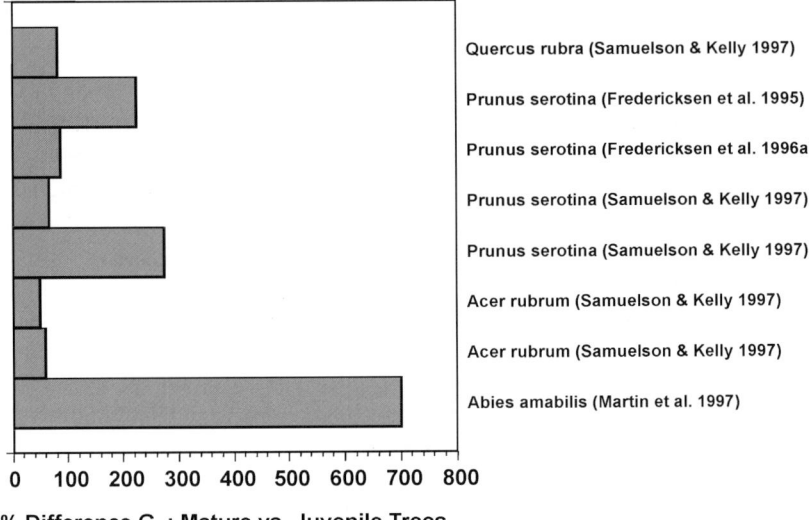

% Difference G$_w$: Mature vs. Juvenile Trees

Figure 8.   Percent difference in stomatal conductance (G_W) between sunlit leaves of mature trees and leaves of shaded, understory juveniles for several tree species. Positive values indicate greater G_W of mature trees versus juveniles. The differences were calculated based on seasonal or daily maximum values where possible; otherwise seasonal mean values were used.

juvenile and mature trees may be relevant towards scaling O_3 impacts to forest ecosystems. The data in Fig. 7 suggests that mature trees often have lower G_W than juveniles for sun leaves. However, because of shading within forests, this comparison applies only to the outer canopy of mature trees and small juveniles growing in openings. When G_W is compared between sun leaves of mature trees and leaves of shaded juveniles of the same species growing in the forest understory, a very different pattern emerges (Fig. 8). In this case, sun leaves of mature trees have much greater G_W than leaves of shaded juveniles. When G_W is measured at ambient light levels, heavily shaded juveniles often have lower G_W than overstory mature trees because light intensity in the understory is below photosynthetic saturation. Also, shade leaves often have lower G_W than sun leaves even when compared at photosynthetically saturating light intensity because of lower leaf N concentration per leaf area and lower LMA (Abrams and Kubiske, 1990; Ellsworth and Reich, 1993; Samuelson and Kelly, 1997). Thus, sun leaves of overstory mature trees should take up more O_3 than leaves of shaded understory juveniles based on differences in G_W.

Differences in water availability to juvenile and mature trees can cause large differences in G_W in some environments. In particular, dry soil can impose greater constraints on G_W for shallow-rooted juveniles than for deep-rooted mature trees that use deep soil or ground water that is not available to small trees (Dawson and Ehleringer, 1991; Donovan and Ehleringer, 1991; Dawson and Ehleringer, 1993; Dawson, 1996). These examples demonstrate that site-specific environment conditions can strongly influence the direction and magnitude of differences in G_W between juvenile and mature trees. Leaves of mature trees may have lower G_W and O_3 uptake than leaves of juveniles under some environmental conditions (e.g., sun leaves for both tree sizes, a large difference in tree height), but they also may have greater G_W and O_3 uptake than leaves of juveniles under other conditions (e.g., sun leaves of mature trees vs. shade leaves of juveniles; Samuelson and Kelly, 1997).

Studies of tree maturation that have compared scions from branches of juveniles with scions from mature trees after grafting onto seedling rootstocks have shown that G_W can also vary with tissue age because of tree maturation processes independent of differences in tree height, light environment, or water availability (Rebbeck et al., 1993; Greenwood, 1995). Limited evidence suggests that some changes in physiological traits during tree maturation result from changes in phytohormones and gene expression (Hutchison et al., 1990; Greenwood, 1995). Hence, tree ontogenetic stage, G_W, and O_3 uptake may be inherently linked.

2.3. Modification of O_3 impact by light environment

Light environment can influence leaf physiological response to O_3. Shade leaves can show greater foliar O_3 injury symptoms than sun leaves (e.g., *Prunus serotina* and *Acer saccharum*, Fig. 9), or vice versa (e.g., *Populus*, Fig. 9). The greater O_3 sensitivity of shade leaves may be related to their low palisade/spongy mesophyll ratio and less compact mesophyll structure that increases O_3 exposure of palisade cells compared with sun leaves (Bennett et al., 1992). Also, exposure of shade leaves to O_3 can cause uncoupling between P_N and G_W (Volin et al., 1993; Tjoelker et al., 1995; Fredericksen et al., 1995; Fredericksen et al., 1996a), which reduces WUE. After chronic exposure to ambient levels of O_3, shaded leaves of *Prunus serotina* seedlings, saplings, and mature trees had a higher ratio of O_3 uptake per P_N (Fig. 10) and greater visible foliar O_3 damage than sun leaves (Fredericksen et al., 1996a, 1996b). This result is consistent with Hypothesis 7 (Table 1).

An additional aspect of the light environment that might be important in scaling leaf-level response to O_3 in forests is the possibility of differences in O_3 uptake at night between juvenile and mature trees. Although darkness is often assumed to severely limit stomatal aperture, G_W can be surprising high at night (Tobiessen, 1982) because of low vapor-pressure deficit and perhaps other factors (Wieser and Havranek, 1993; Matyssek and Innes, 1999). G_W and

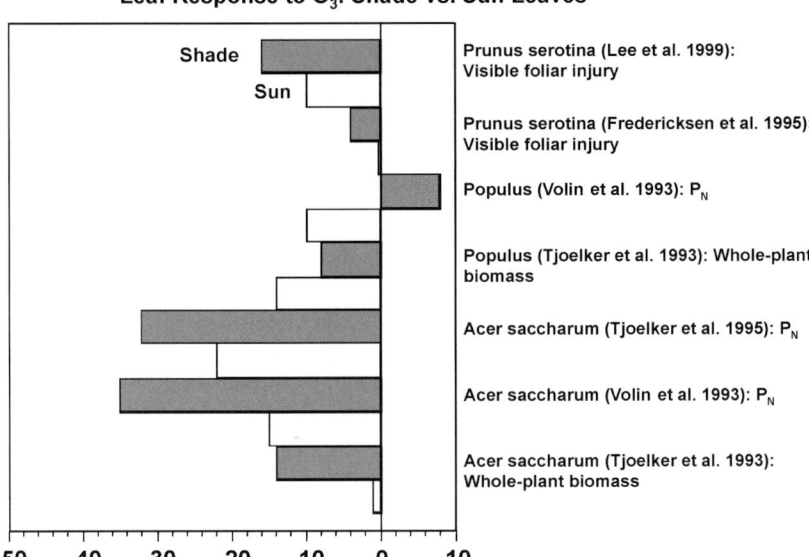

Figure 9. Percent response to O_3 exposure for sun and shade leaves of several tree species. Response variables are listed for each study (P_N = net photosynthetic rate). For studies using controlled exposures (Tjoelker et al., 1993; Volin et al., 1993; Tjoelker et al., 1995), the percent response was calculated for the highest level of exposure compared with the lowest level. For field studies where trees were exposed only to ambient O_3 concentrations (Fredericksen et al., 1996b; Lee et al., 1999), the percent response is the mean amount of visible foliar O_3 injury for each leaf type.

O_3 uptake were high at night in young, well-watered *Betula pendula* cuttings exposed to O_3 in outdoor chambers (Matyssek et al., 1995a). Further, nighttime plus daytime exposure to O_3 reduced biomass growth and altered carbon allocation patterns of these cuttings about two-fold more than daytime exposure alone (Matyssek et al., 1995a). This finding has implications regarding calculations of cumulative O_3 uptake for juvenile and mature trees if nighttime G_W varies with tree age and high concentrations of O_3 occur at night. In a study of different-sized *Prunus serotina* trees, nighttime G_W of mature trees was about 6% of maximum daytime values, whereas nighttime G_W of seedlings was about 18% of maximum daytime values, and seedlings had higher G_W than mature trees during both day and night (Fredericksen et al., 1996c). Given that O_3 concentrations in forests can be high at night under some conditions (Lefohn and Jones, 1986; Krupa and Manning, 1988; Wieser and Havranek, 1993), seedlings may take up more O_3 at night than mature trees.

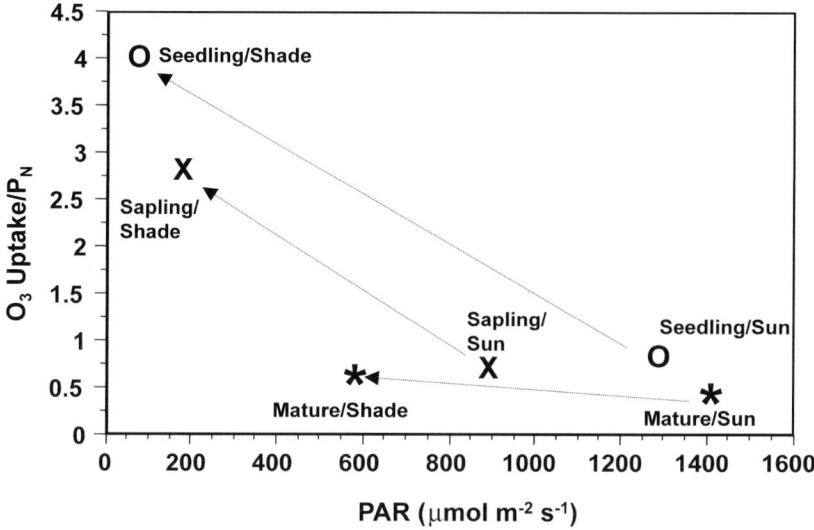

Figure 10. Mean ratio of O_3 uptake to P_N ($nmol\,m^{-2}\,s^{-1}\,O_3\,\mu mol\,m^{-2}\,s^{-1}\,CO_2$) for leaves of seedling, sapling, and mature *Prunus serotina* growing in sun or shade environments in Pennsylvania, and mean daily photosynthetically active radiation (PAR) for each tree size in each environment between June and September, 1993. $N = 10$–30 leaves per tree size for each of four mature trees, five saplings, and five seedlings measured diurnally every two weeks (after data in Fredericksen et al., 1996a).

2.4. Influence of leaf production phenology on O_3 response

Indeterminate growth is a trait of certain tree species and occurs more often for seedlings than for older trees (Kozlowski and Pallardy, 1997). As noted earlier for *Betula pendula*, leaf production phenology of juvenile trees is a plastic trait that can be modified by resource availability (Maurer and Matyssek, 1997) and may be linked to O_3 effects on carbon allocation (Hypothesis 1, Table 1). Differences in phenology of leaf production between juvenile and mature forest trees have been shown to influence cumulative leaf exposure to O_3. Seedlings of *Prunus serotina* grown in an open field had indeterminate growth and produced leaves continuously between May and August whereas mature trees in an adjacent forest had determinate growth and finished leaf production in June (Fig. 11). Despite greater instantaneous rates of O_3 uptake for seedlings because of greater G_W (Fig. 12(A)), many seedling leaves were exposed to O_3 for only part of the growing season, resulting in lower cumulative uptake per average leaf for seedlings than mature trees (Fig. 12(B)). This finding suggests that Hypothesis 4 (Table 1)—stomatal aperture limits O_3 uptake into leaves more for mature than juvenile trees—must also consider the influence of leaf production phenology and leaf longevity on cumulative O_3 uptake.

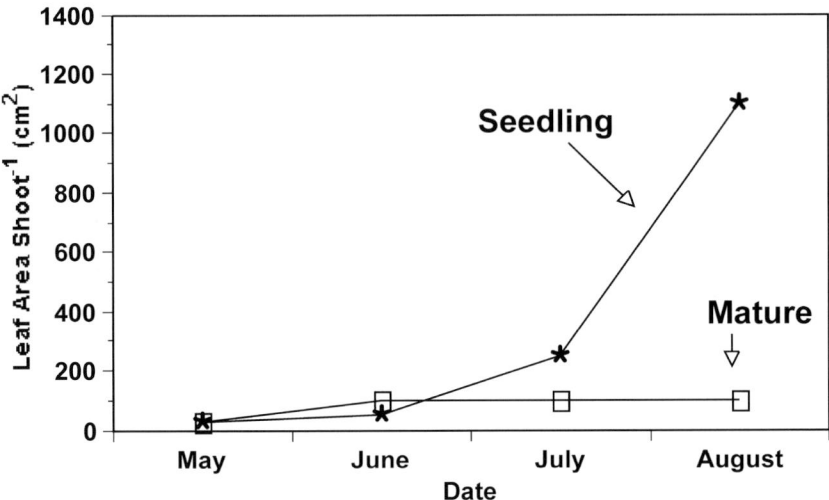

Figure 11. Average leaf area per terminal shoot for seedlings and mature trees of *Prunus serotina* averaged over upper and lower crown positions during 1993 in Pennsylvania. The seedlings were growing in an open field, and the mature trees were growing in an adjacent, natural forest. $N = 20$ shoots per tree for each of 4 mature trees and 12 seedlings. The bars show one standard error of the mean (after data in Fredericksen et al., 1995).

2.5. Changes in O_3 exposure during tree maturation

Spatial variation in O_3 concentration within forests, or variation between forests and areas where test seedlings are grown, could result in differences in O_3 exposure between mature trees and seedlings. Vertical gradients in O_3 concentration in forests may be caused by dense vegetation that is an O_3 sink, resulting in a decrease in O_3 concentration from atmosphere to the forest floor (Kozlowski and Constantinidou, 1986; Fuentes et al., 1992; Taylor and Hanson, 1992; Coe et al., 1995; Skelly et al., 1996; Samuelson and Kelly, 1997; Baumgarten et al., 2000). In addition, nitric oxide (NO) produced by soil microorganisms, which converts O_3 to oxygen, may reduce O_3 concentration near the soil surface (Baumbach and Baumann, 1989). On the other hand, volatile hydrocarbons produced by trees might stimulate the formation of O_3 in the forest because such hydrocarbons are precursors to O_3 formation (Enders et al., 1989; Stockwell et al., 1997).

The interplay among factors that influence local O_3 concentration could result in important spatial variations in O_3 concentration within forests that would need to be considered when comparing responses of juvenile and mature trees. For example, average O_3 concentration was about 33% lower near the forest floor compared with in the upper canopy in a *Fagus sylvatica* forest

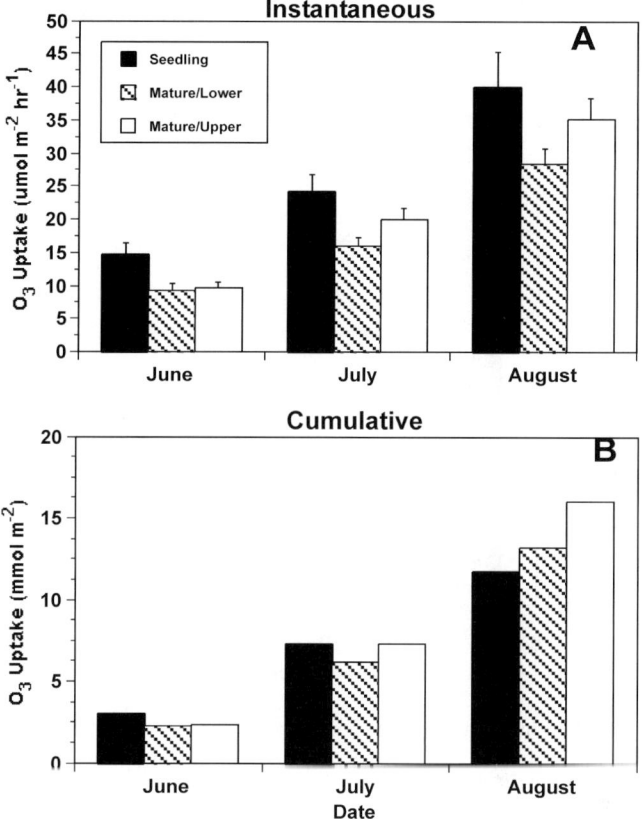

Figure 12. Instantaneous O_3 uptake rate (A) and cumulative O_3 uptake (B) for unshaded seedlings and lower and upper canopy positions of mature forest trees of *Prunus serotina* during June–August 1993 in Pennsylvania. The seedlings were growing in an open field, and the mature trees were growing in an adjacent, natural forest. Instantaneous O_3 uptake rate was calculated from monthly averages of diurnal measurements of O_3 concentration in the canopy and stomatal conductance for each tree size. Cumulative uptake was calculated by summing instantaneous rates of uptake during the study. Because of indeterminate growth of seedlings, cumulative uptake was weighted for this tree age by the proportion of leaf area (monthly mean leaf area/total end-of-season leaf area) exposed to O_3 each month. $N = 30$ to 60 leaves per date for each of 5 seedlings and 4 mature trees. The bars in (A) estimate one standard error of the mean based on pooled variances for O_3 concentration and stomatal conductance (after data in Fredericksen et al., 1995).

in Germany (Fig. 13). Samuelson and Kelly (1997) reported similar vertical gradients in O_3 concentration in a dense hardwood forest in Tennessee. In contrast, only small differences in O_3 concentration occurred for *Prunus serotina* in Pennsylvania among the canopies of mature forest trees, saplings growing in

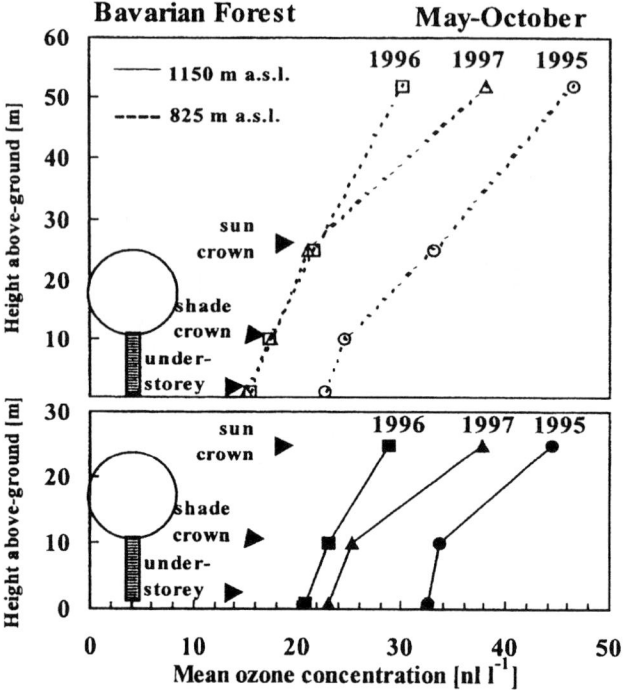

Figure 13. Variation in mean O₃ozone concentration (May–October 1995–1997) within and above a mature forest of *Fagus sylvatica* at altitudes of 1150 m and 825 m in the Bavarian forest (after data in Baumgarten et al., 2000).

a forest opening, and seedlings growing in a field adjacent to the forest during the day when G_W and O_3 uptake were highest (Fig. 14). Obviously, canopy architecture, roughness, and leaf area index may influence spatial patterns of O_3 exposure within stands.

3. Conclusions

Even though short-term effects of O_3 on juvenile trees in controlled experiments may be ameliorated through high supply of CO_2 or mineral nutrients, prevention of chronic injury under forest conditions is questionable because of changes in internal carbon allocation that may compromise resource capture, competitive ability, and defense or tolerance against other stresses (Herms and Mattson, 1992; Bazzaz, 1997). Such controlled experiments on juvenile trees have been valuable in elucidating physiological responses at scales ranging

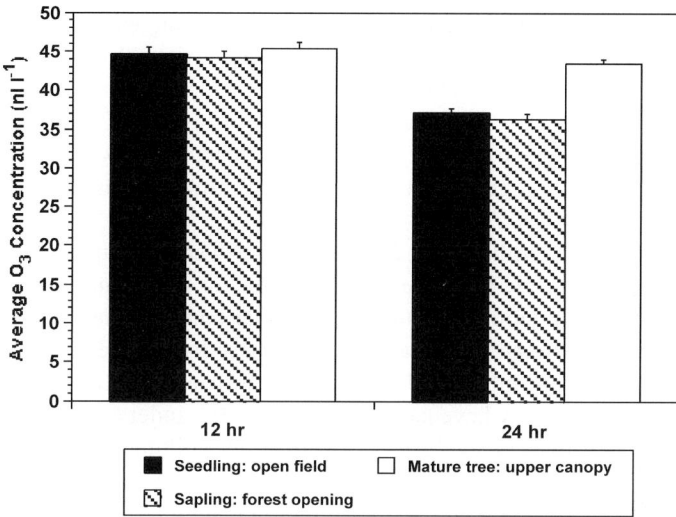

Figure 14. Monthly ambient O_3 concentration (May–August 1994) based on 12 and 24 hour means within the canopies of seedlings, saplings, and mature trees of *Prunus serotina* in a local forest area in Pennsylvania. The seedlings grew in an open field adjacent to the forest containing the mature trees, and the saplings grew in a small forest clearing (after data in Fredericksen et al., 1996b).

from the biochemical level to whole-plant growth. The general similarity between individual leaf and whole-plant responses for water and carbon relations in these experiments have led to scaling diagrams (Fig. 2) that outline probable linkages among molecular, biochemical, tissue, organ, and whole-plant responses. Such linkages form the conceptual basis for scaling units that may be valuable for understanding impacts of O_3 across functional levels of individual trees and during tree ontogeny.

Our analysis of multi-factor studies on plant-internal scaling relationships led to the formulation of seven hypotheses that may be relevant to scaling O_3 responses within forest ecosystems (Table 1). Only three of these hypotheses could be directly evaluated with results from comparisons between juvenile and mature trees, namely Hypotheses 3, 4, and 7. We found inconsistent support in studies with three species for Hypothesis 3—supply of labile carbon available for defense, compensation, and repair responses to O_3 stress will be lower in mature than juvenile trees because of higher respiratory costs for maintaining living tissues and lower photosynthetic rate of mature trees. Obviously, Hypothesis 3 needs to be tested more thoroughly with more species and over a wider range of environmental conditions. We found general support for Hypothesis 4—reduced G_W limits O_3 uptake and mesophyll damage more for mature trees than for juvenile trees—but only for sun leaves. However, this hypothesis does not explain the greater negative impacts of O_3 on the carbon balance of shade leaves (low G_W) compared with sun leaves (high G_W) reported for several species. In this case, high O_3 sensitivity of shade leaves can be conceptually understood to result from an imbalance between O_3 exposure of mesophyll cells and photosynthate available for repair or defense, providing support for Hypothesis 7.

The other four hypotheses (1, 2, 5, 6; Table 1) were difficult to evaluate directly with existing O_3 response data from mature trees, but they should be considered for evaluation in future studies that address quantitative scaling of O_3 effects in forests. Because several of these hypotheses include effects of chemicals that are influenced by human industrial activities (e.g., N compounds, CO_2), they may have particular relevance to understanding impacts of simultaneous changes in levels of O_3 and other air pollutants.

Leaves of juvenile and mature trees show similar qualitative responses in carbon relations and chemical composition to O_3 injury, but quantitative leaf-level responses of juvenile trees to O_3 exposure (e.g., percent reduction in P_N) cannot be assumed to accurately reflect leaf-level responses of mature trees. Factors such as light intensity, vertical profile of soil water availability, rooting depth, spatial variation in light intensity, canopy proportions of sun and shade leaves, and competition from other trees can influence differences in G_W and O_3 uptake between juvenile and mature trees (Kolb, 2002), suggesting problems in generalizing results from leaf-level O_3 responses beyond study site conditions. However, we are encouraged that some of these factors are the subject of current research efforts. These research efforts are especially critical given that leaf-level responses of juvenile trees to O_3 are being used increasingly in process models to predict O_3 impacts on mature forests (e.g., Reich et al., 1990; Taylor et al., 1994; Constable and Retzlaff, 1997; Constable and Taylor, 1997; Ollinger et al., 1997; Retzlaff et al., 1997; Weinstein et al., 1998). The reliability of quantitative predictions of O_3 impact on mature forest trees from process models should increase in the future if such models are refined to include changes in physiological characteristics during tree maturation as highlighted in our review. Certainly, predictions of O_3 impact from such process models are preferable to predictions of impact based on current threshold concepts of damaging O_3 levels (e.g., "Critical Levels for Ozone", formulated by UN-ECE as AOT40, i.e., accumulated O_3 exposure over a threshold of 40 $nl\,l^{-1}$; Skärby et al., 1998; Fuhrer and Achermann, 1999), as the validity of such threshold concepts to mature trees is questionable due to the lack of appropriate databases and conceptual weaknesses in definitions (Matyssek and Innes, 1999). Assessments of O_3 uptake rather than exposure (as adopted by the AOT40 concept) will also improve understanding of O_3 effects on forests.

Tree sensitivity to O_3 is not only determined by O_3 uptake and leaf physiological responses, but also by plant-internal resource allocation to support repair, defense, and compensation mechanisms of coping with O_3 damage. Little is known about controls over internal resource allocation, especially for mature trees, and this is presently a crucial challenge in plant ecophysiological research (Bazzaz, 1997; Lerdau and Gershenzon, 1997). Ontogenetic changes in resource allocation patterns are clearly relevant to understanding differences in O_3 sensitivity between juvenile and old trees, and these patterns likely change during tree ontogeny because of physical (e.g., tree size), environmental (e.g., resource availability, competition), or genetic (e.g., changes in gene expression) factors.

Factorial scenarios that arise from interactions among environment (e.g., light regime, mineral nutrition, atmospheric CO_2), tree genotype, and tree ontogeny likely interact in complex ways to determine tree sensitivity to O_3, thus complicating attempts to scale O_3 impacts to forest trees from responses of juveniles. Free-air O_3 fumigation experiments of forest canopies (Musselman and Hale, 1997; Häberle et al., 1999; Karnosky et al., 2001) alleviate some of these problems, and provide the opportunity to analyze O_3 responses across tree-internal scales and ontogeny. Free-air fumigation experiments in combination with improved assessments of O_3 influx into leaves are likely to resolve some of the remaining uncertainties in understanding O_3 impacts on mature forest trees, and will provide information needed for more credible risk assessments of chronic O_3 stress in forests. Such risk assessments are clearly relevant to concerns about global change, as tropospheric O_3 concentrations will likely stay high in the future (Fowler et al., 1999) and may mitigate the carbon sink strength of forests (Saxe et al., 1998; Ceulemans et al., 1999).

Acknowledgements

This research was supported in part by the US Environmental Protection Agency, by the Swiss 'Bundesamt für Bildung und Wissenschaft' through the 'EUREKA 447 EUROSILVA' program, and by the German Research Foundation (DFG) through the 'Sonderforschungsbereich (SFB) 607: Wachstum oder Parasitenabwehr? Wettbewerb um Ressourcen in Nutzpflanzen aus Land- und Forstwirtschaft'.

References

Abrams, M.D., Kubiske, M.E., 1990. Leaf structural characteristics of 31 hardwood and conifer species in central Wisconsin, Influence of light regime and shade-tolerance rank. Forest Ecol. Manag. 31, 245–253.

Andersen, C.P., Scagel, C.F., 1997. Nutrient availability alters belowground respiration of ozone-exposed ponderosa pine. Tree Physiol. 17, 377–387.

Arp, W.J., 1991. Effects of source-sink relations on photosynthetic acclimation to elevated CO_2. Plant Cell Environ. 14, 869–875.

Baldocchi, D.D., 1993. Scaling water vapor and carbon dioxide exchange from leaves to a canopy: Rules and tools. In: Ehleringer, J.R., Field, C.B. (Eds.), Scaling Physiological Processes, Leaf to Globe. Academic Press, San Diego, CA, pp. 77–114.

Bauerle, W.L., Hinckley, T.M., Cermak, J., Kucera, J., Bible, K., 1999. The canopy water relations of old-growth Douglas-fir trees. Trees 13, 211–217.

Baumbach, G., Baumann, K., 1989. Ozone in forest stands–Examinations to its occurrence and degradation. In: Georgii, H.W. (Ed.), Mechanisms and Effects of Pollutant Transfer into Forests. Kluwer, New York, pp. 37–44.

Baumgarten, M., Werner, H., Häberle, K.-H., Emberson, L.D., Fabian, P., Matyssek, R., 2000. Seasonal ozone response of mature beech trees (*Fagus sylvatica*) at high altitude in the Bavarian forest (Germany) in comparison with young beech grown in the field and in phytotrons. Environ. Pollut. 109, 431–442.

Bazzaz, F.A., 1997. Allocation of resources in plants: State of the science and critical questions. In: Bazzaz, F.A., Grace, J. (Eds.), Plant Resource Allocation. Academic Press, San Diego, CA, pp. 1–38.

Becker, P., Meinzer, F.C., Wullschleger, S.D., 2000. Hydraulic limitation of tree height: A critique. Funct. Ecol. 14, 4–11.

Bennett, J.P., Rassat, P., Berrang, P., Karnosky, D.F., 1992. Relationships between leaf anatomy and ozone sensitivity of *Fraxinus pennsylvanica* Marsh. and *Prunus serotina* Ehrh. Environ. Exp. Bot. 32, 33–41.

Beyers, J.L., Riechers, G.H., Temple, P.J., 1992. Effects of long-term ozone exposure and drought on the photosynthetic capacity of ponderosa pine (*Pinus ponderosa* Laws.). New Phytol. 122, 81–90.

Bond, B.J., 2000. Age-related changes in photosynthesis of woody plants. Trends Plant Sci. 5, 349–353.

Bond, B.J., Ryan, M.G., 2000. Comment on 'Hydraulic limitation of tree height: a critique' by Becker, Meinzer, and Wullschleger. Funct. Ecol. 140, 135–140.

Ceulemans, R., Janssens, I.A., Jach, M.E., 1999. Effects of CO_2 enrichment on trees and forests: Lessons to be learned in view of future ecosystem studies. Ann. Bot. 84, 577–590.

Ceulemans, R., Mousseau, M., 1994. Effects of elevated atmospheric CO_2 on woody plants. New Phytol. 17, 425–446.

Chappelka, A.H., Samuelson, L.J., 1998. Ambient ozone effects on forest trees of the eastern United States: A review. New Phytol. 139, 91–108.

Coe, H., Gallagher, M.W., Choularton, T.W., Dore, C., 1995. Canopy scale measurements of stomatal and cuticular O_3 uptake by Sitka spruce. Atmos. Environ. 29, 1413–1423.

Constable, J.V.H., Retzlaff, W.A., 1997. Simulating the response of mature yellow popular and loblolly pine trees in peak ozone periods during the growing season using the TREEGRO model. Tree Physiol. 17, 627–635.

Constable, J.V.H., Taylor Jr., G.E., 1997. Modeling the effects of elevated tropospheric O_3 on two varieties of *Pinus ponderosa*. Can. J. Forest Res. 27, 527–537.

Cregg, B.M., Halpin, J.E., Dougherty, P.M., Teskey, R.O., 1989. Comparative physiology and morphology of seedling and mature forest trees. In: Noble, D., Martin, J.L., Jensen, K.F. (Eds.), Proceedings of the US-USSR Symposium on Air Pollution Effects on Vegetation, pp. 111–118.

Curtis, P.S., Wang, X., 1998. A meta-analysis of elevated CO_2 effects on wood plant mass, form, and physiology. Oecologia 113, 299–313.

Dawson, T.E., 1996. Determining water use by trees and forests from isotopic, energy balance and transpiration analyses: The role of tree size and hydraulic lift. Tree Physiol. 16, 263–272.

Dawson, T.E., Ehleringer, J.R., 1991. Streamside trees that do not use stream water. Nature 350, 335–337.

Dawson, T.E., Ehleringer, J.R., 1993. Gender-specific physiology, carbon isotope discrimination, and habitat distribution in boxelder, *Acer negundo*. Ecology 74, 798–815.

Dixon, M., Le Thiec, D., Garrec, J.P., 1998. Reactions of Norway spruce and beech trees to two years of ozone exposure and episodic drought. Environ. Exp. Bot. 40, 77–91.

Donovan, L.A., Ehleringer, J.R., 1991. Ecophysiological differences among juvenile and reproductive plants of several woody species. Oecologia 86, 594–597.

Drake, B.G., Azcon-Bieto, J., Berry, J., Bunce, J., Dahlman, R., Dijkstra, P., Farrar, J., Gifford, R.M., Gonzalez-Meler, M.A., Koch, G., Lambers, H., Siedow, J., Wullschleger, S., 1999. Does elevated atmospheric CO_2 concentration inhibit mitochondrial respiration in green plants? Plant Cell Environ. 22, 649–657.

El Kohen, A., Venet, L., Mousseau, M., 1993. Growth and photosynthesis of two deciduous forest species at elevated carbon dioxide. Funct. Ecol. 7, 480–486.

Ellenberg, H., 1996. Vegetation Mitteleuropas mit den Alpen. 5. Auflage. Ulmer Verlag, Stuttgart.

Ellsworth, D.S., Reich, P.B., 1993. Canopy structure and vertical patterns of photosynthesis and related leaf traits in a deciduous forest. Oecologia 96, 169–178.

Enders, G., Teichmann, U., Kramm, G., 1989. Profiles of ozone and surface layer parameters over a mature spruce stand. In: Georgii, H.W. (Ed.), Mechanisms and Effects of Pollutant Transfer into Forests. Kluwer Academic, Dordrecht, pp. 21–35.

Ericsson, T., Rytter, L., Vapaavuori, E., 1996. Physiology and allocation in trees. Biomass Bioenerg. 11, 115–127.

Farquhar, G.D., Ehleringer, J.R., Hubick, K.T., 1989a. Carbon isotope discrimination and photosynthesis. Annu. Rev. Plant Physiol. Plant Mol. Biol. 40, 503–537.

Farquhar, G.D., Hubick, K.T., Condon, A.G., Richards, R.A., 1989b. Carbon isotope fractionation and plant water-use efficiency. In: Rundel, P.W., Ehleringer, J.R., Nagy, K.A. (Eds.), Stable Isotopes in Ecological Research. Springer-Verlag, Berlin, pp. 21–40.

Fowler, D., Cape, J.N., Coyle, M., Flechard, C., Kuylenstienra, J., Hicks, K., Derwent, D., Johnson, C., Stevenson, D., 1999. The global exposure of forests to air pollutants. Water Air Soil Pollut. 116, 5–32.

Fredericksen, T.S., Joyce, B.J., Skelly, J.M., Steiner, K.C., Kolb, T.E., Kouterick, K.B., Savage, J.E., Snyder, K.R., 1995. Physiology, morphology, and ozone uptake of leaves of black cherry seedlings, saplings, and canopy trees. Environ. Pollut. 89, 273–283.

Fredericksen, T.S., Kolb, T.E., Skelly, J.M., Steiner, K.C., Joyce, B.J., Savage, J.E., 1996a. Light environment alters ozone uptake per net photosynthetic rate in black cherry trees. Tree Physiol. 16, 485–490.

Fredericksen, T.S., Skelly, J.M., Steiner, K.C., Kolb, T.E., Kouterick, K.B., 1996b. Size-mediated foliar injury response in black cherry trees. Environ. Pollut. 91, 53–63.

Fredericksen, T.S., Steiner, K.C., Skelly, J.M., Joyce, B.J., Kolb, T.E., Kouterick, K.B., Ferdinand, J.E., 1996c. Diel and seasonal patterns of leaf gas exchange and xylem water potentials in different-sized *Prunus serotina* Ehrh. trees. Forest Sci. 42, 359–365.

Friend, A.L., Tomlinson, P.T., 1992. Mild ozone exposure alters ^{14}C dynamics in foliage of *Pinus taeda* L. Tree Physiol. 11, 214–227.

Fuentes, J.D., Gillespie, T.J., den Hartog, G., Neumann, H.H., 1992. Ozone deposition onto a deciduous forest during dry and wet conditions. Agric. Forest Meteorol. 62, 1–18.

Fuhrer, J., Achermann, B., 1999. Critical Levels for Ozone—Level II. Environmental Documentation 115 Swiss Agency for the Environment, Forests and Landscape, Berne, Switzerland, pp. 333.

Gehring, C.A., Cobb, N.S., Whitham, T.G., 1997. Three way interactions among ectomycorrhizal mutualists, scale insects, and resistant and susceptible pinyon pines. Am. Nat. 149, 824–841.

Gerant, D., Podor, M., Grieu, P., Afif, D., Cornu, S., Morabito, D., Banvoy, J., Robin, C., Dizengremel, P., 1996. Carbon metabolism, enzyme activities and carbon partitioning in *Pinus halepensis* Mill. to mild drought and ozone. J. Plant Physiol. 148, 142.

Grams, T.E.E., Anegg, S., Häberle, K.-H., Langebartels, C., Matyssek, R., 1999. Interactions of chronic exposure to elevated CO_2 and O_3 levels in the photosynthetic light and dark reactions of European beech (*Fagus sylvatica*). New Phytol. 144, 95–107.

Greenwood, M.S., 1995. Juvenility and maturation in conifers: Current concepts. Tree Physiol. 15, 433–438.

Greitner, C.S., Pell, E.J., Winner, W.E., 1994. Analysis of aspen foliage exposed to multiple stresses: Ozone, nitrogen deficiency and drought. New Phytol. 127, 579–589.

Gruber, F., 1998. Präformierte und neoformierte Syllepsis sowie Prolepsis bei der Buche (*Fagus sylvatica* L.). Flora 193, 369–385.

Grulke, N.E., Miller, P.R., 1994. Changes in gas exchange characteristics during the life span of giant sequoia—Implications for response to current and future concentrations of atmospheric ozone. Tree Physiol. 14, 659–668.

Gunthardt-Goerg, M.S., Matyssek, R., Scheidegger, C., Keller, T., 1993. Differentiation and structural decline in the leaves and bark of birch (*Betula pendula*) under low ozone concentration. Trees 7, 104–114.

Gunthardt-Goerg, M.S., McQuattie, C.J., Scheidegger, C., Rhiner, C., Matyssek, R., 1997. Ozone-induced cytochemical and ultrastructural changes in leaf mesophyll cell walls. Can. J. Forest Res. 27, 453–463.

Häberle, K.-H., Werner, H., Fabian, P., Pretzsch, H., Reiter, I., Matyssek, R., 1999. 'Free-air' ozone fumigation of mature forest trees: a concept for validating AOT40 under stand conditions. In: Fuhrer, J., Achermann, B. (Eds.), Critical Level for Ozone–Level II. Swiss Agency for the Environment, Forests and Landscape (SAEFL), Berne, pp. 133–137.

Hanson, P.J., Samuelson, L.J., Wullschleger, S.D., Tabberer, T.A., Edwards, G.S., 1994. Seasonal patterns of light-saturated photosynthesis and leaf conductance for mature and seedling *Quercus rubra* L. foliage—differential sensitivity to ozone exposure. Tree Physiol. 14, 1351–1366.

Heath, J., Kerstiens, G., 1997. Effects of elevated CO_2 on gas exchange in beech and oak at two levels of nutrient supply: consequences for sensitivity to drought in beech. Plant Cell Environ. 20, 57–67.

Herms, D.A., Mattson, W.J., 1992. The dilemma of plants: To grow or defend. Quart. Rev. Biol. 67, 283–335.

Hubbard, R.M., Bond, B.J., Ryan, M.G., 1999. Evidence that hydraulic conductance limits photosynthesis in old *Pinus ponderosa* trees. Tree Physiol. 19, 165–172.

Hunt Jr., E.R., Lavigne, M.B., Franklin, S.E., 1999. Factors controlling the decline of net primary production with stand age for balsam fir in Newfoundland assessed using an ecosystem simulation model. Ecol. Modell. 122, 151–164.

Hutchison, K.W., Sherman, C.D., Weber, J., Smith, S.S., Singer, P.B., Greenwood, M.S., 1990. Maturation in larch II. Effects of age on photosynthesis and gene expression in developing foliage. Plant Physiol. 94, 1308–1315.

Jarvis, P.G., 1993. Prospects of bottom-up models. In: Ehleringer, J.R., Field, C.B. (Eds.), Scaling Physiological Processes, Leaf to Globe. Academic Press, San Diego, CA, pp. 117–126.

Kangasjärvi, J., Talvinen, J., Utrianen, M., Karjalainen, R., 1994. Plant defense systems induced by ozone. Plant Cell Environ. 17, 783–794.

Karnosky, D.F., Gielen, B., Ceulemans, R., Schlesinger, W.H., Norby, R.J., Oksanen, E., Matyssek, R., Hendrey, G.R., 2001. FACE systems for studying the impacts of greenhouse gases on forest ecosystems. In: Karnosky, D.F., Scarascia-Mugnozza, G., Ceulemans, R., Innes,

J.L. (Eds.), The Impacts of Carbon Dioxide and Other Greenhouse Gases on Forest Ecosystems. CABI Press, pp. 297–324.

Kelly, J.M., Taylor Jr., G.E., Edwards, N.T., Adams, M.B., Edwards, G.S., Friend, A.L., 1993. Growth, physiology, and nutrition of loblolly pine seedlings stressed by ozone and acid precipitation: A summary of the ROPIS-south project. Water Air Soil Pollut. 69, 363–391.

Kerstiens, G., Townsend, J., Heath, J., Mansfield, T.A., 1995. Effects of water and nutrient availability on physiological responses of woody species to elevated CO_2. Forestry 68, 303–315.

Kolb, T.E., 2002. Ageing as an influence on tree response to ozone: Theory and observations. In: Matyseek, R., Huttunen, S. (Eds.), Trends in European Forest Tree Physiology Research (in press).

Kolb, T.E., Fredericksen, T.S., Steiner, K.C., Skelly, J.M., 1997. Issues in scaling tree size and age responses to ozone: A review. Environ. Pollut. 98, 195–208.

Kolb, T.E., Stone, J.E., 2000. Differences in leaf gas exchange and water relations among species and tree sizes in an Arizona pine–oak forest. Tree Physiol. 20, 1–12.

Kozlowski, T.T., Constantinidou, H.A., 1986. Environmental pollution and tree growth. Part II. Factors affecting responses to pollution and alleviation of pollution effects. Forestry Abstr. 47, 105–132.

Kozlowski, T.T., Pallardy, S.G., 1997. Physiology of Woody Plants. Academic Press, New York.

Krupa, S.V., Manning, W.J., 1988. Atmospheric ozone formation and effects on vegetation. Environ. Pollut. 50, 101–137.

Kuppers, M., 1994. Canopy gaps: competitive light interception and economic space filling—a matter of whole-plant allocation. In: Caldwell, M.M., Pearcy, R.W. (Eds.), Exploitation of Environmental Heterogeneity by Plants—Ecophysiological Processes Above and Below ground. Academic Press, San Diego, CA, pp. 111–144.

Landolt, W., Gunthardt-Goerg, M.S., Pfenninger, I., Einig, W., Hampp, R., Maurer, S., Matyssek, R., 1997. Effect of fertilization on ozone-induced changes in the metabolism of birch leaves (*Betula pendula*). New Phytol. 137, 389–397.

Langebartels, C., Ernst, D., Heller, W., Lutz, C., Payer, H.-D., Sandermann Jr., H., 1997. Ozone responses of trees: Results from controlled chamber exposures at the GSF phytotron. In: Sandermann Jr., H., Wellburn, A.R., Heath, R.L. (Eds.), Forest Decline and Ozone: A Comparison of Controlled Chamber and Field Experiments. In: Ecological Studies, Vol. 127. Springer-Verlag, Berlin, pp. 163–200.

Laurence, J.A., Amundson, R.G., Friend, A.L., Pell, E.J., Temple, P.J., 1994. Allocation of carbon in plants under stress: An analysis of the ROPIS experiments. J. Environ. Quality 23, 412–417.

Lee, J.C., Skelly, J.M., Steiner, K.C., Zhang, J.W., Savage, J.E., 1999. Foliar response of black cherry (*Prunus serotina*) clones to ambient ozone exposure in central Pennsylvania. Environ. Pollut. 105, 325–331.

Lefohn, A.S., Jones, C.K., 1986. The characterization of ozone and sulfur dioxide air quality for assessing possible vegetation effects. J. Air Pollut. Control Assoc. 36, 1123–1129.

Lerdau, M., Gershenzon, J., 1997. Allocation theory and chemical defense. In: Bazzaz, F.A., Grace, J. (Eds.), Plant Resource Allocation. Academic Press, San Diego, CA, pp. 265–277.

Lippert, M., Steiner, K., Payer, H.-D., Simons, S., Langebartels, C., Sandermann Jr., H., 1996. Assessing the impact of ozone on photosynthesis of European beech (*Fagus sylvatica* L.) in environmental chambers. Trees 10, 268–275.

Luethy-Krause, B., Pfenninger, I., Landolt, W., 1990. Effects of ozone on organic acids in needles of Norway spruce and Scots pine. Trees 4, 198–204.

Luxmoore, R.J., Oren, R., Sheriff, D.W., Thomas, R.B., 1995. Source–sink–storage relationships of conifers. In: Smith, W.K., Hinckley, T.M. (Eds.), Resource Physiology of Conifers: Acquisition, Allocation, and Utilization. Academic Press, New York, pp. 179–216.

Maier, C.A., Teskey, R.O., 1992. Internal and external control of net photosynthesis and stomatal conductance of mature eastern white pine (*Pinus strobus*). Can. J. Forest Res. 22, 1387–1394.

Manes, F., Vitale, M., Donato, E., Paoletti, E., 1998. O_3 and $O_3 + CO_2$ effects on a Mediterranean evergreen broadleaf tree, Holm Oak (*Quercus ilex* L.). Chemosphere 36, 801–806.

Martin, T.A., Brown, K.J., Cermack, J., Ceulemans, R., Kucera, J., Meinzer, F.C., Rombold, J.S., Sprugel, D.G., Hinckley, T.M., 1997. Crown conductance and tree and stand transpiration in a second-growth *Abies amabilis* forest. Can. J. Forest Res. 27, 797–808.

Matyssek, R., Gunthardt-Goerg, M.S., Keller, T., Schneidegger, C., 1991. Impairment of gas exchange and structure in birch leaves (*Betula pendula*) caused by low ozone concentrations. Trees 5, 5–13.

Matyssek, R., Gunthardt-Goerg, M.S., Maurer, S., Keller, T., 1995a. Nighttime exposure to ozone reduces whole-plant production in *Betula pendula*. Tree Physiol. 15, 159–165.

Matyssek, R., Gunthardt-Goerg, M.S., Saurer, M., Keller, T., 1992. Seasonal growth, ^{13}C in leaves and stem, and phloem structure of birch (*Betula pendula*) under low ozone concentrations. Trees 6, 69–76.

Matyssek, R., Havranek, W.M., Wieser, G., Innes, J.L., 1997. Ozone and the forests in Austria and Switzerland. In: Sandermann Jr., H., Wellburn, A.R., Heath, R.L. (Eds.), Forest Decline and Ozone: A Comparison of Controlled Chamber and Field Experiments. In: Ecological Studies, Vol. 127. Springer-Verlag, Berlin, pp. 95–134.

Matyssek, R., Innes, J.L., 1999. Ozone—A risk factor for trees and forests in Europe? Water Air Soil Pollut. 116, 199–226.

Matyssek, R., Reich, P.B., Oren, R., Winner, W.E., 1995b. Response mechanisms of conifers to air pollutants. In: Smith, W.K., Hinckley, T.H. (Eds.), Physiological Ecology of Coniferous Forests. In: Physiological Ecology Series. Academic Press, New York, pp. 255–308.

Maurer, S., Matyssek, R., 1997. Nutrition and the ozone sensitivity of birch (*Betula pendula*), II. Carbon balance, water-use efficiency and nutritional status of the whole plant. Trees 12, 11–20.

Maurer, S., Matyssek, R., Gunthardt-Goerg, M.S., Landolt, W., Einig, W., 1997. Nutrition and the ozone sensitivity of birch (*Betula pendula*), I. Responses at the leaf level. Trees 12, 1–10.

McLaughlin, S.B., Percy, K., 1999. Forest health in North America: Some perspectives on actual and potential roles of climate and air pollution. Water Air Soil Pollut. 116, 151–197.

Mencuccini, M., Magnani, F., 2000. Comment on 'Hydraulic limitation of tree height: A critique' by Becker, Meinzer, and Wullschleger. Funct. Ecol. 140, 135–140.

Miller, P.M., Eddleman, L.E., Miller, J.M., 1992. The seasonal course of physiological processes in Juniperus occidentalis. Forest Ecol. Manag. 48, 185–215.

Momen, B., Anderson, P.B., Helms, J.A., Houpis, J.L.J., 1997. Acid rain and ozone effects on gas exchange of *Pinus ponderosa*: A comparison between trees and seedlings. Int. J. Plant Sci. 158, 617–621.

Momen, B., Helms, J.A., Criddle, R.S., 1996. Foliar metabolic heat rate of seedlings and mature trees of Pinus ponderosa exposed to simulated acid rain and elevated ozone. Plant Cell Environ. 19, 747–753.

Mooney, H.A., Winner, W.E., 1991. Partitioning response of plants to stress. In: Mooney, H.A., Winner, W.E., Pell, E. (Eds.), Response of Plants to Multiple Stresses. Academic Press, San Diego, CA, pp. 129–141.

Mooney, H.A., Winner, W.E., Pell, E.J. (Eds.), 1991. Response of Plants to Multiple Stresses. Academic Press, San Diego, CA, p. 422.

Mousseau, M., Dufrene, E., El Kohen, A., Epron, D., Godard, D., Liozon, R., Pontailler, J.Y., Saugier, B., 1996. Growth strategy and tree responses to elevated CO_2: A comparison of beech (*Fagus sylvatica*) and sweet chestnut (*Castanea sativa* Mill.). In: Koch, G.W., Mooney, H.A. (Eds.), Carbon Dioxide and Terrestrial Ecosystems. Academic Press, San Diego, CA, pp. 71–86.

Musselman, R.C., Hale, B.A., 1997. Methods for controlled and field ozone exposures of forest tree species in North America. In: Sandermann Jr., H., Wellburn, A.R., Heath, R.L. (Eds.), Forest Decline and Ozone a Comparison of Controlled Chamber and Field Experiments. In: Ecological Studies, Vol. 127. Springer-Verlag, Berlin, pp. 277–315.

Niinemets, U., Kull, O., Tenhunen, J.D., 1999. Variability in leaf morphology and chemical composition as a function of canopy light environment in coexisting deciduous trees. Int. J. Plant Sci. 160, 837–848.

Ollinger, S.V., Aber, J.D., Reich, P.B., 1997. Simulating ozone effects on forest productivity: Interactions among leaf-, canopy-, and stand-level processes. Ecol. Applic. 7, 1237–1251.

Pell, E.J., Temple, P.J., Friend, A.L., Mooney, H.A., Winner, W.E., 1994. Compensation as a plant response to ozone and associated stresses: An analysis of ROPIS experiments. J. Environ. Quality 23, 429–436.

Polle, A., Matyssek, R., Gunthardt-Goerg, M.S., Maurer, S., 2000. Defense strategies against ozone in trees: The role of nutrition. In: Agrawal, S.B., Agrawal, M. (Eds.), Environmental Pollution and Plant Responses. Lewis Publishers, Boca Raton, FL, pp. 223–245.

Rebbeck, J., Jensen, K.F., Greenwood, M.S., 1992. Ozone effects on the growth of grafted mature and juvenile red spruce. Can. J. Forest Res. 22, 756–760.

Rebbeck, J., Jensen, K.F., Greenwood, M.S., 1993. Ozone effects on grafted mature and juvenile red spruce: Photosynthesis, stomatal conductance, and chlorophyll concentration. Can. J. Forest Res. 23, 450–456.

Reich, P.B., 1987. Quantifying plant response to ozone: A unifying theory. Tree Physiol. 3, 63–91.

Reich, P.B., Ellsworth, D.S., Kloeppel, B.D., Fownes, J.H., Gower, S.T., 1990. Vertical variation in canopy structure and CO_2 exchange of oak-maple forests: Influences of ozone, nitrogen, and other factors on simulated canopy carbon gain. Tree Physiol. 7, 329–345.

Rennenberg, H., Herschbach, C., Polle, A., 1996. Consequences of air pollution on shoot-root interaction. J. Plant Physiol. 148, 296–301.

Retzlaff, W.A., Weinstein, D.A., Laurence, J.A., Gollands, B., 1997. Simulating the growth of a 160-year-old sugar maple (*Acer saccharum*) tree with and without ozone exposure using the TREGROW model. Can. J. Forest Res. 27, 783–789,

Rey, A., Jarvis, P.G., 1997. Growth response of young birch trees (*Betula pendula* Roth.) after four and a half years of CO_2 exposure. Ann. Bot. 80, 809–816.

Reynolds, J.F., Hilbert, D.W., Kemp, P.R., 1993. Scaling ecophysiology from the plant to the ecosystem: A conceptual framework. In: Ehleringer, J.R., Field, C.B. (Eds.), Scaling Physiological Processes, Leaf to Globe. Academic Press, San Diego, CA, pp. 127–140.

Roloff, A., 1985. Morphologie der Kronenentwicklung von *Fagus sylvatica* L. (Rotbuche) unter besonderer Berücksichtigung möglicherweise neuartiger Veränderungen. Berichte des Forschungszentrums Waldökosysteme/Waldsterben. Göttingen Bd. 18, 1–177.

Rust, S., Huttl, R.F., 1999. The effect of shoot architecture on hydraulic conductance in beech (*Fagus sylvatica* L.). Trees 14, 39–42.

Ryan, M.G., Bond, B.J., Law, B.E., Hubbard, R.M., Woodruff, D., Cienciala, E., Kucera, J., 2000. Transpiration and whole-tree conductance in ponderosa pines trees of different heights. Oecologia 124, 553–560.

Ryan, M.G., Binkley, D., Fownes, J.H., 1997. Age-related decline in forest productivity: Pattern and process. Adv. Ecol. Res. 27, 213–262.

Ryan, M.G., Waring, R.H., 1992. Maintenance respiration and stand development in a subalpine lodgepole pine forest. Ecology 73, 2100–2108.

Samuelson, L.J., 1994a. The role of microclimate in determining the sensitivity of *Quercus rubra* L. to ozone. New Phytol. 128, 235–241.

Samuelson, L.J., 1994b. Ozone-exposure responses of black cherry and red maple seedlings. Environ. Exper. Bot. 34, 355–362.

Samuelson, L.J., Edwards, G.S., 1993. A comparison of sensitivity to ozone in seedlings and trees of *Quercus rubra* L. New Phytol. 125, 373–379.

Samuelson, L.J., Kelly, J.M., 1996. Carbon partitioning and allocation in northern red oak seedlings and mature trees in response to ozone. Tree Physiol. 16, 853–858.

Samuelson, L.J., Kelly, J.M., 1997. Ozone uptake in *Prunus serotina, Acer rubrum* and *Quercus rubra* forest trees of different sizes. New Phytol. 136, 255–264.

Samuelson, L.J., Kelly, J.M., Mays, P.A., Edwards, G.S., 1996. Growth and nutrition of *Quercus rubra* L. seedlings and mature trees after three seasons of ozone exposure. Environ. Pollut. 91, 317–323.

Sandermann Jr., H., 1996. Ozone and plant health. Annu. Rev. Phytopathol. 34, 347–366.

Sandermann Jr., H., Wellburn, A.R., Heath, R.L. (Eds.), 1997. Forest Decline and Ozone: A Comparison of Controlled Chamber and Field Experiments. In: Ecological Studies, Vol. 127. Springer-Verlag, Berlin, p. 400.

Saurer, M., Maurer, S., Matyssek, R., Landolt, W., Gunthardt-Goerg, M.S., Siegenthaler, U., 1995. The influence of ozone and nutrition on ^{13}C in *Betula pendula*. Oecologia 103, 397–406.

Saxe, H., Ellsworth, D.S., Heath, J., 1998. Tree and forest functioning in an enriched CO_2 atmosphere. New Phytol. 139, 395–436.

Schoettle, A.W., 1994. Influence of tree size on shoot structure and physiology of *Pinus contorta* and *Pinus aristata*. Tree Physiol. 14, 1055–1068.

Schulze, E.-D., Hall, A.E., 1982. Stomatal responses to water loss and CO_2 assimilation rates of plants in contrasting environments. In: Lange, O.L., Nobel, P.S., Osmond, C.B., Ziegler, H. (Eds.), Encyclopedia of Plant Physiology, Water Relations and Carbon Assimilation, Vol. 12B. Springer-Verlag, Berlin, pp. 181–230.

Skärby, L., Ro-Poulsen, H., Wellburn, A.M., Sheppard, L.J., 1998. Impacts of ozone on forests: A European perspective. New Phytol. 139, 109–122.

Skelly, J.M., Fredericksen, T.S., Savage, J.M., Snyder, K.R., 1996. Vertical gradients of ozone and carbon dioxide in a Pennsylvania deciduous forest. Environ. Pollut. 94, 235–240.

Smeulders, S.M., Gorissen, A., Joosten, N.N., Vanveen, J.A., 1995. Effects of short-term ozone exposure on the carbon economy of mature and juvenile Douglas firs [*Pseudotsuga menziesii* (Mirb) Franco]. New Phytol. 129, 45–53.

Spence, D.R., Rykiel Jr., E.J., Sharpe, P.J.H., 1990. Ozone alters carbon allocation in loblolly pine assessment with carbon-11 labeling. Environ. Pollut. 64, 93–106.

Stitt, M., Schulze, E.-D., 1994. Plant growth, storage and resource allocation: From flux control in metabolic chain to the whole-plant level. In: Schultze, E.D. (Ed.), Flux Control in Biological Systems From Enzymes to Populations and Ecosystems. Academic Press, San Diego, CA, pp. 57–118.

Stockwell, W.R., Kramm, G., Scheel, H.-E., Mohnen, V.A., Seiler, W., 1997. Ozone formation, destruction and exposure in Europe and the United States. In: Sandermann Jr., H., Wellburn, A.R., Heath, R.L. (Eds.), Forest Decline and Ozone: A Comparison of Controlled and Field Experiments. In: Ecological Studies, Vol. 127. Springer-Verlag, Berlin, pp. 1–38.

Taylor Jr., G.E., Hanson, P.J., 1992. Forest trees and tropospheric ozone: Role of canopy deposition and leaf uptake in developing exposure-response relationships. Agric. Ecosyst. Environ. 42, 255–273.

Taylor Jr., G.E., Johnson, D.W., Andersen, C.P., 1994. Air pollution and forest ecosystems: A regional to global perspective. Ecol. Applic. 4, 662–689.

Temple, P.J., Riechers, G.H., 1995. Nitrogen allocation in ponderosa pine seedlings exposed to interacting ozone and drought stresses. New Phytol. 130, 97–104.

Tjoelker, M.G., Volin, J.C., Oleksyn, J., Reich, P.B., 1993. Light environment alters response to ozone stress in seedlings of *Acer saccharum* Marsh. and hybrid *Populus* L. I. In situ net photosynthesis, dark respiration and growth. New Phytol. 124, 627–636.

Tjoelker, M.G., Volin, J.C., Oleksyn, J., Reich, P.B., 1995. Interaction of ozone pollution and light effects on photosynthesis in a forest canopy experiment. Plant Cell Environ. 18, 895–905.

Tobiessen, P., 1982. Dark opening of stomata in successional trees. Oecologia 52, 356–359.

Tyree, M.T., Ewers, F.W., 1991. The hydraulic architecture of trees and other woody plants. New Phytol. 119, 345–360.

Volin, J.C., Reich, P.B., Givnish, T.J., 1998. Elevated carbon dioxide ameliorates the effects of ozone on photosynthesis and growth Species respond similarly regardless of photosynthetic pathway or plant functional group. New Phytol. 138, 315–325.

Volin, J.C., Tjoelker, M.G., Oleksyn, J., Reich, P.B., 1993. Light environmental alters responses to ozone stress in seedlings of *Acer saccharum* Marsh. and hybrid Populus L. II Diagnostic gas exchange and leaf chemistry. New Phytol. 124, 637–646.

Waring, R.H., 1993. How ecophysiologists can help scale from leaves to landscapes. In: Ehleringer, J.R., Field, C.B. (Eds.), Scaling Physiological Processes, Leaf to Globe. Academic Press, San Diego, CA, pp. 159–166.

Waring, R.H., Schlesinger, W.H., 1985. Forest Ecosystems, Concepts and Management. Academic Press, San Diego, CA, p. 340.

Waring, R.H., Silvester, W.B., 1993. Variation in foliar ^{13}C values within the crowns of *Pinus radiata* trees. Tree Physiol. 14, 1203–1213.

Weinstein, D.A., Samuelson, L.J., Arthur, M.A., 1998. Comparison of the response of red oak (*Quercus rubra*) seedlings and mature trees to ozone exposure using simulation modeling. Environ. Pollut. 102, 307–320.

Wieser, G., 2002. Effects of ozone on conifers in the timberline ecotone. In: Matyseek, R., Huttunen, S. (Eds.), Trends in European Forest Tree Physiology Research (in press).

Wieser, G., Havranek, W.M., 1993. Ozone uptake in the sun and shade crown of spruce: Quantifying the physiological effects of ozone exposure. Trees 7, 227–232.

Wiskich, J.T., Dry, I.B., 1985. The tricarboxylic acid cycle in plant mitochondria: Its operation and regulation. In: Douce, R., Day, D.A. (Eds.), Higher Plant Cell Respiration. In: Encyclopedia of Plant Physiology, New Series, Vol. 8. Springer-Verlag, Berlin, pp. 281–313.

Yoder, B.J., Ryan, M.G., Waring, R.H., Schoettle, A.W., Kaufmann, M.R., 1994. Evidence of reduced photosynthetic rates in old trees. Forest Sci. 40, 513–527.

Zeuthen, J., Mikkelsen, T.N., Paludan-Muller, G., Ro-Poulsen, H., 1997. Effects of increased UV-B radiation and elevated levels of trophospheric ozone on physiological processes in European beech (*Fagus sylvatica*). Physiol. Plant. 100, 281–290.

Zimmermann, M.H., 1983. Xylem Structure and the Ascent of Sap. Springer, New York.

Air Pollution, Global Change and Forests in the New Millennium
D.F. Karnosky et al., editors

175

Chapter 7

Simulating the growth response of aspen to elevated ozone: A mechanistic approach from leaf-level photosynthesis to complex architecture

M.J. Martin*

Natural Resources Research Institute, University of Minnesota, 5013 Miller Trunk Highway, Duluth, MN 55811, USA

G.E. Host

Natural Resources Research Institute, University of Minnesota, 5013 Miller Trunk Highway, Duluth, MN 55811, USA

K.E. Lenz

Department of Mathematics and Statistics, University of Minnesota, University Drive, Duluth, MN 55812, USA

J.G. Isebrands

Environmental Forestry Consultants, LLC, P.O. Box 54, E7323 Hwy 54, New London, WI 54501, USA

Abstract

Predicting ozone-induced reduction of carbon sequestration of forests under elevated tropospheric ozone concentrations requires robust mechanistic leaf-level models, scaled up to whole tree and stand level. As ozone effects depend on genotype, the ability to predict these effects on forest carbon cycling via competitive response between genotypes will also be required. This study tests a process-based model that predicts the relative effects of ozone on the photosynthetic rate and growth of an ozone-sensitive aspen clone, as a first step in simulating the competitive response of genotypes to atmospheric and climate change.

The resulting composite model simulated the relative above ground growth response of ozone-sensitive aspen clone 259 exposed to square wave variation in ozone concentration. This included a greater effect on stem diameter than on stem height, earlier leaf abscission, and reduced stem and leaf dry matter production at the end of the growing season. Further development of the model to reduce predictive uncertainty is discussed.

*Corresponding author.

DOI:10.1016/S1474-8177(03)03007-9

1. Introduction

Atmospheric and climate changes are affecting the productivity of North American forests (Melillo et al., 1996; McLaughlin and Percy, 1999; Kirschbaum, 2000), and young to middle-aged forests are playing an increasingly important role in long-term carbon sequestration (Melillo et al., 1996). However, due to the complexity and relatively large spatial and temporal scale of forest ecosystems and ecosystem processes, predictions of forest response to future environmental change will require modelling techniques based on the mechanisms that underlie tree growth and dry matter production. The most fundamental process of vegetative growth is photosynthesis, which determines the maximum potential rate of carbon uptake by vegetation (Long, 1994). The photosynthetic response to global environmental changes such as increased concentrations of ozone and carbon dioxide depend on genotype. Therefore, forest models need to be able to predict competitive interactions within a forest stand to determine how changes in the composition and abundance of genotypes might alter potential carbon sequestration (Constable and Friend, 2000).

Atmospheric concentrations of CO_2, the substrate of photosynthesis, have increased from pre-industrial levels of approximately 280 μmol mol^{-1} to current values of more than 360 μmol mol^{-1}, and continue to rise at an estimated rate of 1.5 to 2 μmol mol^{-1} per year (IPCC, 1996). A meta-analysis, based on experimental results on young trees, indicates tree growth rates may be expected to increase under elevated CO_2 (Norby et al., 1999). However, the potential increase in forest biomass production and associated carbon dioxide sequestration resulting from increased growth rates may be moderated by the interacting effects of other changing conditions, such as temperature (Long, 1991), soil nutrient deficiency (Stitt and Krapp, 1999) and increasing concentrations of phytotoxic ozone (Long, 1994; McKee et al., 1995; Schmieden and Wild, 1995).

Tropospheric concentrations of ozone have risen at an estimated annual rate of 1% in the northern hemisphere for the last few decades (PORG, 1993). Increasing ozone concentrations threaten to reduce potential forest dry matter production as background concentrations are already close to harmful levels (PORG, 1993) and ozone, formed by a complex suite of reactions between hydrocarbons and nitrogen oxides in sunlight, may be transported over long distances to relatively pristine environments (Chameides et al., 1994).

Although the phytotoxic effects of highly reactive ozone have been reported for several decades (Heath, 1987; Krupa and Manning, 1988), the complex suite of reactions within the leaf that convert ozone into reactive oxygen intermediaries (ROI), together with the resultant production of a variety of possible protective scavenging mechanisms, has made it difficult to establish the biochemical mechanisms of ozone damage (Heath, 1994; Pell et al., 1994, 1999).

Although concentrations of ambient ozone close to background levels of 20 to 30 $nmol\,m^{-2}\,s^{-1}$ have no significant effects on photosynthesis, concentrations of 60 $nmol\,m^{-2}\,s^{-1}$ and above may impair photosynthetic functioning within the mesophyll (McKee et al., 1995; Farage, 1996). The primary effect of chronic and acute ozone exposure on photosynthesis in wheat is a reduction in the maximum capacity of carboxylation (V_{cmax}), thereby inducing stomatal closure via an increase in intercellular [CO_2] (Farage et al., 1991; Farage and Long, 1999; McKee et al., 1995). Indeed, stomatal closure observed in wheat under the acute ozone experiments of Farage et al. (1991) could be explained solely by the decrease in V_{cmax}, via the predicted change in intercellular CO_2 concentration (Martin et al., 2000). Ozone also reduces the photosynthetic capacity and growth of the ozone-sensitive clone 259 of trembling aspen (*Populus tremuloides*) (Coleman et al., 1995a; Kull et al., 1996; Karnosky et al., 1996, 1999). Other symptoms of ozone damage reported for aspen include black bifacial necrosis, and upper leaf surface black or red stipple, a loss of chlorophyll, accelerated leaf senescence associated with earlier leaf abscission, changes in carbon allocation patterns and reduction in productivity and growth (Coleman et al., 1995b; Kull et al., 1996; Karnosky et al., 1999; Yun and Laurence, 1999).

By scaling a leaf-level model to the whole tree, it is possible to investigate whether the ozone-induced reduction in photosynthetic rates of leaves of varying maturity might be enough to account for the observed change in carbon allocated to roots, stems and leaves, and subsequent changes in measured growth parameters.

Further scaling from individual tree to simulate a 'patch' of trees composed of different genotypes may be used to investigate the interactive effects of elevated [CO_2] and [O_3] on forest growth and composition, under both limiting and non-limiting conditions, via competitive and species distribution response (Host et al., 1996; Kirschbaum, 2000). Future work to produce robust models of such complexity will require the cooperation of interdisciplinary teams of experimentalists, physiologists, modellers and programmers (Isebrands and Burk, 1992). Each stage of the scaling process will require model testing and validation (Jarvis, 1995).

The focus of this paper is to test the scaling of an ozone model from the leaf level to the whole tree level for the ozone-sensitive aspen clone 259. This is a first step in building a model to simulate the effects of atmospheric and climate change on carbon sequestration potential of a forest stand via competitive effects within a stand of trees. To this end, a model developed to predict the effects of ozone on photosynthesis and stomatal conductance has been incorporated into the functional-structural tree growth model ECOPHYS (Rauscher et al., 1990; Host et al., 1996; Isebrands et al., 2000; ECOPHYS web site: http://www.nrri.umn.edu/ecophys).

The process-based ozone model originally used the linear relationship between the maximum capacity of in vivo carboxylation (V_{cmax}) and effective ozone dose, based on an accumulated dose above a threshold flux of ozone entering the leaf, to calculate the relative effect of ozone on wheat leaf photosynthesis (Martin et al., 2000). The model was first developed within WIMOVAC (Windows Intuitive Model of Vegetation response to Atmosphere and Climate Change) (Humphries and Long, 1995) and determined net CO_2 assimilation rates via coupled mechanistic, biochemical model equations of photosynthesis (Farquahar et al., 1980; von Caemmerer and Farquahar, 1981) and stomatal conductance (Ball et al., 1987; adapted by Harley et al., 1992).

Incorporation of the ozone model and the coupled photosynthesis and stomatal conductance model equations into the functional-structural tree growth model ECOPHYS integrates physiological processes with the architectural attributes of a tree canopy and incorporates the numerous feedback mechanisms that occur among individual plant components (Host et al., 1990a, 1999). Simulations from whole systems models, that include detailed component process sub-models, have the potential to be extrapolated beyond the conditions used for model development (Reynolds et al., 1993).

The objectives of the present study are threefold. First, to scale the ozone effect model to the whole tree level by incorporating the recently developed process-based model to predict the effects of ozone on photosynthesis into the object-oriented functional and structural tree growth model ECOPHYS (Isebrands et al., 2000). Second, to parameterize the model to simulate the growth of aspen clone 259 exposed to episodic variation in ozone during 1991, based on data of Karnosky et al. (1996). And, finally, to test the tree growth simulation against field results under square-wave exposure to ozone. The extent to which observed relative ozone response of aspen growth and whole tree carbon allocation can be attributed to effects on photosynthesis is discussed. Further model development, planned to reduce predictive uncertainty, is summarized.

2. Methods

2.1. ECOPHYS-ozone model

ECOPHYS was developed to simulate the growth of hybrid poplar and aspen clones over multiple years under interacting environmental stresses by integrating information on canopy architecture, leaf light interception and photosynthetic rate, root distribution and a dynamic process model of carbon allocation. Environmental inputs to the model include latitude, solar radiation, air temperature, relative humidity and ambient concentrations of both carbon dioxide and ozone. Individual leaves and root segments are defined in a three-dimensional coordinate system, allowing detailed calculations of hourly light interception for each leaf, and differential uptake of water and nutrients by roots in a heterogeneous soil environment. For this study, ECOPHYS is used to simulate the growth of a single tree in the first establishment year, on an hourly time-step, assuming optimal water and nutrient availability.

2.2. Shading

Direct beam solar radiation intercepted by each leaf in a given hour is calculated within ECOPHYS from solar altitude and azimuth, according to leaf angle and leaf position relative to the sun, and shading from other leaves (Isebrands et al., 2000). Each upper-leaf surface is represented by a 2-dimensional quadrilateral with bilateral symmetry (kite shape), with general leaf shape, for example, lance, flaring, square, or stubby, defined by parameters listed in the genetic library.

The shaded fraction of each leaf is calculated at each hourly time step. The four vertices of each leaf are located within an $x–y–z$ coordinate system, where the z axis is parallel to a vector from the earth to the sun. The leaf with the smallest z value is above all the other leaves (Fig. 1). The four vertices of each leaf are projected onto a plane perpendicular to the sun vector, represented by a 600×600 pixel area termed the canvas (Wu, 1999).

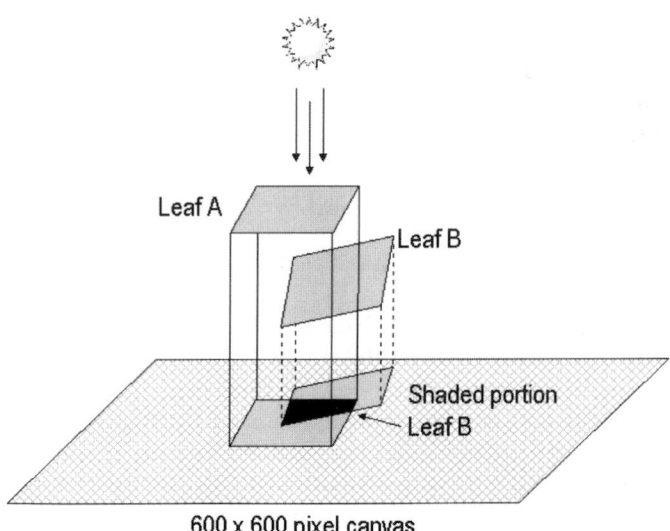

Figure 1. The relationship between the radiation vector, leaves, and the 600×600 pixel canvas (adapted from Zhao, 2000a).

A list of all the leaves is sorted in ascending order according to leaf center point z-values. If a pixel is shaded, the corresponding element of the 600×600 array is set to 1, otherwise, it is 0. Starting from the first leaf in the list, the four vertices of each leaf are projected onto the canvas. In the projected region for the leaf, the number of shaded pixels is compared to the total number of pixels covered by the leaf's projection to determine the fraction of the leaf that is shaded. Then all the pixels covered by the leaf's projection are marked as shaded for calculating the shaded area of the next leaf (Zhao, 2000a).

2.3. Photosynthesis

Photosynthetic rate is calculated in response to hourly inputs of photon flux density, air temperature, relative humidity and ambient concentrations of CO_2 and O_3 for both the sunlit and shaded portions of the leaf. Given that the simulations represent only the first year of growth, and that there is no mutual shading from adjacent trees, foliar nitrogen levels are assumed to be optimum for the purposes of this study. For each leaf, photosynthetic rates for sunlit and shaded areas are multiplied, respectively, by the sunlit and the shaded leaf area and summed to determine total photosynthate production for the leaf at each hourly time step. For this study, ECOPHYS uses the coupled photosynthesis and stomatal conductance model developed for WIMOVAC (Humphries and Long, 1995), parameterized for aspen and modified to account for leaf age and O_3 damage.

Within WIMOVAC, net CO_2 assimilation rates are calculated by combining the well-accepted mechanistic biochemical model equations for photosynthesis developed by Farquahar et al. (1980) and von Caemmerer and Farquahar (1981) and adapted to account for phosphorylation-limited rate (Sharkey, 1985), with the phenomenological model equations of stomatal conductance (Ball et al., 1987; adapted by Harley et al., 1992). Values for net CO_2 assimilation and stomatal conductance are solved through an iterative process.

Although the same equations were used in ECOPHYS to calculate assimilation rates, the equations were solved analytically. The analytical solution (Zhao, 2000b) was derived as in Baldocchi (1994), although mathematical expressions differ somewhat from those in Baldocchi (1994). In particular, a phosphate limitation on photosynthetic rate was included. As in Baldocchi (1994), the system of equations is solved analytically by expressing photosynthetic rate as the minimum among roots of quadratic and cubic polynomials. The expressions for the roots are complicated, but are evaluated quickly by computer. To verify that the analytical method was implemented correctly, it was extensively tested against the iterative method (Zhao, 2000b). The two were found to coincide both at the leaf-level and at the whole-tree level when simulations were run under normal ranges of environmental conditions.

The analytical calculations require a fixed amount of time whereas the time taken by the iterative method depends on operating conditions. Under typical environmental conditions the iterative method was found to be approximately as fast as the analytical method. However, for some environmental conditions many iterations are needed and outside certain ranges the iterative method may not converge. Also, performance of the iterative method depends on parameter values that vary with plant species and ozone damage.

2.4. Ozone model

The ozone model incorporated into ECOPHYS was adapted from the process-based model developed to predict the photosynthetic response of wheat to acute ozone exposure, based on the data from Farage et al. (1991) (Martin et al., 2000). Model equations are listed in Table 1 and symbols defined in Table 2. The underlying mechanism of the ozone-inhibited photosynthesis model is that damage occurs once the protective scavenging detoxification system is overloaded, above a critical flux of ozone entering the leaf (Heath, 1994). The model uses the linear relationship between the relative reduction in V_{cmax} and the 'effective ozone dose' (F'_{O_3eff}), that is, the accumulated amount of ozone entering the leaf above the threshold flux ($F_{O_3(0)}$), to calculate the effect of ozone exposure on leaf photosynthesis. The threshold flux is related to the maximum capacity of scavenging protective metabolism within the leaf. The slope coefficient of the linear function, K_z, reflects the sensitivity of the photosynthetic apparatus to ozone above the threshold flux. This occurs once the maximum rate of protective metabolism against active oxygen radicals has been exceeded. This linear relationship is then used to determine the depen-

Table 1. Model equations to simulate the effects of ozone exposure on rates of photosynthesis in aspen clone 259

	Number of equation
$F'_{O_3eff} = \int_0^t ([O_3]g_z) - F_{O_3(0)} \, dt$	(1)
$g_z = \dfrac{g_s}{1.67}$	(2)
$g_s = g_{(0)} + g_{(1)} \dfrac{AR_H}{C_a}$	(3)
$\Delta V_{cmax} = K_z F'_{O_3eff}$	(4)
$\Delta J_{max} = K_z F'_{O_3eff}$	(5)

Table 2. Definition of symbols and parameter values of the combined ozone/photosynthesis model

Term	Value	Units	Definition and source
A		$\mu\mathrm{mol\,m^{-2}\,s^{-1}}$	Net leaf rate of CO_2 uptake per unit leaf area
C_a	350	$\mu\mathrm{mol\,mol^{-1}}$	Atmospheric concentration of CO_2
$F'_{O_3\mathrm{eff}}$		$\mathrm{mmol\,m^{-2}}$	Effective ozone dose
$F_{O_3(0)}$	9	$\mathrm{nmol\,m^{-2}\,s^{-1}}$	Threshold flux of ozone entering the leaf
g_s		$\mathrm{mmol\,m^{-2}\,s^{-1}}$	Stomatal conductance to water
g_z		$\mathrm{mmol\,m^{-2}\,s^{-1}}$	Stomatal conductance to ozone (Laisk et al., 1989)
g_0	81.1	dimensionless	Minimum stomatal conductance to water when $A = 0$ at light compensation point
g_1	9.58	dimensionless	Empirical coefficient of stomatal conductance sensitivity to A, C_a, and R_H
$J_{\max}$	162	$\mu\mathrm{mol\,m^{-2}\,s^{-1}}$	Light saturated rate of potential rate of electron transport
K_z	8.35	dimensionless	Empirical coefficient of sensitivity of V_{cmax} to $F'_{O_3\mathrm{eff}}$
R_H		%	Relative humidity
V_{cmax}	101	$\mu\mathrm{mol\,m^{-2}\,s^{-1}}$	Maximum RuBP saturated rate of carboxylation

dence of ozone-induced stomatal closure on V_{cmax}, via intercellular $[CO_2]$, (c_i).

Thus, the wheat ozone model predicts stomatal closure caused by ozone exposure, via its effect on V_{cmax} (Eqs. (1)–(4), Table 1), where g_z is the stomatal conductance to ozone and g_s is the stomatal conductance to water, calculated by Eq. (3), using the method of Ball et al. (1987), as adapted by Harley et al. (1992). Ozone enters both wheat and aspen leaves via the stomata and, like wheat, the V_{cmax} of ozone-sensitive aspen (clone 259) is reduced by exposure to ozone (Kull et al., 1996). However, ozone was also found to reduce the light-saturated rate of electron transport, $J_{\max}$, in leaves of clone 259 (Eq. (5), Table 1) (Kull et al., 1996). Evidence of this additional effect of high doses of ozone on photosynthesis has also been found in other woody species, such as oak (Farage and Long, 1995). The proposed leaf-level model of ozone effects on aspen is outlined in Fig. 2.

Over extended periods of ozone exposure visible ozone symptoms occur. For this study, it is assumed that any loss of green leaf area due to necrosis and stippling observed in aspen leaves results from the loss of photosynthetic capacity and is accounted for as a loss of photosynthetic capacity in the model.

2.5. Carbon allocation and growth

Translocation of photosynthate to various parts of the tree is achieved according to detailed carbon allocation matrices based on carbon tracing measure-

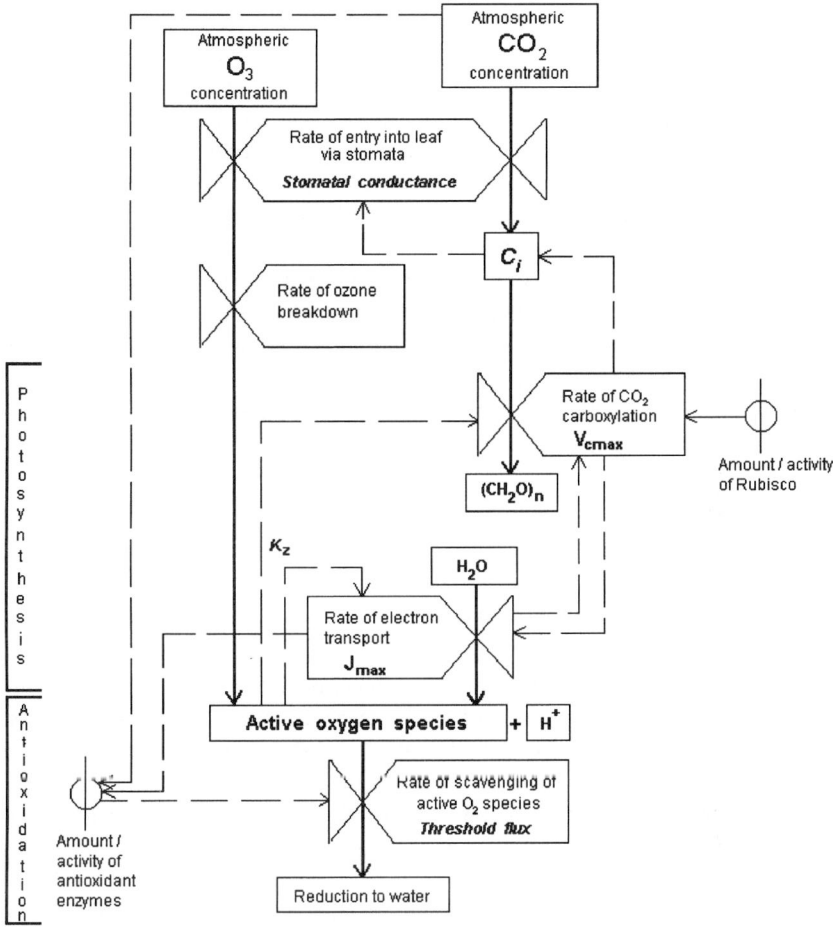

Figure 2. Schematic of the ozone model for aspen, including effect on J_{max}.

ments (Rauscher et al., 1990; Host et al., 1990a, 1996). Each leaf in the canopy has a carbon transport pattern based on LPI (leaf plastochron index) (Larson and Isebrands, 1971). Newly emerged leaves (LPI 1–4) retain all of their photosynthate, while leaves in the expanding leaf zone (LPI 5–9) transport most of their photosynthate upward to leaves and stem internodes, and mature leaves (LPI > 10) transport most of their photosynthate downward to stem internodes, trunk, and root. The carbon transport coefficients for active, unstressed growth were determined by ^{14}C tracer studies in controlled environments and in the field (Larson, 1977; Isebrands and Nelson, 1983; Dickson, 1986; Host et al., 1990a).

After bud set, carbon allocation patterns change as active growth gives way to preparation for winter. This is simulated in ECOPHYS by a linear interpolation of active growth carbon allocation matrices to end of season storage carbon allocation matrices (Host et al., 1990b). During the ten days centered around bud set, linear interpolation is used to gradually decrease the upward leaf and internode transportation coefficients from 100% to 20% of their active growth values, with the corresponding increase in carbon allocated to lower stem and root tissues (Zang, 1999).

SOILPSI simulates the movement of water in a 3-dimensional space as a function of soil water potential, and allows for differential uptake by a heterogeneous 3-dimensional root growth submodel (Theseira et al., 2003). Root architecture is based on a relaxed fractal algorithm for carbon allocation coupled with genetically determined branching rules. Roots are tracked by position and order using a tree structure (Host et al., 1996). The growth process is simulated by incremental increases in photosynthate allocated to the leaf, internode and root sinks on an hourly basis, accounting for temperature effects and respiratory losses.

2.6. Growing season leaf-drop algorithm

Increasing levels of ozone exposure can increase leaf senescence during the growing season. The simulation of leaf senescence during the growing season is based on the productivity of each leaf. For each leaf the net photosynthate per unit leaf area remaining after respiration and reallocation is averaged over ten days. If this average falls below a threshold value, then with a certain probability, the leaf drops. Full details of the leaf-drop algorithm are presented in Appendix A.

ECOPHYS model outputs are: leaf count, leaf area (cm^2), stem height (cm), stem diameter (cm), leaf dry weight (g), stem dry weight (g), cumulative whole-tree dry matter production (g), root length (cm) and dead leaf matter production (g).

2.7. Model development and parameterisation

2.7.1. Clonal library

The clone library file of ECOPHYS contains values for genetic input parameters, originally determined for hybrid poplar (*Populus × euramericana*) 'Eugenei', but altered for this work to represent aspen clone 259 (Host et al., 1990b, 1996). The clonal library parameter values are listed in Table 3.

Table 3. Clonal library parameter values for aspen clone 259

Parameter description	Value (units)
Bud break date	128 (Julian day)
Bud set date	233 (Julian day)
Leaf senescence date	268 (Julian day)
Leaf initiation rate	36 (Number of days)
Leaf senescence rate	3 (Number of days)
Initial expanding leaf zone specific leaf area	0.2222 $(cm^2\,mg^{-1})$
Bud set expanding leaf zone specific leaf area	0.159 $(cm^2\,mg^{-1})$
Internode specific gravity	0.46 $(g\,cm^{-3})$
Ratio of leaf width to leaf length	1
Growth respiration rate	0.250 (ratio)
Maintenance respiration rate	0.015 (day^{-1})

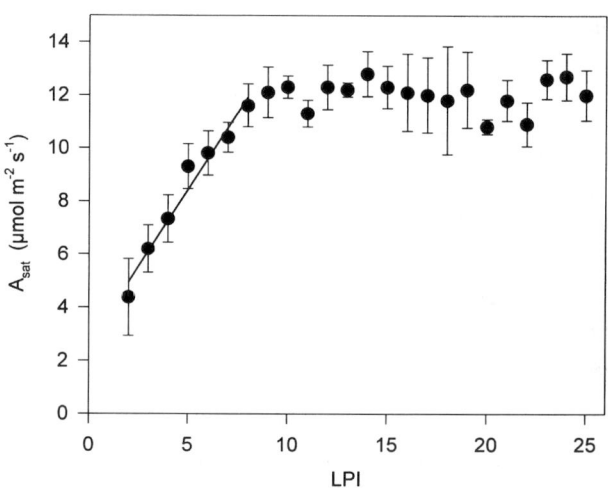

Figure 3. Relationship between light-saturated rate of photosynthesis and leaf plastochron index for aspen clone 259 (after Kull et al., 1996).

2.7.2. Leaf maturity class effects on photosynthesis

Kull et al. (1996) found that for leaves of aspen clone 259, light-saturated rates of photosynthesis reached maximum values only above LPI 11, and that ozone had no effect on leaves of LPI less than 11 (Fig. 3). Therefore, the leaf maturity class effects on photosynthetic rate and ozone response were adapted in ECO-PHYS according to the reported data, whereby the maximum photosynthetic rates for leaves of LPI less than 11 was linearly related to LPI by Eq. (6), and the ozone model was not used for leaves with an LPI value of less than 11 (Fig. 3).

$$A_{\text{ratiotoLPI}} = 1.161\,\text{LPI} + 2.625 \tag{6}$$

2.7.3. Model input and parameter values

Air temperature and relative humidity values recorded in Karnosky's 1991 experiments were used as inputs to the simulation (Karnosky et al., 1996). Photon flux density was simulated by model equations using the latitude 46°N, and ambient $[CO_2]$ was set at 350 μmol m^{-2} s^{-1}. Input variation of ozone concentrations were those values targeted by the experiments of Karnosky et al. (1996). Ozone treatments simulated consisted of charcoal filtered (CF) and an episodic 'two times' ozone concentration (2X), based on a doubling of hourly ambient ozone concentrations measured in Michigan's Lower Peninsula (Hoggsett et al., 1988).

Carboxylation efficiency and Ac_i curve data for aspen clone 259 grown in open-top chambers under various ozone treatments (Kull et al., 1996) were used to determine mean V_{cmax} and J_{max} values, and to provide a first estimate of ozone parameter values. The value of the ozone sensitivity coefficient (K_z) was adjusted from that found using the data of Kull et al. (1996) to reflect the relative reduction in stem dry matter production observed between 2X and CF treatments measured in experiments conducted in 1991 by Karnosky et al. (1996). Parameter values are listed in Table 2. The model would then be tested against separate growth data of aspen grown in 1991 under the square wave ozone exposure regime used by Karnosky et al. (1996).

2.8. Model testing

To test the model, the results of an ECOPHYS simulation of growth response of aspen clone 259 grown in open-top chambers in 1991 by Karnosky et al. (1996) under square wave ozone profile were compared with the observed changes in growth parameters, including stem diameter, stem height, and total leaf area. The square-wave (SQ) variation of ozone concentration consisted of exposure to $[O_3]$ of 100 nmol mol^{-1} for 6 hours a day, for 4 days a week, for a total of 12 weeks.

3. Results

Model simulations of the relative growth response of ozone-sensitive aspen clone 259 under ozone exposure, compared with the charcoal filtered treatment, reflect the general trends reported by Karnosky and co-workers. In particular, the model showed that:

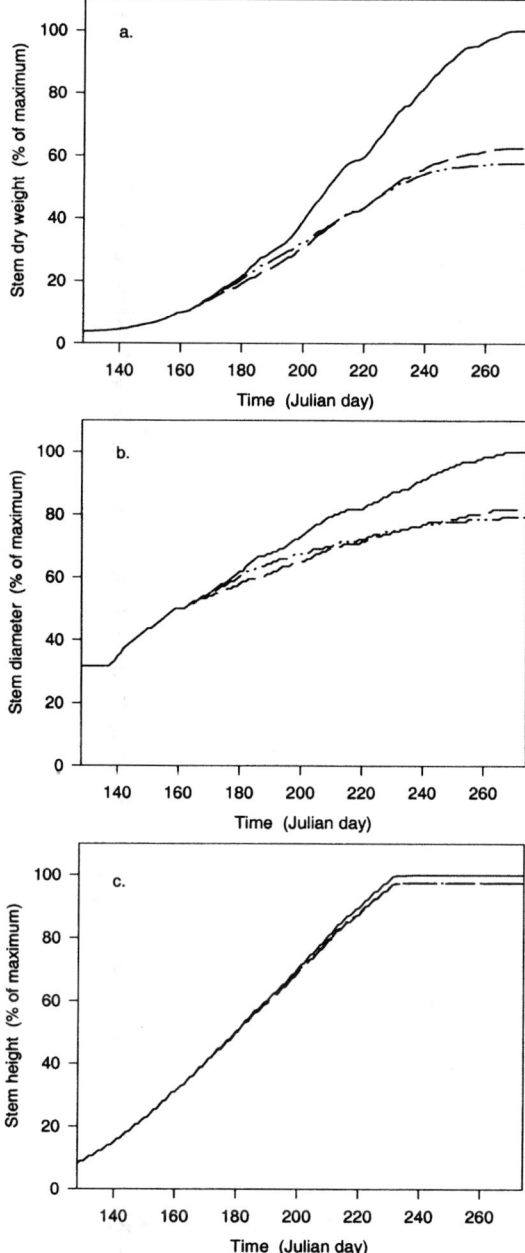

Figure 4. Simulations of aspen clone 259 stem (a) dry matter production, (b) diameter and (c) height under exposure to different ozone treatments: CF (—), 2X (– –) and SQ (–··–··).

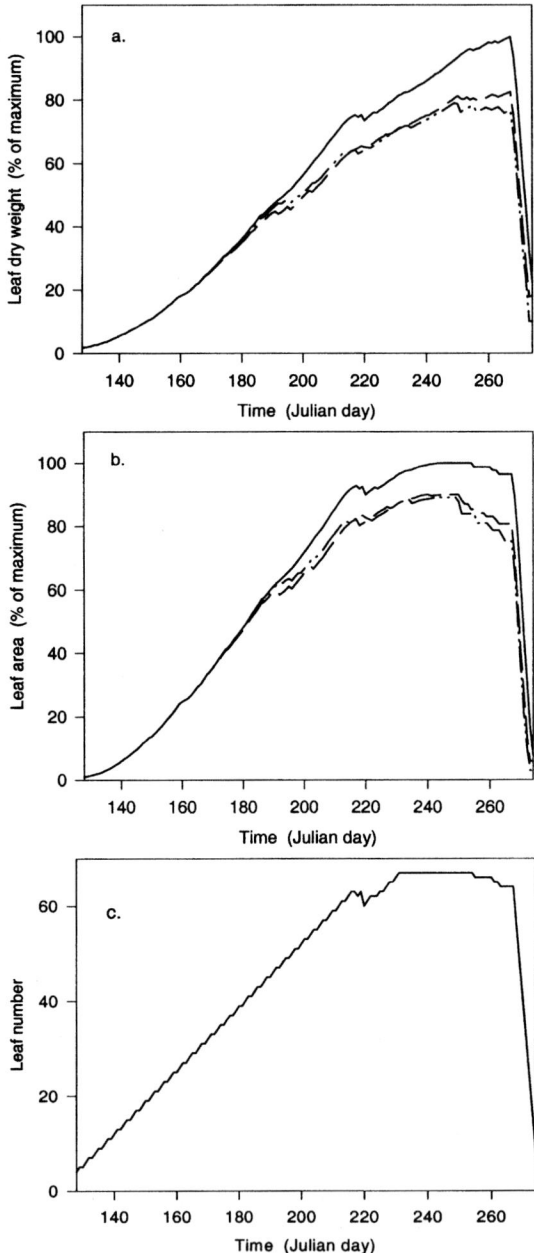

Figure 5. Simulations of aspen clone 259 leaf (a) dry matter production, (b) area and (c) number of leaves initiated, under exposure to different ozone treatments: CF (—), 2X (– –) and SQ (–··–··).

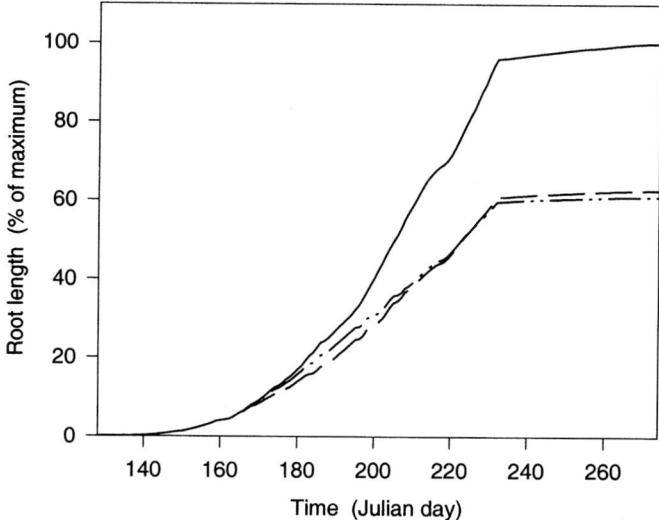

Figure 6. Simulations of aspen clone 259 root length under exposure to different ozone treatments, CF (—), 2X (– –) and SQ (– ·· – ··).

(1) Stem dry matter production and stem diameter are dramatically reduced by ozone, while the effect on stem height was small (Fig. 4(a), (b) and (c));

(2) Ozone-induced earlier leaf abscission dramatically reduces both leaf dry matter production and retained leaf area, but with little or no effect on the number of leaves initiated (Fig. 5(a), (b) and (c));

(3) Decline in root growth is one of the most sensitive indicators of chronic ozone exposure (Fig. 6) (Karnosky et al., 1996).

The predicted reduction in stem dry matter production was within 2% of that measured by Karnosky et al. (1996). A drop of 40% in total root length was simulated (Fig. 6), whereas Karnosky et al. (1996) measured a 55% reduction in root dry matter production, with exposure to 2X. The simulated percentage reduction in retained leaf biomass under 2X ozone treatment depended on the date (Fig. 5(a)) and varied between 17% and 30%. The measured reduction was 36%.

The square wave ozone profile was used to test the model. Simulated responses to square wave treatment are of a similar order to those of the 2X treatment, even though the total ozone supplied for the whole experimental period of the 2X treatment was calculated by Karnosky et al. (1996) to be nearly twice as high as that of the SQ treatment. This agrees with the similar response between square wave and 2X treatments measured by Karnosky et al. (1996), although slightly lower reductions in response were measured under square wave exposure, simulated as slightly greater reductions by ECOPHYS (Figs. 4–6). Once again, only a 40% drop in root length was simulated (Fig. 6), although a 53% reduction in root biomass was measured by Karnosky et al. (1996) at the end of the growing season. Leaf dry matter production was predicted to drop by between 20 and 37%, depending on date, compared with a measured reduction of 32% under ozone regime SQ.

4. Discussion

The model presented here combines validated leaf-level sub-models within a functional and a structural tree growth model to increase our understanding of scaling processes from leaf to whole tree level. The model incorporates a process-based model to predict photosynthetic response to ozone (Martin et al., 2000), the combined stomatal conductance (Ball et al., 1987; Harley et al., 1992) and mechanistic models of CO_2 assimilation (Farquahar et al., 1980; von Caemmerer and Farquahar, 1981; Sharkey, 1985), within a detailed light interception model of tree growth (Rauscher et al., 1990; Host et al., 1990a, 1996, 1999; Isebrands et al., 2000). Working on an hourly time step, this composite, mechanistic approach allows this revised version of ECOPHYS to simulate the observed relative response of above ground growth of ozone-sensitive aspen clone 259 to one season's exposure to ozone. Model simulations concur with observations that ozone affects aspen growth parameters differentially. Stem dry matter production and stem diameter are reduced dramatically, whereas ozone has a relatively small effect on stem height (Fig. 4). Also, while both leaf biomass and leaf area are sensitive to ozone, the number of leaves initiated is unaffected (Fig. 5). These findings support the scaling method employed here, whereby process-based models of ozone effects and photosynthesis at the leaf level are incorporated into a structural-functional tree growth model to simulate the observed effects on above ground dry matter production of an ozone-sensitive aspen clone.

A direct comparison between simulated and observed effects on aspen clone root growth is not possible at this stage, as the present version of ECOPHYS simulates root growth in terms of root length, and not root dry matter production, as measured by Karnosky et al. (1996). A linkage between the 3-dimensional root growth model of ECOPHYS and the SOILPSI water redistribution model (Theseira et al., 2003) is under development. Meanwhile, the predicted 40% reduction in root length may be interpreted as a reasonable simulation of the measured 55% reduction in root biomass when uncertainties due to the practical problems of root biomass measurement, possible sink restrictions by growth in pots and growth chamber effects are taken into account.

When tested against independent data for growth response under the square wave [O_3] profile, ECOPHYS simulates a similar magnitude of reduction in ozone-sensitive growth parameters to those measured under the 2X episodic exposure treatment, despite the difference between ozone regimes. The total seasonal ozone exposure under the square wave ozone profile is nearly twice that under the 2X episodic ozone treatment, and the pattern of exposure and peak values of [O_3] also differed markedly. This ability of the model to predict the similar magnitude of effects of the two ozone treatments supports the threshold flux concept when scaling results to the whole tree. The threshold flux, below which no damage occurs, reflects the maximum capacity of the protective oxidant scavenging system. The hourly time step of leaf level calculations and input values for [O_3], [CO_2], temperature, light and relative humidity allow CO_2 assimilation rates, stomatal conductance and ozone effects to be simulated on a physiologically realistic time scale (for a more detailed discussion of appropriate time steps for model simulations, see Constable and Friend, 2000).

The aim of this modelling exercise was to test the scaling of the model for predicting ozone effects on whole tree photosynthesis, and thus to investigate how much the observed ozone-induced reduction in carbon allocated to leaves, stems and roots may be accounted for solely by reduced photosynthetic rates. Coleman et al. (1995a) reported the ozone-induced reduction in total carbon translocated to sink tissue in aspen clone 259 to be controlled by reduced photosynthetic rates. ECOPHYS simulations that followed observed trends in the response of above ground dry matter production of ozone-sensitive aspen to one season's exposure to ozone support this. The model predicts root length to be dramatically reduced (Fig. 6). The effects of ozone on the more mature leaves of the lower canopy, which are exposed to ozone for longer periods than less mature leaves and export most of their assimilate down to the roots (Coleman et al., 1995a), helps to explain the ozone-sensitivity of root dry matter production.

Observed ozone-induced changes in allocation in other species have been attributed to other processes affecting allocation, such as phloem translocation, in addition to effects on CO_2 assimilation rates, as found in Pima cotton when exposed to acute ozone (see Grantz and Farrar, 1999). Grantz and Farrar suggest effects on phloem translocation may be mediated by oxidation damage to membranes within intercellular air spaces, such as the plasmalemma and plasmodesmata of mesophyll and phloem companion cells. Therefore, although this model can simulate effects on growth of ozone-sensitive aspen via effects on photosynthesis, additional routines describing the ozone effects on carbon allocation might be needed to simulate growth response in other species. Perhaps the relationship between the ozone threshold flux $F_{O_3(0)}$ and the maximum capacity of the scavenging potential could also be useful for calculating ozone-induced reduced rates of carbon translocation in species where ozone is known to damage membranes within intercellular spaces.

Many uncertainties are introduced when scaling models, both temporally, from hours and days to months, and spatially, from a leaf to a whole tree. Uncertainties associated with the change in temporal scale include possible recovery from ozone damage and adaptation, and whether the mechanism for adaptation might occur by natural selection over several generations, an adjustment from one season's growth to the next, or a more immediate response, for example, by an increase in the capacity of protective scavenging mechanisms. This study concentrates on just one year's growth of aspen clone 259 under conditions where water and nitrogen are assumed to be non-limiting. Although a possible change in maximum capacity of the oxidative scavenging system with time of ozone exposure and leaf position should be investigated, initial evidence suggests the production of the anti-oxidants chloroplastic and cytosolic Cu/Zn superoxide dismutase, does not increase in ozone-sensitive clone 259 following O_3 exposure, contrary to the increase reported in ozone tolerant clones (Karnosky et al., 1998).

Therefore, while the results of this study support the suggestion that relative changes in above ground carbon allocation under ozone can be accounted for solely by changes in photosynthetic rates at the leaf level, via differential leaf age effects, earlier leaf abscission and resulting changes in carbon allocation, uncertainties in simulating and measuring effects on root carbon gain need to be resolved, particularly when further scaling the model to predict ozone effects on a stand of trees, over multiple years. Not only will the effects of ozone over several successive years need to be taken into account, but also the effects of mutual shading of leaves within the stand, and competition for light, water and nitrogen effects on ozone uptake.

Understanding how much the response of forest biomass production and carbon sequestration capacity to climate and atmospheric change might be influenced by effects on competitive and species distribution within a forest stand (Kirschbaum, 2000), will require robust detailed models to compare simulated changes in biomass via effects on genotype composition, with simulations of biomass changes from more generic models of a forest canopy. Care must also be taken when using models based on seedling studies to try to predict effects on mature trees (Kolb et al., 1997). Future work is planned to scale the model further, from the individual tree to the level of a patch of trees of known age and of known genotype composition (Host et al., 1996). ECOPHYS is currently being adapted to enable model simulations of tree growth over several seasons. Not only will this require adaptation to incorporate the possible over-winter effects of ozone on bud break and branching, but it will also require adjustment to simulate season-to-season variability in branch budding under non-stressed conditions. To overcome limitations imposed by running such large simulations on processor time, the use of component object modelling (COM) is being developed for ECOPHYS, to allow simulations to be conducted across parallel computers (Isebrands et al., 2000).

Meanwhile, the results of this study support the method of incorporating process-based leaf level models into whole tree growth models to further understand processes at the whole tree level. The findings support Coleman's et al. (1995a) suggestion that the above ground growth response of ozone-sensitive aspen to ozone exposure can be accounted for by the direct effects of ozone on photosynthesis.

Acknowledgements

We would like to extend our thanks to M.D. Coleman, A. Sober, E. McDonald and A. Noormets for useful discussions on aspen data and W. Zhao for programming support. We acknowledge Steve Long and Steve Humphries for background information on the formulation of the model WIMOVAC. Work on the original ozone model was funded by the Natural Environmental Research Council, UK, under grant GT4/92/16/L; initial parameterization of the model for aspen clones was funded by a grant to the University of Essex, from Brookhaven National Laboratory. Development of the ECOPHYS project was funded jointly by the Computational Biology Program of the National Science Foundation, Grant No. DBI-972395, the Northern Global Change Program of the USDA Forest Service, and the U.S. Department of Energy under interagency agreement No. DE-A105-800R20763. Additional funding came from the NSF/DOE/NASA/USDA Joint Program on Terrestrial Ecology and Global Change through a cooperative agreement with Michigan Technological University. This is Contribution No. 296 of the Center for Water and the Environment, Natural Resources Research Institute, University of Minnesota, Duluth, MN.

Appendix A Growing season leaf-drop algorithm

For each leaf on any given day, d denotes gross photosynthate produced by the leaf by $P_g(d)$, leaf maintenance respiration by $R_m(d)$, photosynthate transported out of the leaf by $P_t(d)$, photosynthate received from other leaves by $P_r(d)$, leaf growth respiration by $R_g(d)$, and leaf area by $A_L(d)$. Net photosynthate produced per unit leaf area for a given leaf on day d, $P_{n/a}(d)$, is computed as follows:

$$P_{n/a}(d) = \frac{((P_g(d) - R_m(d)) - P_t(d) + P_r(d)) - R_g(d)}{A_L(d)}$$

The parentheses are necessary because $P_t(d)$ is based on $(P_g(d) - R_m(d))$ as well as on leaf age. Likewise, $R_g(d)$ is based on $((P_g(d) - R_m(d)) - P_t(d) + P_r(d))$ as well as on leaf age.

For each leaf for each day, beginning on the 10th day since the leaf's emergence, the net photosynthate per unit leaf area is averaged over the current day together with the previous nine days. That is, $Pa_{n/a}(d) = (P_{n/a}(d) + P_{n/a}(d - 1) + \cdots + P_{n/a}(d - 9))/10$ denotes the threshold value for $Pa_{n/a}(d)$, below which the leaf might drop, by τ.

Define $K = (Pa_{n/a}(d) - \tau)\delta$. Here τ is a measure of leaf starvation and δ is a scaling factor affecting the likelihood that a starving leaf will drop. In this study $\tau = 0$ and $\delta = 3$, although the values of τ and δ vary among genotypes. Let ξ denote a uniform random number such that $0 \leqslant \xi \leqslant 1$. Whether or not a given leaf drops on a given day, d is determined by the following algorithm:

If $K \geqslant 0$ the leaf remains, else if $\xi > 1 - e^K$ the leaf remains, else the leaf drops.

Here ξ is used to simulate uncertainty and variability due to unmodelled dynamics. The probability that a leaf drops when $Pa_{n/a}(d) < K$ is $1 - e^K$.

References

Baldocchi, D., 1994. An analytical solution for coupled leaf photosynthesis and stomatal conductance models. Tree Physiol. 14, 1069–1079.

Ball, J.T., Woodrow, I.E., Berry, J.A., 1987. A model predicting stomatal conductance and its contribution to the control of photosynthesis under different environmental conditions. In: Biggins, I. (Ed.), Proceedings of the International Congress on Photosynthesis. In: Progress in Photosynthesis Research, Vol. IV. Nihjoff, Dordrecht, pp. 221–224.

Chameides, W.L., Kasibhatla, P.S., Yienger, J., Levy II, H., 1994. Growth of continental-scale metro-agro-plexes, regional ozone pollution and world food production. Science 264, 74–77.

Coleman, M.D., Isebrands, J.G., Dickson, R.E., Karnosky, D.F., 1995a. Photosynthetic productivity of aspen clones varying in sensitivity to tropospheric ozone. Tree Physiol. 15, 585–592.

Coleman, M.D., Dickson, R.E., Isebrands, J.G., Karnosky, D.F., 1995b. Carbon allocation and partitioning in aspen clones varying in sensitivity to tropospheric ozone. Tree Physiol. 15, 593–604.

Constable, J.V.H., Friend, A.L., 2000. Suitability of process-based tree growth models for addressing tree response to climate change. Environ. Pollut. 110, 47–59.

Dickson, R.E., 1986. Carbon fixation and distribution in young *Populus* trees. In: Fujimori, T., Whitehead, D. (Eds.), Crown and Canopy Structure in Relation to Productivity Proceedings. Forestry and Forest Products Research Institute, Ibaraki, Japan, pp. 409–426.

Farage, P.K., Long, S.P., Lechner, E.G., Baker, N.R., 1991. The sequence of change within the photosynthetic apparatus of wheat following short-term exposure to ozone. Plant Physiol. 95, 529–535.

Farage, P.K., Long, S.P., 1995. An in vivo analysis of photosynthesis during short-term O_3 exposure in three contrasting species. Photosynth. Res. 43, 11–18.

Farage, P.K., 1996. The effect of ozone fumigation over one season on photosynthetic processes of *Quercus robur* seedlings. New Phytol. 134, 279–285.

Farage, P.K., Long, S.P., 1999. The effects of O_3 fumigation during leaf development on photosynthesis of wheat and peas: An in vivo analysis. Photosynth. Res. 59, 1–7.

Farquahar, G.D., von Caemmerer, S., Berry, J.A., 1980. A biochemical model of photosynthetic CO_2 assimilation in leaves of C_3 species. Planta 149, 78–90.

Grantz, D.A., Farrar, J.F., 1999. Acute exposure to ozone inhibits rapid carbon translocation from source leaves of Pima cotton. J. Exp. Bot. 50, 1253–1262.

Harley, P.C., Thomas, R.B., Reynolds, J.F., Strain, B.R., 1992. Modelling photosynthesis of cotton grown in elevated CO_2. Plant Cell Environ. 15, 271–282.

Heath, R.L., 1987. Biochemical mechanisms of pollutant stress. In: Heck, W.W., Taylor, O.C., Tingey, D.T. (Eds.), Assessment of Crop Loss from Air Pollutants. Elsevier Science, London, pp. 259–286.

Heath, R.L., 1994. Possible mechanisms for the inhibition of photosynthesis by ozone. Photosynth. Res. 39, 439–451.

Hoggsett, W.E., Tingey, D.T., Lee, E.H., 1988. Ozone exposure indices: concepts for development and evaluation of their uses. In: Heck, W.W., Taylor, O.C., Tingey, D.T. (Eds.), Assessment of Crop Loss from Air Pollutants. Elsevier Science, London, pp. 107–137.

Host, G.E., Rauscher, H.M., Isebrands, J.G., Michael, D.A., 1990a. Validation of photosynthate production in ECOPHYS, an ecophysiological growth process model of *Populus*. Tree Physiol. 7, 283–296.

Host, G.E., Rauscher, H.M., Isebrands, J.G., Dickmann, D.I., Dickson, R.E., Crow, T.R., Michael, D.A., 1990b. The microcomputer scientific software series #6: the ECOPHYS user's manual. USDA Forest Service General Technical Report, NC-141, 50 p.

Host, G.E., Isebrands, J.G., Theseira, G.W., Kiniry, J.R., Graham, R.L., 1996. Temporal and spatial scaling from individual trees to plantations: a modeling strategy. Biomass Bioenerg. 11, 233–243.

Host, G.E., Theseira, G.W., Heim, C., Isebrands, J.G., Graham, R., 1999. EPIC-ECOPHYS: A linkage of empirical and process models for simulating poplar plantation growth. In: Amaro, A., Tome, M (Eds.), Empirical and Process Models for Forest Tree and Stand Growth Simulation. Edicos Salamandra, pp. 419–429.

Humphries, S.W., Long, S.P., 1995. WIMOVAC: a software package for modelling the dynamics of plant leaf and canopy photosynthesis. CABIOS Comput. Applic. Biosci. 11, 361–371.

IPCC, 1996. Technical summary. In: Houghton, J.T., Meira Filho, L.G., Callander, B.A., Harris, N., Kattenberg, A., Maskell, K. (Eds.), IPCC: Climate Change 1995. Cambridge Univ. Press, Cambridge, UK, pp. 9–51.

Isebrands, J.G., Burk, T.E., 1992. Ecophysiology growth process models of short rotation forest crops. In: Mitchell, C.P. (Ed.), Ecophysiology of Short Rotation Forest Crops. Elsevier Applied Science, London, pp. 231–266.

Isebrands, J.G., Nelson, N.D., 1983. Distribution of [14C]-labeled photosynthates within intensively cultured *Populus* clones during the establishment year. Physiol. Plant. 59, 9–18.

Isebrands, J.G., Host, G.E., Lenz, K.E., Wu, G., Stech, H.W., 2000. Hierarchical, parallel computing strategies using Component Object Model for process modeling responses of forest plantations to interacting multiple stresses. In: Ceulemans, R.J.M., Veroustraete, F., Gond, V., Van Rensbergen, J.B.H.F. (Eds.), Forest Ecosystem Modeling, Upscaling, and Remote Sensing. SPB Academic Publishing, The Hague, The Netherlands, pp. 123–135.

Jarvis, P.J., 1995. Scaling processes and problems. Plant Cell Environ. 18, 1079–1089.

Karnosky, D.F., Gagnon, Z.E., Dickson, R.E., Coleman, M.D., Lee, E.H., Isebrands, J.G., 1996. Changes in growth, leaf abscission, and biomass associated with seasonal tropospheric ozone exposures of *Populus tremuloides* clones and seedlings. Can. J. Forest Res. 26, 23–37.

Karnosky, D.F., Podila, G.K., Gagnon, Z., Pechter, P., Akkapeddi, A., Sheng, Y., Riemenschneider, D.E., Coleman, M.D., Dickson, R.E., Isebrands, J.G., 1998. Genetic control of responses to interacting ozone and CO_2 in *Populus tremuloides*. Chemosphere 36, 807–812.

Karnosky, D.F., Mankovska, B., Percy, K., Dickson, R.E., Podila, G.K., Sober, J., Noormets, A., Hendry, G., Coleman, M.D., Kubiske, M., Pregitzer, K.S., Isebrands, J.G., 1999. Effects of tropospheric O_3 on trembling aspen and interaction with CO_2: results from an O_3-gradient and a FACE experiment. Water Air Soil Pollut. 116, 311–322.

Kirschbaum, M.U.F., 2000. Forest growth and species distribution in a changing climate. Tree Physiol. 20, 309–322.

Kolb, T.E., Fredericksen, T.S., Steiner, K.C., Skelly, J.M., 1997. Issues in scaling tree size and age responses to ozone: a review. Environ. Pollut. 98, 195–208.

Krupa, S.V., Manning, W.J., 1988. Atmospheric ozone: Formation and effects on vegetation. Environ. Pollut. 50, 101–137.

Kull, O., Sober, A., Coleman, M.D., Dickson, R.E., Isebrands, J.G., Gagnon, Z., Karnosky, D.F., 1996. Photosynthetic responses of aspen clones to simultaneous exposures of ozone and CO_2. Can. J. Forest Res. 26, 639–648.

Laisk, A., Kull, O., Moldau, H., 1989. Ozone concentration in leaf intercellular air spaces is close to zero. Plant Physiol. 90, 1163–1167.

Larson, P.R., 1977. Phyllotactic transitions in the vascular system of *Populus deltoides* bartr. as determined by ^{14}C labelling. Planta 134, 241–249.

Larson, P.R., Isebrands, J.G., 1971. The plastochron index as applied to developmental studies of cottonwood. Can. J. Forest Res. 1, 1–11.

Long, S.P., 1991. Modification of the response of photosynthetic productivity to rising temperature by atmospheric CO_2 concentrations: Has its importance been underestimated? Plant Cell Environ. 14, 729–739.

Long, S.P., 1994. Increases in temperature, CO_2 and O_3 on net photosynthesis, as mediated by Rubisco. In: Alscher, A.G., Wellburn, A.R. (Eds.), Plant Responses to the Gaseous Environment: Molecular, Metabolic and Physiological Aspects. Chapman and Hall, London, pp. 21–38.

McLaughlin, S., Percy, K., 1999. Forest health in North America: Some perspectives on actual and potential roles of climate and air pollution. Water Air Soil Pollut. 116, 151–197.

McKee, I.F., Farage, P.K., Long, S.P., 1995. The interactive effects of elevated CO_2 and O_3 on photosynthesis in spring wheat. Photosynth. Res. 45, 111–119.

Martin, M.J., Farage, P.K., Humphries, S.W., Long, S.P., 2000. Can the stomatal changes caused by acute ozone exposure be predicted by changes occurring in the mesophyll? A simplification for models of vegetation response to the global increase in tropospheric elevated ozone episodes. Aust. J. Plant Physiol. 27, 211–219.

Melillo, J.M., Prentice, I.C., Farquhar, G.D., Schulze, E.-D., Sala, O.E., 1996. Terrestrial Biotic Responses to Environmental Change and Feedbacks to Climate. In: Houghton, J.T., Meira Filho, L.G., Callander, B.A., Harris, N., Kattenberg, A., Maskell, K. (Eds.), IPCC: Climate Change 1995. Cambridge Univ. Press, Cambridge, UK, pp. 445–481.

Norby, R.J., Wullschleger, S.D., Gunderson, C.A., Johnson, D.W., Ceulemans, R., 1999. Tree responses to rising CO_2 in field experiments: implications for the future forest. Plant Cell Environ. 22, 683–714.

Pell, E.J., Eckhart, N.A., Glick, R.E., 1994. Biochemical and molecular basis for impairment of photosynthetic potential. Photosynth. Res. 39, 453–462.

Pell, E.J., Sinn, J.P., Brendley, B.W., Samuelson, L., Vinten-Johansen, C., Tien, M., Skillman, J., 1999. Differential response of four tree species to ozone-induced acceleration of foliar senescence. Plant Cell Environ. 22, 779–790.

PORG, 1993. Ozone in the United Kingdom. In: Third Report of the United Kingdom Photochemical Oxidants Review Group. AEA Harwell Laboratory, Didcot, UK, pp. 59–66.

Rauscher, H.M., Isebrands, J.G., Host, G.E., Dickson, R.E., Dickmann, D.I., Crow, T.R., Michael, D.A., 1990. ECOPHYS: An ecophysiological growth process model for juvenile poplar. Tree Physiol. 7, 255–281.

Reynolds, J.F., Hilbert, D.W., Kemp, P.R., 1993. Scaling ecophysiology from the plant the ecosystem: a conceptual framework. In: Ehleringer, J.R., Field, C.B. (Eds.), Scaling Physiological Processes: Leaf to Globe. Academic Press, San Diego, CA, pp. 127–140.

Schmieden, U., Wild, A., 1995. The contribution of ozone to forest decline. Physiol. Plant. 94, 371–378.

Sharkey, T.D., 1985. O_2-insensitive photosynthesis in C_3 plants. Plant Physiol. 78, 71–75.

Stitt, M., Krapp, A., 1999. The interaction between elevated carbon dioxide and nitrogen nutrition: the physiological and molecular background. Plant Cell Environ. 22, 583–621.

Theseira, G., Host, G.E., Isebrands, J.G., Whisler, F.D., 2003. SOILPSI: A potential-driven three-dimensional soil water redistribution model: description and comparative evaluation. Environ. Software and Modeling 18, 13–23.

von Caemmerer, S., Farquahar, G.D., 1981. Some relationships between the biochemistry of photosynthesis and the gas exchange of leaves. Planta 153, 376–387.

Wu, G., 1999. A parallel implementation of numerical experiments investigating the shading characteristics of *Populus eugeneii*. Master's Thesis in Applied and Computational Mathematics, University of Minnesota, Duluth, MN.

Yun, S.-C., Laurence, J.A., 1999. The response of clones of *Populus tremuloides* differing in sensitivity to ozone in the field. New Phytol. 141, 411–421.

Zang, G., 1999. Development of regulatory components for ECOPHYS. Master's Thesis in Applied and Computational Mathematics, University of Minnesota, Duluth, MN.

Zhao, W., 2000a. University of Minnesota, Duluth, MN, Technical report 2000-3.

Zhao, W., 2000b. An analytical solution for a mechanistic photosynthesis model, Master's Thesis in Applied and Computational Mathematics, University of Minnesota, Duluth, MN.

Chapter 8
Ozone affects the fitness of trembling aspen

D.F. Karnosky*

*School of Forest Resources and Environmental Science, Michigan Technological University,
101 U.J. Noblet Forestry Building, 1400 Townsend Drive, Houghton, MI 49931, USA
E-mail: karnosky@mtu.edu*

K.E. Percy

*Natural Resources Canada, Canadian Forest Service-Atlantic Forestry Centre, P.O. Box 4000,
Fredericton, New Brunswick, E3B 5P7 Canada*

B. Mankovska

Forest Research Institute, T.G. Masarykova Street 2195, 960 92 Zvolen, Slovakia

T. Prichard

*Wisconsin Department of Natural Resources, 101 S. Webster, P.O. Box 7921,
Madison, WI 53707, USA*

A. Noormets

The University of Toledo, Department EEES, LEES Lab, Mail Stop 604, Toledo, OH 42606, USA

R.E. Dickson

*USDA Forest Service, North Central Forest Experiment Station, Forestry Sciences Laboratory,
5985 Highway K, Rhinelander, WI 54501, USA*

E. Jepsen

*Wisconsin Department of Natural Resources, 101 S. Webster, P.O. Box 7921,
Madison, WI 53707, USA*

J.G. Isebrands

Environmental Forestry Consultants, LLC, P.O. Box 54, E7323, New London, WI 54961, USA

Abstract

Trembling aspen (*Populus tremuloides* Michx.) is sensitive to tropospheric
ozone (O_3) as determined by visible foliar symptoms, accelerated foliar senescence and premature abscission, degradation and change in composition of
epicuticular waxes, decreased photosynthesis and chlorophyll, and decreased

---

*Corresponding author.

DOI:10.1016/S1474-8177(03)03008-0

aboveground and belowground growth. The species is highly variable in O_3 responses as some clones are similar in sensitivity to Bel W3 tobacco and other clones are tolerant to moderate levels of O_3. We have, therefore, hypothesized and presented evidence for natural selection of O_3-tolerance in aspen populations. This hypothesis has been criticized, however, as improbable because O_3 is thought to be a fairly weak selection force and because changes in tree population are thought to occur over very long time periods, longer than tropospheric O_3 has been known to be a problem. To shed more light on this argument, in 1994 we established a set of research field trials using clones of known origin and previously determined O_3 sensitivity at three sites in the Lake States region with differing O_3 profiles (Rhinelander, Wisconsin—low O_3; Kalamazoo, Michigan—moderate O_3; and Kenosha, Wisconsin—high O_3). In this paper, we present evidence of changes in the relative volume d^2h growth of clone 259 (O_3-sensitive) compared to clone 216 (O_3-tolerant) of -0.1%, -44.2%, and -62.8% at the low, medium and high O_3 sites at age 5. In addition, relative survival of the clone 259 compared to 216 was -11.0%, -6.8%, and -38.4% at the low, moderate, and high O_3 sites. Actual survival rates at the high O_3 site were 78.2% for clone 216 and only 48.2% for clone 259. Our results suggest that very rapid and significant changes in competitive ability and fitness can occur under ambient levels of O_3 in the lower Great Lakes region for aggrading forests. These results are consistent with the hypothesis that O_3 is inducing natural selection for O_3 tolerance in aspen.

1. Introduction

Ozone (O_3), a widespread and highly phytotoxic secondary air pollutant, has long been known to impact the growth of trembling aspen (*Populus tremuloides* Michx.) (Wang et al., 1986) by degrading chlorophyll (Gagnon et al., 1992), decreasing rubisco (Noormets et al., 2001), decreasing photosynthesis (Coleman et al., 1995), altering carbon allocation (Coleman et al., 1996), causing visible foliar symptoms and premature senescence (Karnosky et al., 1996), altering leaf epicuticular waxes (Mankovska et al., 1998), and predisposing trees to disease pests (Karnosky et al., 2002; Percy et al., 2002). It is also well known that trembling aspen is highly variable in response to O_3 (Karnosky, 1976) and that the response variation is under strong genetic control (Karnosky, 1977). Genetic control appears to be exerted on gas exchange rates (Noormets et al., 2001) and an antioxidant formation in response to O_3 exposure (Sheng et al., 1997; Karnosky et al., 1998; Noormets et al., 2000; Wustman et al., 2001).

Evidence for differences in O_3 tolerances among various populations differing in O_3 exposures has been presented by Berrang et al. (1986, 1989, 1991) who made the prediction that O_3 was altering the competitive ability of trembling aspen genotypes in areas of high O_3 exposure in North America. This

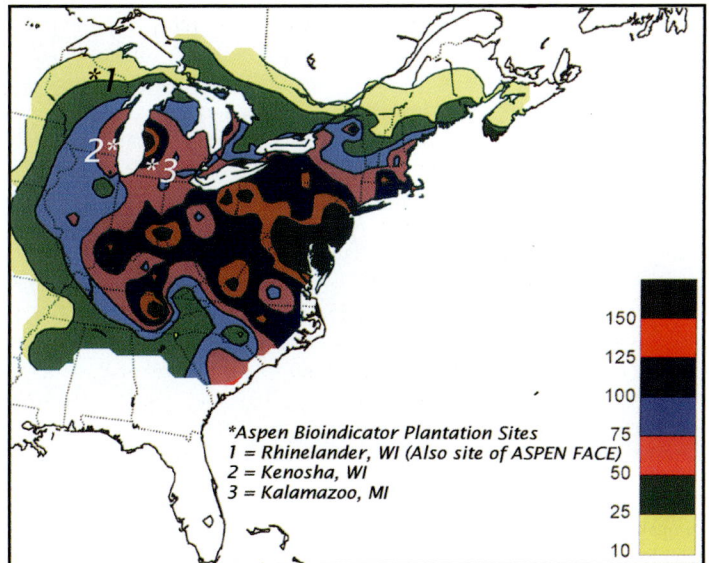

Figure 1.   Spatial distribution of average hours with ozone > 82 ppb (1986–1993) (from Dann and Summers, 1997) and locations of aspen plantations (1 = Rhinelander, WI; 2 = Kenosha, WI; 3 = Kalamazoo, MI).

research was challenged by other researchers who speculated that O_3 was not a strong enough selective pressure in most areas (Taylor et al., 1994; Taylor and Pitelka, 1992) and that trees had not been exposed to O_3 for a long enough time period to evolve O_3 tolerance (Barrett and Bush, 1991).

This study was designed to address the question as to whether or not O_3 could impact the growth and survival of trembling aspen genotypes differing in O_3 tolerance and growing in a competitive environment over the first five years after planting.

2. Methods

In 1994, we established three types of aspen plantations at each of three locations (Rhinelander, Wisconsin—low O_3; Kalamazoo, Michigan—moderate O_3; and Kenosha, Wisconsin—high O_3) along a natural O_3 gradient in the Lake States (Fig. 1).

At each location, we established three experiments (Karnosky et al., 1999). The first was a "common garden" experiment of 7 aspen clones differing in O_3 sensitivity (clones 1, 253 and 259 = O_3-sensitive; clones 10, 216, 221, and

Figure 2. The Kenosha, WI site is shown here with the Wisconsin Department of Natural Resources air quality monitoring shed (foreground), the growth and yield plot (background) and the competition plot (right).

$271 = O_3$-tolerant (Karnosky, 1976; Karnosky et al., 1992)). This experiment consisted of individual trees of each clone planted at 2 m × 2 m spacing and there were 10 replicates at each site. In the second experiment, we established a "growth and yield" trial with O_3-tolerant clone 216 and O_3-sensitive clone 259 planted in 16-tree (4 × 4) blocks at 2 m × 2 m spacing between trees and with six replicates. The final experiment at each site was a "competition trial" between clones 216 and 259 where trees were planted at 0.5 m × 0.5 m between trees and 100-tree (10 × 10) blocks were established in either pure clonal blocks or mixed clonal blocks (the O_3-sensitive and tolerant clones were alternately spaced in this part of the plantation). Again, six replicates were used at each site. Two border rows of clone 271 were established around each of the three plantations at each site. Fig. 2 shows the Kenosha, WI, site.

Weeds were controlled around all trees by herbicide application in the first season and then mowing in subsequent years until weed competition was no longer a problem. All trees were measured annually and also observed at least once per season for visible foliar symptoms. All three sites were old field sites that had not been in agricultural use for several years previously. The sites all have sandy loam soils with relatively high fertility. Temperature and rainfall patterns from the three sites suggest that they are not widely different in summer climate (Table 1). Each site was enclosed in a 3 m tall deer fence.

O_3 has been monitored independently and continuously at the Kenosha site and Rhinelander site by the Wisconsin Department of Natural Resources. The

Table 1. Comparisons of monthly mean temperatures (°C) and seasonal total rainfall (cm) at our three aspen sites along a natural gradient (from: http://www.acdc.noaa.gov/online)

	Average monthly temperatures (°C)														
	1995			1996			1997			1998			1999		
	Rhine	Kala	Keno	Rhine	Kala	Keno	Rhine	Kala	Keno	Rhine	Kala	Keno	Rhine	Kala	Keno
June	21.98	23.38	21.57	18.00	22.38	17.95	19.60	21.97	17.98	16.89	19.67	18.72	18.17	20.78	19.06
July	21.83	25.37	23.37	17.88	22.95	20.75	19.55	23.05	20.61	20.33	22.33	22.22	22.22	23.72	25.44
Aug	23.92	25.89	23.92	20.08	24.35	22.33	16.81	21.68	19.26	20.56	22.33	22.67	18.72	19.89	21.22

Date	June through August rainfall (cm)		
	Rhinelander	Kalamazoo	Kenosha
1995	20.17	26.58	19.64
1996	35.43	30.48	39.10
1997	24.12	29.58	38.17
1998	20.60	16.0	26.80
1999	34.70	25.17	30.43

Table 2. Summary of O_3 values at the three locations where we have aspen bioindicator plots for 1995 to 1999

Site	1995	1996	1997	1998	1999	5-year
	Max 1-h (ppb)					
Rhinelander	79	87	80	74	93	82.2
Kalamazoo	125	106	96	109	103	107.8
Kenosha	120	122	130	119	123	122.8
	Sum 06 (ppm h 8 a.m.–8 p.m. for June 1 to August 31)					
Rhinelander	4.1	2.5	3.6	2.7	4.6	3.5
Kalamazoo	20.6	17.3	24.4	21.4	21.4	21.0
Kenosha	29.8	15.1	–	17.7	27.3	22.5

Kalamazoo site was monitored for O_3 by Pharmacia about 0.5 km from our aspen plots. The O_3 data was quality assured and was part of the AIRS (Aerometric Information and Retrieval Systems of the US Environmental Protection Agency) network. A summary of the O_3 values for 1995–1999 is shown in Table 2.

We evaluated foliar injury on all leaves on a subset of the trees representing all the clones at the O_3 gradient sites.

Heights (±10 cm), diameters (±1 mm), and survival were recorded yearly.

Table 3. Age 5 heights, diameters, and estimated volumes for an O_3-tolerant clone (216) and an O_3-sensitive clone (259) from three sites along an O_3 gradient (Rhinelander, WI = low O_3; Kalamazoo, MI = moderate O_3; Kenosha, WI = high O_3). Values are means ± Standard error

Variable	Clone	Rhinelander	Kalamazoo	Kenosha
Height (cm)	216	248 ± 6	383 ± 8	531 ± 10
	259	239 ± 4	326 ± 5	374 ± 7
Diameter (cm)	216	1.71 ± 0.08	2.26 ± 0.06	2.98 ± 0.06
	259	1.89 ± 0.07	1.86 ± 0.05	2.21 ± 0.05
Volume (D^2H)	216	1014 ± 114	2589 ± 122	5872 ± 309
	259	1012 ± 74	1444 ± 76	2182 ± 131
Ratio	216/259	1.001	1.793	2.691

3. Results

Maximum 1-hr O_3 concentrations averaged 82.2 ppb (Rhinelander), 107.8 ppb (Kalamazoo) and 122.8 (Kenosha) over the 5-year (1994–1999) period (Table 2). During five years (1994–1999) after establishment, we have documented significant, repeatable and consistent site-to-site and clone-to-clone differences in visible O_3 symptoms, degradation of epicuticular waxes and changes in wax composition, and occurrence of pests. The Kenosha site has always had the most severe O_3 impacts.

3.1. Visible foliar symptoms

Visible foliar symptoms were seen on O_3-sensitive aspen clones each growing season by late July to early August at the Kalamazoo and Kenosha sites (Fig. 3). Visible foliar symptoms occurred on 5–8% of clone 216 leaves and 34.7–43.1% of clone 259 leaves (Karnosky et al., 1999). No visible foliar symptoms were detected at the Rhinelander site. Visible foliar symptoms consisted of black bifacial necrosis, chlorosis, or upper leaf surface black stipple and premature leaf abscission. These are all classic O_3 symptoms on aspen (Karnosky, 1976; Wang et al., 1986; Karnosky et al., 1996). Visible symptoms were also seen on other O_3 bioindicator plants, including understory black cherry volunteer seedlings and on nearby milkweed plants, at the Kalamazoo and Kenosha sites, but not at Rhinelander.

3.2. Growth

The trees grew at different rates at the three sites with the trees being the largest at the Kenosha site (Table 3). The ratio of the volume growth of the O_3-tolerant clone (216) to the volume growth of the O_3-sensitive clone changed from 1.001

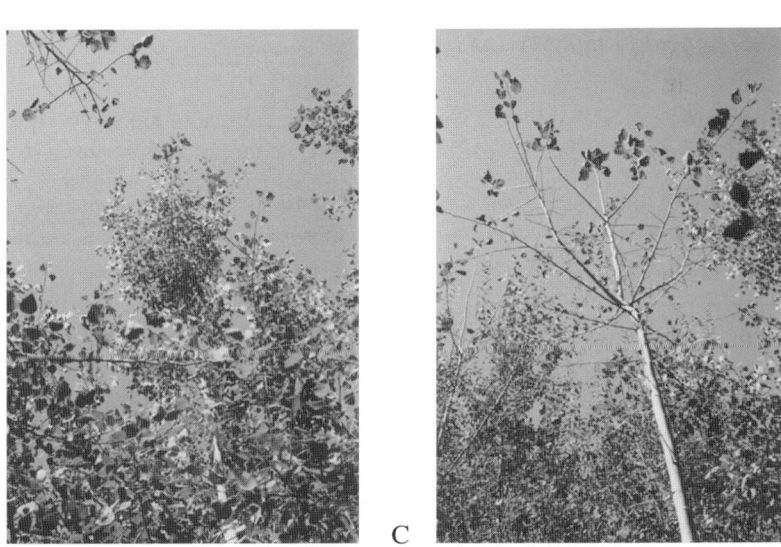

Figure 3. Examples of visible foliar symptoms seen at the Kenosha, WI site. A: Black bifacial necrosis on clone 259. B, C: Clones 216 (B) and 259 (C) showing the differential premature leaf abscission due to O_3 following a late August O_3 event.

at Rhinelander to 1.793 (Kalamazoo) and 2.691 (Kenosha) indicating the larger detrimental effect of O_3 on the O_3-sensitive clone 259 than on the O_3-tolerant clone 216.

3.3. Fitness

Fitness involves the ability of plants to survive and to produce offspring. The trends in survival can be seen in Fig. 4. Survival at the low O_3 site was 95%

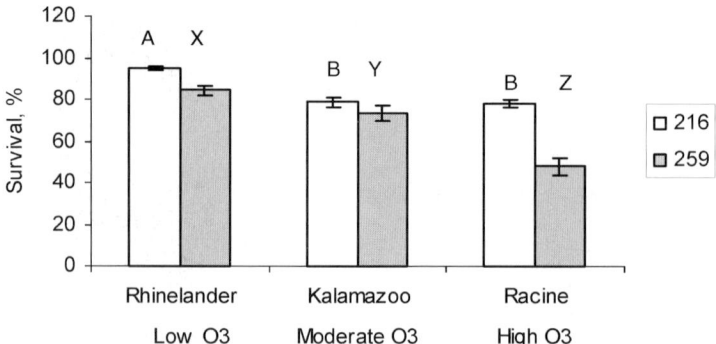

Figure 4. Survival at age 5 of two aspen clones differing in O_3 tolerance clone 216 = O_3 tolerant and clone 259 = O_3 sensitive. Rhinelander, WI = low O_3; Kalamazoo, MI = moderate O_3; Kenosha, WI = high O_3. The values shown with similar letters are not different at the 0.0001 level (Duncan's Multiple Range test via SAS).

(clone 216) and 84% (clone 259). Significant decreases in the survival of both clones were seen at the elevated O_3 sites and the largest difference in survival between the O_3-tolerant and O_3-sensitive clones came at the highest O_3 site.

4. Discussion

In this paper we present evidence linking elevated levels of ambient O_3 to differential survival of O_3-sensitive and O_3-tolerant aspen clones grown for five years along a natural O_3 gradient in the Lake States region. Age 5 survival data shows the same statistical trends as to the findings of visible foliar symptoms and height and diameter growth in that the O_3-tolerant clone (216) had less visible foliar symptoms, less growth impact and higher survival rates than did the O_3^--sensitive clone (259).

These results are consistent with previous visible symptoms and growth studies in open-top chambers (Karnosky et al. 1992, 1996) and in Free-air CO_2 Enrichment (FACE) studies (Karnosky et al., 1999; Isebrands et al., 2001; McDonald et al., 2002) where clone 216 has always been more tolerant than clone 259.

We present evidence of a change in the ratio of volume growth of clone 216 from 1.01 at the low O_3 site to a maximum of 2.36 at the highest O_3 site. In short-term open-top chamber studies, we found an increase in the ratio of total biomass from 1.015 in charcoal-filtered air to 1.324 in 2 × ambient O_3 (Karnosky et al., 1996).

The impacts of O_3 on evolutionary processes of plants are poorly understood. Dunn (1959) reported effects of O_3 on southern California *Lupinus* populations. Reiling and Davison (1992) found evidence of O_3-induced population changes in native herbaceous species. Karnosky (1981, 1989) presented evidence for a 10-fold higher mortality rate among O_3-sensitive *Pinus strobus* genotypes as compared to O_3-tolerant genotypes over a 15-year time frame in southern Wisconsin. Berrang et al. (1986, 1989, 1991) found evidence of population differences in O_3 tolerance and these were related to the O_3 concentrations of each population.

The question if evolutionary changes are occurring under ambient O_3 in natural populations remains a key genecological question (Pitelka, 1988; Davison and Barnes, 1998; Winner et al., 1991). In this paper, we provide evidence for a reduced fitness of an O_3-sensitive clone as determined by a 36% decrease in survival compared to the low O_3 site within five years from establishment. The spacing conditions (0.5 m × 0.5 m) are characteristic of aspen stands regenerating after a clear-cut or fire and the highest ambient O_3 levels reported in this study are typical of those found over much of the eastern one-half of the United States.

Acknowledgements

This research was supported in part by the USDA Forest Service Global Change Program, the USDA Forest Service North Central Research Station, the National Council of the Paper Industry for Air and Stream Improvement (NCASI), Michigan Technological University, the Slovakian Forest Research Institute, Pharmacia, and the Canadian Forest Service Forest Health Network. The authors would like to thank the University of Wisconsin-Parkside for the use of their land for this study and the Wisconsin Department of Natural Resources Air Quality Division for operating and maintaining the Air Quality Monitoring at the Kenosha site.

References

Barrett, C.H., Bush, E.J., 1991. Population processes in plants and the evolution of resistance to gaseous air pollutants. In: Taylor Jr., G.E., Pitelka, L.F., Clegg, M.T. (Eds.), Ecological Genetics and Air Pollution. Springer-Verlag, New York, pp. 137–165.

Berrang, P.C., Karnosky, D.F., Mickler, R.A., Bennett, J.P., 1986. Natural selection for ozone tolerance in *Populus tremuloides*. Can. J. Forest Res. 16, 1214–1216.

Berrang, P.C., Karnosky, D.F., Bennett, J.P., 1989. Natural selection for ozone tolerance in *Populus tremuloides*. II. Field verification. Can. J. Forest Res. 19, 519–522.

Berrang, P.C., Karnosky, D.F., Bennett, J.P., 1991. Natural selection for ozone tolerance in *Populus tremuloides*: An evaluation of nationwide trends. Can. J. Forest Res. 21, 1091–1097.

Coleman, M.D., Dickson, R.E., Isebrands, J.G., Karnosky, D.F., 1995. Photosynthetic productivity of aspen clones varying in sensitivity to tropospheric ozone. Tree Physiol. 15, 585–592.

Coleman, M.D., Dickson, R.E., Isebrands, J.G., Karnosky, D.F., 1996. Root growth and physiology of potted and field-grown trembling aspen exposed to tropospheric ozone. Tree Physiol. 16, 145–152.

Dann, T., Summers, P., 1997. Ground-level ozone and its precursors, 1980–1993. Canadian 1996 NO_x/VOC Science Assessment Report. 295 p.

Davison, A.W., Barnes, J.D., 1998. Effects of ozone on wild plants. New Phytol. 139, 135–151.

Dunn, D.B., 1959. Some effects of air pollution on *Lupinus* in the Los Angeles area. Ecology 40, 621–625.

Gagnon, Z.E., Karnosky, D.F., Dickson, R.E., Isebrands, J.G., 1992. Effect of ozone on chlorophyll content in *Populus tremuloides*. Am. J. Bot. 79 (6), 107 (Abstract).

Isebrands, J.G., McDonald, E.P., Kruger, E., Hendrey, G., Pregitzer, K., Percy, K., Sober, J., Karnosky, D.F., 2001. Growth responses of *Populus tremuloides* clones to interacting carbon dioxide and tropospheric ozone. Environ. Pollut. 115, 359–371.

Karnosky, D.F., 1976. Threshold levels for foliar injury to *Populus tremuloides* Michx. by sulfur dioxide and ozone. Can. J. Forest Res. 6, 166–169.

Karnosky, D.F., 1977. Evidence of genetic control of response to sulfur dioxide and ozone in *Populus tremuloides* Michx. Can. J. Forest Res. 7, 437–440.

Karnosky, D.F., 1981. Changes in eastern white pine stands related to air pollution stress. Mitteilungen der Forstlichen Bundesversuchsanstalt, Wien 137, 41–45.

Karnosky, D.F., 1989. Air pollution induced population changes in North American forests. In: Bucher, J.B., Bucher-Wallin, I. (Eds.), Air Pollution and Forest Decline, Proc. 14th Int. Meeting for Specialists in Air Pollution Effects on Forest Ecosystems, Interlaken, Switzerland. Birmensdorf, pp. 315–317.

Karnosky, D.F., Gagnon, Z.E., Reed, D.D., Witter, J.A., 1992. Growth and biomass allocation of symptomatic and asymptomatic *Populus tremuloides* clones in response to seasonal ozone exposures. Can. J. Forest Res. 22, 1785–1788.

Karnosky, D.F., Gagnon, Z.E., Dickson, R.E., Coleman, M.D., Lee, E.H., Isebrands, J.G., 1996. Changes in growth, leaf abscission, and biomass associated with seasonal tropospheric ozone exposures of *Populus tremuloides* clones and seedlings. Can. J. Forest Res. 26, 23–37.

Karnosky, D.F., Podila, G.K., Gagnon, Z., Pechter, P., Akkapeddi, A., Coleman, M., Dickson, R.E., Isebrands, J.G., 1998. Genetic control of responses to interacting O_3 and CO_2 in *Populus tremuloides*. Chemosphere 36, 807–812.

Karnosky, D.F., Mankovska, B., Percy, K., Dickson, R.E., Podila, G.K., Sober, J., Noormets, A., Hendrey, G., Coleman, M.D., Kubiske, M., Pregitzer, K.S., Isebrands, J.G., 1999. Effects of tropospheric O_3 on trembling aspen and interaction with CO_2: Results from an O_3-gradient and a FACE experiment. In: Sheppard, L.J., Cape, J.N. (Eds.), Forest Growth Responses to the Pollution Climate of the 21st Century. Kluwer Academic, London, pp. 311–322.

Karnosky, D.F., Percy, K.E., Xiang, B., Callan, B., Noormets, A., Mankovska, B., Hopkin, A., Sober, J., Jones, W., Dickson, R.E., Isebrands, J.G., 2002. Interacting CO_2-tropospheric O_3 and predisposition of aspen (*Populus tremuloides* Michx.) to infection by *Melampsora medusae* rust. Global Change Biol. 8, 329–338.

Mankovska, B., Percy, K., Karnosky, D.F., 1998. Impact of ambient tropospheric O_3, CO_2, and particulates on the epicuticular waxes of aspen clones differing in O_3 tolerance. Ekológia (Bratislava) 18 (2), 200–210.

McDonald, E.P., Kruger, E.L., Riemenschneider, D.E., Isebrands, J.G., 2002. The role of competition in modifying growth responses of trembling aspen to elevated CO_2 and tropospheric ozone. Funct. Ecol. 16, 792–801.

Noormets, A., Podila, G.K., Karnosky, D.F., 2000. Rapid response of antioxidant enzymes to O_3-induced oxidative stress in *Populus tremuloides* clones varying in O_3 tolerance. Forest Gen. 7, 335–338.

Noormets, A., Sober, A., Pell, E.J., Dickson, R.E., Podila, G.K., Sober, J., Isebrands, J.G., Karnosky, D.F., 2001. Stomatal and non-stomatal control of photosynthesis in trembling aspen (*Populus tremuloides* Michx.) exposed to elevated CO_2 and/or O_3. Plant Cell Environ. 24, 327–336.

Percy, K.E., Awmack, C.S., Lindroth, R.L., Kubiske, M.E., Kopper, B.J., Isebrands, J.G., Pregitzer, K.S., Hendrey, G.R., Dickson, R.E., Zak, D.R., Oksanen, E., Sober, J., Harrington, R., Karnosky, D.F., 2002. Altered performance of forest pests under CO_2- and O_3-enriched atmospheres. Nature 420, 403–407.

Pitelka, L.F., 1988. Evolutionary responses of plants to anthropogenic pollutants. Trends Ecol. Evolut. 3, 233–236.

Reiling, K., Davison, A.W., 1992. The response of native, herbaceous species to ozone: growth and fluorescence screening. New Phytol. 120, 29–37.

Sheng, Y., Podila, G.K., Karnosky, D.F., 1997. Differences in O_3-induced SOD and glutathione antioxidant expression in tolerant and sensitive aspen (*Populus tremuloides* Michx.) clones. Forest Gen. 4, 25–33.

Taylor Jr., G.E., Pitelka, L.F., 1992. Genetic diversity of plant populations and the role of air pollution. In: Barker, J.R., Tingey, D.T. (Eds.), Air Pollution Effects on Biodiversity. Van Nostrand Reinhold, New York, pp. 111–130.

Taylor, G.E., Johnson, D.W., Anderson, C.P., 1994. Air pollution and forest ecosystems: A regional to global perspective. Ecol. Applic. 4, 662–689.

Wang, D., Karnosky, D.F., Bormann, F.H., 1986. Effects of ambient ozone on the productivity of *Populus tremuloides* Michx. grown under field conditions. Can. J. Forest Res. 16, 47–55.

Winner, W., Coleman, J., Gillespie, C., Mooney, H., Pell, E., 1991. Consequences of evolving resistance to air pollutants. In: Taylor, G.E., Pitelka, L.F., Clegg, M.T. (Eds.), Ecological Genetics and Air Pollution. Springer-Verlag, New York, pp. 177–202.

Wustman, B.A., Oksanen, E., Karnosky, D.F., Sober, J., Isebrands, J.G., Hendrey, G.R., Pregitzer, K.S., Podila, G.K., 2001. Effects of elevated CO_2 and O_3 on aspen clones varying in O_3 sensitivity: Can CO_2 ameliorate the harmful effects of O_3? Environ. Pollut. 115, 473–481.

Air Pollution, Global Change and Forests in the New Millennium
D.F. Karnosky et al., editors
© 2003 Elsevier Ltd. All rights reserved.

211

Chapter 9

Responses of Aleppo pine to ozone

R. Alonso

*Ecotoxicology of Air Pollution, CIEMAT-DIAE, Ed. 70. Avda. Complutense 22,
Madrid-28040, Spain*

S. Elvira, R. Inclán, V. Bermejo

*Ecotoxicology of Air Pollution, CIEMAT-DIAE, Ed. 70. Avda. Complutense 22,
Madrid-28040, Spain*

F.J. Castillo

*Departamento Ciencias del Medio Natural, Universidad Pública de Navarra,
Pamplona-31006, Spain*

B.S. Gimeno*

*Ecotoxicology of Air Pollution, CIEMAT-DIAE, Ed. 70. Avda. Complutense 22,
Madrid-28040, Spain*
E-mail: benjamin.gimeno@ciemat.es

Abstract

Tropospheric ozone (O_3) has become a pollutant of major concern in southern Europe. Aleppo pine (*Pinus halepensis*) exhibits O_3-induced visible injury across the Mediterranean Basin. Therefore, two experiments were carried out in open-top chambers to assess the influence of O_3 on growth and physiology of this species and potential interactions between O_3 and the summer environment. Elevated levels of O_3 were found to alter some antioxidant enzyme activities and reduce chlorophyll content, photosynthetic activity, and stomatal conductance of Aleppo pine seedlings, although no significant effects on growth rates were detected after 3 years' exposure. These effects were observed even when O_3 exposure was restricted to the summer, i.e., the time of year with the lowest O_3 uptake. Combined exposure to both O_3 and water stress resulted in reductions in needle biomass and net photosynthesis rates, as well as disturbances in the response of plant defense systems. Our results suggest that O_3 may be a factor contributing to reductions in the vitality of Aleppo pine forests across the Mediterranean.

---

*Corresponding author.

DOI:10.1016/S1474-8177(03)03009-2

1. Introduction

Tropospheric ozone (O_3) is the most widespread air pollutant in the Mediterranean area. In this region, meteorological conditions are especially favorable to its formation and persistence (Millán et al., 1996). Much work has focused on O_3-induced damage on agricultural crops in this area, and visible injury, yield losses, and reductions in fruit and grain quality have been reported (see reviews by Lorenzini, 1993; Schenone, 1993; Gimeno et al., 1994; Velissariou et al., 1996; Fumagalli et al., 2001). However, natural vegetation has received much less attention, although concerns have been expressed about the impacts of the pollutant on southern European forests (Bussotti and Ferretti, 1998; Barnes et al., 2000). In the early 1980s, Naveh et al. (1980) described O_3-induced visible injury on some pine species, highlighting the potentially important threat that photochemical pollution could represent to Mediterranean ecosystems. Aleppo pine (*Pinus halepensis* Mill.) is considered to be especially susceptible to O_3 damage because the O_3-induced visible symptoms have been reported in forests across the Mediterranean Basin (Velissariou et al., 1992; Gimeno et al., 1992; Davison et al., 1995; Barnes et al., 2000; Sanz et al., 2000). These symptoms have been replicated under experimental fumigation conditions (Wellburn and Wellburn, 1994; Anttonen et al. 1995, 1998; Manninen et al., 1999).

In recent years, concern over the health of Aleppo pine forests has increased prompted by reports of declines in tree condition in eastern Spain (Montoya, 1995; Sanz et al., 2000). It has been proposed that O_3 might be one factor contributing to the reduction in vitality of Aleppo pine forests, a phenomenon that could have significant ecological implications because of the important role this species plays in soil stabilization in the Mediterranean region.

This paper presents an overview of the results of two experiments performed within the context of a pan-European study on the interactive effects of O_3 and other environmental factors, such as water stress, on *Pinus halepensis* physiology.

2. Materials and methods

The experiments were carried out in field-based, open-top chambers located at the Ebro Delta in northeastern Spain. This facility is described in detail elsewhere (Reinert et al., 1992; Pujadas et al., 1997; Elvira et al., 1998). Two-year-old Aleppo pine seedlings were planted in 18-dm^3 pots containing 50% peat, 30% sand, 20% vermiculite and a slow–release fertilizer (Osmocote Plus; NPK 15 : 18 : 11) to prevent nutrient limitations. Seedlings were exposed to the following O_3 treatments: charcoal-filtered air (CFA), non-filtered air (NFA), and non-filtered air plus 40 ppb O_3 (NFA+40). In the latter treatment, 40 ppb O_3 was added from 8 to 18 hrs daily to the ambient levels of the pollutant. Three chambers per treatment were used and three additional chamberless plots (Ambient Air, AA) were established to evaluate chamber effects.

In the first experiment, 20 plants per plot (60 plants per treatment) were exposed to the different O_3 treatments just during summer months. In the second experiment, 24 seedlings per plot were continuously exposed to elevated O_3 over 3 years; then after 20 months, a water stress treatment was introduced. Two watering regimes were established in order to assess the interactive effects of O_3 and water stress. In the well-watered treatment (WW), water was supplied according to requirement, while water-stressed (WS) seedlings received half the water supplied to WW seedlings. After WS plants had been exposed to 10 weeks' water shortage, water was withheld altogether for 11 days. Well-watered seedlings were irrigated daily to field capacity over the same period. The water deficit resulted in a loss of 7–10% in needle water content; pre-dawn needle water potential (Scholander pressure-bomb, SKPM 1400, Skye Instruments Ltd., UK) fell to -3.8 MPa in the WS plants, while it remained in the range of -1.6 to -2.5 MPa in the WW seedlings.

Ozone, nitrogen oxides (NO_x) and sulfur dioxide (SO_2) were continuously monitored, as well as meteorological variables such as air temperature, air relative humidity, and photosynthetic active radiation (PAR). Stem height and diameter of trees in the different treatments were assessed seasonally, and aboveground biomass was determined at the end of the experiment. Net photosynthesis (A) and stomatal conductance to water vapor (g_s) were measured under the prevailing environmental conditions using a LICOR-6200 infrared gas analysis system (Licor Inc., Licoln, NB, USA). Gas exchange rates were calculated according to von Caemmerer and Farquhar (1981) and expressed on the basis of projected needle area. Both previous and current-year needles were sampled on a seasonal basis to determine their levels of photosynthetic pigments and antioxidant enzyme activities. Five replicates of needles from different trees were sampled per chamber on each sampling date to determine pigments. Regarding xanthophyll analyses, needle samples were always collected at midday and immediately frozen in liquid nitrogen. Chlorophyll and carotenoid concentrations were measured spectrophotometrically (Barnes et al., 1992) and xanthophylls were determined by HPLC according to Val et al. (1994). Antioxidant enzyme activity determination involved collecting current and previous-year needles between 09:00 and 10:00 hours (local time). Two pooled samples from six individual trees were taken per chamber on each sampling date. Cell extractions (see Alonso et al., 1999) were performed to analyze the following antioxidant enzyme activities: catalase (CAT; Aebi, 1983); glutathione reductase (GR; Castillo and Greppin, 1988), guaiacol peroxidase (POD) and superoxide dismutase (KCN-resistant SOD and CuZnSOD) (Elvira et al., 1998). Also, POD activity was assessed in extracellular fluid isolated as described elsewhere (Elvira et al., 1998).

Data were subjected to ANOVA and differences between treatments were assessed using the LSD calculated at the 5% level (STATISTICA v5.1. StatSoft Inc., USA)

3. Results

3.1. Air quality

Ozone was the most abundant air pollutant in ambient air at the field site, with SO_2 and NO_x concentrations within the range of the detection limits of the monitors used. Ozone levels showed a typical diurnal profile superimposed on a 20-ppb background level. The highest concentrations were recorded during the midday (10:00–17:00 h GMT), and the lowest concentrations were experienced around dawn (4:00–6:00 h GMT) (see Elvira et al., 1998; Alonso et al., 1999). Moreover, strong seasonal variations were also found, with the highest O_3 concentrations recorded during spring and early summer and the lowest values during winter (Table 1). Ozone concentrations were compared with the guidelines defined in the Directive 92/72/EC of the European Commission (EC) for the protection of vegetation. Thus, hourly average ambient O_3 concentrations never exceeded 100 ppb, but 24-h average concentrations were greater than 33 ppb almost every month over the experimental period. Each year, ambient levels of O_3 exceeded current United Nations (UN)—Economic Commission for Europe (ECE) critical-level guidelines for the protection of forest trees (see Table 2).

3.2. Gas exchange and ozone uptake

Values of A and g_s of Aleppo pine seedlings showed clear seasonal variations. Moreover, diurnal gas exchange patterns varied depending on the season (Fig. 1). During winter, A and g_s increased during the morning, attaining maximum values during the central part of the day, which coincided with maximum levels of solar radiation and temperature. In contrast, during summer, maximum A and g_s were recorded early in the morning (6:00–8:00 GMT) and decreased around midday, while temperature, irradiance, and vapor pressure deficit continued to increase. A partial recovery in the gas exchange values was observed during the afternoon on some summer days.

Ozone uptake rates were estimated from the relationship between g_s and time of day in the different seasons. During winter, O_3 concentrations were lower than during spring and early summer (Table 1), but the higher rates of g_s

Table 1. Monthly ozone concentrations in the different treatments during sampling dates. M10h = 10-hour average expressed as ppb calculated from 7:00 to 17:00 h GMT. M24h = 24-hour average expressed as ppb. AOT40 = Accumulated exposure over threshold of 40 ppb during daylight hours (expressed as ppb h)

		AA			NFA			NFA+40			CFA	
	M10h	M24h	AOT40	M10h	M24h	AOT40	M10h	M24h	AOT40	M10h	M24h	AOT40
1994 Feb	31	28	224	30	27	196	45	33	2410	20	18	0
May	45	42	2714	43	41	2267	71	55	11998	4	3	0
Jul	46	35	2457	40	34	1071	64	46	9514	7	6	0
Sep	37	28	830	33	27	350	68	45	10534	5	4	0
1995 Feb	39	32	1188	40	33	1150	61	44	6576	19	15	0
May	60	54	7674	56	56	6390	85	69	17139	17	14	0
Jul	43	38	2585	43	40	2388	67	52	10799	16	13	0
Sep	41	34	1501	40	33	1182	68	47	9755	13	11	0
1996 Mar	42	37	1755	40	36	1153	64	48	8509	13	11	0

Table 2. Cumulative ozone exposure during daylight hours for the period April to September expressed as AOT40 (Accumulated exposure over threshold of 40 ppb) in the different treatments

Year	AA	NFA+40	NFA	CFA
1994	11060	54231	7879	0
1995	23422	70042	19128	0
1996	12362	66699	9667	0

resulted in greater values of O_3 uptake (Fig. 2) than during summer months, when stomatal closure reduced O_3 uptake to a minimum. In trees exposed to the highest O_3 levels (NFA+40), O_3 uptake was usually higher in current-year than in 1-year-old needles, especially during winter. These differences were not apparent in plants exposed to non-filtered air (Fig. 2).

Maximum daily values of A and g_s also changed throughout the year, with lower values recorded during the summer compared with winter and spring (Fig. 3). These seasonal variations were detectable even when plants were not subjected to water stress. Aleppo pine seedlings exposed to elevated O_3 concentrations (NFA+40) exhibited significantly lower values of A and g_s than plants grown in charcoal-filtered air. In trees where O_3 exposure was restricted to the summer months (Fig. 3(a)), a 30% reduction in A and g_s was detected in 1-year-old needles. In contrast, trees exposed to O_3 throughout the year (Fig. 3(b)) exhibited a comparable reduction in A and g_s in current-year and 1-year-old needles.

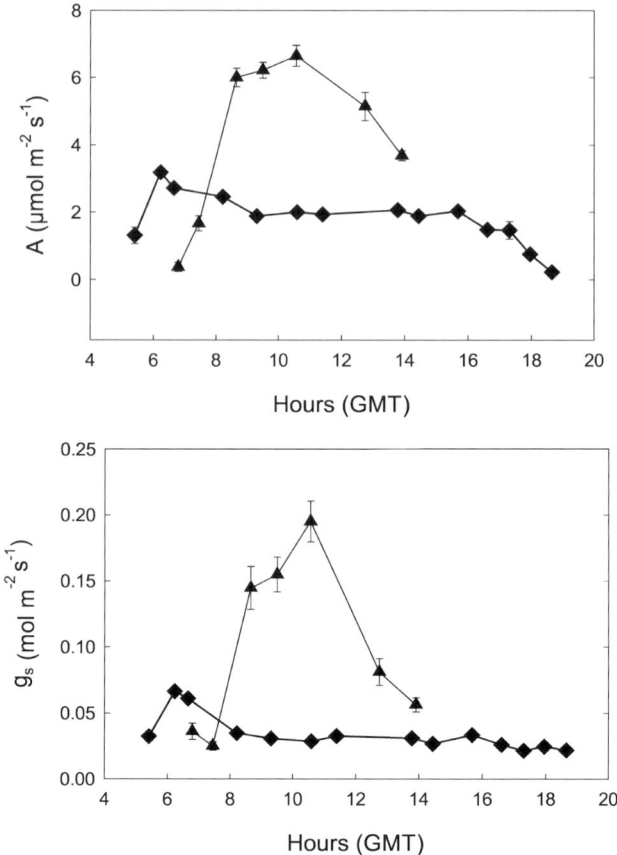

Figure 1. Daily profile of gas exchange parameters of Aleppo pine seedlings growing in char-coal-filtered air during summer (◆) and winter (▲). Means ± SE.

Soil water deficit induced a stronger depression (Fig. 3(b)) in A (up to 76%) and g_s (up to 37%) in 1-year-old needles than O_3 (around 20–23%). Water shortage, in combination with ozone, did not induce additive reductions in net photosynthesis. However, maximum stomatal conductance values for seedlings exposed to O_3 plus water stress were higher than observed in trees exposed to water stress alone.

3.3. Photosynthetic pigments

Photosynthetic pigments of Aleppo pine also exhibited marked seasonal fluctu-ations. During summer, the activation of the xanthophyll cycle (de-epoxidation

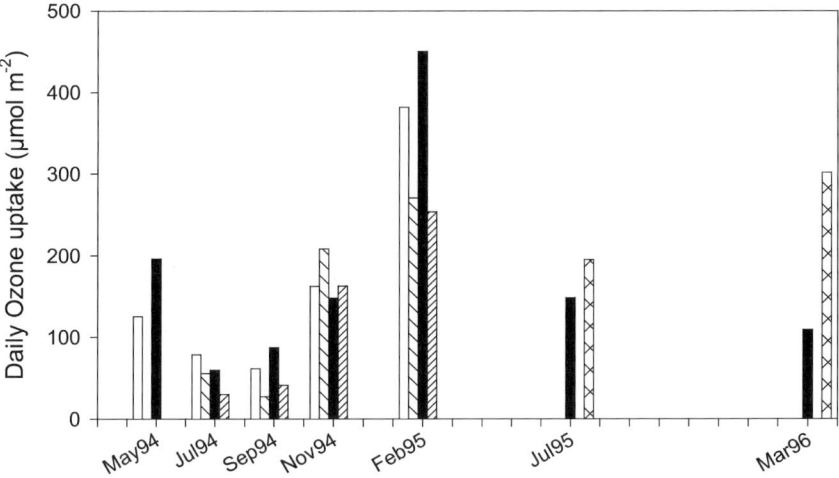

Figure 2. Daily ozone uptake on sampling dates in Aleppo pine seedlings exposed to ozone. □ 1993-needles exposed to NFA+40; \\\\ 1993-needles exposed to NFA; ■ 1994-needles exposed to NFA+40; //// 1994-needles exposed to NFA; ××× 1995-needles exposed to NFA+40.

state of xanthophyll cycle pool, DPS index) was higher ($p < 0.001$) at midday in both current and 1-year-old needles of trees exposed to elevated levels of O_3 than in seedlings raised in clean air (Fig. 4(a), (b)). In winter, an ozone-induced increment ($p < 0.005$) in DPS index was recorded in 1-year-old needles during the morning and the evening (Fig. 4(c), (d)).

Soil water deficit also caused a significant increase in the conversion state of xanthophyll cycle pool intermediates in both needle age classes, regardless of O_3 treatment (Fig. 4(a), (b)). No interactive effects of O_3 and water stress on DPS were detected at midday, but water stress induced a smaller activation ($p < 0.001$) of the xanthophyll cycle during the morning and the evening in 1-year-old needles of plants exposed to O_3. In contrast, current-year needles exposed to O_3 and water stress exhibited greater DPS during the evening than trees exposed to charcoal-filtered air.

3.4. Antioxidant enzyme activities

Current-year needles of Aleppo pine generally exhibited higher antioxidant enzyme activities than 1-year-old needles (Table 3) plus a greater capacity to react to stress. Ozone exposure resulted in alterations in the activity of antioxidant enzymes on some sampling dates. Effects were not related to accumulated O_3 exposure or to episodic O_3 concentrations before sampling, but appeared to be influenced by needle age tree ontogeny. The most significant

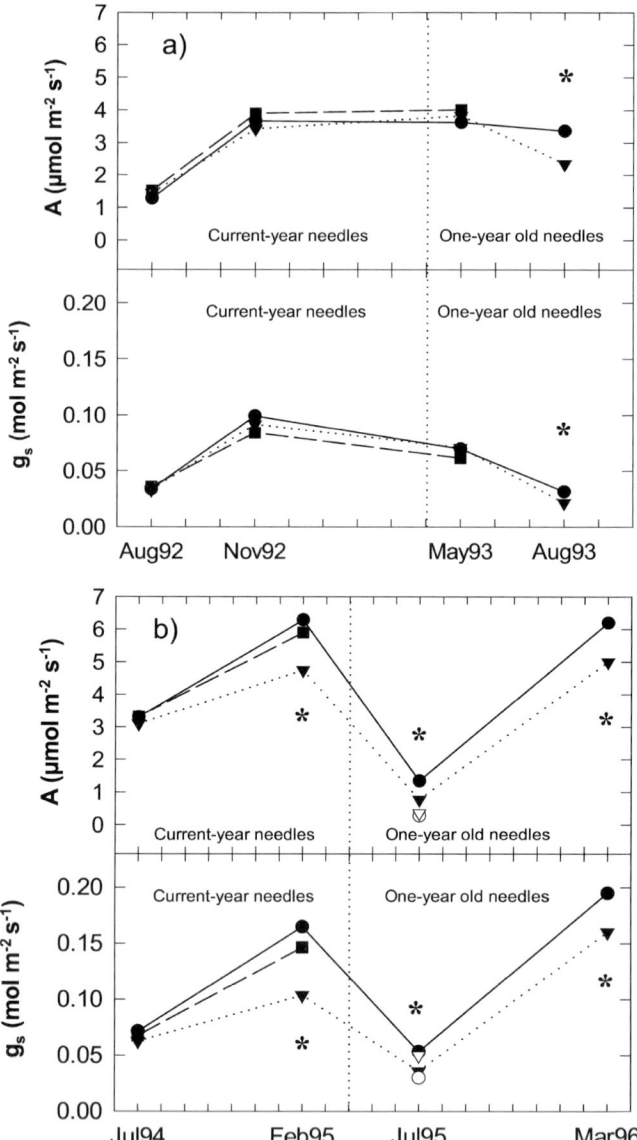

Figure 3. Maximum daily net photosynthesis (A) and stomatal conductance (g_S) measured on Aleppo pine seedlings exposed to the different ozone treatments: ● CFA WW; ○ CFA WS; ■ NFA; ▼ NFA+40 WW; ▽ NFA+40 WS. (a) 1992-needles exposed to O_3 during summer; (b) 1994-needles exposed to O_3 throughout the year. ∗ indicates significant differences between treatments.

Table 3. Antioxidant enzyme activities in Aleppo pine seedlings exposed continuously to the different ozone treatments. Means ± SE, $n = 3$–6. ex.POD = extracellular POD; cel.POD = cellular POD; KCNr.SOD = KCN-resistant SOD; POD activities expressed as ΔAbs_{470} min^{-1} g^{-1} fwt. Other activities expressed as Unit g^{-1} fwt. Different letters indicate significant differences among O_3 treatments on each sampling date

		ex.POD	cel.POD	GR	CAT	KCNr. SOD	CuZnSOD
1993 needles							
May 94	CFA	11.1 ± 2.2	178 ± 9b*	1170 ± 109	95 ± 23	22.6 ± 2.2a*	90 ± 7
	NF	10.5 ± 3.1	183 ± 17b	1269 ± 150	125 ± 18	31.9 ± 3.4b	87 ± 7
	NFA+40	10.8 ± 1.7	138 ± 4a	984 ± 89	92 ± 3	20.5 ± 2.3a	82 ± 7
Jul 94	CFA	22.7 ± 3.1b*	187 ± 9	420 ± 70	35 ± 7a†	19.7 ± 5.8	57 ± 5
	NF	11.5 ± 0.6a	157 ± 13	473 ± 15	38 ± 3a	14.3 ± 3.1	68 ± 5
	NFA+40	10.9 ± 1.2a	167 ± 8	535 ± 40	75 ± 8b	15.5 ± 3.5	60 ± 2
Sep 94	CFA	10.6 ± 3	185 ± 13b†	676 ± 106	84 ± 15	18.9 ± 2.3	86 ± 12
	NF	6.4 ± 1.1	103 ± 8a	883 ± 42	121 ± 9	19.7 ± 3.1	95 ± 10
	NFA+40	6.6 ± 1.8	152 ± 14b	744 ± 108	101 ± 8	13.2 ± 1.1	82 ± 3
Nov 94	CFA	14.1 ± 2.7b†	186 ± 11	641 ± 96	50 ± 3	28.4 ± 2.7	72 ± 5
	NF	6.4 ± 0.9a	161 ± 11	861 ± 132	78 ± 11	42.9 ± 5	81 ± 5
	NFA+40	4.8 ± 1a	155 ± 14	997 ± 100	96 ± 14	27.1 ± 3.7	64 ± 4
Feb 95	CFA	8.8 ± 1.1	198 ± 13	1060 ± 66	111 ± 10	25 ± 1.8	64 ± 6
	NF	9.3 ± 0.7	254 ± 38	1013 ± 34	106 ± 10	22.3 ± 1.1	68 ± 7
	NFA+40	11.7 ± 1	203 ± 15	1115 ± 29	130 ± 13	25.3 ± 2.5	53 ± 5
1994 needles							
Sep 94	CFA	$2 \times 10^{-4} \pm 8 \times 10^{-5}$	82 ± 7	1177 ± 61	154 ± 12	55.2 ± 4.3	144 ± 16
	NF	$3 \times 10^{-4} \pm 8 \times 10^{-5}$	90 ± 5	1093 ± 91	143 ± 16	57.7 ± 3.4	159 ± 6
	NFA+40	$4 \times 10^{-4} \pm 1 \times 10^{-4}$	91 ± 4	1208 ± 68	164 ± 9	53.5 ± 3.6	147 ± 8
Nov 94	CFA	1.2 ± 0.3	188 ± 5	2373 ± 102	280 ± 10	82.1 ± 8.6	147 ± 8
	NF	1.7 ± 0.4	174 ± 8	2089 ± 108	255 ± 11	69.6 ± 6.9	136 ± 6
	NFA+40	2.5 ± 0.7	153 ± 22	1920 ± 125	244 ± 22	79 ± 2.4	144 ± 10
Feb 95	CFA	7.1 ± 0.6a‡	209 ± 17	2043 ± 115	231 ± 12	83.3 ± 7	151 ± 13a*
	NF	16.4 ± 2.4b	242 ± 22	2438 ± 239	250 ± 15	71.7 ± 8.6	136 ± 6a
	NFA+40	36.3 ± 5.4c	253 ± 10	2784 ± 354	333 ± 44	76.7 ± 13.6	228 ± 33b

(continued on next page)

Table 3. (Continued from previous page)

		ex.POD	cel.POD	GR	CAT	KCNr. SOD	CuZnSOD
May95	CFA	12 ± 2.1a‡	287 ± 26	1534 ± 62a*	156 ± 23	56.1 ± 4.3	94 ± 12
	NF	21.8 ± 3.4b	257 ± 22	1810 ± 35b	234 ± 3	69.6 ± 6.6	124 ± 18
	NFA+40	14.1 ± 1a	292 ± 10	1442 ± 45a	204 ± 14	63.4 ± 3.7	94 ± 5
Jul 95	CFA	8.5 ± 0.3ab*	132 ± 10a‡	641 ± 67a*	53 ± 6a*	47.3 ± 6.2	90 ± 7
	NF	11.2 ± 1.6b	202 ± 5b	820 ± 15b	82 ± 10b	44.9 ± 2.6	70 ± 6
	NFA+40	4.9 ± 0.7a	140 ± 7a	668 ± 40a	59 ± 5a	53.9 ± 1.4	86 ± 11
Sep 95	CFA	14.7 ± 3	183 ± 15ab*	886 ± 97b‡	41 ± 8a†	37.4 ± 3.1a*	79 ± 6
	NF	12.8 ± 1	142 ± 10a	665 ± 37a	67 ± 17ab	53.8 ± 5.6b	82 ± 10
	NFA+40	9.4 ± 1.8	204 ± 12b	1092 ± 55c	86 ± 6b	53.8 ± 4.6b	70 ± 4
Mar 96	CFA	6.8 ± 1.2	170 ± 9	904 ± 63a*	73 ± 7	46.5 ± 3.7	73 ± 5
	NF	8.8 ± 1.3	167 ± 10	1007 ± 99ab	43 ± 9	45.5 ± 4.1	49 ± 3
	NFA+40	6 ± 1.1	160 ± 7	1152 ± 49b	72 ± 8	51.6 ± 3.8	62 ± 5
1995 needles							
Jul 95	CFA	–	39 ± 3	882 ± 17b*	107 ± 5b*	71.7 ± 4.9	89 ± 8
	NF	–	–	–	–	–	–
	NFA+40	–	41 ± 3	549 ± 85a	70 ± 7a	60.6 ± 6.9	62 ± 15
Sep 95	CFA	–	68 ± 4	1672 ± 90	119 ± 15	68.2 ± 3.1	72 ± 5
	NF	–	–	–	–	–	–
	NFA+40	–	72 ± 4	1478 ± 50	109 ± 8	63.8 ± 5	85 ± 7
Mar 96	CFA	13.8 ± 2.2	175 ± 8	2067 ± 81ab*	170 ± 11	87.9 ± 9ab†	103 ± 11b†
	NF	9 ± 1.7	184 ± 25	1795 ± 93a	145 ± 26	62 ± 9.7a	52 ± 6a
	NFA+40	16.6 ± 1.8	219 ± 12	2251 ± 98b	197 ± 18	89.1 ± 7.6b	125 ± 10b

* $p \leqslant 0.05$.
† $p \leqslant 0.01$.
‡ $p \leqslant 0.001$.

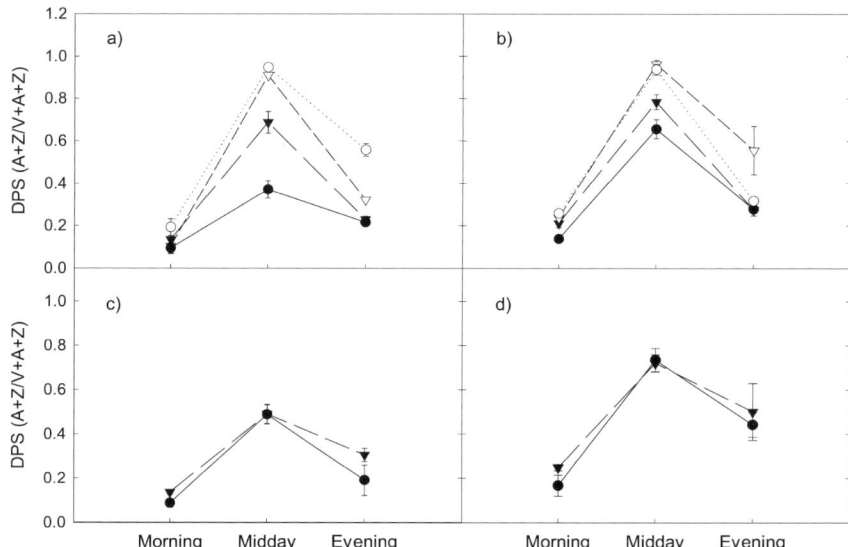

Figure 4. De-epoxidation state of the xanthophyll cycle pool $(A+Z)/(V+A+Z)$ in 1-year-old needles (a), (c) and current-year needles (b), (d) during summer (a), (b) and winter (c), (d) in trees exposed to ozone and/or soil water deficit: ● CFA WW; ○ CFA WS; ▼ NFA+40 WW; ▽ NFA+40 WS. Means ± SD, $n = 4$–12.

effect of O_3 on antioxidant systems was recorded during winter in 10-month-old needles (1994 needles measured in February 1995, Table 3). At this stage, seedlings previously exposed to the highest O_3 levels exhibited significantly higher extracellular-POD and CuZn SOD activity and a trend, though not statistically significant ($p < 0.1$), toward an increase in CAT, GR, and cellular-POD activity compared with plants grown in clean air. Similar responses were observed when antioxidant enzyme activities were expressed on a chlorophyll basis.

Extracellular-POD activity showed the greatest increase in response to O_3 exposure, up to fivefold in the NFA+40 treatment and twice in the NFA treatment during the winter (February 1995). The increase in extracellular-POD activity was due to an increase in the activity of acid (1.5 times higher activity in NFA+40 compared with CFA), but especially basic (3.3 times higher) isoperoxidases (Fig. 5). Although acid extracellular-POD activity was similar throughout the year, the activity of basic isozymes increased during spring and summer in control trees. In contrast, the basic POD activity of the trees exposed to O_3 increased sharply during wintertime before the activation of these activities was detected in CFA plants.

+9.5   pH   +3.5

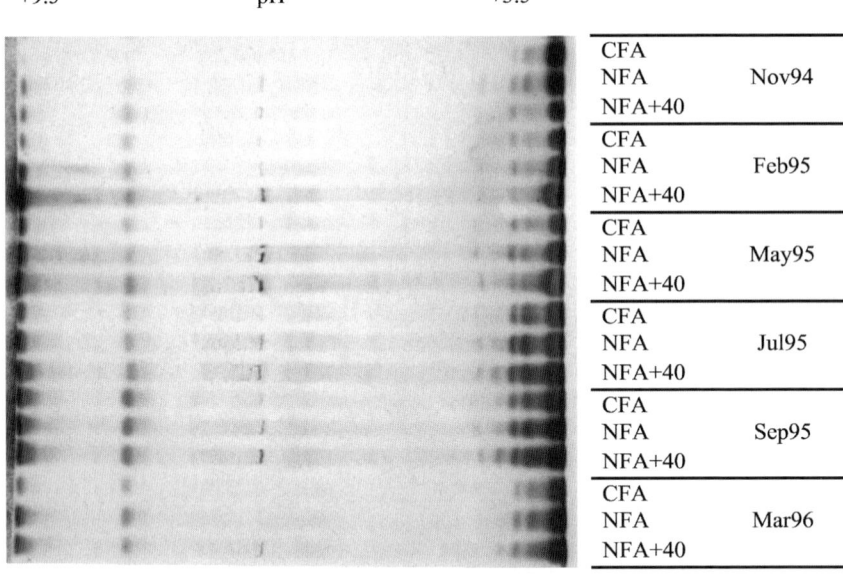

CFA		
NFA		Nov94
NFA+40		
CFA		
NFA		Feb95
NFA+40		
CFA		
NFA		May95
NFA+40		
CFA		
NFA		Jul95
NFA+40		
CFA		
NFA		Sep95
NFA+40		
CFA		
NFA		Mar96
NFA+40		

Figure 5. Extracellular isoperoxidase pattern of 1994-needles exposed to the different ozone treatments. Isoperoxidases were separated by isoelectricfocusing (IEF) on a 3.5–9.5 pH gradient following Elvira et al. (1998).

When Aleppo pine seedlings were subjected to water stress, 1-year-old needles exhibited a decrease in cellular-POD and KCN-resistant SOD activities, while current-year needles exhibited a significant increase in GR and CuZn SOD (Data presented in Alonso et al., 2001). Ozone exposure in combination with water stress resulted in significant decreases in POD, GR, CAT and KCN-resistant SOD activities in current-year needles.

3.5. Growth

Aleppo pine seedlings exhibited active growth throughout the year with the highest rates recorded during springtime (Fig. 6). No significant effects of O_3 were observed on height and stem diameter growth after 3 years of continuous exposure to the respective O_3 treatments. In contrast, water stress significantly decreased stem and diameter growth up to 16%. The combination of O_3 and water stress resulted in no additional decreases in relative growth rate.

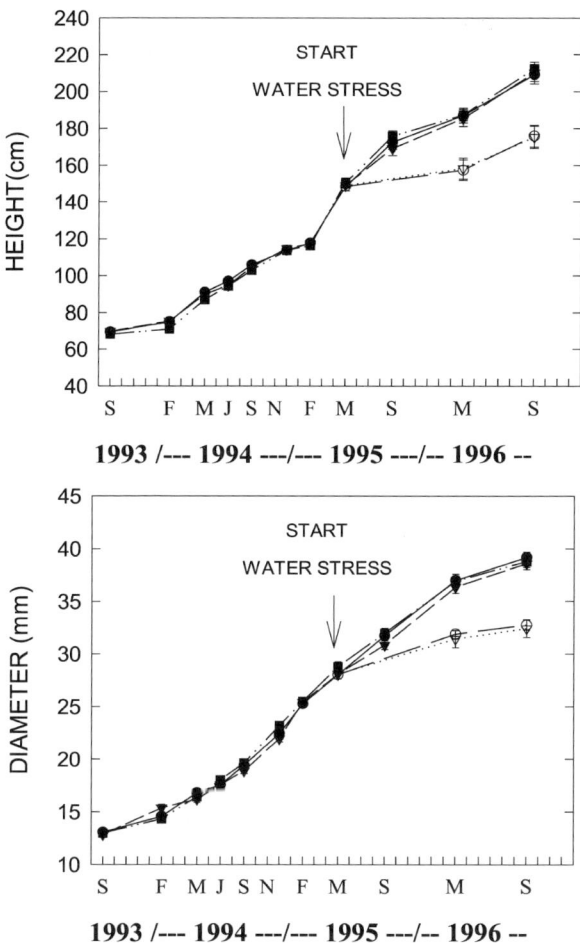

Figure 6. Incremental stem and height growth of Aleppo pine seedlings grown in the different ozone treatments: ● CFA WW; ○ CFA WS; ▼ NFA+40 WW; ▽ NFA+40 WS; ■ NFA WW. Means ± SD. $n = 25$–35 for WW treatments and $n = 12$ for WS treatments.

4. Discussion

The European Commission (EC) Directive 92/72/EC established two guidelines to protect vegetation from O_3 injury: 100 ppb as an hourly average and 33 ppb as a 24-h average. Ambient O_3 concentrations at the Ebro Delta over the experimental period were almost continuously above the 33 ppb 24-h average value. Ambient O_3 levels each year exceeded the UN-ECE critical level for the protection of forest trees (10 000 ppbh in 6 months, April to September).

The experiments involved realistic O_3 exposures, as concentrations in the same range of those reported in NFA+40 treatment have been recorded at different sites on the Spanish Mediterranean coast (Millán et al., 1996; Gimeno et al., 1996; Sanz and Millán, 1998).

The mild temperatures at the field site, along with the fact that trees were irrigated for experimental purposes, enabled Aleppo pine seedlings to keep growing throughout the year. A similar year-round pattern of growth is also observed in natural forests, where trees frequently show several flushes during the same year, depending on weather conditions (Gil et al., 1996). Despite the fact that trees showed continuous growth throughout the year, most of the physiological parameters analyzed showed clear seasonal variations. Chlorophyll content, net photosynthesis, and stomatal conductance decreased during the summer even when plants were not subjected to water stress. For instance, chlorophyll levels exhibited a 40–50% reduction during the summer, followed by recovery during autumn and winter regardless of O_3 treatment (data presented in Elvira and Gimeno, 1996). A similar pattern was also observed for changes in carotenoid content. Summer chlorosis has also been observed in natural forests of Aleppo pine (Davison et al., 1995) and in other Mediterranean species (Kyparissis et al., 1995). The recovery in chlorophyll levels observed during the winter suggests that the reduction of photosynthetic pigments during summer might afford a protection mechanism to avoid the absorption of potentially damaging excitation energy (Kyparissis et al., 1995; Elvira et al., 1998).

Reductions in net photosynthesis and stomatal conductance during the summer have frequently been reported for other Mediterranean species in their natural environment, usually associated with water stress (Epron and Dreyer, 1993; Damesin and Rambal, 1995; Faria et al., 1998). In our experiment, Aleppo pine seedlings exhibited seasonal variations in gas exchange independent of water status. Factors other than soil water deficit, such as vapor pressure deficit, high temperatures, or high levels of irradiance, presumably contributed to the decline in photosynthesis (Faria et al., 1998).

Changes in physiological parameters induced by seasonal weather variations and/or water stress, were more important than changes induced by O_3. Nevertheless, O_3 decreased the rate of net photosynthesis, stomatal conductance, and the levels of photosynthetic pigments (data presented in Elvira and Gimeno, 1996) and triggered alterations in detoxification systems. These effects were detectable even when exposure to the pollutant was restricted to the summer months (i.e., the season with the lowest O_3 uptake because of the low stomatal conductance values).

Ozone-induced declines in A and g_s have been reported previously for Aleppo pine (Anttonen et al., 1998), as in other tree species (Chappelka and Samuelson, 1998). Despite these results, long-term effects of O_3 exposure on photosynthesis are not very conclusive and increases, decreases, or no effects could be found. Although some data suggest that photosynthesis can be lowered without a concomitant fall in stomatal conductance (Heath and Taylor, 1997), Aleppo pine seedlings exhibited reductions in g_s in the same range as net photosynthesis. However, the higher maximum stomatal conductance exhibited by plants subjected to combined O_3 and water stress compared with those grown in charcoal-filtered air might be indicative of poorer stomatal control. Such "sluggish" behavior of stomata in O_3-exposed foliage has been reported previously (Tjoelker et al., 1995; Grulke, 1999). This loss of stomatal regulation might be particularly dangerous for Mediterranean species that have to cope with harsh environmental conditions during the summer.

Ozone-induced reductions in gas exchange were observed before changes in chlorophyll content. Similar findings have been reported for Aleppo pine (Manninen et al., 1999) and other conifers (Takemoto et al., 1997). Thus, chlorosis does not seem to be a primary effect of O_3 exposure, but rather a secondary effect due to impaired photosynthetic capacity (Heath and Taylor, 1997). Moreover, when CO_2 fixation is reduced without reducing light capture, there is the potential for photoinhibition (Demmig-Adams and Adams III, 1992). On the one hand, this may explain the decline in chlorophyll levels, while on the other hand, photoinhibition also affects defense mechanisms. In this sense, O_3 altered some of the photoprotective capabilities of Aleppo pine. The activation of the xanthophyll cycle pool at midday was higher in seedlings exposed to O_3 than in trees raised in charcoal-filtered air. This could be linked with the reduction in CO_2 assimilation rates and was also observed in plants exposed to O_3 only during summer months (Elvira et al., 1998). Other tree species have also exhibited an activation of the xanthophyll cycle induced by O_3 (Mikkelsen et al., 1995; Reichenauer and Bolhar-Nordenkampf, 1999). These findings suggest the involvement of photooxidative stress in the development of O_3-induced injury (see Heath and Taylor, 1997; Tausz et al., 1999). When plants were exposed to a combination of ozone and water stress, a further increase of DPS index was recorded in current-year needles while 1-year-old needles showed lower activation of xanthophyll cycle components than plants raised in charcoal-filtered air. This behavior of 1-year-old needles was observed even though chlorophyll and net photosynthetic rates were similar in water-stressed plants regardless of the O_3 treatment (data presented in Inclán et al., 1998a and Alonso et al., 2001). Thus, effects observed on xanthoplyll cycle intermediates indicate a greater capacity to respond to stress in current-year needles than in 1-year-old needles.

Ozone exposure also caused alterations in the antioxidant systems of Aleppo pine needles. The most significant increase in the activity of the antioxidant-related enzymes coincided with the period of highest O_3 uptake: during winter. At this time of year, O_3 concentrations were lower than in spring and early summer but high stomatal conductance resulted in a higher rate of uptake of the pollutant. During the summer months, stomatal closure reduced O_3 uptake to a minimum. However, a trend toward an increase in some antioxidant activities in plants exposed to elevated O_3 was also detected. Ozone effects on antioxidant enzymes also depended on needle age, as 1-year-old needles did not show the same increase in antioxidant-related enzyme activities as was observed in current-year needles.

The greatest change induced by O_3 in Aleppo pine was found to be the activity of extracellular POD. Similarly, when plants were exposed to O_3 only during summer months, extracellular-POD activation was the first effect recorded (Elvira et al., 1998). Extracellular-POD activity thus appeared to be more O_3 responsive than cellular POD. Similar findings have been reported for other species (Castillo et al., 1984, 1987). These results suggest that the principal effects of O_3 are located in the apoplast with the plasma membrane as the chief target (Kangasjärvi et al., 1994; Alscher et al., 1997).

Ozone exposure did not induce new extracellular-POD isozymes, although the response of basic isoperoxidases was more pronounced than that in acid isozymes. The function of isoperoxidases in conifer physiology is not clear. Acid isoperoxidases seem more related to lignification processes, while basic forms are related to ethylene and indolacetic acid metabolism (Gaspar et al., 1991). Also, basic extracellular isoperoxidases have been linked to a detoxification role under O_3 exposure, using ascorbate as an electron donor (Gaspar et al., 1985; Castillo and Greppin, 1986).

When Aleppo pine seedlings were exposed to both O_3 and water stress, some antioxidant activities (POD, GR, CAT and KCN-resistant SOD) decreased in current-year needles compared with plants grown in charcoal-filtered air, which differed in the response induced by water stress alone. This reduction of antioxidant activities could mean that the protection capacity of these needles was overwhelmed when plants were exposed to both stresses.

No reduction in tree relative growth rates was detectable in either experiment, although small decreases in net photosynthesis compounded over many years might produce significant growth reductions (Teskey, 1995). In fact, Barnes et al. (2000) reported growth decreases in this species after 4 years of repeated O_3 exposure. Although O_3 did not reduce tree growth, seedlings exposed to the highest levels (NFA+40) throughout the year exhibited significantly lower (up to 10%) needle biomass and total aboveground biomass accumulation compared with plants grown in charcoal-filtered air (data in Inclán et al., 1998b). Similarly, water stress resulted in significant decreases in both needle weight (23%) and total aboveground biomass (27%). Ozone and water stress were found to affect biomass additively. Future studies need to address the combined action of O_3 and other environmental stresses on Mediterranean tree species.

5. Conclusions

Ambient O_3 concentrations on the eastern coast of Spain regularly exceed international guidelines for the protection of forest trees. Aleppo pine physiology was affected by O_3 exposure, but the response was strongly modified by environmental conditions. Ozone exposure caused alterations in the protective systems and decreased CO_2 assimilation rates, stomatal conductance, photosynthetic pigments, and biomass accumulation. These effects were also observed when O_3 exposure was restricted to the summer months, i.e., the period when low stomatal conductance resulted in minimal rates of pollutant uptake. Based on the high values of O_3 uptake observed during winter and the fact that the tree exhibited continuous growth, international guidelines to protect Mediterranean trees should consider O_3 concentrations throughout the year. The interactive effects of O_3 and water stress did not induce a further decrease in gas exchange or growth rates, but their combined effects produced detrimental effects in some protective systems of Aleppo pine, such as xanthophyll cycle and antioxidant enzymes, and decreased needle biomass accumulation. These responses suggest that O_3 exposure may impair the ability of this species to withstand water stress in its natural environment, as suggested by previous authors (Wellburn and Wellburn, 1994; Gerant et al., 1996; Barnes et al., 2000). The reduction in net photosynthesis induced by O_3 during the winter and spring (the most active period), along with a weaker capacity for photoprotection during the summer months (the most stressful period), could be considered important threats to the vitality of Aleppo pine forests, and are worthy of detailed consideration to predict the future of Mediterranean vegetation under changing climate conditions.

Acknowledgements

This research was funded by the projects LIFE97/ENV/000336 and CICYT CLI97-0735-C0302. The authors are especially grateful to Sonia Sánchez, José Manuel Gil, and Modesto Mendoza for their excellent help in the laboratory and fieldwork.

References

Aebi, H., 1983. Catalase. In: Bergmeier, H. (Ed.), Methods of Enzymatic Analysis. Verlag Chemie, Weinheim, pp. 121–126.
Alonso, R., Elvira, S., Castillo, F.J., Gimeno, B.S., 1999. Antioxidative defense and photoprotection in *Pinus halepensis* induced by Mediterranean conditions and ozone exposure. Free Radical Res. 31, 59–65.

Alonso, R., Elvira, S., Castillo, F.J., Gimeno, B.S., 2001. Interactive effects of ozone and water stress on pigments and activities of antioxidative enzymes in *Pinus halepensis*. Plant Cell Environ. 24 (9), 905–916.

Alscher, R.G., Donahue, J.L., Cramer, C.L., 1997. Reactive oxygen species and antioxidants: relationships in green cells. Physiol. Plant. 100, 224–233.

Anttonen, S., Herranen, J., Peura, P., Kärenlampi, L., 1995. Fatty acids and ultrastructure of ozone-exposed Aleppo pine (*Pinus halepensis* Mill.) needles. Environ. Pollut. 87, 235–242.

Anttonen, S., Kittilä, M., Kärenlampi, L., 1998. Impacts of ozone on Aleppo pine needles: visible symptoms, starch concentrations and stomatal response. Chemosphere 36, 663–668.

Barnes, J.D., Balaguer, L., Manrique, E., Elvira, S., Davison, A.W., 1992. A reappraisal of the use of DMSO for the extraction and determination of chlorophylls *a* and *b* in lichens and higher plants. Environ. Exp. Bot. 32, 85–100.

Barnes, J.D., Gimeno, B.S., Davison, A.W., Dizengremel, P., Gerant, D., Bussotti, F., Velissariou, D., 2000. Air pollution impacts on pine forests in the Mediterranean basin. In: Ne'eman, G., Traband, L. (Eds.), Ecology, Biogeography and Management of Pinus Halepensis and P. Brutia Forest Ecosystems in the Mediterranean Basin. Backhuys Publishers, Leiden, The Netherlands, pp. 391–404.

Bussotti, F., Ferretti, M., 1998. Air pollution, forest condition and forest decline in Southern Europe: an overview. Environ. Pollut. 101, 49–65.

Castillo, F.J., Penel, C., Greppin, H., 1984. Peroxidase release induced by ozone in *Sedum album* leaves. Plant Physiol. 74, 846–851.

Castillo, F.J., Greppin, H., 1986. Balance between anionic and cationic extracellular peroxidase activities in *Sedum album* leaves after ozone exposure. Analysis by high-performance liquid chromatography. Physiol. Plant. 68, 201–208.

Castillo, F.J., Miller, P.R., Greppin, H., 1987. Extracellular biochemical markers of photochemical oxidant air pollution damage to Norway spruce. Experientia 43 (1), 111–115.

Castillo, F.J., Greppin, H., 1988. Extracellular ascorbic acid and enzyme activities related to ascorbic acid metabolism in *Sedum album* L. leaves after ozone exposure. Environ. Exp. Bot. 28 (3), 231–238.

Chappelka, A.H., Samuelson, L.J., 1998. Ambient ozone effects on forest trees of the eastern United States: a review. New Phytol. 139, 91–108.

Damesin, C., Rambal, S., 1995. Field study of leaf photosynthetic performance by a Mediterranean deciduous oak tree (*Quercus pubescens*) during a severe summer drought. New Phytol. 131, 159–167.

Davison, A.W., Velissariou, D., Barnes, J.D., Inclán, R., Gimeno, B.S., 1995. The use of Aleppo pine, *Pinus halepensis* Mill. as a bioindicator of ozone stress in Greece and Spain. In: Munavar, M.S.P.B. (Ed.), Bioindicators of Environmental Health. Ecovision World Monograph Series, Academic Publishing, pp. 63–72.

Demmig-Adams, B., Adams III, W.W., 1992. Photoprotection and other responses of plants to high light stress. Annu. Rev. Plant Physiol. Plant Mol. Biol. 43, 599–626.

Elvira, S., Gimeno, B.S., 1996. A contribution to the set-up of ozone critical levels for forest trees in Mediterranean areas. Results from the exposure of Aleppo pine (*Pinus halepensis* Mill.) seedling in open-top chambers. In: Kärenlampi, L., Skärby, L. (Eds.), Critical Levels for Ozone in Europe: Testing and Finalizing the Concepts. University of Kuopio, pp. 169–182.

Elvira, S., Alonso, R., Castillo, F.J., Gimeno, B.S., 1998. On the response of pigments and antioxidants of *Pinus halepensis* seedlings to Mediterranean climatic factors and long-term ozone exposure. New Phytol. 138, 419–432.

Epron, D., Dreyer, E., 1993. Long-term effects of drought on photosynthesis of adult oak trees [*Quercus petraea* (Matt.) Liebl. and Quercus robur L.] in a natural stand. New Phytol. 125, 381–389.

Faria, T., Silverio, D., Breia, E., Cabral, R., Abadía, A., Abadía, J., Pereira, J.S., Chaves, M.M., 1998. Differences in the response of carbon assimilation to summer stress (water deficits, high light and temperature) in four Mediterranean tree species. Physiol. Plant. 102, 419–428.

Fumagalli, I., Gimeno, B.S., Velissariou, D., Mills, G., De Temmerman, L., Fuhrer, J., 2001. Evidence of ozone-induced adverse effects on Mediterranean vegetation. Atmos. Environ. 35, 2583–2587.

Gaspar, T., Penel, C., Castillo, F.J., Greppin, H., 1985. A two-step control of basic and acidic peroxidases and its significance for growth and development. Physiol. Plant. 64, 418–423.

Gaspar, T., Penel, C., Hagege, D., Greppin, H., 1991. Peroxidases in plant growth, differentiation and development processes. In: Lobarzewski, J., Greppin, H., Penel, C., Gaspar, T. (Eds.), Biochemical, Molecular and Physiological Aspects of Plant Peroxidases. University of Geneva.

Gerant, D., Podor, M., Grieu, P., Afif, D., Cornu, S., Morabito, D., Banvoy, J., Robin, C., Dizengremel, P., 1996. Carbon metabolism enzyme activities and carbon partitioning in *Pinus halepensis* Mill. exposed to mild drought and ozone. J. Plant Physiol. 148, 142–147.

Gil, L., Díaz-Fernández, P.M., Jiménez, M.P., Roldán, M., Alía, R., Agúndez, D., De Miguel, J., Martín, S., De Tuero y Reina, M., 1996. Regiones de procedencia. *Pinus halepensis*. Organismo Autónomo de Parques Nacionales. Ministerio de Medio Ambiente, España.

Gimeno, B.S., Velissariou, D., Barnes, J.D., Inclán, R., Peña, J.M., Davison, A.W., 1992. Daños visibles por ozono en acículas de *Pinus halepensis* Mill. en Grecia y España. Ecología 6, 131–134.

Gimeno, B.S., Velissariou, D., Schenone, G., Guardans, R., 1994. Ozone effects on the Mediterranean region: an overview. In: Fuhrer, J., Achermann, B. (Eds.), Critical Levels for Ozone. A UN-ECE workshop report. FAC Liebefeld, pp. 122–136.

Gimeno, B.S., Cabal, H., Barquero, C.G., Artiñano, B., Vilaclara, E., Guardans, R., 1996. Ozone exceedance maps in Catalunya–problems and criteria. In: Kärenlampi, L., Skärby, L. (Eds.), Critical Levels for Ozone in Europe: Testing and Finalizing the Concepts. University of Kuopio, pp. 228–233.

Grulke, N.E., 1999. Physiological responses of ponderosa pine to gradients of environmental stressors. In: Miller, P.R., McBride, J.R. (Eds.), Oxidant Air Pollution Impacts in the Montane Forests of Southern California. Springer-Verlag, pp. 126–163.

Heath, R.L., Taylor Jr., G.E., 1997. Physiological processes and plant responses to ozone exposure. In: Forest Decline and Ozone. Springer-Verlag, pp. 317–368.

Inclán, R., Alonso, R., Pujadas, M., Terés, J., Gimeno, B.S., 1998a. Ozone and drought stress: Interactive effects on gas exchange in Aleppo pine (*Pinus halepensis* Mill.). Chemosphere 675, 685–690.

Inclán, R., Alonso, R., Gimeno, B.S., 1998b. Interactions of ozone exposure and water stress on *Pinus halepensis* performance. In: De Kok, L.J., Stulen, I. (Eds.), Responses of Plant Metabolism to Air Pollution and Global Change. Backhuys Publishers, Leiden, The Netherlands, pp. 337–340.

Kangasjärvi, J., Talvinen, J., Utriainen, M., Karjalainen, R., 1994. Plant defense systems induced by ozone. Plant Cell Environ. 17, 783–794.

Kyparissis, A., Petropoulou, Y., Manetas, Y., 1995. Summer survival of leaves in a soft-leaved shrub (*Phlomis fruticosa* L., *Labiatae*) under Mediterranean field conditions: avoidance of photoinhibitory damage through decreased chlorophyll contents. J. Exp. Bot. 46 (293), 1825–1831.

Lorenzini, G., 1993. Towards an ozone climatology over the Mediterranean basin: environmental aspects. Medit. 2, 53–59.

Manninen, S., Le Thiec, D., Rose, C., Nourrison, G., Radnai, F., Garrec, J.P., Huttunen, S., 1999. Pigment concentrations and ratios of Aleppo pine seedlings exposed to ozone. Water Air Soil Pollut. 116, 333–338.

Mikkelsen, T.N., Dodell, B., Lütz, C., 1995. Changes in pigment concentration and composition in Norway spruce induced by long-term exposure to low levels of ozone. Environ. Pollut. 87, 197–205.

Millán, M.M., Salvador, R., Mantilla, E., Artiñano, B., 1996. Meteorology and photochemical air pollution in southern Europe: experimental results from EC research projects. Atmos. Environ. 30, 1909–1924.

Montoya, R., 1995. Red de seguimiento de daños en los montes. Daños originados por la sequia en 1994. Cuadernos de la S.E.C.F. 2, 83–97.

Naveh, Z., Steinberger, E.H., Chaim, S., Rotmann, A., 1980. Photochemical oxidants–A threat to Mediterranean forest and upland ecosystems. Environ. Cons. 7, 301–309.

Pujadas, M., Gimeno, B.S., Terés, J., 1997. La experiencia española en el diseño de sistemas experimentales para el estudio de efectos producidos por contaminantes gaseosos sobre especies vegetales. Boletín de Sanidad Vegetal-Plagas 23, 39–54.

Reichenauer, T.G., Bolhar-Nordenkampf, H.R., 1999. Does ozone exposure protect from photoinhibition? Free Radical Res. 31, 193–198.

Reinert, R., Gimeno, B.S., Salleras, J.M., Bermejo, V., Ochoa, M.J., Tarruel, A., 1992. Ozone effects on watermelon plants at the Ebro Delta (Spain): symptomatology. Agric. Ecosyst. Environ. 38, 41–49.

Sanz, M.J., Millán, M.M., 1998. The dynamics of aged air masses and ozone in the western Mediterranean: relevance to forest ecosystems. Chemosphere 36, 1089–1094.

Sanz, M.J., Calatayud, V., Calvo, E., 2000. Spatial pattern of ozone injury in Aleppo pine related to air pollution dynamics in a coastal-mountain region of eastern Spain. Environ. Pollut. 108, 239–247.

Schenone, G., 1993. Air pollution and vegetation in the Mediterranean area. Medit. 2, 49–52.

Takemoto, B.K., Bytnerowicz, A., Dawson, P.J., Morrison, C.L., Temple, P.J., 1997. Effects of ozone on *Pinus ponderosa* seedlings: comparison of responses in the first and second growing seasons of exposure. Can. J. Forest Res. 27, 23–30.

Tausz, M., Bytnerowicz, A., Weidner, W., Arbaugh, M.J., Padgett, P., Grill, D., 1999. Pigments and photoprotection in needles of *Pinus ponderosa* trees with and without symptoms of ozone injury. Phyton 39, 219–224.

Teskey, R.O., 1995. Synthesis and conclusions from studies of southern commercial pines. In: Fox, S., Mickler, R.A. (Eds.), Impacts of Air Pollutants on Southern Pine Forests. Springer-Verlag, pp. 467–490.

Tjoelker, M.G., Volin, J.C., Oleksyn, J., Reich, P.B., 1995. Interaction of ozone pollution and light effects on photosynthesis in a forest canopy experiment. Plant Cell Environ. 18, 895–905.

Val, J., Monge, E., Baker, N.R., 1994. An improved HPLC method for rapid analysis of the xanthophyll cycle pigments. J. Chromatogr. Sci. 32, 286–289.

Velissariou, D., Davison, A.W., Barnes, J.D., Pfirrmann, T., McClean, D.C., Holevas, C.D., 1992. Effects of air pollution on *Pinus halepensis* (Mill.): pollution levels in Attica, Greece. Atmos. Environ. 26, 373–380.

Velissariou, D., Gimeno, B.S., Badiani, M., Fumagalli, I., Davison, A.W., 1996. Records of O_3 visible injury in the ECE mediterranean region. In: Kärenlampi, L., Skärby, L. (Eds.), Critical Levels for Ozone in Europe: Testing and Finalizing the Concepts. University of Kuopio, pp. 343–350.

von Caemmerer, S., Farquhar, G.D., 1981. Some relationships between the biochemistry of photosynthesis and the gas exchange of leaves. Planta 153, 376–387.

Wellburn, F.A.M., Wellburn, A.R., 1994. Atmospheric ozone affects carbohydrate allocation and winter hardiness of *Pinus halepensis* (Mill.). J. Exp. Bot. 45 (274), 607–614.

Air Pollution, Global Change and Forests in the New Millennium
D.F. Karnosky et al., editors
© 2003 Elsevier Ltd. All rights reserved.

Chapter 10

Ozone affects Scots pine phenology and growth

S. Manninen*

*Department of Ecology and Systematics, University of Helsinki, P.O. Box 65 (Viikinkaari 1),
FIN-00014 Helsinki, Finland*

Botanical Institute, University of Gothenburg, P.O. Box 461, SE-40530 Gothenburg, Sweden

R. Sorjamaa

*Department of Applied Physics, University of Kuopio, P.O. Box 1627,
FIN-70211 Kuopio, Finland*

*Department of Physics (Biophysics), University of Oulu, P.O. Box 3000,
FIN-90014 Oulu, Finland*

S. Kurki, N. Pirttiniemi, S. Huttunen

Department of Biology, University of Oulu, P.O. Box 3000, FIN-90014 Oulu, Finland

Abstract

An open-top chamber (OTC) fumigation experiment with 10- to 15-year-old
Scots pines (*Pinus sylvestris* L.) was started in autumn 1997 at the University of
Oulu ($65°$N, $25°$E). There were six non-filtered air (NF) and six open-field (AA)
control pines. The six NF + O_3 pines were exposed from summer 1998 onwards
to target ozone (O_3) concentrations of ambient air +40 ppb in May, ambient air
+30 ppb in June, ambient air +20 ppb in July, ambient air +10 ppb in August
and ambient air in September. The accumulated O_3 exposure over a threshold
of 40 ppb (AOT40) in the NF + O_3 OTCs was 12.9 ppmh in the summer of
1998, but only 1.1 and 1.8 ppmh in the summers of 1999 and 2000, respectively,
because fumigation started late. The respiration of previous-year needles was
increased by exposure to O_3; and they also showed a decreasing trend in net
photosynthesis and an increasing trend in the internal CO_2 concentration with
increasing O_3 exposure in the summer of 1998. The results on needle carbon
(C) contents suggested O_3-related changes in C allocation, and the chlorophyll
a + b/carotenoid ratio in the current-year needles of the NF + O_3 pines was also
lower than that in the current-year needles of the NF controls in November 1999.
The slightly elevated O_3 concentrations caused clear physiological responses in
Scots pine needles, which may, over a longer period, result growth reductions,
as was suggested by the non-significant changes in the current-year shoot (18%

---

*Corresponding author.

DOI:10.1016/S1474-8177(03)03010-9

increase in main shoots vs. 19% decrease in branches) and needle (15% and 10% decrease in main shoots and branches, respectively) growth of the $NF + O_3$ trees recorded in late July 2000. It seems that peak O_3 episodes during early summer are harmful to subarctic Scots pines.

1. Introduction

The critical level concept based on the accumulated ozone (O_3) exposure over a threshold of 40 ppb (AOT40) has been developed in Europe to protect vegetation from O_3 damage. In the Level I approach, a critical O_3 level of 10 ppmh has been proposed for European forests, on the basis of a few dose-response studies (Kärenlampi and Skärby, 1996). The major uncertainties in defining the critical level values for O_3 relate to the choice of response parameters and species (Fuhrer et al., 1997). There is also need for a longer term perspective. Trees are long-lived organisms, and it may take years for impacts to become apparent (Fuhrer and Achermann, 1999).

Among conifers, Scots pine (*Pinus sylvestris* L.) and Norway spruce (*Picea abies* [L.] Karst.) are the most important species in Scandinavia, both ecologically and economically. The O_3 sensitivity of Scots pine has been studied much less than that of Norway spruce. Recent O_3 studies on Scots pine seedlings include open-top chamber (OTC) experiments by Broadmeadow and Jackson (2000), Landolt et al. (2000) and Utriainen et al. (2000), for example. In the Liphook Forest Fumigation Experiment, seedlings were exposed under field conditions (McLeod and Skeffington, 1995), as was also done by Utriainen and Holopainen (1999). Kellomäki and Wang (1997, 1998) studied the O_3 response of 30-year-old Scots pines using OTCs, whereas Skärby et al. (1987) exposed current-year shoots of 20-year-old Scots pines *in situ*. In the present study, 10-to-15-year-old Scots pines were exposed during the growing seasons 1998, 1999, and 2000 to elevated O_3 levels in OTCs in northern Finland. Their responses were measured by gas exchange, growth, morphology, injury, pigment, and nutrient parameters during the summer and autumn to detect both direct and delayed O_3 effects. The hypothesis was that ambient or slightly elevated O_3 concentrations affect local subarctic Scots pines. One of the special aims of the project was to assess the role of high O_3 concentrations in late spring and early summer as a factor modifying the response of Scots pine to O_3.

2. Materials and methods

2.1. The experiment

The Scots pine (*Pinus sylvestris* L.) used in this study were 10 to 15 years old when they were moved from Kempele, near the city of Oulu, to the University of Oulu's experimental field (65°N, 25°E) in September 1997. The trees were balled transplants of seed origin. Twelve of the trees were planted directly in the OTCs (one tree/OTC) and the rest in an open field. Six of the trees, that had overwintered in the open field, were planted in open-field control plots in May and June 1998. Four of the six trees that had been planted in the control OTCs in September 1997 were also replaced by new trees in May–June 1998, as they seemed to have suffered some winter damage. In summary, there were six pines in the OTCs supplied with non-filtered ambient air + supplemental ozone (NF + O_3), six pines in the non-filtered control OTCs (NF), and six pines in the open-field control plots (AA). The soil around the pine roots was a mixture of humus and sand (3 : 1) from a dryish heath forest. The trees were watered from a nearby lake, but not fertilized.

A detailed description of the experimental system is given by Hirvijärvi et al. (1993). Ozone was produced from pure oxygen by an O_3 generator (Fischer Mod. 502). In the summer of 1998, the O_3 concentrations were measured with a Monitor Labs O_3 analyzer (model 8810). In the summer of 1999, an API 400 O_3 analyzer (No. 066) was used because there were problems with the old analyzer. In the summer of 2000, we hired a Dasibi Environmental Corporation O_3 analyzer (model 1008-RS). Furthermore, the walls of the OTCs were replaced by new polycarbonate ones in June 1999. Subsequently, the average values of photosynthetically active radiation (PAR) in the OTCs were only 15% lower compared with conditions in the open field. In the summer of 1998, they were 28% lower on average.

The NF + O_3 pines were exposed to ambient air +40 ppb O_3 in May 1998, after which the O_3 exposure was decreased by 10 ppb each month to mimic the natural variation in O_3 concentrations during the growing season in northern areas. As a result, the pines were exposed to only ambient air in September 1998. In the summers of 1999 and 2000, fumigation did not start until 23 June and 7 June, respectively, and the NF + O_3 pines were exposed to the same target O_3 concentrations as in the summer of 1998, i.e., ambient air +30 ppb in June, ambient air +20 ppb in July, ambient air +10 ppb in August and ambient air in September. Fumigation was carried out between 08.00 and 16.00 hours for 5 days a week. This episodic approach was chosen to mimic natural O_3 occurrence under northern conditions. A cumulative O_3 exposure index AOT40 was calculated as a sum of the hourly O_3 concentrations above the cut off of 40 ppb. The average AOT40 amounted to 12.9 ppmh (calculated for 24 hours/day) in the NF + O_3 OTCs from May–September 1998, whereas between late June and the end of September 1999, it only reached 1.1 ppmh and, between early June and the end of July 2000, 1.8 ppmh (Table 1).

Table 1. AOT40s (ppbh) for the treatments in May–September 1998, June–September 1999, and June and July 2000 on a 24-hour basis. The 1-hour minimum and maximum O_3 concentrations are given in parentheses

Year	Month	Treatment		
		NF	NF + O_3	AA
1998	May[a]	56 (1–50)	7890 (1–154)	95 (1–47)
	June	0 (0–41)	3706 (0–113)	2 (0–42)
	July	4 (0–50)	1266 (0–97)	1 (0–42)
	August	2 (0–42)	68 (1–112)	0 (0–30)
	September[b]	7 (0–44)	7 (0–44)	7 (0–44)
1999	June[a]	0 (0–41)	526 (0–72)	2 (0–42)
	July	0 (0–39)	553 (0–110)	0 (0–39)
	August	0 (0–41)	40 (0–52)	0 (0–40)
	September[b]	0 (1–38)	0 (1–38)	0 (1–38)
2000	June[a]	0 (0–34)	1244 (0–83)	0 (0–31)
	July	0 (0–30)	584 (0–71)	0 (0–35)
Total	1998	69	12937	105
	1999	0	1119	2
	2000	0	1828	0

[a]Fumigation was started on 4 May 1998, 23 June 1999, and 7 June 2000.
[b]Ozone concentrations measured in the open field.

2.2. Growth recording

Budburst, current-year needle growth, and current-year shoot growth were recorded from the end of May or the beginning of June onwards, each summer. Observations were made twice weekly on the main shoot and one of the branches on the 1997 whorl. Five randomly chosen needles were measured for length each time.

2.3. Gas exchange measurements

Measurements of net photosynthesis (P_n), respiration (R), stomatal conductance for water vapor (g_{H_2O}), and internal CO_2 concentration (c_i) were carried out five times between 16 July and 20 September 1998. Gas exchange was not measured in 1999 and 2000. Three trees were chosen from each treatment, and from each tree, a branch was chosen from the 1996 whorl. The gas exchange of current- (c) and previous-year ($c + 1$) shoots was measured using a portable photosynthesis system (ADC gas analyzer model LCA 2 and ADC PLC(N) cuvette). Photosynthetically active radiation (PAR) was measured with a Li-Cor meter equipped with a SPQA 2260 sensor. Five needles from each age class on each branch were measured for length to calculate total needle surface area ($A_t = 4.2235 \times$ length $- 15.6835$) (Flower-Ellis and Olsson, 1993).

2.4. Pigment and glutathione reductase analyses

Current-year and $c + 1$ needles for chlorophyll (chl) a and b and carotenoid (car) analyses were collected on 15 November 1999. The needles were taken from a 1997 whorl and stored in a freezer ($-72\,°C$) until analysis. The pigments were extracted with DMSO according to Hiscox and Israelstam (1979) and the absorbances were measured at 470, 646, and 663 nm with a Beckman DU$^®$ -64 spectrophotometer. The pigment concentrations were calculated according to Wellburn (1994).

Glutathione reductase (GR) activity was determined from c and/or $c + 1$ needles collected on 11 August and 8 September 1998 by a modification of the method of Polle et al. (1990). A Beckman DU$^®$ -64 spectrophotometer was used to measure the decrease in absorbance at 340 nm, and the activity was calculated using an extinction coefficient of $6.22\ mM^{-1}\ cm^{-1}$ for NADPH.

2.5. Microscopic studies

One millimeter sections were removed from healthy looking needles (three needles/tree) from the 1997 whorl on 1 September 1998 for morphological observations. The pieces were fixed, dehydrated, and embedded according to Soikkeli (1980) and Reinikainen and Huttunen (1989). The samples were stained with toluidine blue and examined under a Nikon Optiphot-2 light microscope connected to a digital image analyzer (Microscale TM/TC, Digithurst Ltd.) by a video camera (Hitachi CCD KP-C571). The following variables were measured: needle width (vertical thickness) and thickness (horizontal thickness), total cross-sectional area, epidermal and hypodermal area, and mesophyll area. The damage in mesophyll cells was classified according to Soikkeli (1981). The samples (two or three needles/tree) for transmission electron microscopic (TEM) studies were stained with lead citrate and uranyl acetate. Because one of the earliest O_3 symptoms observed in plants is the deformation and shrinking of chloroplasts, the chloroplast ultrastructure, i.e., chloroplast size, number of plastoglobuli and swelling of thylakoids, was recorded. Thylakoid swelling was assessed according to the following classes: $0 =$ not swollen, $1 =$ slightly swollen, $2 =$ somewhat swollen, and $3 =$ markedly swollen.

2.6. Elemental analyses

The samples for elemental analyses were collected by taking c and $c + 1$ needles from several branches on the 1996 and 1997 whorls on 18 November 1999. Total foliar concentration of sulfur (S), phosphorus (P), magnesium (Mg), potassium (K), and calcium (Ca) was analyzed using a SRS 303 As X-ray fluorescence spectrometer with an Rh anode; nitrogen (N) and carbon (C) concentrations were analyzed using CE Instrument's EA 1110 CHNS-O Elemental Analyzer supplied with Eager 200 for Windows[TM] (Manninen and Huttunen, 2000).

2.7. Statistical analyses

Differences between treatments were assessed using ANOVA, Fischer's PLSD (Protected Least Significant Difference) as a *post hoc* test, Kruskal–Wallis test and Mann–Whitney test (STATVIEW 4.1, Abacus Concepts Inc.). The data were tested for normality. No data transformation was carried out. The tree (i.e., OTC chamber and AA plot) means were calculated for each parameter before testing; they numbered $n = 6$ for each treatment, except in the case of the gas exchange measurements, where $n = 3$.

3. Results

3.1. Photosynthesis, respiration, and stomatal responses

The gas exchange measurements did not show any marked O_3 effects on current-year (c) needles due to the large variation between individual trees within the treatments, although the $NF + O_3$ pines had lower average net photosynthesis (P_n) and higher average respiration (R) than the NF ones, especially in mid-August 1998 (Fig. 1).

Previous-year ($c + 1$) needles showed an increasing trend in the P_n of the NF pines and a decreasing trend in that of the $NF + O_3$ pines (Fig. 2). The $NF + O_3$ pines had higher R than the NF pines. The difference in R between the $NF + O_3$ and NF trees was statistically significant ($Z = -1.964$, $p = 0.0495$) in mid-August 1998. The internal CO_2 concentration (c_i) of the $NF + O_3$ pines seemed to increase with decreasing P_n, whereas no trends were seen in stomatal conductance (g_{H_2O}). The AA pines always had the highest g_{H_2O}, however.

3.2. Growth

In late July 2000, the current-year main shoots of the $NF + O_3$ pines were, on average 18% longer than those of the NF pines ($Z = -1.281$, $p = 0.2002$), whereas the average current-year growth of branches in the $NF + O_3$ pines was 19% less than that in the NF pines ($Z = -1.922$, $p = 0.0547$) (Fig. 3). The c needles of the $NF + O_3$ pines were also shorter than those of the NF

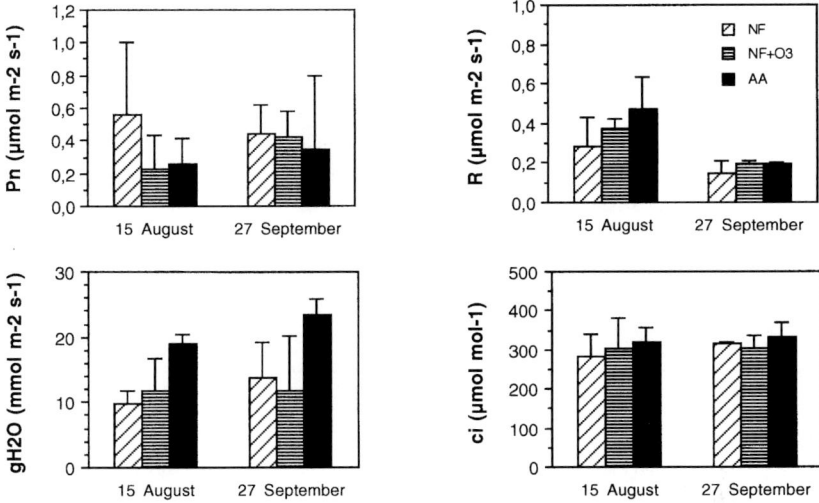

Figure 1. Net photosynthesis (P_n, $\mu mol\,m^{-2}\,s^{-1}$), respiration (R, $\mu mol\,m^{-2}\,s^{-1}$), and stomatal conductance for water vapor (g_{H_2O}, $mmol\,m^{-2}\,s^{-1}$) per total needle area and internal CO_2 concentration (c_i, $\mu mol\,mol^{-1}$) of current-year (c) needles in August and September 1998. Values are means $\pm SD$, $n = 3$.

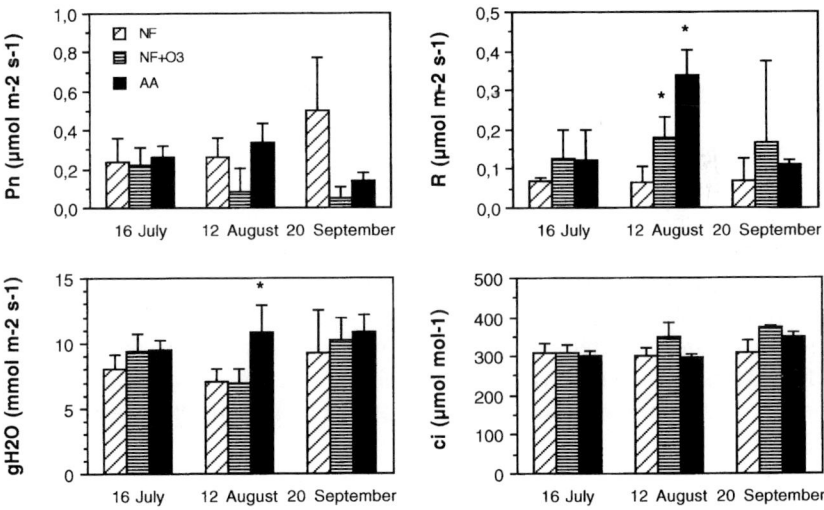

Figure 2. Net photosynthesis (P_n, $\mu mol\,m^{-2}\,s^{-1}$), respiration (R, $\mu mol\,m^{-2}\,s^{-1}$), and stomatal conductance for water vapor (g_{H_2O}, $mmol\,m^{-2}\,s^{-1}$) per total needle area and internal CO_2 concentration (c_i, $\mu mol\,mol^{-1}$) of previous-year ($c+1$) needles in July, August, and September 1998. Asterisks indicate statistically significant differences (Mann–Whitney U, $p < 0.05$) between the NF pines and the pines with the other treatments. Values are means $\pm SD$, $n = 3$.

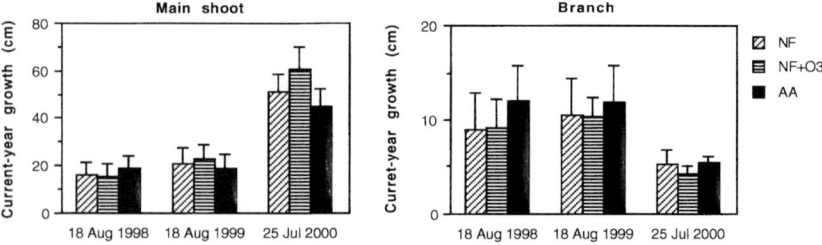

Figure 3. Current-year main shoot and branch growth in the summers of 1998–2000. Values are means ±SD, $n = 6$.

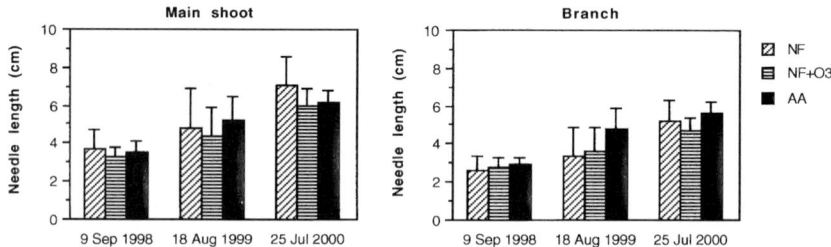

Figure 4. Current-year needle length of main shoots and branches in the summers of 1998–2000. Values are means ±SD, $n = 6$.

pines, but these differences (15% in main shoots and 10% in branches) were not statistically significant either (Fig. 4).

3.3. Microscopic studies

The microscopic studies showed no statistically significant differences in the measured parameters (Table 2). However, epidermal and hypodermal area in the c needles of the NF + O_3 pines was smaller than that in the *c* needles of the NF pines in the summer of 1998. There was no visible O_3 damage in the needles, but more injured mesophyll cells were observed in the *c* needles of the NF + O_3 pines than in those of the NF pines (data not shown). Under TEM, the NF + O_3 pines had the highest percentage of both healthy (class 0) and markedly swollen (class 3) thylakoids.

3.4. Pigment concentrations and glutathione reductase activity

The results suggested an O_3-induced decrease in the (chlorophyll a + b)/carotenoid ratio ((chl a + b)/car) of the *c* needles of the NF + O_3 pines ($p = 0.0547$) in November 1999 as a result of a decrease in the chl a + b concentration and an increase in the car concentration (Fig. 5). The difference in GR activ-

Table 2. Morphology of current-year needles in September 1998[a]

	Treatment		
Variable	NF	NF + O_3	AA
Width (mm)	1.45 ± 0.20	1.45 ± 0.19	1.43 ± 0.08
Thickness (mm)	0.62 ± 0.05	0.63 ± 0.06	0.61 ± 0.05
Cross-sectional area (mm^2)	0.78 ± 0.18	0.76 ± 0.17	0.75 ± 0.12
Epiderm + hypoderm area (mm^2)	0.096 ± 0.016	0.089 ± 0.012	0.090 ± 0.013
Mesophyll area (mm^2)	0.48 ± 0.11	0.47 ± 0.11	0.47 ± 0.08

[a]Treatment means ±SD, $n = 6$.

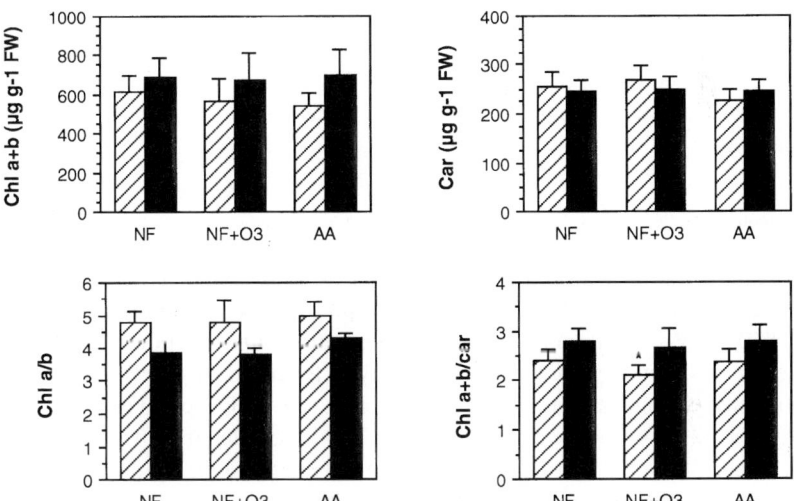

Figure 5. Chlorophyll a + b and carotenoid concentrations and ratios of chl a/b and (chl a + b)/car in current (*c*) and previous-year (*c* + 1) needles in November 1999. Hatched bars = *c* needles, black bars = *c* + 1 needles. The asterisk indicates a statistically significant difference ($p < 0.05$) between the NF + O_3 and NF pines. Values are means ±SD, $n = 6$.

ity between treatments was not statistically significant, although in the *c* and *c* + 1 needles of the NF + O_3 pines showed slightly lower average GR activity compared with the *c* and *c* + 1 needles of the NF pines (Table 3).

3.5. Carbon and nutrient concentrations

There was an O_3 effect on the needle C content, as shown by the higher C content in the *c* + 1 needles of the NF + O_3 pines than in those of either the NF

Table 3. Glutathione reductase activity (nkat g^{-1} FW) of current (c) and previous-year ($c + 1$) needles in August and September 1998[a]

		Treatment		
Month	Needle age class	NF	NF + O$_3$	AA
August	$c + 1$	5.6 ± 1.8	6.0 ± 2.8	5.4 ± 0.8
September	c	12.6 ± 3.0	11.0 ± 5.1	8.5 ± 3.2
	$c + 1$	11.2 ± 4.5	10.1 ± 2.2	10.5 ± 3.3

[a]Treatment means ±SD, $n = 6$.

pines or the AA controls Table 4. All the elemental concentrations (except the Ca concentration) depended on needle age: c needles had higher N, S, P, Mg, and K concentrations than $c + 1$ needles, which in turn had a higher C content than the c needles. The interaction between treatment and needle age in the case of the needle S concentrations could be explained by the differences in the foliar S concentrations of the c vs. $c + 1$ needles between the OTC and AA pines.

4. Discussion

Broadmeadow and Jackson (2000) exposed seedlings of oak (*Quercus petraea* L.), ash (*Fraxinus excelsior* L.), and Scots pine to 20–80 ppb O$_3$ (annual 24 h AOT40s 47.8–74.1 ppmh) in a 3-year experiment. Oak was the most responsive species, with a 30% reduction in growth followed by a 15% reduction in Scots pine. Ozone had no detectable effect on ash. Chlorophyll degradation in response to O$_3$ was only observed in oak and it correlated with stomatal conductance (Broadmeadow et al., 1999). Landolt et al. (2000), in turn, studied the O$_3$ response of seedlings during one growing season (50% ambient +30 ppb; daylight hour AOT40 19.7 ppmh, 24 h AOT40 29.3 ppmh) and calculated a biomass loss of 25.5% for ash, 17.4% for beech (*Fagus sylvatica* L.), 9.9% for Scots pine, and 5.6% for Norway spruce per AOT40 increment of 10 ppmh. Utriainen and Holopainen (1999) and Utriainen et al. (2000) exposed 3-year-old Scots pine to 1.3–1.5 × ambient O$_3$ for three growing seasons in open-field (annual 24 h AOT40s 15.4–38.0 ppmh) and to 1.5 × ambient O$_3$ for two growing seasons in OTCs (growing season 24 h AOT40s 33.3 and 39.9 ppmh). Ozone had a growth-depressing effect on the current-year main shoot length after the third year (19% under elevated O$_3$ during the growing season and 41% under elevated O$_3$ during the springtime and growing season) (Utriainen and Holopainen, 1999). Slight O$_3$-induced yellowing and/or chlorotic mottling was observed in the $c + 1$ needles in the OTC experiment (Utriainen et al., 2000). In the Liphook Project, which lasted for nearly 4 years, no major effects of O$_3$

Table 4. Carbon and nutrient concentrations in the c and c + 1 needles of Scots pine in November 1999[a]

Treatment	Needle age	C (%)	N (%)	S (μg g^{-1})	P (μg g^{-1})	Mg (μg g^{-1})	K (μg g^{-1})	Ca (μg g^{-1})
NF	c	49.9 ± 0.4	1.39 ± 0.12	1303 ± 71	1615 ± 130	1866 ± 299	6268 ± 777	2729 ± 336
	c+1	51.4 ± 0.4	1.25 ± 0.13	1142 ± 100	1338 ± 47	1296 ± 210	5263 ± 399	2279 ± 748
NF + O$_3$	c	50.3 ± 0.4	1.46 ± 0.17	1318 ± 103	1573 ± 91	1791 ± 352	5745 ± 690	2775 ± 428
	c+1	51.9 ± 0.4	1.23 ± 0.13	1076 ± 106	1293 ± 52	1167 ± 180	4877 ± 429	2122 ± 437
AA	c	50.1 ± 0.5	1.28 ± 0.11	1205 ± 99	1587 ± 49	1571 ± 223	5743 ± 433	2606 ± 439
	c+1	51.2 ± 0.4	1.25 ± 0.11	1198 ± 103	1456 ± 113	1286 ± 316	5208 ± 637	2601 ± 771
Treatment		*	ns	ns	ns	ns	ns	ns
Needle age		***	***	***	***	***	***	ns
Treatment × needle age		ns	ns	*	ns	ns	ns	ns

[a]Treatment means ± SD.
Significances of ANOVA ($n = 6$) for the of treatment effects and needle age, or the interaction between them, on the element concentrations are given as follows: *** $p < 0.001$, ** $p < 0.01$, * $p < 0.05$, ns = non-significant.

were observed on Scots pine, Norway spruce, or Sitka spruce (*Picea sitchensis* [Bong.] Carr.) at average AOTs of 28.8–30.7 ppmh (McLeod and Skeffington, 1995).

Scots pine may be considered moderately sensitive to O_3 based on the seedling studies cited above. The marked variation in the growth parameters between individual young trees in late July 2000 makes it difficult to demonstrate any systematic O_3 response. Furthermore, the increasing trends in needle and main shoot length from the summer of 1998 to the summer of 2000 may only reflect the rooting and adaptation of the trees to the experimental site. If so, then the growth results from the 1998 vs. 2000 summers may suggest that high O_3 levels have little effect on Scots pines when the trees have root damage or soil water availability is limited due to climatic factors (McLaughlin and Downing, 1995), whereas under normal conditions Scots pine is sensitive to O_3. Actually, marked differences might have been found merely by increasing the number of needles studied. The results of Laakso (2001), based on ten needles/tree rather than five needles/tree, as in this study, showed a statistically significant decrease in the length of c needles attached to the branches of our NF + O_3 pines in the summer of 2000 ($p = 0.037$). Given the formidable costs of free-air and mature-tree fumigation and the number of replications needed to detect small biological changes over short-term experimentation, Samuelson and Kelly (2001) recommend that research on cause-effect relationships in forest trees should apply the rigorous statistical and monitoring protocol developed by Schreuder and Thomas (1991).

At any rate, the reduced growth of c needles may be attributed to the decreased net photosynthesis and increased respiration under elevated O_3. Kellomäki and Wang (1997, 1998) exposed naturally grown 30-year-old Scots pines to doubled ambient O_3 in OTCs. In the third year of O_3 treatment, the doubled ambient O_3 (69 ppb; average daylight hour AOT40 19.2 ppmh) significantly reduced the photosynthetic rate, the specific growth rate of needles undergoing early expansion, and the needle N concentration in the late stage, but increased the apparent respiration rates in the late stage. An increase in dark respiration was also found by Skärby et al. (1987), who exposed current-year shoots of 20-year-old pines to 60–200 ppb in branch chambers for 1 month. The TREGRO simulations by Constable and Retzlaff (1997) showed that, regardless of O_3 exposure and peak O_3 episode occurrence, a peak O_3 episode in August caused the greatest reduction in C gain in yellow poplar, whereas a peak O_3 episode in July caused the greatest reduction in the C gain of loblolly pine. In other words, maximum O_3 response was observed when the peak O_3 episode occurred at or near the completion of the annual foliage production phenophase.

Age-dependent differences in O_3 uptake, anatomy, and detoxification as well as injury repair appear to be of paramount importance (Matyssek and Innes,

1999). Older needles of Norway spruce have been found to react more negatively to O_3 stress than young needles (Skärby et al., 1995), as was also suggested by the present results. The average net photosynthetic rate of the $c + 1$ needles in the $NF + O_3$ pines was only 10% that of the $c + 1$ needles in the NF pines in September 1998, and their average respiration rate was 2.3-fold compared to the NF control. This means that the O_3 effect is small in young trees and becomes greater in old trees, mainly because of the different proportions of the needle age classes (Skärby et al., 1995). It also means that a loss of 2- to 3-year-old needles in Scots pine influences the remaining crown more significantly than a loss of 7- to 10-year-old needles in Norway spruce (Langebartels et al., 1997).

Glutathione reductase activity did not suggest any changes in the antioxidative status of the $NF + O_3$ pines in September 1998. According to Foyer et al. (1994), glutathione, ascorbate, and superoxide dismutase defenses are often not responsive until visible injury occurs. There were no visible foliar O_3 injuries in the $NF + O_3$ pines. On the other hand, the reduced chl a + b concentration and the increased car concentration, i.e., the reduced (chl a + b)/car ratio, in c needles of the $NF + O_3$ pines in November 1999 compared with pigment concentrations and ratio in c needles of the NF pines point to O_3-related oxidative stress. Carotenoids provide one line of defence against oxidative stress (Young and Britton, 1990; Polle and Rennenberg, 1994). The lower (chl a + b)/car ratio in c needles of the $NF + O_3$ pines than in those of the NF pines in November 1999 may be considered a memory effect that develops in early autumn in Scots pine (Langebartels et al., 1997). It has been suggested that a reduction in the proportion of surface structures is an acclimation reaction to elevated O_3 levels (Günthardt-Goerg et al., 1993; Pääkkönen et al., 1993, 1995). It may, however, merely indicate restricted resources as a result of increased respiration and decreased net photosynthesis.

Foliar nutrient concentrations do not explain any of the observed changes. The higher C content of $c + 1$ needles of the $NF + O_3$ pines compared with $c + 1$ needles of the NF pines in November 1999 may indicate changes in carbohydrate allocation. It seems that O_3 sensitivity is strongly affected by (genetic) variation in stomatal conductance for water vapor (g_{H_2O}) and C allocation to fine root biomass (Constable and Taylor, 1997; Skärby et al., 1998; Samuelson and Kelly, 2001). The possible decrease in carbohydrate allocation to roots and the change in the root/shoot ratio can be verified after final harvesting (in summer 2001). The average g_{H_2O} of $c + 1$ needles of the $NF + O_3$ pines was higher than that of $c + 1$ needles of the NF pines in the summer of 1998.

Samuelson and Kelly (2001) point out that is unclear to what degree the higher O_3 uptake rates in seedlings are offset by the production of new foliage, as there is some variation in the shoot phenology of *Pinus* species between juvenile and mature trees (Clark et al., 1995). Scots pine has a determinate growth pattern, and as the O_3 concentrations decreased towards the end of the growing season, the developing needles always experienced peak O_3 episodes. This may explain the observed effects of low AOT40s in this study. This also means that elevated O_3 concentrations, especially occurring in early summer in northern areas, may have harmful effects on Scots pine.

Acknowledgements

The study was financed by the Jenny and Antti Wihuri Foundation and the University of Oulu. Mr. Jorma Mäkilä and Mr. Jouni Määttä are thanked for their help with the summer 1998 fumigations. Ms. Tuulikki Pakonen, M.Sc., and Mr. Seppo Sivonen are acknowledged for their help with the elemental analyses. The English was revised by Ms. Sirkka-Liisa Leinonen.

References

Broadmeadow, M.J.S., Heath, J., Randle, T.J., 1999. Environmental limitations to O_3 uptake—some key results from young trees growing at elevated CO_2 concentrations. Water Air Soil Pollut. 116, 299–310.
Broadmeadow, M.J.S., Jackson, S.B., 2000. Growth responses of *Quercus petraea, Fraxinus excelsior* and *Pinus sylvestris* to elevated carbon dioxide, ozone and water supply. New Phytol. 146, 437–451.
Clark, C.S., Weber, J.A., Lee, E.H., Hogsett, W.E., 1995. Accentuation of gas exchange gradients in flushes of ponderosa pine exposed to ozone. Tree Physiol. 15, 181–189.
Constable, J.V.H., Retzlaff, W.A., 1997. Simulating the response of mature yellow poplar and loblolly pine trees to shifts in peak ozone periods during the growing season using the TREGRO model. Tree Physiol. 17, 627–635.
Constable, J.V.H., Taylor, G.E., 1997. Modeling the effects of elevated tropospheric O_3 on two varieties of *Pinus ponderosa*. Can. J. Forest. Res. 27, 527–537.
Flower-Ellis, J.G.K., Olsson, L., 1993. Estimation in volume, total and projected area of Scots pine needles from the regression on length. Studia Forest. Suecica 190, 1–19.
Foyer, C., Lelandais, M., Kunert, K.J., 1994. Photooxidative stress in plants. Physiol. Plant. 92, 696–717.
Fuhrer, J., Achermann, B. (Eds.), 1999. Critical Levels for Ozone—Level II. Environmental Documentation No. 115. Swiss Agency for Environment, Forest and Landscape, Bern, Switzerland.
Fuhrer, J., Skärby, L., Ashmore, M.R., 1997. Critical levels for ozone effects on vegetation in Europe. Environ. Pollut. 97, 91–106.
Günthardt-Goerg, M.S., Matyssek, R., Scheidegger, C., Keller, T., 1993. Differentiation and structural decline in the leaves and bark of birch (*Betula pendula*) under low ozone concentrations. Trees 7, 104–114.
Hirvijärvi, E., Huttunen, S., Rankka, N., 1993. Description of the experimental system used in a northern open-top fumigation project in Oulu. Aquilo Ser. Bot. 32, 1–7.
Hiscox, J.D., Israelstam, G.F., 1979. A method for the extraction of chlorophyll from leaf tissue without maceration. Can. J. Bot. 57, 1332–1334.

Kärenlampi, L., Skärby, L. (Eds.), 1996. Critical Levels for Ozone in Europe: Testing and Finalizing the Concepts. UN-ECE Workshop Report, University of Kuopio, Department of Ecology and Environmental Science.

Kellomäki, S., Wang, K., 1997. Effects of elevated O_3 and CO_2 on chlorophyll fluorescence and gas exchange in Scots pine during the third growing season. Environ. Pollut. 101, 263–274.

Kellomäki, S., Wang, K., 1998. Growth, respiration and nitrogen content in needles of Scots pine exposed to elevated ozone and carbon dioxide in the field. Environ. Pollut. 97, 17–27.

Laakso, K., 2001. Leaf and needle characteristics of two climax tree species, holm oak (*Quercus ilex* L.) and Scots pine (*Pinus sylvestris* L.), in elevated-ozone environments in Italy and Finland. Master's thesis. Department of Biology, University of Oulu.

Landolt, W., Bühlmann, U., Bleuler, P., Bucher, J.B., 2000. Ozone exposure-response relationships for biomass and root/shoot ratio of beech (*Fagus sylvatica*), ash (*Fraxinus excelsior*), Norway spruce (*Picea abies*) and Scots pine (*Pinus sylvestris*). Environ. Pollut. 109, 473–478.

Langebartels, C., Ernst, D., Heller, W., Lütz, C., Payer, H.-D., Sandermann, H., 1997. Ozone responses in trees: Results from controlled chamber exposures at the GSF phytotron. In: Sandermann, H., Wellburn, A.R., Heath, R.L. (Eds.), Forest Decline and Ozone. In: Ecological Studies, Vol. 127. Springer-Verlag, Berlin, pp. 163–200.

Manninen, S., Huttunen, S., 2000. Response of needle sulphur and nitrogen concentrations of Scots pine versus Norway spruce to SO_2 and NO_2. Environ. Pollut. 107, 421–436.

Matyssek, R., Innes, J., 1999. Ozone–a risk factor for trees and forests in Europe? Water Air Soil Pollut. 116, 199–226.

McLaughlin, S.B., Downing, D.J., 1995. Interactive effects of ambient ozone and climate measured on growth of mature forest trees. Nature 374, 252–254.

McLeod, A.R., Skeffington, R.A., 1995. The Liphook Forest Fumigation project: an overview. Plant Cell Environ. 18, 327–335.

Pääkkönen, E., Holopainen, T., Kärenlampi, L., 1995. Ageing-related anatomical and ultrastructural changes in leaves of birch (*Betula pendula* Roth.) clones as affected by low ozone exposure. Ann. Bot. 75, 285–294.

Pääkkönen, E., Paasisalo, S., Holopainen, T., Kärenlampi, L., 1993. Growth and stomatal responses of birch (*Betula pendula* Roth.) clones to ozone in open-air and chamber fumigations. New Phytol. 125, 615–623.

Polle, A., Chakrabarti, K., Schürmann, W., Rennenberg, H., 1990. Composition and properties of hydrogen peroxide decomposition systems in extracellular and total extracts from needles of Norway spruce (*Picea abies* L. Karst.). Plant Physiol. 94, 312–319.

Polle, A., Rennenberg, H., 1994. Photooxidative stress in trees. In: Foyer, C., Mullineaux, P.M. (Eds.), Causes of Photooxidative Stress and Amelioration of Defense Systems in Plants. CRC Press, Tokyo, pp. 191–218.

Reinikainen, J., Huttunen, S., 1989. The level of injury and needle ultra structure of acid rain-irrigated pine and spruce seedlings after low temperature treatment. New Phytol. 112, 29–39.

Samuelson, L., Kelly, J.M., 2001. Scaling ozone effects from seedlings to forest trees. New Phytol. 149, 21–41.

Schreuder, H.T., Thomas, C.E., 1991. Establishing cause-effect relationships using forest survey data. Forest Sci. 37, 1497–1512.

Skärby, L., Ro-Poulsen, H., Wellburn, F.A.M., Sheppard, L.J., 1998. Impacts of ozone on forests: a European perspective. New Phytol. 139, 109–122.

Skärby, L., Troeng, E., Boström, C.-Å., 1987. Ozone uptake and effects on transpiration, net photosynthesis, and dark respiration in Scots pine. Forest Sci. 33, 801–808.

Skärby, L., Wallin, G., Selldén, G., Karlsson, P.E., Ottosson, S., Sutinen, S., Grennfelt, P., 1995. Tropospheric ozone—a stress for Norway spruce in Sweden. Ecol. Bull. 44, 133–146.

Soikkeli, S., 1980. Ultrastructure of the mesophyll in Scots pine and Norway spruce: Seasonal variation and molarity of the fixative buffer. Protoplasma 103, 241–252.

Soikkeli, S., 1981. Comparison of cytological injuries in conifer needles from several polluted industrial environments in Finland. Ann. Bot. Fenn. 18, 47–61.

Utriainen, J., Holopainen, T., 1999. Impact of increased springtime O_3 exposure on Scots pine (*Pinus sylvestris*) seedlings in central Finland. Environ. Pollut. 109, 479–487.

Utriainen, J., Janhunen, S., Helmisaari, H.-S., Holopainen, T., 2000. Biomass allocation, needle structural characteristics and nutrient composition in Scots pine seedlings exposed to elevated CO_2 and O_3 concentrations. Trees 14, 475–484.

Wellburn, A.R., 1994. The spectral determination of chlorophylls a and b, as well as total carotenoids, using various solvents with spectrophotometers of different resolution. J. Plant Physiol. 144, 307–313.

Young, A., Britton, G., 1990. Carotenoids and stress. In: Alscher, R.G., Cumming, J.R. (Eds.), Stress Responses in Plants: Adaptation and Acclimation Mechanisms. In: Plant Biology, Vol. 12, pp. 87–112.

Air Pollution, Global Change and Forests in the New Millennium
D.F. Karnosky et al., editors
© 2003 Elsevier Ltd. All rights reserved.

Chapter 11

Ozone affects leaf surface–pest interactions

K.E. Percy*

Natural Resources Canada, Canadian Forest Service, Atlantic Forestry Centre, P.O. Box 4000, Fredericton, New Brunswick E3B 5P7, Canada

B. Mankovska

Forest Research Institute, T.G. Masarykova Street 2195, 960 92 Zvolen, Slovakia

A. Hopkin

Natural Resources Canada, Canadian Forest Service, Great Lakes Forestry Centre, 1219 Queen Street, Sault Ste. Marie, Ontario P6A 5E2, Canada

B. Callan

Natural Resources Canada, Canadian Forest Service, Pacific Forestry Centre, 506 West Burnside Rd., Victoria, BC, V8Z 1M5 Canada

D.F. Karnosky

School of Forest Resources and Environmental Science, Michigan Technological University, 101 U.J. Noblet Forestry Building, 1400 Townsend Drive, Houghton, MI 49931, USA

Abstract

Tropospheric ozone (O_3) levels are increasing around the world and damaging concentrations now occur in 25% of the world forests. This study was conducted at an open air CO_2 and O_3 exposure (Aspen FACE) facility in northern Wisconsin. Here, we present evidence for a link between long-term, low-level O_3 exposure and alterations in trembling aspen (*Populus tremuloides* Michx.) epicuticular waxes resulting in consequential changes to leaf surface properties. In turn, these changes have resulted in increased incidence of natural infection by the aspen leaf rust (*Melampsora medusae* Thuem. f. sp. *tremuloidae*). These results have been consistent over 3 years varying in natural rust occurrence on three trembling aspen clones differing in O_3 sensitivity. The presence of elevated CO_2 did not alleviate the O_3 effects.

---

*Corresponding author.

DOI:10.1016/S1474-8177(03)03011-0

1. Introduction

The global atmospheric CO_2 concentration has risen by nearly 30% since the preindustrial time period (IPCC, 2001). This increase is likely largely due to increasing industrial emissions and global deforestation (Keeling et al., 1995). Similarly, emissions of oxidized nitrogen (NO_x) and volatile organic compounds from fossil fuel emissions related to human activities, such as industrial production and transportation, have resulted in large increases in background tropospheric O_3 concentrations (Finlayson-Pitts and Pitts, 1997; Stevenson et al., 1998; IPCC, 2001). Fowler et al. (1999a, 1999b) suggest that nearly one-quarter of the earth's forests are currently at risk from tropospheric O_3 where the July peak concentrations exceed 60 ppb.

Elevated CO_2 and O_3 impact aspen (*Populus tremuloides* Michx.) trees in diametrically opposite ways. Elevated CO_2 stimulates photosynthesis (Tjoelker et al., 1998; Noormets et al., 2001a, 2001b), delays foliar senescence in autumn (Karnosky et al., 2003), and stimulates aboveground (Norby et al., 1999) and belowground (King et al., 2001; Kubiske and Godbold, 2001) growth. Trees grown under elevated CO_2 generally have lower nitrogen concentrations in their foliage, lower Rubisco (ribulose biphosphate carboxylase) concentrations (Moore et al., 1999), altered defense compounds (Lindroth et al., 1993, 1997) and decreased levels of antioxidants (Wustman et al., 2001).

In contrast to the largely beneficial effects of CO_2, O_3 is generally detrimental to aspen growth and productivity. Ozone has been shown to induce foliar injury (Karnosky, 1976), decrease foliar chlorophyll content (Gagnon et al., 1992), accelerate leaf senescence (Karnosky et al., 1996), decrease photosynthesis (Coleman et al., 1995a), alter carbon allocation (Coleman et al., 1995b), alter epicuticular wax production (Percy et al., 2002) and chemical composition (Karnosky et al., 2002; Percy et al., 2002), and decrease above- and belowground growth (Wang et al., 1986a, 1986b; Karnosky et al., 1996; Isebrands et al., 2001). Extrapolation of open-top chamber data from O_3 impacts on aspen to the natural aspen range suggests that 14–33% biomass loss may be occurring over 50% of their distribution in the eastern US (Hogsett et al., 1997). Of significance to industry, is the fact that the aspen resource is now almost fully utilized in many parts of North America (Tuskan and Walsh, 2001).

Current climate change scenarios predict further increases in CO_2 (Stott et al., 2000; IPCC, 2001) and O_3 (Stevenson et al., 1998; Fowler et al., 1999a, 1999b) over the next century. Because little research has been done on the impacts of these interacting pollutants and conflicting results have been found even for the same species, it is difficult to predict how future forests will respond to these interacting pollutants. What is known, particularly from our work on Aspen FACE, is that the negative roles of insects and disease are likely to become more important under co-exposure to both greenhouse gases (Percy et al., 2002). In this paper, we present evidence of the impact of O_3 on aspen leaf surfaces and we link these changes to increased occurrence of aspen leaf rust caused by *Melampsora medusae*.

2. Methods

2.1. The aspen FACE experiment

This experiment was conducted at the FACTS-II (Aspen FACE) site located on USDA Forest Service land near Rhinelander, WI (Karnosky et al., 1999; Dickson et al., 2000). The experiment includes a full factorial with 12 30 m diameter treatment rings with three control rings, three rings with elevated O_3, three rings with elevated CO_2, and three rings with elevated O_3 + elevated CO_2. All rings are a minimum of 100 m apart. The rings were planted in late summer 1997 and treatments ran from aspen bud-break to bud-set in 1998, 1999, 2000, 2001, and 2002. The eastern half of each ring was randomly planted at 1 m × 1 m spacing in two tree plots with five aspen clones differing in O_3 tolerance (8L, 216 and 271 = relatively tolerant and 42E and 259 = relatively sensitive). The northwestern quadrant of each ring was planted at the same spacing with alternating aspen clone 216 and sugar maple seedlings, and the southwestern quadrant of each ring was planted as above with aspen clone 216 and paper birch seedlings.

CO$_2$ and O_3 were administered during the daylight hours with elevated CO_2 being targeted for 560 ppm, which is 200 ppm above the daylight ambient CO_2 concentration. O_3 is administered at a target of 1.5× ambient and is not administered during periods of cold weather, or when plants are wet from fog, dew, or rain events. Additional details on the experimental design and pollutant generation and monitoring can be found in Karnosky et al. (1999), Dickson et al. (2000), and Karnosky et al. (2003).

2.2. The aspen clones

Essential information on the three aspen clones investigated is listed below in Table 1.

2.3. Epicuticular wax physicochemical characteristics

To characterize epicuticular wax (EW) production, five recently mature (LPI 8–12) leaves from major lateral branches from each of five trees per clone, per FACE ring, were collected and pooled for analysis. Wax was removed with

Table 1.   *Populus tremuloides* clones used in this study

Clone	Origin state (county)	O$_3$ tolerance[a]	Rust susceptibility[b] (low O$_3$)	Rust susceptibility[b] (high O$_3$)	O$_3$-induced stomatal occlusion[c]
216	Wisconsin (Bayfield)	tolerant	low	low	moderate
259	Indiana (Porter)	susceptible	moderate	high	high
271	Indiana (Porter)	tolerant	low	moderate	moderate

[a]Ozone tolerance by visible foliar injury (Berrang et al., 1991), photosynthesis responses (Coleman et al., 1995a, 1995b; Noormets et al., 2001a, 2001b) and growth responses (Isebrands et al., 2001).
[b]From Karnosky et al., 2002.
[c]From Karnosky et al., 1999, and Percy et al., 2002.

chloroform (Karnosky et al., 2002) and amount of wax deposit per unit leaf area calculated. Quantitative wax chemical composition was then determined using high-temperature capillary gas chromatography (Percy et al., 2002).

For examination of EW structure, leaf segments were excised, mounted on a stub, gold-coated on a cold stage and examined under a JSM 6400 SEM. Leaf surface EW and fungal (hyphae, spores, germtubes) structures and density were quantified on abaxial leaf surfaces. Stomatal occlusion was quantified as described by Mankovska et al. (2003).

2.4. Leaf surface wettability

Aspen leaf rust occurs on abaxial leaf surfaces. Aspen clone leaf surface wettability was assessed by measuring abaxial leaf surface droplet contact angle. Segments from leaves used for SEM examination were excised and mounted onto glass slides using double-faced tape. Small volume (0.2 μl; ~ 700 μm diam.) deionized water droplets were detached from a glass syringe and equilibrium contact angles measured (1 degree) using an NRL Contact Angle Goniometer (Rame-Hart, Mountain Lakes, New Jersey).

2.5. Rust occurrence

Melampsora medusae Thuem. f. sp. *tremuloidae* was examined by observations of urediniospores with scanning electron microscopy and/or light microscopy. The percentage of infected leaves per tree for every tree in the core of each Aspen FACE ring ($n = 1839$ trees total) was estimated in mid to late August and the severity of rust occurrence was scored on an average of 20 leaves per tree from 1 to 5 with 1 = 1–20% of the leaf area covered with urediniospores, 2 = 21–40%, 3 = 41–60%, 4 = 61–80%, and 5 = 81–100%. An index of mean severity of infection will then be calculated as percent leaves injured x severity of rust occurrence (Karnosky et al., 2002; Percy et al., 2002).

3. Results and discussion

3.1. Epicuticular wax production and chemical composition

The leaf cuticle, and in particular the outermost epicuticular wax layer, forms an important defensive barrier against pests (Jeffree, 1996). The role of the cuticle as a physical barrier to protect leaves from fungal infection is well established (Mengden, 1996). Exposure of aspen clones to O_3 greatly altered epicuticular wax physicochemical characteristics. Under O_3, leaves developed a significantly larger wax deposit than in the control treatment. Amount of EW recovered from leaves exposed to O_3 averaged 45.5 $\mu g\,cm^{-2}$, or 23% greater than in the control (36.8 $\mu g\,cm^{-2}$) treatment (Fig. 1).

The structure of epicuticular wax is normally conferred by its chemical composition. In other words, alpihatic compounds that crystallize onto the leaf surface do so in a form determined by their inherent chemistry. For instance,

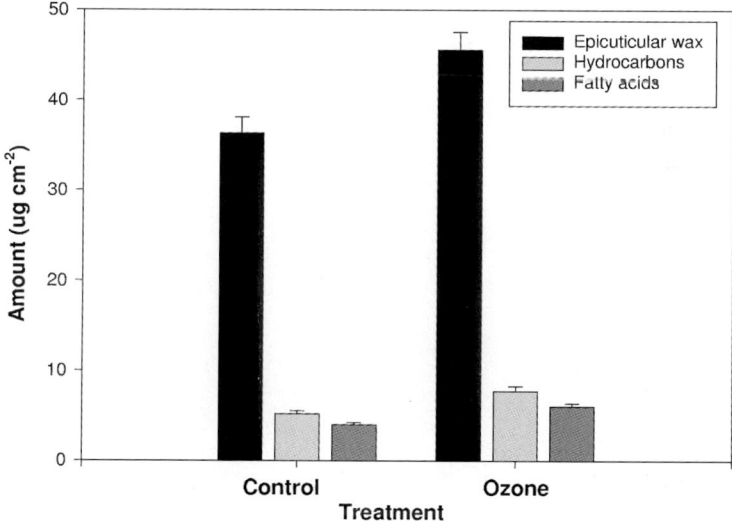

Figure 1. Effect of O_3 on *Populus tremuloides* epicuticular wax physicochemical characteristics. Amount of epicuticular wax recovered from aspen leaf surfaces and relative proportions of two major wax classes which influence surface wettability. Modified from Karnosky et al. (2002) and Percy et al. (2002).

primary alcohols crystallize into small irregular plates, secondary alcohols into hollow tubes (Jeffree et al., 1975). In a number of tree species, O_3 has been reported to act upon the two major biosynthetic pathways producing epicuticular wax and causing shifts in C allocation between them, thus altering chemical composition of the leaf surface, and by default, wax structure (Percy et al., 1994). In aspen, synthesis of the two largest wax classes (hydrocarbons, fatty acids) was stimulated under O_3 fumigation. Amount of hydrocarbons (carbon chains 24–33) was increased 48% and fatty acids (carbon chains 16–30) 51% relative to the controls (Fig. 1). Although hydrocarbons form crystalline plates conferring microroughness and greater hydrophobicity, fatty acids form amorphous films that increase leaf water retention. Also of interest was the reported (Karnosky et al., 2002) carbon chain lengthening in hydrocarbon leaf wax deposits under O_3, however, this is likely of little importance in terms of leaf surface properties and fungal performance.

3.2. Epicuticular wax structure

To qualify the degree of epicuticular wax alteration following seasonal O_3 exposure we examined aspen leaf surfaces using SEM. When examining the O_3 sensitive clone 259, clear and consistent differences between the control and O_3 treated abaxial leaf surfaces were seen. Particularly in areas around stomata, a change from fine crystalline to amorphous (and more hydrophillic) structure was observed (Fig. 2).

 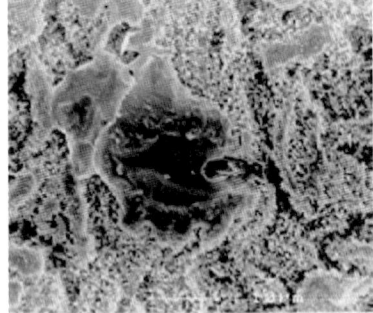

Figure 2. Effect of O_3 on epicuticular wax structure as seen under the SEM. On the left is a SEM of the lower leaf surface waxes and stomate of clone 259 under low O_3 conditions. On the right is a SEM of the same clone grown under $1.5\times$ ambient O_3. Note the difference in wax structure and high level of stomatal occlusion (from Karnosky et al., 2003).

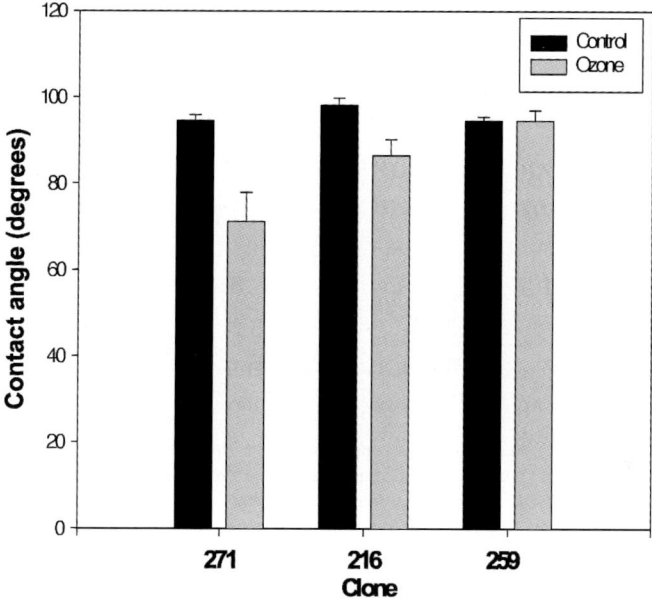

Figure 3. Leaf droplet contact angle (DCA) on abaxial leaf surfaces of three *Populus tremuloides* clones. Note the large decrease in DCA in the two O_3 tolerant clones. Modified from Karnosky et al. (2002).

3.3. Leaf surface wettability

Wettability of the leaf cuticle has a well-known role in influencing fungal spore survival (Kerstiens, 1996). As aspen rust is usually found on abaxial leaf surfaces, abaxial surface wettability has a particular significance in *Melampsora* uredinospores. As we reported earlier (Karnosky et al., 2002), aspen exposure to O_3 caused the abaxial leaf surfaces to become more wettable (decreased contact angle) in two of the three clones studied. Contact angles in the O_3 treatment were 25% lower in clone 271 and 12% lower in clone 216 relative to the controls Fig. 3). Both clones are considered to be O_3 tolerant (see Table 1) based on expression of visual symptoms. Interestingly, there was no difference in contact angle due to treatment in clone 259, which is known to be very sensitive to O_3.

3.4. Rust occurrence

Allen (1991) has reported on the important role of topographic signals from the *Populus deltoides* (Bartr.) leaf surface on development of appressoria from

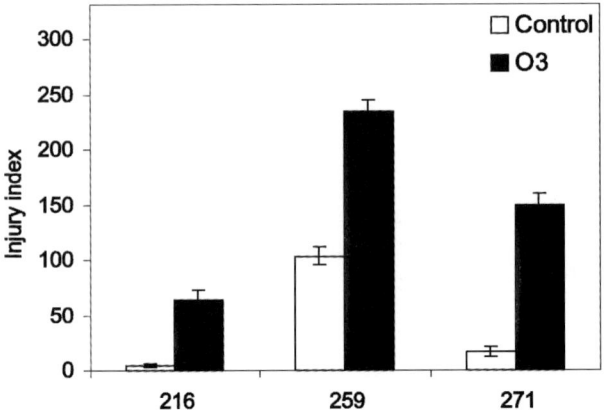

Figure 4.   Effect of O_3 on aspen leaf rust infection index calculated for three aspen clones. Significantly higher (3–5×) levels of rust occurrence were seen in two clones due to O_3. Modified
from Karnosky et al. (1999) and Karnosky et al. (2002).

urediniospores of *Melampsora medusae*. The optimum morphology would be
expected to be similar to that we observed for the amorphous wax deposits
(see Fig. 2(b)) developed under O_3 in this study. In our experiment, significant
differences in rust infection index occurred due to treatment and clone (Fig. 4).
Seasonal exposure to O_3 dramatically increased rust infection three- to fivefold in all clones. Increases were greatest for clone 271 and least in clone 216.

Our data support those of Beare et al. (1999) who also reported that O_3 increases *Melampsora* rust infection in *Populus*. In our clones, the increase in
proportion of amorphous wax deposits (Fig. 2) and the observed alteration towards less crystalline wax deposits noted previously (Mankovska et al., 1998;
Karnosky et al., 1999) on these aspen clones could have reduced ridge heights
towards the optimum preferred for appressorium formation. Also, fatty acid
deposits tend to be situated in areas of the leaf with reduced topography, so the
concurrent increase in amount of more hydrophobic (Holloway, 1969) hydrocarbons we noted (Fig. 1) may not have been able to counteract the increased
wettability due in large part to the increase in fatty acids following O_3 exposure.

4. Conclusions

Of special relevance to this investigation is the fact that modifications to epicuticular wax physicochemical characteristics and leaf surface properties we
have documented above were induced by O_3 at a scale relevant to organisms

such *as Melampsora*. Our data show that O_3 can modify aspen leaf surfaces. Leaf morphology, microroughness and physicochemical characteristics of the outermost epicuticular wax layer in direct contact with the atmosphere influence key leaf surface properties such as wettability, water retention and foliar uptake (Percy and Baker, 1987, 1988). The microenvironment at the phylloplane is critical to the process of fungal infection (Mengden, 1996). In our opinion, one critical impact of low-level O_3 exposure over the long term may be an increased predisposition of aspen to diseases such as aspen leaf rust.

Acknowledgements

This research was supported in part by the US Department of Energy (Office of Biological and Environmental Research, BER), USDA CSREES Grant 2001-35100-10624, the USDA Forest Service Global Change Program, the USDA Forest Service North Central Research Station, Natural Resources Canada (Canadian Forest Service), Canadian Federal Panel on Energy Research and Development (PERD), the National Council of the Paper Industry for Air and Stream Improvement (NCASI), Michigan Technological University, and the Slovakian Forest Research Institute.

References

Allen, E.A., 1991. Appressorium formation in response to topographical signals by 27 rust species. Phytopathology 81, 323–331.

Beare, J.A., Archer, S.A., Bell, J.N.B., 1999. Effects of *Melampsora* leaf rust disease and chronic ozone exposure on poplar. Environ. Pollut. 105, 419–426.

Berrang, P.C., Karnosky, D.F., Bennett, J.P., 1991. Natural selection for ozone tolerance in *Populus tremuloides*: An evaluation of nationwide trends. Can. J. Forest Res. 21, 1091–1097.

Coleman, M.D., Dickson, R.E., Isebrands, J.G., Karnosky, D.F., 1995a. Photosynthetic productivity of aspen clones varying in sensitivity to tropospheric ozone. Tree Physiol. 15, 585–592.

Coleman, M.D., Dickson, R.E., Isebrands, J.G., Karnosky, D.F., 1995b. Carbon allocation and partitioning in aspen clones varying in sensitivity to tropospheric ozone. Tree Physiol. 15, 593–604.

Dickson, R.E., Lewin, K.F., Isebrands, J.G., Coleman, M.D., Heilman, W.E., Riemenschneider, D.E., Sober, J., Host, G.E., Hendrey, G.R., Pregitzer, K.S., Karnosky, D.F., 2000. Forest atmosphere carbon transfer storage-II (FACTS II)—The aspen free-air CO_2 and O_3 enrichment (FACE) project in an overview. USDA Forest Service North Central Experiment Station. General Tech. Rep. NC-214.

Finlayson-Pitts, B.J., Pitts Jr., J.N., 1997. Tropospheric air pollution: ozone, airborne toxics, polycyclic aromatic hydrocarbons, and particulates. Science 276, 1045–1051.

Fowler, D., Cape, J.N., Coyle, M., Flechard, C., Kuylenstierna, J., Hicks, K., Derwent, D., Johnson, C., Stevenson, D., 1999a. The global exposure of forests to air pollutants. J. Water Air Soil Pollut. 116, 5–32.

Fowler, D., Cape, J.N., Coyle, M., Smith, R.I., Hjellbrekke, A.G., Simpson, D., Derwent, R.G., Johnson, C.E., 1999b. Modelling photochemical oxidant formation, transport, deposition, and exposure of terrestrial ecosystems. Environ. Pollut. 100, 43–55.

Gagnon, Z.E., Karnosky, D.F., Dickson, R.E., Isebrands, J.G., 1992. Effects of ozone on chlorophyll content in *Populus tremuloides*. Amer. J. Bot. 79 (Suppl.), 107.

Hogsett, W.E., Weber, J.E., Tingey, D., Herstrom, A., Lee, E.H., Laurence, J.A., 1997. An approach for characterizing tropospheric ozone risk to forests. Environ. Manag. 21, 105–120.

Holloway, P.J., 1969. Chemistry of leaf waxes in relation to wetting. J. Sci. Food Agric. 20, 124–128.

IPCC (Intergovernmental Panel on Climate Change), 2001. A report of working group I of the Intergovernmental Panel on Climate Change. Available at http://www.ipcc.ch/.

Isebrands, J.G., McDonald, E.P., Kruger, E., Hendrey, G., Pregitzer, K., Percy, K., Sober, J., Karnosky, D.F., 2001. Growth responses of *Populus tremuloides* clones to interacting carbon dioxide and tropospheric ozone. Environ. Pollut. 115, 359–371.

Jeffree, C.E., 1996. Structure and ontogeny of plant cuticles. In: Kerstiens, G. (Ed.), Plant Cuticles. Bios Scientific Publishers, pp. 175–188 and pp. 33–82.

Jeffree, C.E., Baker, E.A., Holloway, P.J., 1975. Ultrastructure and recrystallization of plant epicuticular waxes. New Phytol. 75, 539–549.

Karnosky, D.F., 1976. Threshold levels for foliar injury to *Populus tremuloides* Michx. by sulfur dioxide and ozone. Can. J. Forest Res. 6, 166–169.

Karnosky, D.F., Gagnon, Z.E., Dickson, R.E., Coleman, M.D., Lee, E.H., Isebrands, J.G., 1996. Changes in growth, leaf abscission, and biomass associated with seasonable tropospheric ozone exposures of *Populus tremuloides* clones and seedlings. Can. J. Forest. Res. 16, 23–27.

Karnosky, D.F., Mankovska, B., Percy, K., Dickson, R.E., Podila, G.K., Sober, J., Noormets, A., Hendrey, G., Coleman, M.D., Kubiske, M., Pregitzer, K.S., Isebrands, J.G., 1999. Effects of tropospheric O_3 on trembling aspen and interaction with CO_2: results from an O_3-gradient and a FACE experiment. J. Water Air Soil Pollut. 116, 311–322.

Karnosky, D.F., Percy, K.E., Xiang, B., Callan, B., Noormets, A., Mankovska, B., Hopkin, A., Sober, J., Jones, W., Dickson, R.E., Isebrands, J.G., 2002. Interacting CO_2–tropospheric O_3 and predisposition of aspen (*Populus tremuloides* Michx.) to infection by *Melampsora medusae* rust. Global Change Biol. 8, 329–338.

Karnosky, D.F., Zak, D.R., Pregitzer, K.S., Awmack, C.S., Bockheim, J.G., Dickson, R.E., Hendrey, G.R., Host, G.E., King, J.S., Kopper, B.J., Kruger, E.L., Kubiske, M.E., Lindroth, R.L., Mattson, W.J., McDonald, E.P., Noormets, A., Oksanen, E., Parsons, W.F.J., Percy, K.E., Podila, G.K., Riemenschneider, D.E., Sharma, P., Thakur, R., Sober, A., Sober, J., Jones, W.S., Anttonen, S., Vapaavuori, E., Mankovska, B., Heilman, W.E., Isebrands, J.G., 2003. Low levels of tropospheric O_3 moderate responses of temperate hardwood forests to elevated CO_2: a synthesis of results from the Aspen FACE project. Funct. Ecol. 17, 289–304.

Keeling, C.M., Whort, T.P., Wahlen, M., Vander Plict, J., 1995. International extremes in the rate of rise of atmospheric carbon dioxide since 1980. Nature 375, 666–670.

Kerstiens, G., 1996. Barrier properties of the cuticle to water, solutes and pest and pathogen penetration in leaves of plants grown in polluted atmospheres. In: Yunus, M., Iqbal, M. (Eds.), Plant Response to Air Pollution. Wiley, pp. 167–178.

King, J.S., Pregitzer, K.S., Zak, D.R., Karnosky, D.F., Isebrands, J.G., Dickson, R.E., Hendrey, G.R., Sober, J., 2001. Fine root biomass and fluxes of soil carbon in young stands of paper birch and trembling aspen is affected by elevated CO_2 and tropospheric O_3. Oecologia 128, 237–250.

Kubiske, M.E., Godbold, D.L., 2001. Growth and function of roots and root systems. In: Karnosky, D., Scarascia-Mugnozza, G., Ceulemans, R., Innes, J. (Eds.), The Impact of Carbon Dioxide and Other Greenhouse Gasses on Forest Ecosystems. CABI Publishing, New York, pp. 325–340.

Lindroth, R.L., Kinney, K.K., Platz, C.L., 1993. Responses of deciduous trees to elevated atmospheric CO_2: Productivity, phytochemistry and insect performance. Ecology 74, 763–777.

Lindroth, R.L., Roth, S., Kruger, E.L., Volin, J.C., Koss, P.A., 1997. CO_2-mediated changes in aspen chemistry: effects on gypsy moth performance and susceptibility to virus. Global Change Biol. 3, 279–289.

Mankovska, B., Percy, K., Karnosky, D.F., 1998. Impact of ambient tropospheric O_3, CO_2, and particulates on the epicuticular waxes of aspen clones differing in O_3 tolerance. Ekologia (Bratislava) 18 (2), 200–210.

Mankovska, B., Percy, K., Karnosky, D.F., 2003. Impact of greenhouse gases on epicuticular waxes of *Populus tremuloides* Michx: results from an open-air exposure and a natural O_3 gradient. Ekologia (in press).

Mengden, K., 1996. Fungal attachment and penetration. In: Kerstiens, G. (Ed.), Plant Cuticles. Bios Scientific Publishers, pp. 175–188.

Moore, B.D., Cheng, S.H., Sims, D., Seemann, J.R., 1999. The biochemical and molecular basis for photosynthetic acclimation of elevated atmospheric CO_2. Plant Cell Environ. 22, 567–582.

Noormets, A., Sober, A., Pell, E.J., Dickson, R.E., Podila, G.K., Sober, J., Isebrands, J.G., Karnosky, D.F., 2001a. Stomatal and nonstomatal control of photosynthesis in trembling aspen (*Populus tremuloides* Michx.) exposed to elevated CO_2 and O_3. Plant Cell Environ. 24, 327–336.

Noormets, A., McDonald, E.P., Kruger, E.L., Isebrands, J.G., Dickson, R.E., Karnosky, D.F., 2001b. The effect of elevated CO_2 and/or O_3 on potential plant level carbon gain in aspen. Trees 15, 262–270.

Norby, R.J., Wullschleger, S.D., Gunderson, C.A., Johnson, D.W., Ceulemans, R., 1999. Tree responses to rising CO_2 in field experiments: implications for the future forest. Plant Cell Environ. 22, 683–714.

Percy, K.E., Baker, E.A., 1987. Effects of simulated acid rain on production, morphology and composition of epicuticular wax and on cuticular membrane development. New Phytol. 107, 577–589.

Percy, K.E., Baker, E.A., 1988. Effects of simulated acid rain on leaf wettability, rain retention and uptake of some inorganic ions. New Phytol. 108, 75–82.

Percy, K.E., McQuattie, C.J., Rebbeck, J.A., 1994. Effects of air pollutants on epicuticular wax chemical composition, pp. 67–79. In: Percy, K.E., Cape, J.N., Jagels, R., Simpson, C.M. (Eds.), Air Pollutants and the Leaf Cuticle. In: NATO ASI Series, Vol. G-36. Springer-Verlag, Heidelberg.

Percy, K.E., Awmack, C.S., Lindroth, R.L., Kubiske, M.E., Kopper, B.J., Isebrands, J.G., Pregitzer, K.S., Hendrey, G.R., Dickson, R.E., Zak, D.R., Oksanen, E., Sober, J., Harrington, R., Karnosky, D.F., 2002. Altered performance of forest pests under CO_2- and O_3-enriched atmospheres. Nature 420, 403–407.

Stevenson, D.S., Johnson, C.E., Collins, W.J., Derwent, R.G., Shine, K.P., Edwards, J.M., 1998. Evolution of tropospheric ozone radiative forcing. Geophys. Res. Lett. 25, 3819–3822.

Stott, P.A., Tett, S.F.B., Jones, G.S., Allen, M.R., Mitchell, J.F.B., Jenkins, G.J., 2000. External control of 20th century temperature by natural and anthropogenic forcing. Science 290, 2133–2137.

Tjoelker, M.G., Oleksyn, J., Reich, P.B., 1998. Seedlings of five boreal tree species differ in acclimation of net photosynthesis to elevated CO_2 and temperature. Tree Physiol. 18, 715–726.

Tuskan, G.A., Walsh, M.E., 2001. Short-rotation woody crop systems, atmospheric carbon dioxide and carbon management: A U.S. case study. The Forest Chron. 77, 259–264.

Wang, D., Bormann, F.H., Karnosky, D.F., 1986a. Regional tree growth reductions due to ambient ozone: evidence from field experiments. Environ. Sci. Technol. 20, 1122–1125.

Wang, D., Karnosky, D.F., Bormann, F.H., 1986b. Effects of ambient ozone on the productivity of *Populus tremuloides* Michx. grown under field conditions. Can. J. Forest Res. 16, 47–55.

Wustman, B.A., Oksanen, E., Karnosky, D.F., Sober, J., Isebrands, J.G., Hendrey, G.R., Pregitzer, K.S., Podila, G.K., 2001. Effects of elevated CO_2 and O_3 on aspen clones varying in O_3 sensitivity: can CO_2 ameliorate the harmful effects of O_3? Environ. Pollut. 115, 473–481.

Air Pollution, Global Change and Forests in the New Millennium
D.F. Karnosky et al., editors
© 2003 Elsevier Ltd. All rights reserved.

Chapter 12

Ultrastructural response of a Mediterranean shrub species to O₃

F. Bussotti*, E. Gravano, P. Grossoni, C. Tani, B. Mori

*Dipartimento di Biologia Vegetale, Università di Firenze, Piazzale delle Cascine 28,
I-50144 Firenze, Italy*

Abstract

Mediterranean vegetation (especially evergreen shrubs) is regarded as resistant to ozone (O_3) because of its avoidance mechanisms (summer stomatal closure, low stomatal conductance, low gaseous exchange rates, slow growth). Nevertheless, visible symptoms of ozone damage were observed on some species (*Arbutus unedo, Pistacia lentiscus*) both in natural and controlled conditions. This paper reports the results of a survey on the ultrastructural characterization of the visible symptoms (red stippling) occurring on strawberry trees (*Arbutus unedo*) in controlled conditions (fumigation chambers), i.e., treated with a chronic exposure to two realistic O_3 doses (50 and 100 ppb for 21 days, 5 h day^{-1}) on recently mature leaves. The findings, obtained with light and transmission electron microscopy, show that the most important responses are in plants treated with 100 ppb O_3, and are localized in the epidermis-cuticle complex. The thickness of the cuticle increases and changes also occur in the reticular structure of its lower layer. In the epidermal cells of 100 ppb O_3-treated leaves, tannins embed the outer primary wall. In the mesophyll, we observe the alteration of tannins contained in the vacuoles. Results suggest that *Arbutus unedo* increases its defenses as an active response to ozone stress.

1. Introduction

Ozone (O_3) is one of the most widely spread pollutants in Europe and North America, and its concentration in the troposphere is increasing (Allegrini and Brocco, 1995; Fowler et al., 1999). Ambient levels of O_3 are already known to be high enough to cause extensive visible injury in forest vegetation, both on conifers (Miller and McBride, 1999) and broadleaved trees (Skelly et al., 1987). In Europe, ambient concentrations of this pollutant are known to be

---

*Corresponding author.

DOI:10.1016/S1474-8177(03)03012-2

particularly high in the Mediterranean region because of the high summer temperatures and radiation levels (Butkovic et al., 1990; Gimeno et al., 1994; Millán et al., 1996, 1997). In many sites around the Mediterranean basin, the 24-h mean of the O_3 concentration from May to September was, in recent years, above 40 ppb with peaks (hourly means) that occasionally exceeded 100–150 ppb (Velissariou and Skretis, 1999; Chaloulakou et al., 1999; Soda et al., 2000). During the hottest and driest periods of the year, Mediterranean forest species have low gas exchange rates, thus avoiding the foliar uptake of O_3 (Rhizopoulou and Mitrakos, 1990; Tretiach, 1993; Gucci et al., 1999), but visible symptoms on leaves were recorded in several species. Ozone-induced symptoms occur primarily on *Pinus halepensis* Mill. (chlorotic mottle, see Gimeno et al., 1992; Velissariou et al., 1992; Barnes et al., 2000), but some evergreen sclerophyllous shrubs are also known to exhibit stippling and/or change in color. Among these, *Arbutus unedo* L. (strawberry tree) is probably the most sensitive species (Skelly et al., 1999; Sanz and Millán, 2000).

The present study reports structural effects in epidermal cells on the upper leaf surface of recently mature leaves of *A. unedo* caused by chronic exposure to ozone (50 and 100 ppb for 21 days, 5 h day^{-1}).

2. Materials and methods

Two-year-old seedlings of strawberry tree derived from a single mother plant were grown in pots containing a fertilized compost of peat, perlite, and natural soil in natural conditions under a shade awning until the time of the experimental fumigation. All containers were regularly provided with optimal water supply by means of an automatic drip irrigation system. Plants, selected by phenotypical homogeneity, were pre-adapted to greenhouse conditions a week before the treatment.

Fumigations were performed in a set of Perspex chambers, each measuring $0.90 \times 0.90 \times 0.65$ m, continuously ventilated with charcoal-filtered air (two complete air changes/min). Ozone was produced by electric discharge with an air-cooled generator (Fischer 500, Zurich, CH) supplied with pure oxygen, and was mixed with the inlet air entering the fumigation chambers. The concentration of O_3 at plant height was continuously monitored with a photometric ML8810 analyzer (Monitor Labs, San Diego, USA). More details are reported elsewhere (Lorenzini et al., 1994). The target doses were 50 and 100 ppb for 21 days (5 h day $^{-1}$, from 09.00 to 14.00, solar time). Control plants were exposed only to charcoal-filtered air (< 3 ppb). Five individuals were used, each randomly allocated to a fumigation chamber and the exposure was performed in the summer of 1996.

For light microscopy observations, five leaves were examined from each plant. From the median intervenial zone of each leaf, four samples 3 × 5 mm wide were collected and fixed in 4% formalin. Of these, two samples per leaf were cut using a freezing microtome (semithin cross-sections 30 μm thick) and stained with Floral Yellow 088 ($C_{22}H_{16}O$, Sigma, Italy) to highlight lipids (Brundrett et al., 1991). The other two samples per leaf were dehydrated in an ethanol series, embedded in LR White Resin (London Resin Company Ltd.) and cut using a Reichert OM-U3 Ultramicrotome glass blade to a thickness of 1 μm. These sections were stained with Schiff reagent plus Calcofluor (Mori and Bellani, 1996), to display the cellulosic matrix of the wall and subsequent modifications (caused, for example, by apposition of phenolic substances and lignification process—see Bussotti et al., 1997).

For transmission electron microscopy, a further two samples from each leaf were fixed in 2.5% glutaraldehyde and 4% paraformaldehyde in phosphate buffer, pH 7.2. After 20 h at 5 °C, the samples were rinsed twice, each time for 10 min in the same buffer, postfixed in 2% OsO_4 in the same buffer for 2 h and, later, dehydrated in an acetone series, spending 10 min at each stage of the dehydration series. Finally, after two 5-min rinses in propylene oxide, the samples were embedded in resin, according to Spurr's procedure (Spurr, 1969). A Reichert Ultracut S microtome with a diamond knife was used to cut ultra-fine sections (0.09 μm). These sections were stained with uranyl acetate and lead citrate and observed under a Carl Zeiss EM9-S2 microscope. Electron-microscopy observations show phenolic compounds as electron-dense structures, due to the osmiophilic properties of these compounds (Parham and Kaustinen, 1976; Bussotti et al., 1998).

3. Results

After 15 days of O_3 exposure, reddish intervenial spots (Fig. 1(A), (B)) were observed on the adaxial surface of leaves treated with 100 ppb O_3. In plants fumigated with 50 ppb O_3, visible injury was scarce and appeared only at the end of experimental period (21 days).

Observations with LM and TEM were performed in sections of O_3-treated leaves from both symptomatic and asymptomatic patches. The results described in this paper refer to observations in symptomatic patches, because the green areas did not differ from controls. *Arbutus* leaves characteristically have a 2-layer palisade parenchyma. The layer next to the upper leaf surface shows vacuoles filled with phenolic content (identified as tannins by Gravano et al., 2000). Vacuoles of epidermal cells do not stain as phenolic substances. Outer walls and cuticles of the abaxial and adaxial epidermal cells are thick (see Gravano et al., 2000).

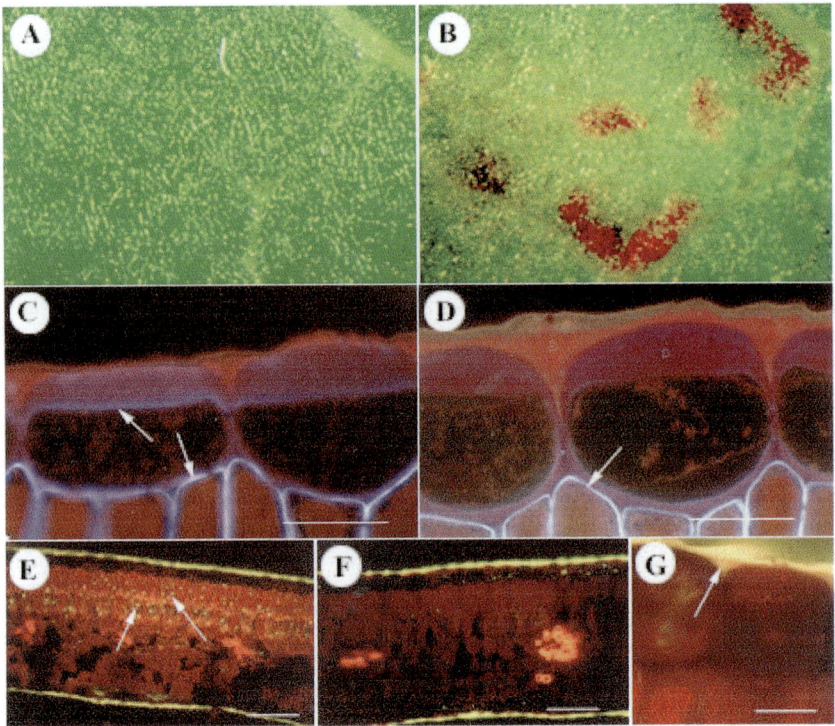

Figure 1. (A)–(G): Macroscopic and LM observations. (A), (B): Upper surface of control (A) and 100 ppb O_3-treated leaves (B). Typical ozone symptoms (reddish spots) are visible in the latter. (C), (D): Samples stained with Calcofluor + Schiff reagent; arrows indicate the cellulosic matrix of the wall (A = Control; B = 100 ppb O_3-treated sample), bars = 20 μm. (E)–(G): Samples stained with Floral Yellow. (E), (F): arrows indicate the pattern of lipidic bodies (E = Control; F = 100 ppb O_3-treated sample, bars = 0.1 mm); (G) (100 ppb O_3-treated sample): arrows indicate the cuticular nails, bar = 20 μm.

Reactions to the 100 ppb O_3 treatment were observed with a light microscope (Fig. 1(A)–(D)), whereas no changes were detected in the 50 ppb O_3 treatment compared with the filtered air control. Fig. 1(C), (D) (Calcofluor + Schiff reagent) illustrates the effects of the 100 ppb O_3 treatment in the walls of the adaxial epidermal cells. The cellulose-pectic component (that stains light blue) was clearly visible in the control leaves (Fig. 1(C), arrow), but in the 100 ppb O_3 treatment (Fig. 1(D)), in the outer wall, the fluorescent response of cellulose disappeared and walls stained red. This indicated the presence of modification processes, probably apposition of phenolic substances (Bussotti et al., 1997). In 100 ppb O_3-treated leaves the typical cellulose response was evident only in the walls of the mesophyll cells (Fig. 1(D), arrow).

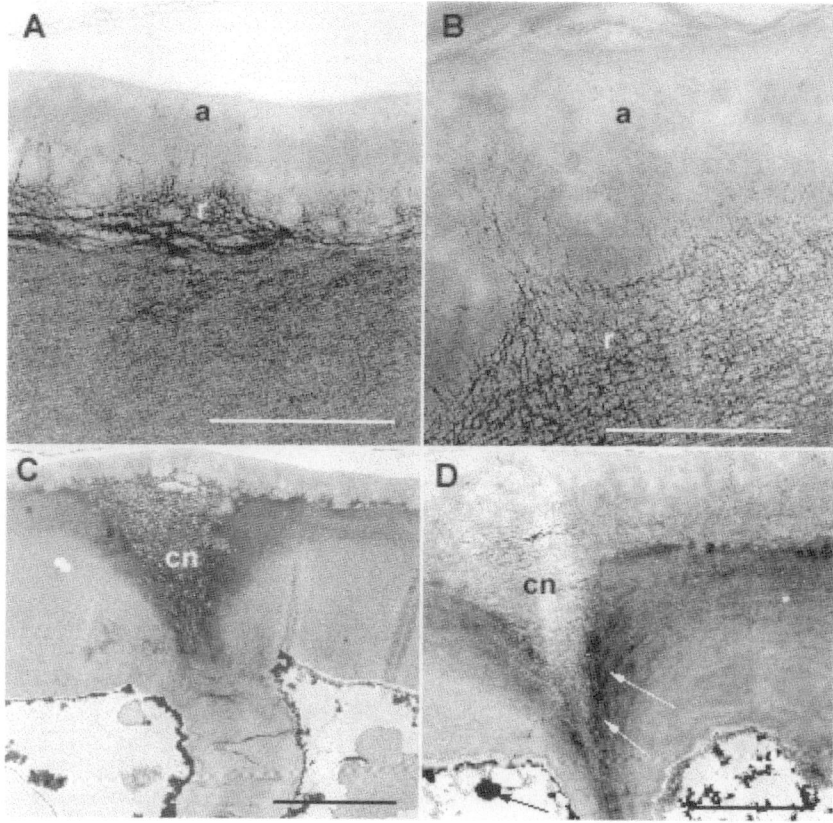

Figure 2.   (A)–(F): Electron microscope observations. (A), (B): Bi-layered cuticle, (a) amorphous and (r) reticulate layer (A = Control; B = 100 ppb O₃-treated sample), bars = 0.5 µm. (C), (D): Cuticular nail (cn) (C = Control; D = 100 ppb O₃-treated sample). Arrows indicate the phenolic substances embedding the wall, bars = 2 µm.

Fig. 1(E)–(G) shows the pattern of lipids: lipidic bodies were notably present in the palisade cells of filtered-air control leaves (Fig. 1(E)), and they appeared localized in the epidermal cells in those subjected to fumigation (Fig. 1(F)). In Fig. 1(G), lipidic substances filled the cuticular pegs in fumigated leaves.

The results of TEM observations (means) are shown in Figs. 2 and 3. Only the controls and 100 ppb-treated leaves are shown, because the 50 ppb-treated leaves did not differ from controls. Fig. 2(A), (B) shows the condition of the cuticle of the adaxial surface in controls (Fig. 2(A)) and in samples from the 100 ppb O₃ treatment (Fig. 2(B)). The cuticle of *A. unedo* is composed of a reticulate and an amorphous layer (*Pyrus* type: see Gouret et al., 1993), and

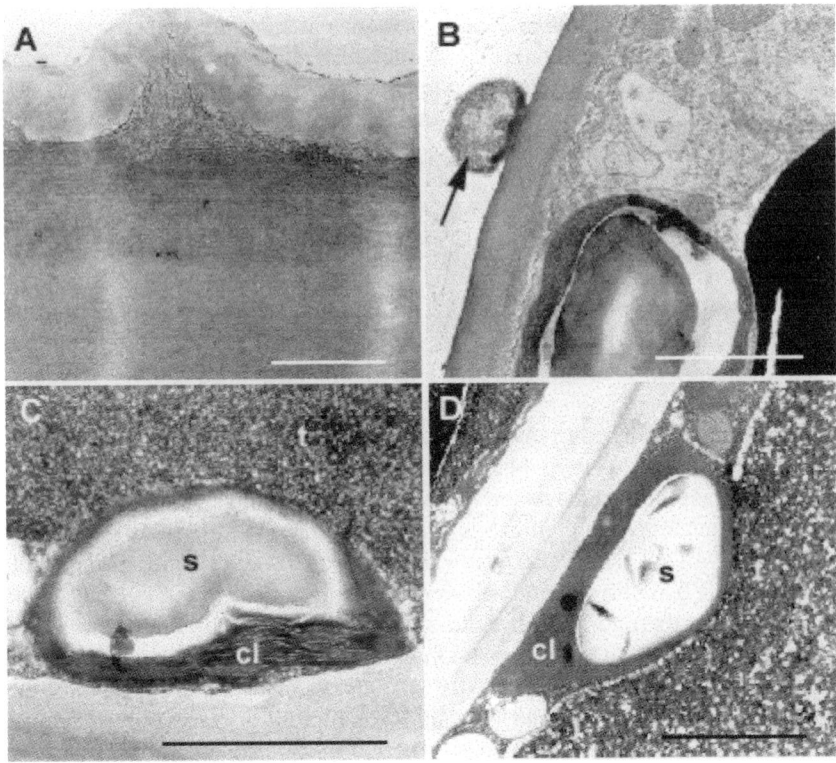

Figure 3. (A)–(D): Electron microscope observations. (A), (B): 100 ppb O₃-treated samples. A: Cuticular ridge; B: Cell wall projection in the mesophyll cell (arrow), bar = 1 μm. (C), (D): Condition of the mesophyll cells (C = Control; D = 100 ppb O₃-treated sample): (t) tannins; (cl) chloroplast; (s) starch grain, bars = 1 μm.

both of them had developed considerably by the end of this treatment. Cuticular nails (Fig. 2(C), (D)) had a finely reticulated structure in the controls (Fig. 2(C)), whereas in the treated leaves (Fig. 2(D)) that structure appeared amorphous. In the 100 ppb O₃ treatment, amorphous electron-dense material (probably tannins: Fig. 2(D) arrow) was found embedding the outer part of the primary cell wall. Cuticular ridges (Fig. 3(A)) and probably cell-wall projections (or exudates) at the mesophyll cell surface facing the intercellular space (Fig. 3(B)) were observed in the samples from the 100 ppb O₃ treatment. Fig. 3(C), (D) shows electron-dense vacuolar material particularly in the cells from the adaxial palisade layer, which is more homogeneous in the control (Fig. 3(A)) compared with the granulated appearance in the O₃-treated samples. Chloroplasts (Fig. 3(D)) show a normal lenticular shape both in controls and in the 100 ppb O_3 treatment.

4. Discussion and conclusions

The behavior we observed suggests the presence of strong active responses located primarily in the epidermal layer. The epidermis represents the interface between plant and environment, so it is the seat of important biochemical protection processes (Bell, 1981) that include an additional synthesis of cutine with the thickening of the cuticular layer, phenolic compounds embedded in the cell wall, and an increased synthesis of polysaccharides that help to build up the cell wall. Ozone is known to be involved in all these processes. Percy et al. (1992) observed the increase of the cuticular membrane thickness in red spruce (*Picea rubens* Sarg.) exposed to O_3 and acid fog. Phenols are involved in several protection processes against biotic and abiotic stress factors (Bennett and Wallsgrove, 1994; Rosemann et al., 1991; and Kangasjärvi et al., 1994) have reported on the role of O_3 in the formation of enzymes as precursors of phenolic compounds. Tannins filling the adaxial epidermis wall were found in *Fagus sylvatica* L. (Bussotti et al., 1998) and *A. unedo* (Gravano et al., 2000) in response to water stress, and also in *Fraxinus excelsior* L. as response to the influence of ozone (Günthardt-Goerg et al., 2000).

In the mesophyll, the main alterations related to the effects of O_3 were found in the vacuolar content of tannins and the cell walls, and in the lipids pattern. The observed granulation of tannins in response to the treatment is consistent with the process described by Kärenlampi (1986) and Pääkkönen and Holopainen (1995) in pines and birch as a response to O_3 fumigation. Wall excrescences and protuberances were observed by Günthardt-Goerg (1996) in ozone-treated broadleaves. As far as lipids are concerned, previous studies by our group (Soda et al., 2000) report an accumulation of these substances in ozone-exposed leaves, this behaviour was associated with the degeneration of organelle membranes. In the present study, lipidic bodies did not increase in the mesophyll of the treated leaves but, on the contrary, they disappeared. Their disappearance may be related to a mobilization of reserves necessary to build up the epidermal structures.

This experiment focused on the early active responses at the epidermis level and did not investigate the mesophyll alterations, namely the condition of chloroplasts. In much of the literature (Sutinen et al., 1990; Selldén et al., 1996), these organelles are considered the structures most sensitive to ozone and the fact that we observed no change in these organelles cannot allow us to rule out damage entirely. It simply suggests that, in *A. unedo,* the active responses occur earlier than the organic damage, and symptoms are not necessarily associated with mesophyll injuries (as was reported by Moss et al., 1998; Evans and Miller, 1972). On the other hand, active responses imply a different metabolism of the reserves, so that they are employed to increase defense mechanisms rather than growth (Heath, 1999).

Acknowledgements

Authors are grateful to Giacomo Lorenzini and Cristina Nali (DCDSL, University of Pisa) for the experiments in fumigation chambers.

References

Allegrini, I., Brocco, D., 1995. The historical development of atmospheric pollution and its effects in Italy. In: Lorenzini, G., Soldatini, G.F. (Eds.), Responses of Plants to Air Pollution. Agric. Medit. s.v., pp. 11–22.

Barnes, J.D., Gimeno, B.S., Davison, A.W., Dizengremel, P., Gerant, D., Bussotti, F., Velissariou, D., 2000. Air pollution impacts on pine forests in the Mediterranean basin. In: Ne'eman, G., Traband, L. (Eds.), Ecology, Biogeography and Management of Pinus Halepensis and Pinus Brutia Forest Ecosystems in the Mediterranean Basin. Backhuys, Leiden, pp. 391–404.

Bell, A.A., 1981. Biochemical mechanisms of disease resistance. Annu. Rev. Plant Physiol. 32, 21–81.

Bennett, R.N., Wallsgrove, R.M., 1994. Secondary metabolites in plant defense mechanisms. New Phytol. 127, 617–633.

Brundrett, M.C., Kendrick, B., Peterson, C., 1991. Efficient lipid staining in plant material with Sudan Red 7B or Floral Yellow 088 in Polyethylene Glycol-Glycerol. Biotech. Histochem. 66, 111–116.

Bussotti, F., Bottacci, A., Grossoni, P., Mori, B., Tani, C., 1997. Cytological and structural changes in *Pinus pinea* L. needles following the application of an anionic surfactant. Plant Cell Environ. 20, 513–520.

Bussotti, F., Gravano, E., Grossoni, P., Tani, C., 1998. Occurrence of tannins in leaves of beech trees (*Fagus sylvatica* L.) along an ecological gradient, detected with an histochemical and ultrastructural analyses. New Phytol. 138, 469–479.

Butkovic, V., Cvitas, T., Klasing, L., 1990. Photochemical ozone in the Mediterranean. Sci. Tot. Environ. 99, 145–151.

Chaloulakou, A., Assimacopoulos, D., Lekkas, T., 1999. Forecasting daily maximum ozone concentrations in the Athens basin. Environ. Monitor. Assess. 56, 97–112.

Evans, L.S., Miller, P.R., 1972. Comparative needle anatomy and relative ozone sensitivity of four pine species. Can. J. Bot. 50, 1067–1071.

Fowler, D., Cape, J.N., Cyle, M., Flechard, C., Kuylenstierna, J., Hicks, K., Derwent, D., Johnson, C., Stevenson, D., 1999. The global exposure of forests to air pollutants. Water Air Soil Pollut. 116, 5–32.

Gimeno, B.S., Velissariou, D., Barnes, J.D., Inclán, R., Peña, J.M., Davison, A., 1992. Daños visibles por ozono en aciculas de *Pinus halepensis* Mill. en Grecia y España. Ecologia 6, 131–134.

Gimeno, B.S., Velissariou, D., Schenone, G., Guardans, R., 1994. Ozone effects in the Mediterranean region: an overview. In: Fuhrer, J., Achermann, B. (Eds.), Critical Levels for Ozone. UNECE Workshop Publ. Federal Research Station for Agricultural Chemistry and Environmental Hygiene (FAC), CH-3097 Liebefeld-Bern, Switzerland, pp. 122–137.

Gouret, E., Rohr, R., Chamel, A., 1993. Ultrastructure and chemical composition of some isolated plant cuticles in relation to their permeability to the herbicide, diuron. New Phytol. 124, 423–431.

Gravano, E., Desotgiu, R., Bussotti, F., Tani, C., Grossoni, P., 2000. Structural adaptations in leaves of two Mediterranean evergreen shrubs under different climatic conditions. J. Medit. Ecol. 1, 165–170.

Gucci, R., Massai, R., Casano, S., Mazzoleni, S., 1999. Seasonal changes in the water relations of Mediterranean co-occurring woody species. Plant Biosyst. 133, 117–128.

Günthardt-Goerg, M.S., 1996. Different responses to ozone of tobacco, poplar, birch and alder. J. Plant Physiol. 148, 207–214.

Günthardt-Goerg, M.S., McQuattie, C.J., Maurer, S., Frey, B., 2000. Visible and microscopical injury in leaves of five deciduous tree species related to current critical ozone levels. Environ. Pollut. 109, 489–500.

Heath, R., 1999. Biochemical processes in an ecosystem: how should they be measured? Water Air Soil Pollut. 116, 279–298.

Kangasjärvi, J., Talvinen, J., Utriainen, M., Karjalainen, R., 1994. Plant defence systems induced by ozone. Plant Cell Environ. 17, 783–794.

Kärenlampi, L., 1986. Relationship between macroscopic symptoms of injury and cell structural changes in needles of ponderosa pine exposed to air pollution in California. Ann. Bot. Fenn. 23, 255–264.

Lorenzini, G., Nali, C., Panicucci, A., 1994. Surface ozone in Pisa (Italy): a six-year study. Atmos. Environ. 38, 51–59.

Millán, M., Salvador, R., Mantilla, E., Artiñano, B., 1996. Meteorology and photochemical air pollution in Southern Europe: experimental results from EC research projects. Atmos. Environ. 12, 1909–1924.

Millán, M.M., Salvador, R., Mantilla, E., Kallos, G., 1997. Photooxidant dynamics in the Mediterranean basin in summer: results from European research projects. J. Geophys. Res. 102, 8811–8823.

Miller, P.R., McBride, J. (Eds.), 1999. Oxidant air pollution impacts in the montane forests of southern California. In: Ecological Studies, Vol. 134. Springer, New York.

Mori, B., Bellani, L., 1996. A differential staining for cellulosic and modified plant cell walls. Biotech. Histochem. 71, 71–72.

Moss, D.M., Rock, B.N., Bogle, A.L., Bilkova, J., 1998. Anatomical evidence of the development of damage symptoms across a growing season in needles of red spruce from central New Hampshire. Environ. Exp. Bot. 39, 247–262.

Pääkkönen, E., Holopainen, T., 1995. Influence of nitrogen supply on the response of clones of birch (*Betula pendula* Roth.) to ozone. New Phytol. 129, 595–603.

Parham, R.A., Kaustinen, H.M., 1976. Differential staining of tannin in sections of exposy-embedded plant cells. Stain Technol. 51, 237–240.

Percy, K.E., Jensen, K.F., McQuattie, C.J., 1992. Effects of ozone and acidic fog on red spruce needle epicuticular wax production, chemical composition, cuticular membrane ultrastructure and needle wettability. New Phytol. 122, 71–80.

Rhizopoulou, S., Mitrakos, K., 1990. Water relations of evergreen sclerophylls. I. Seasonal changes in the water relations of eleven species from the same environment. Ann. Bot., 171–178.

Rosemann, D., Heller, W., Sandermann Jr., H., 1991. Biochemical plant response to ozone. II. Induction of stilbene biosynthesis in Scots pine (*Pinus sylvestris* L.) seedlings. Plant Physiol. 97, 1280–1286.

Sanz, M.J., Millán, M.M., 2000. Ozone in the Mediterranean region: evidence of injury to vegetation. In: Innes, J.L. (Ed.), Forest Dynamics in Heavily Polluted Regions. In: IUFRO Research Series, Vol. 1. CABI Publishing, Wallingford, UK, pp. 165–192.

Selldén, G., Sutinen, S., Skärby, L., 1996. Controlled ozone exposure and field observations in Fennoscandia. In: Sandermann, H., Wellburn, A.R., Heath, R.L. (Eds.), Forest decline and ozone. A comparison of controlled chamber and field experiments. In: Ecological Studies, Vol. 127. Springer, Berlin, pp. 249–276.

Skelly, J.M., Davis, D.D., Merrill, W., Cameron, E.A., Brown, H.D., Drummond, D.B., Dochinger, L.S. (Eds.), 1987. Diagnosing Injury to Eastern Forest Trees. University Park, Pennsylvania State University, College of Agriculture, Pennsylvania.

Skelly, J.M., Innes, J.L., Savage, J.E., Snyder, K.R., Vanderheyden, D., Zhang, J., Sanz, M.J., 1999. Observation and confirmation of foliar ozone symptoms of native plant species of Switzerland and Southern Spain. Water Air Soil Pollut. 116, 227–234.

Soda, C., Bussotti, F., Grossoni, P., Barnes, J.D., Mori, B., Tani, C., 2000. Impacts of urban levels of ozone on *Pinus halepensis* foliage. Environ. Exp. Bot. 44, 69–82.

Spurr, A., 1969. A low viscosity epoxy resin embedding medium for electron microscopy. J. Ultrastruct. Res. 26, 31–43.

Sutinen, S., Skärby, L., Wallin, G., Selldén, G., 1990. Long-term exposure of Norway spruce, *Picea abies* (L.) Karst., to ozone in open-top chambers. II. Effects on the ultrastructure of needles. New Phytol. 115, 345–355.

Tretiach, M., 1993. Photosynthesis and transpiration of evergreen Mediterranean and deciduous trees in an ecotone during a growing season. Acta Oecol. 14, 341–360.

Velissariou, D., Davison, A.W., Barnes, J.D., Pfirmann, T., MacLean, D.C., Holevas, C.D., 1992. Effects of air pollution on *Pinus halepensis* Mill.: pollution levels in Attica, Greece. Atmos. Environ. 26, 373–380.

Velissariou, D., Skretis, L., 1999. Critical levels exceedances and ozone biomonitoring in the Greek fir forest (*Abies cephalonica* Loud.) at the Parnis mountain national park in Attica, Greece. In: Fuhrer, J., Achermann, B. (Eds.), Critical Levels for Ozone. UNECE Workshop Publ., Federal Research Station for Agricultural Chemistry and Environmental Hygiene (FAC), CH-3097 Liebefeld-Bern, Switzerland, pp. 205–208.

Air Pollution, Global Change and Forests in the New Millennium
D.F. Karnosky et al., editors
© 2003 Elsevier Ltd. All rights reserved.

Chapter 13

Ozone injury symptoms on vegetation in an Alpine valley, North Italy

F. Bussotti*

Dipartimento di Biologia Vegetale, Università di Firenze, Piazzale delle Cascine 28, 50144 Firenze, Italy

C. Mazzali

Fondazione Lombardia per l'Ambiente, Piazza Diaz 7, 20123 Milano, Italy

A. Cozzi, M. Ferretti

Linnaea-ambiente, Via Sirtori 37, 50137 Firenze, Italy

E. Gravano

Dipartimento di Biologia Vegetale, Università di Firenze, Piazzale delle Cascine 28, 50144 Firenze, Italy

G. Gerosa

DMF, Department of Mathematics and Physics, Università Catolica del Sacro Cuore, Via Musei 2, 25121 Brescia, Italy

A. Ballarin-Denti

Department of Mathematics and Physics, Catholic University of Brescia, via Musei 41, 25121 Brescia, Italy

Abstract

In recent years multidisciplinary surveys examining the impact of pollutants on forest ecosystems have been carried out in the Valtellina (Northern Italy, Alpine region). A large part of the activity of these surveys has involved the study of ozone and its distribution in the area, development of the exceedances maps (AOT40) and their validation by means of field observation of visible foliar symptoms in the indigenous and sensitive forest vegetation. The present paper reports the results of a field survey carried out in 1998 on the visible symptoms expressed by several native species (trees, shrubs and herbs). A non-systematic grid of 81 plots was assessed throughout the valley. Due to the high variability of the physical (altitude, slope, exposure, etc.) and vegetational (species assemblage, tree age, forest structure) features, the comparability among the plots was

---

*Corresponding author.

DOI:10.1016/S1474-8177(03)03013-4

found to be very weak, so only the presence or absence of leaf injuries was assessed. Results do not show any correlation between the distribution of foliar symptoms in Valtellina and ozone levels. Rather, leaf injuries follow the same distribution of environmental modifying factors (such as depth and moisture of the soil) and tree sensitivity.

1. Introduction

High concentrations of ozone have been recorded in the transnational region between the north of Lombardy, Italy, and Ticino, southern Switzerland (Bacci et al., 1990; Staffelbach et al., 1997; Gerosa et al., 1999). This pollution is attributed to the influence of the wide urban and industrial area of Milan (Italy) and to the presence of the mountains (Alps and Prealps) that provide an obstacle to the spreading of the precursors to the north and enhance their accumulation in the area south of Alps, i.e., the Po plain (Staffelbach et al., 1997). On the other hand, the mountains lend the landscape a complex orography, and the distribution of ozone is irregular.

A research project, aimed at investigating the behaviour of ozone in Alpine areas, was carried out in a valley of Lombardy, the Valtellina (see Fig. 1(A)). Results were already partially reported by Ballarin-Denti et al. (1998a, 1998b) Dell'Era et al. (1998), and Gerosa et al. (1999).

Recently, a large part of the research activities was devoted to studying the distribution of ozone in the area. Fig. 1(B) (Mazzali, unpublished) shows the map of the exceedances for forests (based on the concept of AOT40, see Kärenlampi and Skärbi, 1996). Data were provided by a campaign with passive samplers from May to August 1998, and processed applying methods of spatial statistics (Cressie, 1991; Loibl et al., 1994). In synthesis, the AOT40 (May–August) levels estimated in a territory covering 80×42 km^2, ranged from 5000 to more than 25 000 ppb h, and most of the area showed levels higher than the critical threshold for forest vegetation (10 000 ppbh, see Fuhrer et al., 1997). The highest exposures were reached in the southern part of the valley, i.e., the northern slopes of the *Orobiche* Alps.

As part of the above-mentioned project, the distribution of the ozone-like symptoms on the foliage of sensitive native forest vegetation was also investigated. It is a well known fact that ambient concentrations of ozone in Canton Ticino (CH) may produce visible injuries (Skelly et al., 1987, 1998; Vander-Heyden et al., 2000).

The aim of this investigation was to answer the following questions:

- Is it possible to distinguish areas with different ozone exposures by monitoring symptoms in wild vegetation?

- What are some of the problems which may be connected to a spatial survey for O_3 induced injuries to vegetation in alpine and prealpine conditions?

1.1. Feasibility survey

The rationale of the survey in Valtellina was based on the open-top chamber experiments made in the Lattecaldo nursery, Canton Ticino in southern Switzerland (Skelly et al., 1998; VanderHeyden et al., 2000), where foliar symptoms were reproduced on several native plant species with ambient concentrations of ozone. The Lattecaldo nursery is located very close to Valtellina (Fig. 1); both sites are located in the same region subjected to the pollution from Milan (Gerosa and Ballarin-Denti, unpublished data), and show very similar ecological and vegetational features.

Preliminary inspections enabled us to recognize in the native vegetation of the Valtellina region the same kind of symptoms recorded in Lattecaldo (Cozzi et al., 2000). These symptoms were further validated by means of microscopic observations (Gravano et al., 2000). The main symptomatic species recorded were: *Ailanthus altissima* (Mill.) Swingle, *Corylus avellana* L., *Fraxinus excelsior* L., *Laburnum alpinum* (Mill.) Berchtold & J. Presler, *Parthenocissus quinquefolia* (L.) Planch., *Prunus avium* L., *Robinia pseudoacacia* L., *Rubus* spp., *Ulmus glabra* Hudson and *Vitis vinifera* L.

Other important features of the study area were: scattered plants with visible symptoms, seedlings and young trees most severely affected, visible symptoms generally affecting only some plants in a population and even only some leaves on an individual plant. Finally the species assemblage varied with the altitude, with the broadleaved trees in the lower areas and mainly conifers at higher elevations.

1.2. Survey design

The survey was performed at the end of the summer 1998 on 81 observation plots distributed throughout the whole area (Fig. 1(B)), but depending on the conditions of accessibility. The first problem was the comparability between the plots. This problem has usually been solved by choosing areas with similar ecological conditions, with similar species assemblage and of a similar age; however this was not possible in this specific case. In fact, the Alps are characterized by a great variability in the physical properties of the landscape over short distances (altitude, slope, exposure, previous soil use, soil conditions and forest management); consequently the conditions of the vegetation (species composition, intra-specific genetic variability, tree age) are quite variable, so each plot was representative only of its specific conditions and no between-plots comparison were possible. For these reasons we chose to determine only the "presence" or "absence" of symptoms on any species (trees and shrubs) without any further discrimination. That meant that we were unable to find any relationship between the severity of the symptoms and the AOT40 levels, but only the possible minimum exposure thresholds for the presence of the symptoms.

2. Results

Foliar symptoms were detected on 52 of 81 plots (64%) and were distributed fairly evenly throughout the study area (Fig. 1(B)). Symptomatic plants were to be found at all altitudes: relatively rare at the lower elevations (28% of plots between 200 and 400 m asl); more frequent at intermediate levels (between 62 and 66% of plots between 400 and 1000 m asl); and their frequencies decreased again at altitudes above 1000 m asl (50%). Other morphological and topographical parameters (exposure, position) did not appear to affect the distribution of symptomatic plots.

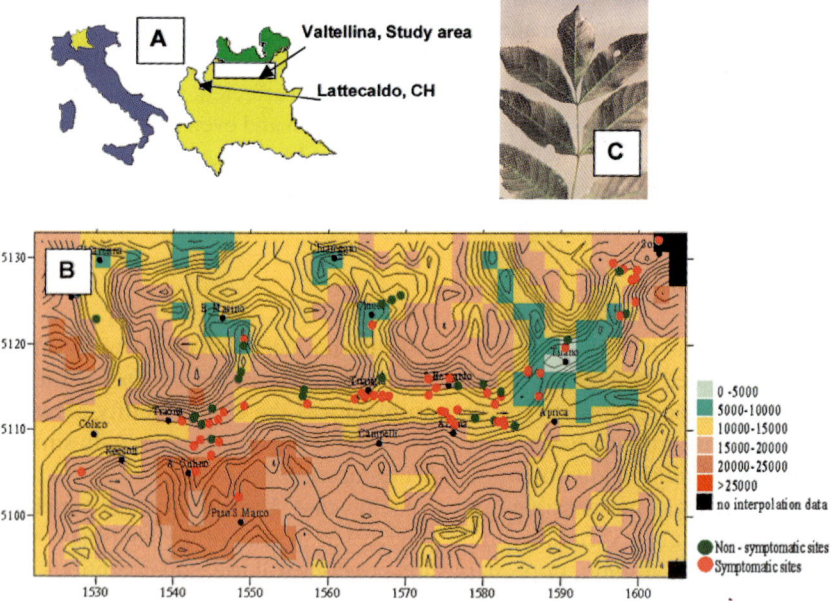

Figure 1. (A) The location of the study area; (B) Map of the exceedances (May–August 1998) and location of symptomatic and non-symptomatic plots. The values are expressed as ppbh; (C) Typical symptoms on *Fraxinus excelsior* leaves.

Symptomatic plots were especially distributed in areas with considerable anthropogenic disturbance or those with irregular forest structures (in these conditions no asymptomatic plots were found at all), where we can find more frequently ruderal sensitive species (*Ailanthus altissima* or *Rubus* spp.) or small trees, more sensitive than the big ones. As far as the stand features are concerned, the main parameter linked to the presence of symptoms was soil depth (73% of the symptomatic plants were found more frequently on deep and moist soil).

The species that was most frequently found to be symptomatic was *Fraxinus excelsior* (26 plots), followed by *Ailanthus altissima* (19), *Prunus avium* (7) and *Rubus* spp. (5). The other species display symptoms with less frequency. The typical symptoms of ozone injury on ash leaves are presented on Fig. 1(C).

Symptomatic sites were scattered throughout the whole area, and no relationship was found with any of the estimated AOT40 values.

3. Discussion

It is usually recognized that the occurrence of foliar symptoms, as well as any physiological injury, does not depend totally on the environmental concentrations of ozone, but rather on the stomatal uptake of this pollutant (Matyssek and Innes, 1999). Thus, factors limiting stomatal gas exchanges are also believed to reduce the harmful effects of ozone. Among these factors those considered most important in the scientific literature are:

- Soil moisture (Schaub et al., 1999). In field conditions symptoms are less evident because water constraints and dry soils enhance stomatal closure (Davison and Barnes, 1998);
- The age of the trees (Kolb et al., 1998). Seedlings and young trees usually have a greater stomatal conductance than mature trees. Aging induces a loss in the efficiency of water transport (Magnani et al., 2000);
- The genotype. Bennett et al. (1992) and Ferdinand et al. (2000) have shown that sensitive individuals of *Fraxinus pennsylvanica* and *Prunus serotina* have a more mesophytic foliar structure than the tolerant individuals: a greater amount of intercellular spaces enhances the effectiveness of gas exchange;
- Structure of vegetation. Canopies provide an obstacle to the deposition and circulation of ozone within forest ecosystems (Dell'Era et al., 1998), so only the individuals in open areas are fully exposed to the action of ozone; and finally
- The different sensitivity of the various tree species and their assemblage according to the ecological conditions markedly influences the distribution of symptoms.

The influence of the factors mentioned above is confirmed by the results of the present survey. In fact the distribution of the symptoms does not follow any AOT40 gradient, but it is influenced by the distribution (according to an altitudinal gradient) of species with different sensitivity (at the higher altitudes we found only coniferous trees, which are less sensitive) and by the distribution of the "modifying" ecological factors, such as soil moisture and canopy structure.

On the other hand, we can observe that the AOT40 levels across the entire study area are high enough to cause visible symptoms; in fact, the sensitive species show foliar injuries with levels of AOT40 less than 10 000 ppb h (VanderHeyden et al., 2000). The lack of differences along the estimated AOT40 gradient may also be due to the period of the survey (late summer). In late summer the response is "flattened" (since each plot has received a sufficient exposure) and the more severely damaged leaves have been shed. To enhance the differences of ozone exposure, it is probably more suitable to detect the symptoms at their onset, assessing the sample trees several times from the beginning of the summer to the onset of the symptoms.

4. Conclusions

The alpine environment is characterized by marked variability of territorial and vegetational features. In these conditions the most important problem in conducting ozone-foliar injuries surveys is to obtain a homogeneous sample, evenly distributed across the investigated territory. Without such a sample (as in the present investigation), only qualitative observations are possible. Moreover, in field conditions the distribution and intensity of leaf injuries doesn't follow a gradient of ozone concentration but, rather, symptom patterns follow the distribution of the sensitive species and the environmental "modifying" factors which are related to a greater or lesser extent to the uptake of ozone.

Probably, more information can be obtained with same-species-trees and shrubs (belonging to the native vegetation) planted in plots distributed according to the AOT40 gradient (*trend plots*, see Chappelka et al., 1999).

Acknowledgements

This survey was carried out with the financial support of Provincia di Sondrio—Settore Ambiente e Sviluppo Economico.

References

Bacci, P., Sandroni, S., Ventura, A., 1990. Patterns of tropospheric ozone in the pre-alpine region. Sci. Total Environ. 96, 297–312.

Ballarin-Denti, A., Cocucci, S.M., Di Girolamo, F., 1998a. Environmental pollution and forest stress: a multidisciplinary approach study on alpine forest ecosystems. Chemosphere 36, 1049–1054.

Ballarin-Denti, A., Cocucci, S.M., Sartori, F. (Eds.), 1998b. Monitoraggio delle Foreste sotto Stress Ambientale. Fondazione Lombardia per l'Ambiente Publisher, Milano, Italy.

Bennett, J.P., Rassat, P., Berrang, P., Karnosky, D.F., 1992. Relationship between leaf anatomy and ozone sensitivity of *Fraxinus pennsylvanica* Marsh. and *Prunus serotina* Ehrh. Environ. Exp. Bot. 32, 33–41.

Chappelka, A., Somers, G., Renfro, J., 1999. Visible ozone injury on forest trees in Great Smoky Mountains National Park, USA. Water Air Soil Pollut. 116, 255–260.

Cozzi, A., Ferretti, M., Innes, J.L., 2000. Sintomi fogliari attribuibili ad ozono sulla vegetazione spontanea in Valtellina. Monti e Boschi 51 (3/4), 42–49.

Cressie, N.A.C., 1991. Statistics for spatial data. Wiley, New York.

Davison, A.W., Barnes, J.D., 1998. Effects of ozone on wild plants. New Phytol. 139, 135–151.

Dell'Era, R., Brambilla, E., Ballarin-Denti, A., 1998. Ozone and air particulate measurements in mountain forest sites. Chemosphere 36, 1083–1088.

Ferdinand, J.A., Fredericksen, T.S., Kouterick, K.B., Skelly, J.M., 2000. Leaf morphology and ozone sensitivity of two pollinated genotypes of black cherry (*Prunus serotina*) seedlings. Environ. Pollut. 108, 297–302.

Fuhrer, J., Skarby, L., Ashmore, M.R., 1997. Critical levels for ozone effects on vegetation in Europe. Environ. Pollut. 97, 91–106.

Gerosa, G., Spinazzi, F., Ballarin-Denti, A., 1999. Tropospheric ozone in alpine forest sites: air quality monitoring and statistical data analysis. Water Air Soil Pollut. 116, 345–350.

Gravano, E., Bussotti, F., Grossoni, P., Tani, C., 2000. Danni fogliari da ozono: caratterizzazione ultrastrutturale di *Fraxinus excelsior* L. e *Prunus avium* L. In: Bucci, G., Minotta, G., Borghetti, M. (Eds.), Applicazioni e Prospettive per la Ricerca Forestale Italiana. SISEF Atti 2. Edizione Avenue Media, Bologna, pp. 447–452.

Kärenlampi, L., Skarbi, L., 1996. Critical levels for Ozone in Europe. Testing and finalizing the concepts. UN-ECE Workshop report, published by the University of Kuopio, Department of Ecology.

Kolb, T.E., Fredericksen, T.S., Steiner, K.C., Skelly, J.M., 1998. Issues in scaling tree size and age response to ozone: a review. Environ. Pollut. 98, 195–208.

Loibl, W., Winiwater, W., Kopsca, A., Zueger, J., Baumann, R., 1994. Estimating the spatial distribution of ozone concentrations in complex terrain. Atmos. Environ. 28, 2557–2566.

Magnani, F., Mencuccini, M., Grace, J., 2000. Age-related decline in stand productivity: the role of structural acclimation under hydraulic constraints. Plant Cell Environ. 23, 251–263.

Matyssek, R., Innes, J.L., 1999. Ozone—a risk factor for trees and forests in Europe? Water Air Soil Pollut. 116, 199–226.

Schaub, M., Zhang, J., Skelly, J.M., Steiner, K.C., Davis, D.D., 1999. Influence of varying soil moisture on gas exchange and ozone injury in three hardwood species. In: Fuhrer, J., Achermann, B. (Eds.), Critical levels for ozone UNECE Workshop Publ. Federal Research Station for Agricultural Chemistry and Environmental Hygiene (FAC), CH-3097 Liebefeld-Bern, Switzerland, pp. 63–66.

Skelly, J.M., Davis, D.D., Merrill, W., Cameron, E.A., Brown, H.D., Drummond, D.B., Dochinger, L.S., 1987. Diagnosing injury to eastern forest trees. University Park, Pennsylvania State University, College of Agriculture, Pennsylvania.

Skelly, J.M., Innes, J.L., Snyder, K.R., Savage, J.E., Hug, C., Landolt, W., Bleuler, P., 1998. Investigations of ozone-induced injury in forests of Southern Switzerland: field surveys and open-top chamber experiments. Chemosphere 36, 995–1000.

Staffelbach, T., Neftel, A., Blattner, A., Gut, A., Fahrni, M., Stähelin, J., Prévôt, A., Hering, A., lehning, M., Neininger, B., Bäumle, M., Kok, G.L., Dommen, J., Hutterli, M., Anklin, M., 1997. Photochemical oxidant formation over Southern Switzerland. I. Results from summer 1994. J. Geophys. Res. 102, 23345–23362.

VanderHeyden, D.J., Skelly, J.M., Innes, J.L., Hug, C., Zhang, J., Landolt, W., Bleuler, P., 2000. Ozone exposure thresholds and foliar injury on forest plants of Switzerland. Environ. Pollut. 111, 333–348.

Air Pollution, Global Change and Forests in the New Millennium
D.F. Karnosky et al., editors
© 2003 Elsevier Ltd. All rights reserved.

Chapter 14

Forest growth and critical air pollutant loads in Scandinavia

C. Nellemann[*]

*Department of Biology and Nature Conservation, Agricultural University of Norway,
Box 5014, 1432 Ås, Norway*

M.G. Thomsen

Norwegian Institute of Land Inventory, Raveien 9, 1432 Ås, Norway

U. Söderberg

*Department of Forest Resource Management and Geomatics,
Swedish University of Agricultural Sciences, SE-901 83 Umeå, Sweden*

K. Hansen

*Department of Forest Ecology, Danish Forest and Landscape Research Institute, Hørsholm,
Kongevej 11, 2970 Hørsholm, Denmark*

Abstract

We investigated radial increment in relation to natural growing conditions and critical loads exceedance for 8577 Norway spruce forest inventory plots in Norway, Sweden, and Denmark. Tree age, growing season, stand density and site productivity accounted for about 50% of the variation in increment, underlining the importance of natural variation for forest growth. We developed a model for assessing increment in comparable forest stands in relation to exceedance of critical loads for nitrogen. Increment was 8–17% lower in exceeded areas. Increment declined with increasing exceedance of critical loads for nitrogen. These results demonstrate that natural growing conditions may conceal important patterns in forest vitality correlated to pollution loadings. Although forest condition thus varies primarily with natural stressors, air pollution may have weakened forest health in exceeded areas in parts of Scandinavia. As air pollution is likely to enhance natural stressors for tree growth, geographic patterns of forest damage are likely to continue to be highly variable and complex.

[*]Corresponding author.

DOI:10.1016/S1474-8177(03)03014-6

1. Introduction

Strategies to reduce levels of atmospheric sulfur (S) and nitrogen (N) in Europe are currently focused on exceedance of critical loads and acceptance levels for forests. Exceedance of critical loads and levels for forests currently form the basis for abatement strategies on sulfur and nitrogen in Europe (DeVries, 1993; Sverdrup and Warfvinge, 1993; Raitio and Kilponen, 1994; Cronan and Grigal, 1995; Hansson, 1995; Jönsson et al., 1995; Posch et al., 1995; UN, 1996). Effective abatement strategies were implemented under the United Nations I and II. Sulfur protocols have resulted in a steady reduction in the deposition of S in Nordic countries and elsewhere in Europe since 1985 (UN, 1994; Posch et al., 1995; Hansen et al., 1998). However, N emissions in Europe reached relatively high levels in the late 1980s and only a small decrease has been observed since then (Pacyna et al., 1991; Robertson, 1991; Grennfelt et al., 1994; Barrett and Berge et al., 1996).

Cumulative effects of long-range-transported air pollutants have primarily been of concern in relation to increased leaching of calcium (Ca) and magnesium (Mg) (van Breemen et al., 1983; Abrahamsen et al., 1994), reduced availability of phosphorus (P) and a shifting of fine roots to higher, more drought-sensitive horizons (Schulze, 1989; Schulze et al., 1989; Schulze and Freer-Smith, 1991) due to aluminium toxicity in forest soils (Driscoll et al., 1985; Mulder et al., 1990, 1995). High deposition of N to nutrient-limited areas may result in eutrophication, nutrient imbalance, increased growth, and, hence, increased acidification (Nihlgård, 1985; Wright et al., 1995; Gundersen, 1989; Gundersen and Rasmussen, 1995). Furthermore, the combined action of soil acidification, reduced availability of essential mineral elements, increased drought sensitivity, and N-eutrophication is most likely to cause nutrient imbalances and forest damage in the long term (Driscoll et al., 1985; van Breemen et al., 1983; Mulder et al., 1989; Abrahamsen et al., 1994; Nellemann and Frogner, 1994; Bytnerowicz and Fenn, 1996; Wright et al., 1995; Nellemann and Esser, 1998), with subsequent impacts on forest growth (Nellemann and Thomsen, 2001).

Defoliation has increased in most of Europe (Müller-Edzards et al., 1997). The high levels of defoliation in the Nordic countries are, to a large extent, related to factors such as a short growing season, advanced age, and climatic stress (Thomsen and Nellemann, 1994). However, when data on growth are analyzed excluding the confounding effect of natural factors, the degree of defoliation appears to be relatively higher, while growth is lower, in areas where the critical loads for forest soils are exceeded (Nellemann and Frogner, 1994; Thomsen et al., 1995; Nellemann and Thomsen, 2001). Effects of acidification on forest soils have been noted in parts of Scandinavia, including reduced availability of cations and lower concentrations of phosphorous in the southern areas where the highest input of S and N is received (Dahl, 1988; Tamm and Hallbäcken, 1988; Billett et al., 1990; Løbersli et al., 1990; Mulder et al., 1990; Løbersli, 1991; Frogner, 1991; Eriksson et al., 1992; Gustafson and Jacks, 1993; Steinnes et al., 1993; Nellemann and Esser, 1998). However, correlation with forest growth at regional scales is still lacking.

Even so, total forest growth increased in the last half of the 20th century in Europe (Spiecker et al., 1996), possibly in response to N eutrophication (Kauppi et al., 1992). In the Nordic countries, cutting rates have been below the overall increment and, thus, have resulted in increased standing stock, and, therefore, increased total volume growth (Tomter, 1996). However, based on the combined action of soil acidification and potential nutrient imbalances, forest increment has been hypothesized to decline (Nihlgård, 1985). Because of the complexity of the effects of N deposition, such analyses are, however, very complicated.

Cumulative effects of acid rain and N deposition on increment are likely to be confounded with those of other growing conditions, such as age, site productivity, stand density, soil condition, and climate (Thomsen and Nellemann, 1994). To avoid confounding of these effects, investigations of forest increment and air pollution must necessarily include a thorough assessment of forest stands in both polluted and background areas, with particular reference to sites of similar age, growing season, and soil conditions in order to isolate any potential impacts of air pollution (Thomsen and Nellemann, 1994; Klap et al., 1997; Müller-Edzards et al., 1997). This is particularly complicated in Scandinavia where the southern high-deposition areas also have the most favorable climate and longest growing seasons.

Few researchers, if any, have analyzed the radial increment on a regional scale in relation to deposition loadings in Europe. Although a substantial number of experimental studies have predicted effects of air pollution, regional correlations between exceedance of critical loads and forest growth are virtually non-existent. As critical loads form the basis of emission protocol negotiations, it is crucial to know if these models actually reflect forest condition on a more regional scale. In the following, we present an analysis of increment in Scandinavian forests seen in relation to natural growing conditions and exceedance of critical loads for N.

2. Methods

Data for this study were derived from the National Forest Inventories of Norway 1992–1994 (2339 plots) and Sweden 1988–1994 (6206 plots), and from monitoring plots in Denmark 1987–1995 (32 plots), representing the most recent data sets available for each country. Data included 5-year radial increment, stand age, basal area, basal area increment, site productivity (H_{40}-system for Norway, SIS for Sweden, and the Møller-system for Denmark), and growing-season variables. The three countries use different parameters as indicators for growing season, i.e., altitude in Norway (Thomsen and Nellemann, 1994), temperature sum in Sweden (sum of temperature for days $> 5\,°C$) (Odin et al., 1983), and standard growth regime classes in Denmark. All plots are located below $62°$ latitude. Despite differences in terminology of forest for classification systems, all three countries rely on standard techniques for measurements of radial increment, using increment cores (1/10 mm) taken at 1.3 m above ground level. All site productivity indices are based on age-height curves.

Critical loads for N in boreal forests are currently suggested to be 7 kg N $ha^{-1}\,yr^{-1}$ (UN ECE, 1996; Esser and Tomter, 1996). We used a modelled deposition map from Lövblad (in: Strand, 1997) to calculate estimated exceedance of N-critical loads for forests in three deposition zones receiving < 7 kg N $ha^{-1}\,yr^{-1}$; 7–14 kg N $ha^{-1}\,yr^{-1}$, and > 14 kg N $ha^{-1}\,yr^{-1}$, in Norway, Sweden, and Denmark, respectively (Fig. 1). Mean exceedance for each deposition zone was calculated by averaging mean deposition for an EMEP-grid (Posch et al., 1995; Strand, 1997) in relation to a critical load for N for northern boreal forests of 7 kg N $ha^{-1}\,yr^{-1}$ (Werner and Spranger, 1996).

In order to avoid potential bias resulting from differences in forest management, forest types, or methodology among the countries, a separate analysis was conducted for each country. Firstly, the importance of stand age, site productivity (using the three different classification systems available in each country), climate, and stand density for variation in radial increment was investigated for each country using multiple linear regression analysis.

Secondly, an investigation of increment in relation to critical load exceedance was conducted for Norway and Sweden. In order to avoid bias resulting from potential natural differences, data were stratified so that we could obtain increment data for comparable forest types (Thomsen and Nellemann, 1994; Thomsen et al., 1995). Stratification was performed according to tree age, growing season, and site productivity indices (SIS 25–34 in Sweden, H_{40} 11–20 in Norway). Tree age was split into three groups (40–59 years, 60–80 years, and > 80 years) for each country, growing season in one group (< 400 m altitude in Norway and > 1250 day-degrees in Sweden), and site productivity into four groups for Norway (11, 14, 17, and 20) and ten groups for Sweden (25, 26, . . . , 34). This gave a total of 12 strata in Norway and 30 strata in Sweden. A pairwise comparison of increment between the deposition zones was made (Wilcoxon rank sum test) using mean weighted radial increment according to the equation shown below. We thus obtained an unbiased estimate of increment independent of tree age, site productivity, and growing season by comparing similar forest types in background and exceeded areas, respectively.

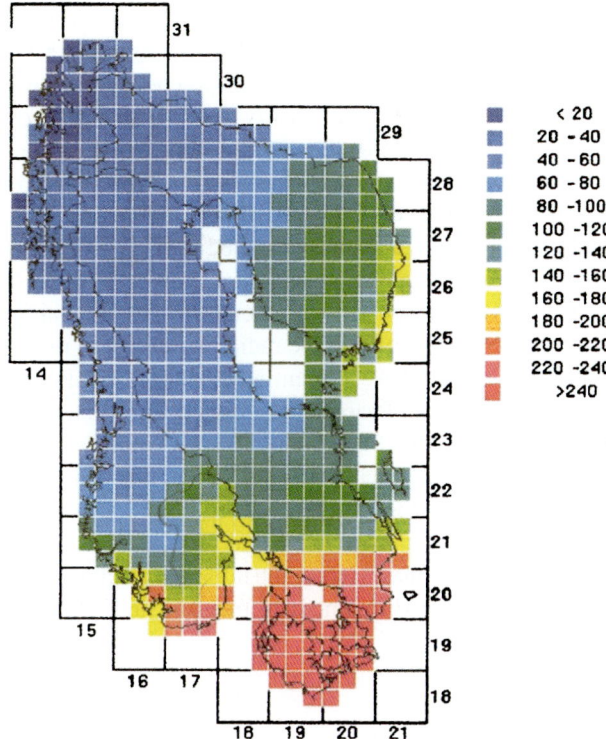

Figure 1. Total N deposition in Norway spruce stands ($cg\,N\,m^{-2}\,yr^{-1}$) in the Scandinavian countries (courtesy of G. Lövblad, IVL, Sweden).

Mean difference in increment was tested by comparing paired samples according to a null-hypothesis of no difference in mean weighted increment (H_0: $\theta_x - \theta_{nx} = 0$) against ($H_1$: $\theta_x - \theta_{nx} \neq 0$), using the following equation (Nellemann and Thomsen, 2001):

$$D_T = {^\wedge}\theta_x - {^\wedge}\theta_{nx}$$

$$= \frac{1}{N}\left[\sum \frac{n_{xi} + n_{nxi}}{2}\mu_{xi} - \sum \frac{n_{xi} + n_{nxi}}{2}\mu_{nxi}\right]$$

where D_T = mean weighted difference in radial increment between exceeded ($> 7\ kg\,N\,ha^{-1}\,yr^{-1}$) and non-exceeded areas ($< 7\ kg\,N\,ha^{-1}\,yr^{-1}$), respectively; θ_x, θ_{nx} = mean weighted radial increment in exceeded (x) and non-exceeded (nx) areas, respectively; n_{xi}, n_{nxi} = number of plots in strata i in exceeded (x) and non-exceeded (nx) areas, respectively; μ_{xi}, μ_{nxi} = mean radial increment for strata i in exceeded (x) and non-exceeded (nx) areas, respectively; and N = number of strata (Nellemann and Thomsen, 2001).

Thirdly, an overall comparison of increment was conducted in relation to deposition loads for the two countries, Norway and Sweden. As these countries have slightly different classification systems for stand age and site indices, the mean radial increment in background areas (< 7 kg N ha^{-1} yr^{-1}) was classified as being 100% for both countries, and then the mean radial increment was calculated as a percentage of that observed in background areas in each country for two additional deposition regions receiving $7-14$ kg N ha^{-1} yr^{-1} and > 14 kg N ha^{-1} yr^{-1}, respectively. A total of six observations (three in Norway and three in Sweden) of relative radial increment in relation to background areas was obtained. These percent measures of radial increment were tested against deposition using Pearson's coefficient of correlation. The analysis was conducted using 60- to 100-year-old forests, stratified by site indices, thus comparing sites of similar stand age and site productivity in different deposition zones. In order to control the confounding factors of tree age and site productivity, this analysis could not be performed on smaller geographic units than these deposition zones. Hence, the number of replicates in increment with deposition adjusted for differences in tree age and site index becomes six.

3. Results

Radial increment varies as a function of tree age (Fig. 2(a)), site productivity (Fig. 2(b)), and growing season (Fig. 2(c)). Tree age, site productivity and growing season explained 58% of the variation in increment in Denmark ($R^2 = 0.58$, $p < 0.01$, $n = 32$), 47% in Norway ($R^2 = 0.47$, $p < 0.01$, $n = 2339$), and 52% in Sweden ($R^2 = 0.52$, $p < 0.01$, $n = 6206$).

On average, weighted and stratified increment was from about 9% (Sweden) ($p < 0.01$) to about 20% (Norway) ($p < 0.01$) lower in exceeded areas compared to background areas for comparable forest stands. A similar analysis was not possible for Denmark, as the entire country was located within one deposition zone, and sample size was too small for this type of analysis.

A combined assessment of increment for comparable forest types for the boreal spruce forest of Norway and Sweden revealed declining increment with increasing pollution loadings ($r_p = -0.92$; $n = 6$) (Fig. 3). This pattern was not attributable to potential differences in growing season, stand age, or site productivity, nor to stand density, which was not significantly different among areas.

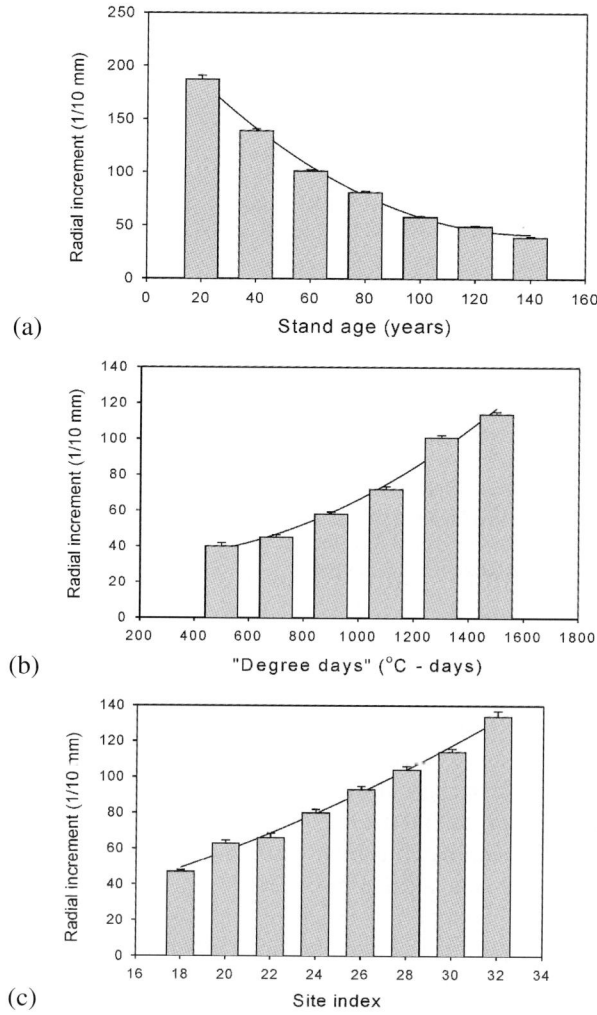

Figure 2. Mean 5-year radial increment (±SE) from 6206 Norway spruce plots in Sweden seen in relation to (a) stand age, (b) growing season, and (c) site index, 1988–1994.

4. Discussion

Radial tree increment varies primarily as a function of tree age, climate, stand density, and variation in site productivity across Scandinavia. However, the analysis of stands having comparable growth conditions clearly demonstrates reduced increment in exceeded areas for Norway and Sweden. The same pat-

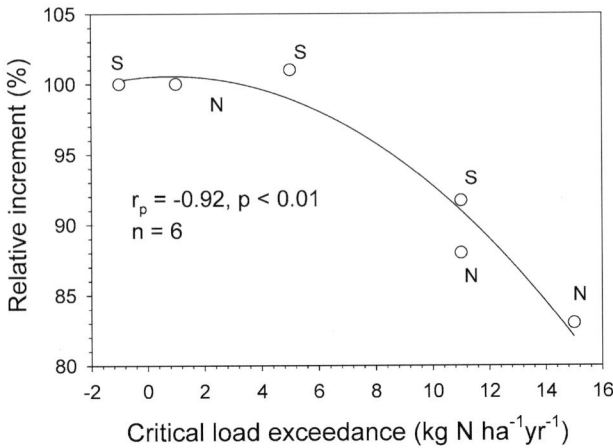

Figure 3. Mean relative radial increment for 8545 Norway spruce plots for three deposition zones (< 7 kg N ha^{-1} yr^{-1}; 7–14 kg N ha^{-1} yr^{-1} and > 14 kg N ha^{-1} yr^{-1}, respectively) with varying exceedance of critical loads for nutrient N. Increment is expressed as percentage of background areas ($= 100\%$), (S = Sweden and N = Norway).

tern was found in Finland (Tomppo and Henttonen, personal commun.). This agrees with previous studies in Norway (Nellemann and Frogner, 1994; Thomsen et al., 1995; Nellemann and Thomsen, 2001). In southern Norway, a study on more than 30 000 forest plots revealed that increment increased with increasing N-deposition throughout 1960–1970, without a similar response in comparable forests in background areas (Nellemann and Thomsen, 2001). This was followed by a substantial decline in exceeded areas, again with no similar decline in adjacent, comparable areas that did not receive N in exceedance of critical loads (Nellemann and Thomsen, 2001).

However, we could not perform such an analysis for Denmark. The country represents a relatively small geographic area with low variation in deposition, and most of the forest plots were found in exceeded areas.

Even though exceedance of critical loads is greatest in the southern part of the boreal region, forest growth is more abundant there than in the non-polluted but climatically marginal areas further north because the growing season is longer (Fig. 2(c)). Indeed, an absolute 75% reduction in mean increment for the areas with the longest growing season (and highest N deposition) would still leave these areas with an increment above that of the climatically growth-limited areas at higher latitudes and altitudes (Fig. 2(c)). Hence, difficulty in correlating increment to N deposition in areas with high defoliation and acid input is due in part to confounding factors.

This type of regional analysis does not allow for site-specific assessments of individual stress factors, but serves to document large-scale, regional patterns associated with natural growth conditions, and, potentially, differences in pollution loadings. Empirical studies suggest that increased soil acidification may lead to antagonistic effects of Al^{3+} on nutrient and water uptake by fine roots. A shifting of fine roots combined with lower amounts of mychorrhizae in exceeded areas may reduce nutrient uptake and increase drought sensitivity (Shafer and Schoeneberger, 1991; Holopainen et al., 1996). As N fertilization may increase frost and drought sensitivity, the cumulative risk of abiotic damages will gradually increase over time (Nilsen, 1994; Chappelka and Freer-Smith, 1995). Indeed, substantially higher levels of needle discoloration have been observed in spruce forest having low humic pH (Nellemann and Esser, 1998). An indirect effect of long-range transported air pollution in these areas may, therefore, be greater annual forest growth in response to different weather conditions (Chappelka and Freer-Smith, 1995), along with a greater likelihood of insect damage. The result of this then is a highly variable damage pattern, depending on local climate and growth conditions. Even relatively moderate levels of N deposition (> 10 $kg\,N\,ha^{-1}\,yr^{-1}$) in these sensitive northern areas may affect forests over time (Wright et al., 1995). European data show that there is a good correlation between N deposition and N leaching (Gundersen, 1995). At N deposition levels of 9–25 $kg\,N\,ha^{-1}\,yr^{-1}$, N leaching has been observed; at levels beyond > 25 $kg\,N\,ha^{-1}\,yr^{-1}$, substantial leaching has occurred well within the observed levels in the areas studied here (Wright et al., 1995). Although nitrogen saturation may occur over the long term, it is far more important to emphasize mechanisms related to nutrient imbalances at earlier stages, which, unfortunately, will also vary substantially according to growing conditions.

Forest growth varies primarily with stand age, growing season, and site index, and other obvious stand parameters. However, cumulative effects of long-range-transported air pollution may be a significant additive factor affecting forest growth in southern Scandinavia. Cause-effect relationships cannot be established from geographic analysis alone, but we conclude that when natural growth factors are controlled, there appears to be a spatial agreement between exceedance of critical loads of N in forest soils and lower radial increment in boreal forests of Scandinavia.

References

Abrahamsen, G., Stuanes, A.O., Tveite, B. (Eds.), 1994. Long-Term Experiments With Acid Rain in Norwegian Forest Ecosystems. In: Ecological Studies, Vol. 104.

Barrett, K., Berge, E., 1996. Transboundary air pollution in Europe. EMEP MSC-W Status Report 1/96. The Norwegian Meteorological Institute, Oslo.

Billett, M.F., FitzPatrick, E.A., Cresser, M.S., 1990. Changes in the carbon and nitrogen status of forest soil organic horizons between 1949/50 and 1987. Environ. Pollut. 66, 67–79.

Bytnerowicz, A., Fenn, M.E., 1996. Nitrogen deposition in California forests: A review. Environ. Pollut. 92, 127–146.

Chappelka, A.H., Freer-Smith, P.H., 1995. Predisposition of trees by air pollutants to low temperatures and moisture stress. Environ. Pollut. 87, 105–117.

Cronan, C.S., Grigal, D.F., 1995. Use of calcium/aluminum ratios as indicators of stress in forest ecosystems. J. Environ. Quality 24, 209–226.

Dahl, E., 1988. Forsuring av jordsmonn i Rondane, Syd-Norge, på grunn av sur nedbør. Økoforsk rapport 1, 1–53.

DeVries, V., 1993. Average critical loads for nitrogen and sulfur and its use in acidification abatement policy in the Netherlands. Water Air Soil Pollut. 68, 399–434.

Driscoll, C.T., Van Breemen, N., Mulder, J., 1985. Aluminum chemistry in a forested spodosol. New Hampshire. Soil Sci. Soc. Amer. J. 49, 437–444.

Eriksson, E., Karltun, E., Lundmark, J.-E., 1992. Acidification of forest soils in Sweden. Ambio 21, 150–154.

Esser, M.J., Tomter, S.M., 1996. Reviderte kart for tålegrenser for nitrogen basert på empiriske verdier for ulike vegetasjonstyper. NIJOS Rapp. 7/96.

Frogner, T., 1991. Skogsjordas næringsinnhold i kystnære strøk i sør-Norge. Overvåkningsprogram for skogskader. Norsk Institutt for Skogforskning, Ås.

Grennfelt, P., Hov, Ø., Derwent, D., 1994. Second generation abatement strategies for NO_x, NH_3, SO_2 and VOCs. Ambio 23, 425–433.

Gundersen, P., 1989. Luftforurening med kvælstofforbindelser—effekter i nåleskov (in Danish). Licentiatafhandling, Lab. for Økologi og Miljølære, Danmarks Tekniske Højskole.

Gundersen, P., 1995. Nitrogen deposition and leaching in European forests—preliminary results from a data compilation. Water Air Soil Pollut. 85, 1179–1184.

Gundersen, P., Rasmussen, L., 1995. Nitrogen mobility in a nitrogen limited forest at Klosterhede, Denmark, examined by NH_4NO_3 addition. For. Ecol. Manag. 71, 75–88.

Gustafson, J.P., Jacks, G., 1993. Sulphur status in some Swedish podzols as influenced by acid deposition and extractable organic carbon. Environ. Pollut. 81, 185–191.

Hansen, K., Hovmand, M.F., Dybkjær, T., Bille-Hansen, J., 1998. De danske skove fik det bedre i 1997. Skoven 4, 154–157 (in Danish).

Hansson, J., 1995. Modelling effects of soil solution BC/Al-ratio and available N on ground vegetation composition. Rep. Ecol. Environ. Eng., Report 3.

Holopainen, T., Heinonen-Tanski, H., Halonen, A., 1996. Injuries to Scots pine mycorrhizas and chemical gradients in forest soil in the environment of a pulp mill in central Finland. Water Air Soil Pollut. 87, 111–130.

Jönsson, C., Warfvinge, P., Sverdrup, H., 1995. Uncertainty in predicting weathering rate and environmental stress factors with the PROFILE model. Water Air Soil Pollut. 81, 1–23.

Kauppi, P., Mielikäinen, K., Kuusela, K., 1992. Biomass and carbon budget of European forests, 1971 to 1990. Science 256, 70–74.

Klap, J.M., de Vries, W., Erisman, J.W., Leeuwen, E.P., 1997. Assessment of the possibilities to derive relationships between stress factors and forest condition. WSC, Wageningen. SC-report 150.

Løbersli, E.M., 1991. Soil acidification and metal uptake in plants. Ph.D. dissertation, University of Trondheim, Norway.

Løbersli, E.M., Steinnes, E., Ødegård, M., 1990. A historical study of mineral elements in forest plants from south Norway. Environ. Monitor. Assess. 15, 111–129.

Mulder, J., Christophersen, N., Hauhs, M., Vogt, R.D., Andersen, S., Andersen, D.O., 1990. Water flow paths and hydrochemical controls in the Birkenes catchment as inferred from a rainstorm high in seasalts. Water Resours. Res. 26, 611–622.

Mulder, J., Christophersen, N., Kopperud, K., Fjeldal, P.H., 1995. Water flow paths and the spatial distribution of soils as a key to understanding differences in streamwater chemistry between three catchments (Norway). Water Air Soil Pollut. 81, 67–91.

Mulder, J., Van Breemen, N., Eijck, H.C., 1989. Depletion of soil aluminium by acid deposition and implications for acid neutralization. Nature 337, 247–249.

Müller-Edzards, C., De Vries, W., Erisman, J.W., 1997. Ten years of monitoring forest condition in Europe. Studies on temporal development, spatial distribution and impacts of natural and anthropogenic stress factors. UN-ICP-report, Geneva.

Nellemann, C., Esser, J., 1998. Crown condition and soil acidification in Norwegian spruce forests. Ambio 27, 143–147.

Nellemann, C., Frogner, T., 1994. Spatial patterns of spruce defoliation: Relation to acid deposition, critical loads and natural growth conditions in Norway. Ambio 23, 255–259.

Nellemann, C., Thomsen, M.G., 2001. Critical loads, nitrogen deposition and long-term changes in forest growth. Water Air Soil Pollut. 128, 197–205.

Nihlgård, B., 1985. The ammonium hypothesis—An additional explanation to the forest dieback in Europe. Ambio 14, 2–8.

Nilsen, P., 1994. In: Proceedings of an international seminar on: Counteractions Against Acidification in Forest Ecosystems, March 3–4, 1994, Mastemyr, Norway. Aktuelt fra Skogfors 14, 1–54.

Odin, H., Eriksson, B., Pertuu, K., 1983. Temperaturkartor för svenskt skogsbruk. Department of Forest Site Research, Univ. for Agric. Sciences. Report 45 (in Swedish).

Pacyna, J.M., Larssen, S., Semb, A., 1991. European survey for NO_x emissions with emphasis on eastern Europe. Atmos. Environ. 25, 425–439.

Posch, M., de Smet, P.A.M., Hettelingh, J.-P., Downing, R.J., 1995. Calculation and mapping of critical thresholds in Europe. Status report 1995. RIVM-report 259101004.

Raitio, H., Kilponen, T., 1994. Critical loads and critical limit values. Proceedings of the Finnish-Swedish Environmental Conference. Oct. 27–28, 1994, Vaasa, Finland. The Finnish Forest Research Institute. Research Papers 513.

Robertson, K., 1991. Emissions of N_2O in Sweden—natural and anthropogenic sources. Ambio 20, 151–155.

Schulze, E.-D., 1989. Air pollution and forest decline in a spruce (*Picea abies*) forest. Science 244, 776–783.

Schulze, E.-D., Lange, O.L., Oren, R., 1989. Air pollution and forest decline. Ecol. Stud. 77, 1–475.

Schulze, E.-D., Freer-Smith, P.H., 1991. An evaluation of forest decline based on field observations focused on Norway spruce, *Picea abies*. Proc. R. Soc. Edinburgh 97B, 155–168.

Shafer, S.R., Schoeneberger, M.M., 1991. Mycorrhizal mediation of plant response to atmospheric change: Air quality concepts and research considerations. Environ. Pollut. 73, 163–177.

Spiecker, H., Mielikäinen, K., Köhl, M., Skovsgaard, J.P., 1996. Growth trends in European forests. European Forest Institute Research Report No. 5. Springer-Verlag.

Steinnes, E., Flaten, T.P., Varskog, P., Låg, J., Bølviken, B., 1993. Acidification status of Norwegian forest soils as evident from large scale studies of humus samples. Scand. J. For. Res. 8, 291–304.

Strand, L. (Ed.), 1997. Monitoring the Environmental Quality of Nordic Forests. NORD, p. 14.

Sverdrup, H., Warfvinge, P., 1993. The effect of soil acidification on the growth of trees, grass and herbs as expressed by the (Ca + Mg + K)/Al ratio. Rep. Ecol. Environ. Eng. 2.

Tamm, C.O., Hallbäcken, L., 1988. Changes in soil acidity in two forest areas with different acid deposition: 1920s to 1980s. Ambio 17, 56–61.

Thomsen, M.G., Nellemann, C., 1994. Isolation of natural factors affecting crown density and crown color in coniferous forest: Implications for monitoring of forest decline. Ambio 23, 251–254.

Thomsen, M.G., Nellemann, C., Frogner, T., Henriksen, A., Tomter, S.T., Mulder, J., 1995. Tilvekst og vitalitet for granskog sett i relasjon til tålegrenser og forurensningsbelastning. Rapport fra Skogforsk 22/95.

Tomter, S.M., 1996. Skog 96. Statistics of forest conditions and resources in Norway. NIJOS Rapp. 19/96.

UN, 1994. Protocol to the 1979 convention on long-range transboundary air pollution on further reduction of sulphur emissions. United Nations.

UN, 1996. UN-ECE. Convention on Long Range Transboundary Air Pollution. Manual on methodologies for mapping critical loads/levels and geographical areas where they are exceeded. Final draft prepared for the 12th Meeting of the Task Force on Mapping. March 22, 1996, Budapest, Hungary.

UN ECE, 1996. UN ECE Task Force on Mapping: Minutes of the 11th Meeting of the Task Force on Mapping. Geneva, 14 December 1995.

van Breemen, N., Mulder, J., Driscoll, C.T., 1983. Acidification and alkalinization of soils. Plant Soil 75, 283–308.

Werner, B., Spranger, T., 1996. Manual on methodologies and criteria for mapping critical loads and levels and geographical areas where they are exceeded. UN-ECE Convention on long range transported air pollution. Federal Environmental Agency, Berlin.

Wright, R.F., Roelofs, J.G.M., Bredemeier, M., Blanck, K., Boxman, A.W., Emmett, B.A., Gundersen, P., Hultberg, H., Kjønaas, O.J., Moldan, F., Tietema, A., van Breemen, N., van Dijk, H.F.G., 1995. NITREX: responses of coniferous forest ecosystems to experimentally changed deposition of nitrogen. Forest. Ecol. Manag. 71, 163–169.

Air Pollution, Global Change and Forests in the New Millennium
D.F. Karnosky et al., editors
© 2003 Elsevier Ltd. All rights reserved.

Chapter 15

Norway spruce mortality and critical air pollutant loads

M.G. Thomsen[*,1]

Norwegian Institute of Land Inventory, P.O. Box 115, 1431 Ås, Norway

C. Nellemann

Norwegian Institute of Nature Research, Pressesenteret, Storhove, N-2624 Lillehammer, Norway

Abstract

We analyzed proportions of dead trees to exceedance of critical loads for acidification at regional scales in 3914 Norway spruce (*Picea abies*) plots (1994–1997) in southeastern Norway. The percentage of dead spruce was significantly higher within exceeded areas than in non-exceeded areas for all stands ($p < 0.001$). Within exceeded areas, both mortality and exceedance levels increased with elevation up to about 400 m a.s.l., and then declined correspondingly. No such covariation was observed in non-exceeded areas among elevation, mortality and exceedance levels. Mortality was also significantly higher in exceeded areas than in comparable stands in background areas having similar stand age, site productivity, and altitude. In areas that were exceeded, mortality was about two times higher than in background areas (21.3 and 11.4%, respectively). There were no significant differences in mortality between exceeded areas and background areas in the 1920s and early 1970s; however, mortality increased three-fold in exceeded areas, compared with a much smaller increase in background areas between 1970–1990s, corresponding closely to exceedance of critical loads. We conclude that, although mortality varies with age, site productivity, and other natural stressors, there is a general geographic relationship between mortality and exceedance of critical loads at regional scales. These results strengthen previous findings, which have shown the same pattern for tree increment, crown condition, and lake acidification. Although natural variation may obscure effects of long-range-transported air pollution at individual sites, there are consistent regional patterns suggesting a greater sensitivity to natural stressors in areas where critical loads are exceeded. More intensive studies should be conducted to clarify potential cause-effect relationships that cannot be established at these regional scales.

---

[*]Corresponding author.

[1]Present address: Hedmark University College, Div. Blæstad, 2322 Ridabu, Norway.

DOI:10.1016/S1474-8177(03)03015-8

1. Introduction

Extensive reductions in sulfur (S) emissions over Europe have occurred over the last decades (Posch et al., 1995). In 1985, the first S protocol under the UN convention on Long-range Transboundary Air Pollution was signed. Since the Convention was signed in 1979, there has been a significant increase in the knowledge and methodologies relating to these problems. Advanced mathematical models have been developed (Cosby et al., 1985; Posch et al., 1995). These models have made it possible to conduct qualified risk assessments by calculating critical load and have become a very important political tool for protocol work on emission reductions in Europe (Cronan and Grigal, 1995; Posch et al., 1995). Critical load is the maximum allowable deposition that does not increase the probability of damage to forest soils and surface waters (Hettelingh et al., 1995).

In 1994, the second sulfur protocol, based on modelled critical loads, was signed; proposed reductions should protect 90% of the ecosystems' area in Europe by 2010 (Posch et al., 1995; Hettelingh et al., 1995). Nitrogen emissions increased from 1960 to 1980; the first NO_x protocol was signed in 1988, but emissions have remained at a high level (Pacyna et al., 1991; Robertsson, 1991; Barrett and Berge, 1996). The second NO_x protocol was signed in 1999, again based on the critical loads concept. Currently, northern Europe has areas with the lowest critical loads (Hettelingh et al., 1995).

The response variable used in critical loads calculations is the calcium : aluminium (Ca : Al) ratio (Sverdrup and Warfvinge, 1993; Raitio and Kilponen, 1994). A ratio of 1 : 0 in the soil solution is estimated to give a 50 : 50 risk of adverse impacts on tree growth (Cronan and Grigal, 1995), with risk increasing with a lower ratio. Potential impacts of acidification damage and deposition of excess nitrogen on the trees include nutritional imbalances, foliar discoloration, needle loss, susceptibility to secondary stress, growth decline, and increased tree mortality (Freer-Smith, 1998).

Tree mortality as a result of air pollution has been related to heavy pollution from point sources (Kandler and Innes, 1995). In countrywide studies, mortality or tree growth is often given as a mean for countries or large geographic areas and the general conclusion is that mortality lies within the normal range of 0.5–3% (Hall, 1995; Nellemann and Esser, 1998), and increment often shows large variations (Spiecker, 1999). However, Sverdrup et al. (1994) concluded that, under continued deposition in the Nordic countries at 1990 levels, tree dieback and stemwood growth loss would increase significantly. Mortality can be caused by numerous factors, including frost, drought, insects, pathogenic fungi, etc., but also as a result of increased sensitivity to such factors indirectly induced by ozone stress, soil acidification, or nutrient imbalances (Schulze et al., 1989). Few studies have considered mortality in relation to exceedance of critical loads at regional scales, even though patterns, if they exist, emerge on a regional scale. In the present study, we hypothesize that tree mortality is greater in areas where critical loads are exceeded.

2. Materials and methods

Data on mortality were derived for a total of 3914 plots from a 3 × 3 km grid system covering southeastern Norway from the National Forest Inventory for the period 1994–1997 (Nellemann and Frogner, 1994). For the most part, only plots dominated by Norway spruce (*Picea abies*) were included. However, data on Scots pine (*Pinus sylvestris*) were used in the analysis of the historical development of mortality. Within each site, cutting (age) class, site productivity, and elevation were recorded. Mortality for Norway spruce was defined as the proportion of dead standing trees compared with live trees within each plot. All trees with $a > 5$ cm diameter were included.

Critical loads were calculated on the basis of soil and surface water chemistry data from 712 grid squares, each 12 × 12 km (Posch et al., 1995), covering the forested areas of Norway. The molar ratio Ca : Al in soil solution was used as the modelling criterion, where critical load is exceeded when Ca : Al < 1. This is the threshold at which root damage is expected (DeVries, 1993; Cronan and Grigal, 1995; Posch et al., 1995). The dynamic soil acidification catchment model MAGIC was used to calculate critical loads and exceedances (Cosby et al., 1985; Wright et al., 1988). The exceedance data from this model have also been found to correlate to forest damage, soil chemistry, and tree increment in this region (Nellemann and Frogner, 1994; Nellemann and Thomsen, 2001).

We compared forest condition in exceeded areas in southeastern Norway with background areas, using the zones specified in Nellemann and Thomsen (2001). The exceeded area primarily includes the southern part of eastern Norway, whereas background areas extend east and north of the exceeded area (Fig. 1). The western coastal and northern regions of Norway were not included as they experience substantially different weather and climatic conditions and, therefore, require more specialized methods for comparison.

Defoliation generally increases with altitude (Thomsen and Nellemann, 1994). To avoid potential confounding effects of altitude, site productivity, and tree age on mortality, we analyzed the proportion of dead trees for plots with comparable elevation (0–99 m; 100–199 m; 200–299 m; 300–399 m, etc.), site productivity indices (6–8 lowest, 11–14 and 17–23 highest productivity), and cutting (age) classes (class III represent young stands typically 15–30 years old, class IV 30–60 years old, and class V mature forest typically 60–120 years old), using data stratification (Nellemann and Frogner, 1994; Nellemann and Thomsen, 2001). As deposition and, thus, pollution loadings and exceedance,

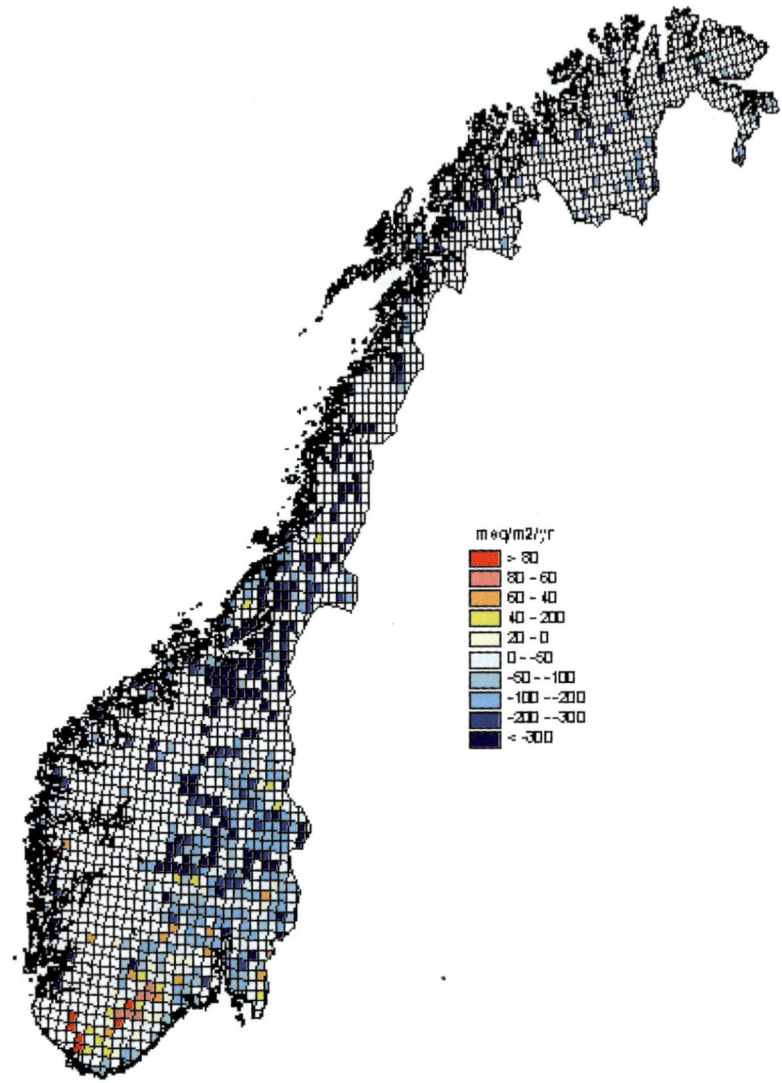

Figure 1. Exceedance of critical loads for acidification calculated by MAGIC in southern Norway. Note the exceeded area in the southeast, a contiguous, elongated exceeded (red) area.

vary substantially with altitude, we compared the variation in exceedance with altitude and the corresponding changes in mortality with elevation for the exceeded and background areas, respectively. This was done using polynomial regression.

We compared volume percentage of dead trees with total live volume on more than 10 000 plots from 1920–1925 (for 1920–1925, only combined conifers, i.e., Norway spruce and Scots pine); 1970–1976; 1986–1992; and 1992–1996. These were the only periods available. Data were accessible from historic county records from the Norwegian Institute of Land Inventory for the years 1920–1996. We compared potential differences in mortality over time between exceeded and background areas from 1920–1996.

Statistical analysis was performed in Sigmastat (Kuo et al., 1992). Data were subjected to a Kolmogorov–Smirnov test of normality. Correlation analyses were performed using Spearman Coefficient of Rank Correlation. Pairwise comparisons were done using Student t-tests or Mann–Whitney tests. In all cases, p-values < 0.05 were considered statistically significant.

3. Results

Critical loads are primarily exceeded in southeastern Norway, 10–50 km from the coast at 200 to 600 m a.s.l. Previous studies have shown that for spruce forests, there are no significant differences in soil depth between the exceeded and background areas (Nellemann, 1997; Nellemann and Thomsen, 2001). Therefore, the higher exceedance of critical loads for acidification in southeastern Norway is primarily due to higher levels of deposition in this region.

Exceedance of critical loads increased with altitude up to approximately 400 m in exceeded areas ($r_s = 0.55$; $p < 0.01$; $n = 113$), and then declined (Fig. 2(a)), whereas there was no pattern evident in background areas ($r_s = -0.03$; $p = 0.71$; $n = 155$) (Fig. 2(b)). To avoid bias of altitude on forest vitality, only forests below 400 m a.s.l. were considered in the following analysis (Thomsen and Nellemann, 1994; Nellemann and Frogner, 1994).

Tree mortality for forest located below 400 m a.s.l., was on average two- to threefold higher in exceeded areas ($23.1 \pm 1.1\%$ and $9.6 \pm 0.3\%$, respectively). These differences were not attributable to potential differences between the areas in site productivity or cutting (age) class distribution ($p < 0.05$).

Mortality varied with both age and site productivity. Old stands or stands experiencing more marginal growth conditions (lower site indices) had, on average, higher mortality. A comparison of mortality between exceeded and non-exceeded areas revealed a significantly higher proportion of dead trees in exceeded areas among all age categories compared with background areas (Fig. 3). This was also true for site productivity (Fig. 4). Forest mortality increased with elevation, corresponding to exceedance patterns (see Fig. 2(a), (b)) in exceeded areas, whereas no such pattern existed in background areas

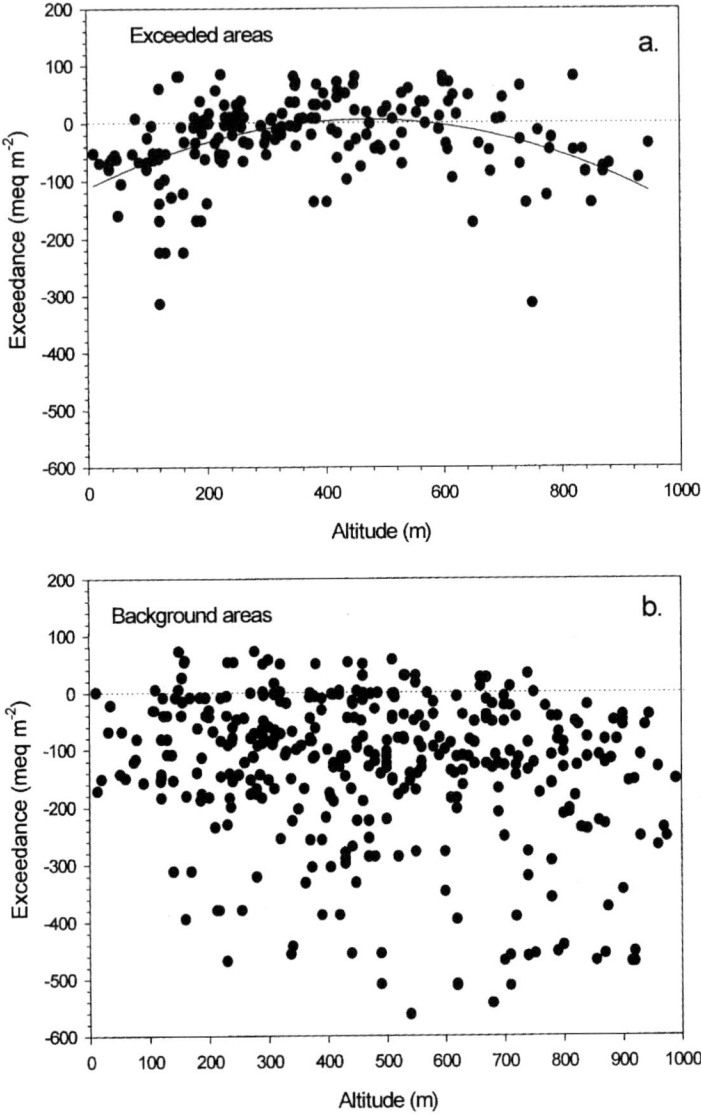

Figure 2. Exceedance of critical loads in relation to altitude for (a) exceeded and (b) background areas.

(Fig. 5). Around 1920 and in the early 1970s, there were no significant differences in mortality between exceeded and background areas. After 1975,
mortality increased in both areas (Fig. 6). However, mortality increased much

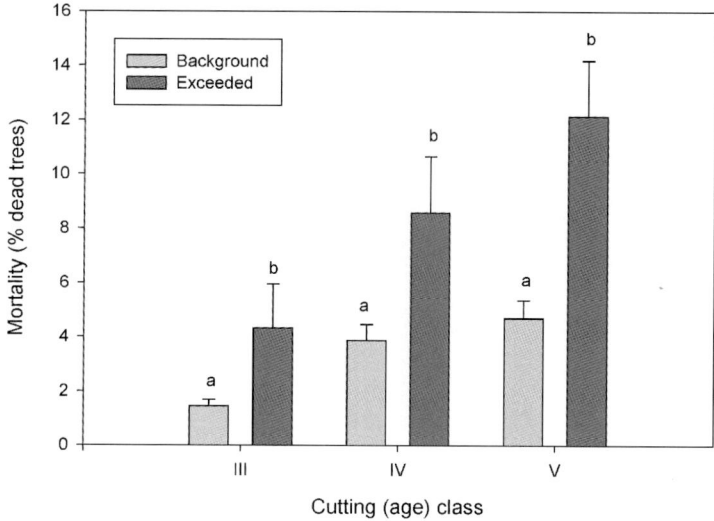

Figure 3. Mortality of Norway spruce in exceeded and background areas for different age classes (class III represent young stands of typically 15–30 years old, class IV 30–60 years old, and class V mature forest, typically 60–120 years old). Different letters for pairs indicate significant difference using Mann–Whitney tests ($p < 0.05$).

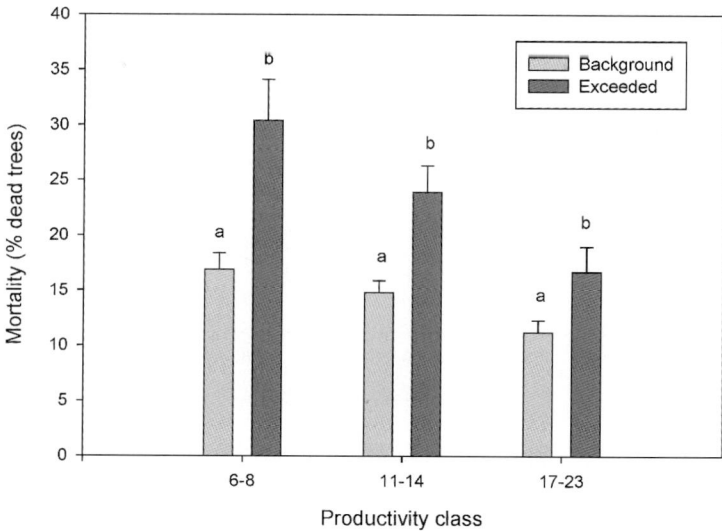

Figure 4. Mortality of Norway spruce in exceeded and background areas in relation to site productivity. Different letters for pairs indicate significant difference using Mann–Whitney tests ($p < 0.05$).

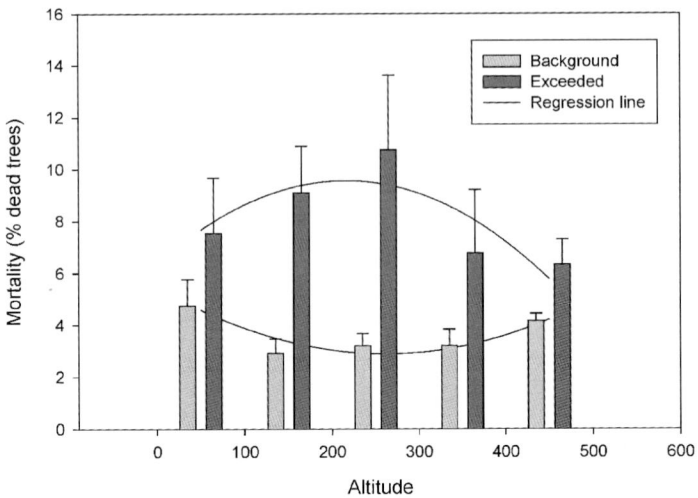

Figure 5. Pairwise comparisons of mortality in Norway spruce in exceeded and non-exceeded areas at different altitudes.

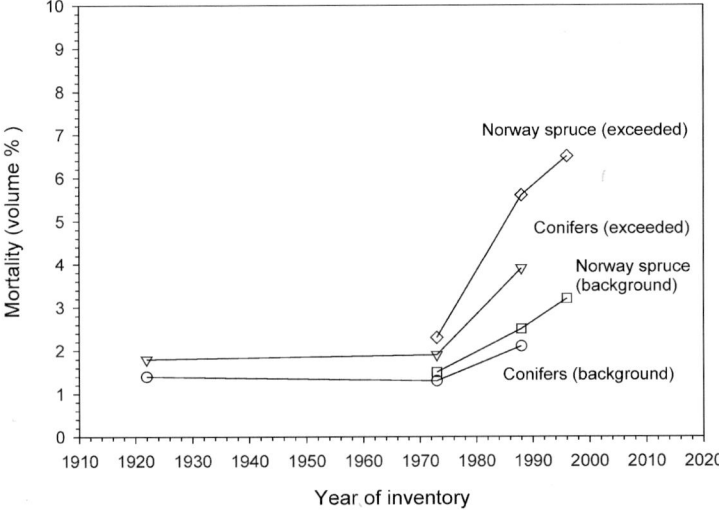

Figure 6. Changes in mortality between background areas and exceeded areas between 1920 and 1996 for conifers and for Norway spruce.

faster in exceeded areas, becoming significantly ($p < 0.001$) higher in exceeded areas, corresponding closely with reduced increment, higher defoliation, and extensive lake acidification in that period (Nellemann and Thomsen, 2001)

4. Discussion

Mortality expressed as the proportion of dead to live trees was significantly higher in exceeded areas. The ratio of dead to live trees varied with stand age and site productivity, but was also, in general, two- to threefold higher in exceeded areas compared with background areas.

The regional covariation between exceedance of critical loads and mortality only provides a geographic indication of a possible correlation. The resolution of our data was not sufficient for more detailed comparisons. There are several indications that the observed pattern corresponding with air pollution loadings may be at least partially causal, although certainly interactive with natural stressors also (Chappelka and Freer-Smith, 1995).

The areas that are classified as exceeding critical loads have a series of characteristics suggesting possible long-term impacts of acidification and N deposition. In humus, base saturation is about 60% lower, and the contents of major base cations and phosphorus are 30–60% lower for comparable soil types and soil depths compared with background areas (Nellemann and Frogner, 1994; Nellemann, 1997; Nellemann and Esser, 1998). Combined with increased soil acidification, these factors may result in reduced forest vitality through increased sensitivity to frost or drought or impeded growth or nutrient imbalances (Nihlgård, 1985; Driscoll et al., 1985; Schulze et al., 1989; Wright et al., 1995; Bytnerowicz and Fenn, 1996).

Forest vitality is consistently poorer in exceeded areas compared with background areas, when forest sites of similar age, site productivity, and altitude are compared. Defoliation is 10–30% greater (Nellemann and Frogner, 1994), on average, and increment correspondingly lower (Thomsen et al., 1995; Solberg and Tveite, 2000; Solberg and Strand, 2000; Nellemann and Thomsen, 2001).

Around 1920, forest surveys indicated little difference in mortality between these regions in southern Norway. The differences in mortality increased throughout 1960–1980 in response to increasing pollution loadings (Nellemann and Thomsen, 2001). Increment followed a remarkably similar pattern. A study of more than 31 000 forest plots from 1954–1996 revealed only minor differences in growth for comparable stands in exceeded and background areas before 1960 (Nellemann and Thomsen, 2001). Increment increased in exceeded areas in the period 1960–1970, closely corresponding to the elevated N deposition. Increases were greatest in exceeded areas, slightly lower in the lowlands, and nearly absent in background areas. During the late 1970s, increment declined following the same pattern, most dramatically in exceeded areas at 200–400 m a.s.l., again without any great changes in background areas. These patterns coincide roughly with exceedance of critical loads and higher mortality at these elevations, as well as with increases in mortality at this time.

Hindcast modelling has shown a distinct covariation in these patterns with exceedance of critical loads (Nellemann and Thomsen, 2001). Similar covariation in comparable stands for exceedance of critical loads and increment has also been observed in the other Scandinavian countries (Tomppo and Henttonen, personal commun.; Nellemann et al., 2002).

We consider it unlikely, that air pollution and nutrient imbalance alone caused these patterns in the forests studied. There is, however, a significant body of studies documenting potential long-term cumulative impacts of air pollution, stressing that most of the damage likely results from elevated sensitivity to abiotic stress as a result of nutrient imbalance, soil acidification or ozone-related impacts (Sverdrup et al., 1994; Chappelka and Freer-Smith, 1995; Cronan and Grigal, 1995; Freer-Smith, 1998). It is likely, therefore, that the observed damage patterns are the result of the combined action of air pollution and natural stressors (climatic changes). We conclude that there is a relatively consistent pattern between forest mortality and exceedance of critical loads at regional levels, and support the use of these criteria for regional risk assessments and, thus, as a basis for emission protocols.

References

Barrett, K., Berge, E., 1996. Transboundary Air Pollution in Europe. EMEP MSC-W Status report 1/96. The Norwegian Meteorological Institute, Oslo.

Bytnerowicz, A., Fenn, M.E., 1996. Nitrogen deposition in California forests: a review. Environ. Pollut. 92, 127–146.

Chappelka, A.H., Freer-Smith, P.H., 1995. Predisposition of trees by air pollutants to low temperatures and moisture stress. Environ. Pollut. 87, 105–117.

Cosby, B.J., Hornberger, G.M., Galloway, J.N., Wright, R.F., 1985. Modelling the effects of acid deposition: assessment of a lumped-parameter model of soil water and stream water chemistry. Water Resour. Res. 21, 51–63.

Cronan, C.S., Grigal, G.F., 1995. Use of calcium/aluminum ratios as indicators of stress in forest ecosystems. J. Environ. Quality 24, 209–226.

DeVries, V., 1993. Average critical loads for nitrogen and sulfur and its use in acidification abatement policy in the Netherlands. Water Air Soil Pollut. 68, 399–434.

Driscoll, C.T., van Breemen, N., Mulder, J., 1985. Aluminum chemistry in a forested spodosol in New Hampshire. Soil Sci. Soc. Am. J. 49, 437–444.

Freer-Smith, P.H., 1998. Do pollution-related forest declines threaten the sustainability of forests? Ambio 27, 123–127.

Hall, J.P., 1995. Forest health monitoring in Canada—how healthy is the boreal forest? Water Air Soil Pollut. 82, 77–85.

Hettelingh, J.P., Posch, M., de Smet, P.A.M., Downing, R.J., 1995. The use of critical loads in emission reduction agreements in Europe. Water Air Soil Pollut. 4, 2381–2388.

Kandler, O., Innes, J.L., 1995. Air-pollution and forest decline in central Europe. Environ. Pollut. 90, 171–180.

Kuo, J., Fox, E., McDonald, S., 1992. SIGMASTAT Statistical Software, User's Manual. Jandel Scientific GmbH, Erkrath, Germany.

Nellemann, C., Frogner, T., 1994. Spatial patterns of spruce defoliation: relation to acid deposition, critical loads and natural growth conditions in Norway. Ambio 23, 255–259.

Nellemann, C., 1997. Forest soil condition in Norway. In: Vanmechelen, L., Groenemans, R., Van Ranst, E. (Eds.), Soil Condition in Europe. UN-ICP-report, Geneva, pp. 174–177.

Nellemann, C., Esser, J., 1998. Crown condition and soil acidification in Norwegian spruce forests. Ambio 27, 143–147.

Nellemann, C., Thomsen, M.G., 2001. Critical loads, nitrogen deposition and long-term changes in forest growth. Water Air Soil Pollut. 128, 197–205.

Nellemann, C., Thomsen, M.G., Söderberg, U., Hansen, K., 2002. Forest growth and critical loads in Scandinavia. This volume.

Nihlgård, B., 1985. The ammonium hypothesis—an additional explanation to the forest dieback in Europe. Ambio 14, 2–8.

Pacyna, J.M., Larssen, S., Semb, A., 1991. European survey for NO_x emissions with emphasis on eastern Europe. Atmos. Environ. 25A, 425–439.

Posch, M., de Smet, P.A.M., Hettelingh, J.-P., Downing, R.J. (Eds.), 1995. Calculation and Mapping of Critical Thresholds in Europe. RIVM, Bilthoven, RIVM-rep., 259101004.

Raitio, H., Kilponen, T., 1994 Critical loads and critical limit values. Proceedings of the Finnish-Swedish Environmental Conference. October 27–28, 1994, Vaasa, Finland. The Finnish Forest Res. Inst., Research paper 513.

Robertsson, K., 1991. Emissions of N_2O in Sweden—natural and anthropogenic sources. Ambio 20, 151–155.

Schulze, E.-D., Lange, O.L., Oren, R. (Eds.), 1989. Air Pollution and Forest Decline. In: Ecological Studies, Vol. 77.

Solberg, S., Strand, G.H., 2000. Comparing the geography of changing crown density from two sampling systems for *Picea abies* in Norway. Scand. J. For. Res. 15, 81–86.

Solberg, S., Tveite, B., 2000. Crown density and growth relationships between stands of *Picea abies* in Norway. Scand. J. For. Res. 15, 87–96.

Spiecker, H., 1999. Overview of recent growth trends in European forests. Water Air Soil Pollut. 116, 33 46.

Sverdrup, H., Warfvinge, P., 1993. The effect of soil acidification on the growth of trees, grass and herbs as expressed by the $(Ca + Mg + K)/Al$ ratio. Rep. Ecol. Environ. Eng. 2.

Sverdrup, H., Warfvinge, P., Nihlgård, B., 1994. Assessment of soil acidification effects on forest growth in Sweden. Water Air Soil Pollut. 78, 1–36.

Thomsen, M.G., Nellemann, C., 1994. Isolation of natural factors affecting crown density and crown color in coniferous forest: implications for monitoring of forest decline. Ambio 23, 251–254.

Thomsen, M.G., Nellemann, C., Frogner, T., Henriksen, A., Tomter, S.T., Mulder, J., 1995. Increment and vitality in Norwegian spruce forests seen in relation to critical loads and pollution loadings. Research paper from Skogforsk 22.

Wright, R.F., Lotse, E., Semb, A., 1988. Reversibility of acidification shown by whole catchment experiments. Nature 334, 670–675.

Wright, R.F., Roelofs, J.G.M., Bredemeier, M., Blanck, K., Boxman, A.W., Emmett, B.A., Gundersen, P., Hultberg, H., Kjønaas, O.J., Moldan, F., Tietema, A., van Breemen, N., van Dijk, H.F.G., 1995. NITREX: responses of coniferous forest ecosystems to experimentally changed deposition of nitrogen. For. Ecol. Manag. 71, 163–169.

Air Pollution, Global Change and Forests in the New Millennium
D.F. Karnosky et al., editors
© 2003 Elsevier Ltd. All rights reserved.

Chapter 16

Beech foliar chemical composition:
A bioindicator of air pollution stress

G. Amores, J.M. Santamaría*

*Department of Chemistry and Soil Science, University of Navarre, Irunlarrea s/n,
31080 Iruña-Pamplona, Spain*

Abstract

During the summers of 1995 and 1997, 238 foliar samples were taken from beech (*Fagus sylvatica* L.) trees in 17 stands belonging to the two most representative vegetation series found in Navarre (western Pyrenees, Spain). Each unwashed sample was analysed for calcium (Ca), copper (Cu), iron (Fe), potassium (K), magnesium (Mg), manganese (Mn), nitrogen (N), sodium (Na), phosphorus (P), sulphur (S), and zinc (Zn). Defoliation of sampled trees was also assessed.

Data analysis showed Ca, Mg, and S concentrations exceed the values reported in literature, while Fe and Cu concentrations were below such references. Main macronutrient ratios also exceeded referenced values, a possible cause of nutritional imbalances in the sampled trees. A low concentration of microelements of anthropogenic origin shows that atmospheric pollution caused by such pollutants is very low in this area, although a decreasing gradient from NW to SE can be observed, probably due to long-range transport from emitter points situated in the Bay of Biscay area.

1. Introduction

The decline of forests induced by air pollutants, global climate change, and their interaction with traditional diseases and pathogens over the last few decades, clearly indicates the need for urgent measures to protect our forests.

In this context, an evaluation of the nutritional status of trees and quantification of pollutants can be one of the most powerful diagnostic tools to determine their condition (Innes, 1993; Bussotti et al., 1995; Anonymous, 1997).

Nevertheless, although several authors have measured element concentrations in beech (*Fagus sylvatica* L.) leaves (Brumme et al., 1992; Fischer et al., 1993; Szarek et al., 1993; DeVries et al., 1995; Maňkovská, 1997, 1998;

---

*Corresponding author.

DOI:10.1016/S1474-8177(03)03016-X

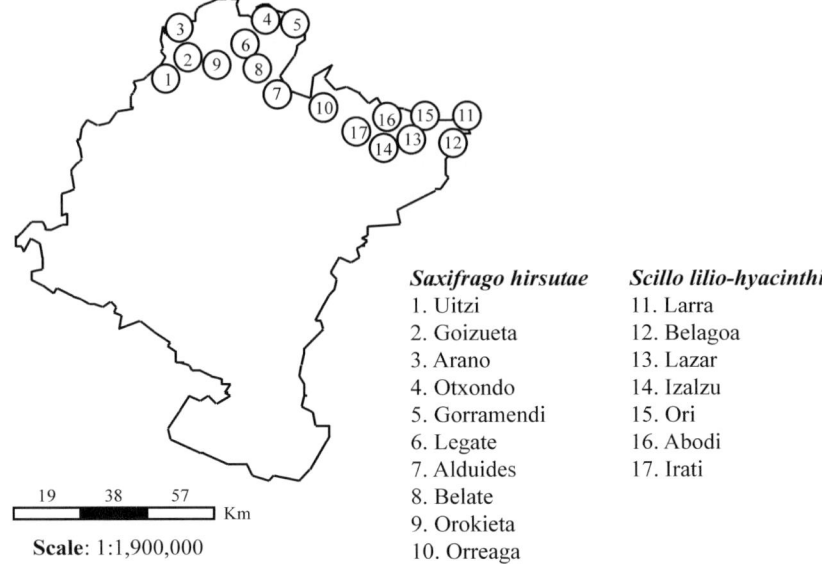

Saxifrago hirsutae	***Scillo lilio-hyacinthi***
1. Uitzi	11. Larra
2. Goizueta	12. Belagoa
3. Arano	13. Lazar
4. Otxondo	14. Izalzu
5. Gorramendi	15. Ori
6. Legate	16. Abodi
7. Alduides	17. Irati
8. Belate	
9. Orokieta	
10. Orreaga	

Figure 1.   Geographic location of selected beech stands in Navarre.

Heinze, 1998; Meiwes et al., 1998), their conclusions when relating their results to the influence of anthropogenic pollution on forest health are varied and
even contradictory.

Beech is the most important deciduous forest species in Navarre, covering
an area of 136 291 ha, or 39% of the forest cover. Therefore, in this study, our
aim was to assess the condition of Navarre's beech forests and establish its
possible relationship to foliar nutrient concentrations and ratios, studying their
validity as potential indicators of forest health.

2. Materials and methods

Seventeen forest stands, distributed throughout Navarre (western Pyrenees,
Spain), were selected. These forests belong to the two most representative vegetation series found in this region (Rivas-Martínez et al., 1991), namely, *Saxifrago hirsutae–Fageto sylvaticae* S. (acidophilic) and *Scillo lilio-hyacinthi–
Fageto sylvaticae* S. (basophilic). Forests belonging to the first vegetation series developed on silicon (Si)-rich oligotrophic soils, and those belonging to
the second series developed on Ca-rich soils (Fig. 1).

Sampling was carried out over the summers of 1995 and 1997, in late August
and early September. In all stands, mean amount of precipitation and mean air
temperature were higher during the 10 months preceding the second sampling

(1997) than during those preceding the first sampling (1995). For precipitation, the differences in June, July, and August were particularly pronounced (Anonymous, 1994–1997).

Seven trees of the same canopy class (dominant or codominant trees according to Kraft classification) were chosen in each forest stand. Foliage samples were taken from the upper third of the crown (wind- and light-exposed), using a telescoping tree pruner. The unwashed samples were dried at 60 °C and then ground in a mill.

To determine the concentrations of Ca, Cu, Fe, K, Mg, Mn, Na, P, S, and Zn, the dried samples were digested with a mixture of concentrated HNO_3 and $HClO_4$ (2 : 1), via wet digestion. A small quantity of concentrated HF was also added, in order to destroy the silica. All chemicals used were of Suprapur Grade. Digestions were gauged with Milli-Q quality water (resistivity $< 16\ M\Omega\ cm$). Ca, Cu, Fe, Mg, Mn, and Zn concentrations were determined by means of atomic absorption spectrophotometry, and K and Na concentrations by atomic emission spectrophotometry, using a Perkin Elmer AAnalyst 800 spectrophotometer. The concentrations of P and S were determined by ICP (Jovin Ibon J-38S). Nitrogen concentration in foliage samples was determined by the Kjeldahl method. The performance of all methods was verified by analyzing certified reference material (CRM-*Fagus sylvatica*) and no method bias was detected.

Crown condition assessment of the selected trees was carried out according to the rules detailed in the UN/ECE (1997) Manual.

Relationships between element concentrations, ratios, and site-specific data were examined with Pearson's r test (Bonferroni's adjustment) and Factor Analysis. The significance of differences between groups of data was assessed by one-way analysis of variance. All calculations were done in SPSS, v. 9.

3. Results

Results corresponding to both sampling years were pooled in the same group for each stand, as conclusive differences were not obtained by statistical analysis of the data.

In general, crown condition of Navarre's beech forests is quite good, and only three of the stands (Otsondo, Orreaga, and Larra) are moderately defoliated (defoliation $> 25\%$). The rest of the stands are only slightly defoliated (defoliation $= 11$–25%).

Concerning element concentrations in foliage, we compared our results to various data cited in the literature for beech (Maňkovská, 1998; Flückiger and Braun, 1999) (Table 1).

For macronutrient concentrations, many of the values exceed the maximums shown in the cited literature. Calcium concentrations were higher in all stands,

Table 1. Yearly rainfall (mm), defoliation (%), and foliage element concentrations (mean ± S.D.) in beech stands (* literature data reported for beech)

Stand	Rainfall	Defoliation	Macronutrients (mg/g)						Microelements (µg/g)				
			Ca	K	Mg	N	P	S	Cu	Fe	Mn	Na	Zn
Uitzi	1852	12.7 ± 6.3	9.7 ± 1.4	6.0 ± 1.7	1.8 ± 0.4	22.9 ± 2.6	1.3 ± 0.3	2.1 ± 0.3	7.0 ± 1.0	107 ± 34	453 ± 199	232 ± 89	35 ± 11
Goizueta	2032	18.1 ± 5.6	10.1 ± 2.4	8.5 ± 2.3	1.1 ± 0.5	21.4 ± 3.6	1.4 ± 0.4	2.0 ± 0.3	6.6 ± 1.7	177 ± 61	419 ± 131	517 ± 250	38 ± 12
Arano	2179	17.5 ± 3.3	11.5 ± 4.7	7.0 ± 1.6	1.7 ± 0.5	21.2 ± 2.1	1.5 ± 0.3	2.1 ± 0.3	6.3 ± 1.0	91 ± 63	409 ± 202	1097 ± 413	48 ± 15
Otsondo	1991	26.1 ± 5.6	12.2 ± 3.6	6.9 ± 2.2	2.1 ± 0.4	20.1 ± 2.8	1.5 ± 0.5	2.6 ± 0.8	5.0 ± 1.2	140 ± 53	667 ± 365	577 ± 246	29 ± 10
Gorramendi	2008	18.6 ± 4.1	8.9 ± 2.2	10.0 ± 2.9	1.3 ± 0.3	21.7 ± 2.5	1.4 ± 0.3	2.8 ± 0.9	7.2 ± 1.9	125 ± 27	871 ± 195	417 ± 138	33 ± 10
Legate	1411	12.9 ± 8.9	9.9 ± 1.5	9.6 ± 1.7	2.2 ± 0.4	22.3 ± 1.4	1.6 ± 0.6	2.6 ± 0.7	6.8 ± 1.4	123 ± 29	1378 ± 445	493 ± 195	28 ± 9
Aldude	1614	16.1 ± 4.0	9.6 ± 1.1	6.2 ± 1.0	2.1 ± 0.2	20.9 ± 2.5	1.4 ± 0.3	2.4 ± 0.7	4.5 ± 0.9	83 ± 31	544 ± 364	162 ± 37	22 ± 9
Belate	1846	13.3 ± 4.4	12.2 ± 2.2	6.6 ± 2.2	1.9 ± 0.4	22.0 ± 1.9	1.3 ± 0.2	2.1 ± 0.5	6.5 ± 0.7	119 ± 18	304 ± 165	507 ± 180	32 ± 8
Orokieta	1560	14.6 ± 6.3	9.7 ± 2.1	7.4 ± 2.3	1.3 ± 0.3	18.6 ± 3.0	1.5 ± 0.5	2.4 ± 0.8	6.3 ± 1.4	105 ± 25	372 ± 166	433 ± 164	24 ± 6
Orreaga	1506	25.0 ± 6.8	9.6 ± 1.4	7.6 ± 2.0	1.2 ± 0.3	22.6 ± 2.5	1.2 ± 0.3	2.2 ± 0.5	5.7 ± 0.9	110 ± 30	1020 ± 417	272 ± 89	24 ± 7
Larra	1158	35.4 ± 9.6	12.6 ± 3.2	11.4 ± 2.1	1.4 ± 0.3	24.4 ± 3.3	1.6 ± 0.6	2.3 ± 0.2	7.1 ± 2.0	83 ± 38	193 ± 79	125 ± 91	24 ± 6
Belagoa	1158	18.9 ± 3.5	17.4 ± 2.8	10.0 ± 2.7	1.1 ± 0.2	21.5 ± 1.9	1.4 ± 0.2	2.3 ± 0.6	5.9 ± 1.2	127 ± 37	127 ± 90	153 ± 72	34 ± 5
Lazar	1270	20.4 ± 5.0	12.2 ± 2.0	9.7 ± 2.2	1.3 ± 0.4	20.6 ± 1.8	1.2 ± 0.4	2.4 ± 0.5	5.5 ± 0.6	127 ± 27	453 ± 193	125 ± 55	25 ± 6
Izalzu	1270	18.8 ± 3.1	12.1 ± 2.2	9.8 ± 3.5	2.1 ± 0.4	21.2 ± 3.0	1.3 ± 0.4	2.4 ± 0.5	6.4 ± 1.5	100 ± 22	740 ± 311	108 ± 81	28 ± 11
Ori	1158	21.8 ± 6.4	12.4 ± 2.4	8.2 ± 1.9	2.0 ± 0.6	19.9 ± 2.4	1.3 ± 0.2	2.3 ± 0.5	5.8 ± 1.0	92 ± 21	1033 ± 261	135 ± 51	25 ± 10
Abodi	1270	23.5 ± 5.9	11.7 ± 1.5	6.9 ± 1.6	1.9 ± 0.6	20.6 ± 3.5	1.4 ± 0.3	2.2 ± 0.5	6.2 ± 1.1	100 ± 31	207 ± 167	193 ± 48	35 ± 10
Irati	1422	20.0 ± 3.9	10.9 ± 2.5	9.1 ± 3.9	1.7 ± 0.4	20.3 ± 3.4	1.4 ± 0.2	2.4 ± 0.6	4.8 ± 1.3	109 ± 30	243 ± 208	136 ± 44	32 ± 6
Mean	1571	19.6 ± 7.7	11.3 ± 3.1	8.2 ± 2.7	1.7 ± 0.5	21.3 ± 2.9	1.4 ± 0.4	2.3 ± 0.6	6.1 ± 1.4	113 ± 41	556 ± 420	338 ± 300	30 ± 11
Reference*		4.0–8.0	4.0–8.0	5.0–10.0	1.0–1.5	18.0–25.0	1.0–1.7	1.3–2.0	6.0–14.0	200–2000	<1000	<100	20–80

as were S concentrations (with the exception of Goizueta). Magnesium concentrations in foliage in more than half the stands were also higher than referenced values. Only N and P concentrations were within the cited range in all the stands. Most of the stands showed K foliage concentrations within the cited range too, except the two eastern-most sampled stands (Larra and Belagoa) where a surplus was observed.

For microelement concentrations, only Zn concentrations were within the referenced range in all stands, although many of the microelements were close to the minimum. However, Na concentrations were higher than the maximum and Fe was lower than the minimum in all stands. Some of the Cu concentrations were lower than the referenced minimum, and even in those stands where they fell within the range, concentrations were close to the minimum. Most stands had Mn concentrations within the range shown in the cited literature (only the concentrations in Legate, Orreaga, and Ori were higher than the maximum).

Comparing the main macronutrient ratios with the literature data reported for beech by Flückiger and Braun (1999), most of the stands (mainly in those belonging to the basophilic vegetation series) had higher N/P values than the referenced maximum. Stands presenting N/Mg and N/K values outside the range were few (Table 2).

Table 2. Main macronutrient ratios (mean ± S.D.) in beech stands

	Macronutrient ratios		
Stand	N/Mg	N/P	N/K
Uitzi	13.3 ± 4.1	18.3 ± 2.8	4.0 ± 0.9
Goizueta	22.1 ± 8.9	16.3 ± 5.0	2.8 ± 1.0
Arano	15.0 ± 9.0	15.1 ± 3.6	3.2 ± 0.9
Otsondo	9.9 ± 1.7	14.2 ± 3.5	3.2 ± 1.0
Gorramendi	18.0 ± 5.0	16.2 ± 4.3	2.3 ± 0.5
Legate	10.3 ± 2.1	16.0 ± 4.9	2.4 ± 0.5
Aldude	10.0 ± 1.9	14.9 ± 3.1	3.4 ± 0.5
Belate	12.1 ± 2.8	17.3 ± 3.7	3.6 ± 1.0
Orokieta	14.3 ± 4.5	13.3 ± 4.5	2.8 ± 1.0
Orreaga	19.4 ± 5.4	20.4 ± 5.0	3.1 ± 0.8
Larra	18.6 ± 6.9	16.2 ± 3.5	2.2 ± 0.4
Belagoa	20.4 ± 4.0	16.1 ± 3.2	2.2 ± 0.5
Lazar	17.5 ± 4.8	19.5 ± 5.9	2.2 ± 0.6
Izalzu	10.9 ± 3.3	16.9 ± 4.0	2.4 ± 0.9
Ori	10.7 ± 3.1	15.3 ± 1.5	2.5 ± 0.6
Abodi	12.7 ± 6.8	15.0 ± 2.3	3.1 ± 0.9
Irati	13.0 ± 5.4	14.8 ± 2.8	2.4 ± 0.8
Mean	14.5 ± 6.3	16.3 ± 4.2	2.8 ± 0.9
Reference	8.0–15.0	8.0–15.0	2.0–3.0

Table 3. Differences in defoliation (%) and foliage element concentrations (mean ± S.D.) between vegetation series. 'a' Not significant, 'b' significant at the 95% probability level, 'c' significant at the 99% probability level

Vegetation series	Defoliation	Macronutrients (mg/g)						Microelements (µg/g)				
		Ca	K	Mg	N	P	S	Cu	Fe	Mn	Na	Zn
Acidophilic	17.5 ± 7.2	10.3 ± 2.7	7.5 ± 2.3	1.7 ± 0.5	21.4 ± 2.7	1.4 ± 0.4	2.3 ± 0.7	6.2 ± 1.5	118 ± 46	647 ± 430	475 ± 322	31 ± 12
Basophilic	22.5 ± 7.6	12.8 ± 3.1	9.2 ± 2.9	1.6 ± 0.6	21.1 ± 3.0	1.3 ± 0.3	2.3 ± 0.5	5.9 ± 1.4	106 ± 33	425 ± 370	140 ± 67	29 ± 9
Significance	c	c	c	a	a	a	a	a	b	c	c	a

As far as differences between vegetation series are concerned, Ca and K concentrations were significantly higher in forest stands belonging to the basophilic vegetation series, whereas acidophilic forest stands had higher Fe, Mn, and Na concentrations. Also, defoliation was significantly greater in basophilic stands than in acidophilic stands, although both values were within the same defoliation class (slight defoliation) (Table 3).

Correlations between foliage element concentrations, when present, were weak (Table 4), so Principal Components Analysis was performed in order to find groups of element concentrations that are somehow joined. VARIMAX

Table 4. Correlation degree (Pearson's r test) between foliage element concentrations

	Ca	Cu	K	Fe	Mg	Mn	N	Na	P	S	Zn
Ca	1.00										
Cu	0.05[a]	1.00									
K	0.15[b]	0.13[b]	1.00								
Fe	−0.06[a]	0.08[a]	−0.03[a]	1.00							
Mg	0.10[a]	0.00[a]	−0.24[c]	−0.13[a]	1.00						
Mn	−0.21[c]	0.06[a]	0.05[a]	0.04[a]	0.23[c]	1.00					
N	0.05[a]	0.30[c]	0.22[c]	0.03[a]	−0.12[a]	0.07[a]	1.00				
Na	−0.09[a]	0.16[b]	−0.22[c]	0.16[b]	0.09[a]	0.02[a]	−0.06[a]	1.00			
P	0.11[a]	0.14[b]	0.09[a]	−0.08[a]	0.03[a]	0.05[a]	0.24[c]	0.05[a]	1.00		
S	−0.13[a]	0.09[a]	0.03[a]	0.07[a]	0.11[a]	0.04[a]	−0.29[c]	0.07[a]	−0.18[c]	1.00	
Zn	0.21[c]	0.30[c]	0.01[a]	0.16[b]	0.03[a]	−0.08[a]	0.22[c]	0.39[c]	0.15[b]	−0.10[a]	1.00

[a] Not significant.
[b] Significant at the 95% probability level.
[c] Significant at the 99% probability level.

Table 5. Groups of foliage element concentrations obtained by Principal Components Analysis

	Component				
	1	2	3	4	5
Na	0.818				
Zn	0.736				
Fe	0.474				
S		−0.753			
N		0.704			
P		0.598			
Mn			0.749		
Ca			−0.718		
K				0.817	
Cu				0.522	
Mg					0.841

Table 6.   Correlation degree (Pearson's *r* test) between geographic coordinates of the stands and foliage element concentrations

	Cu	Na	K	Ca	Mg	N	Zn	Fe	Mn	P	S
Longitude	−0.13[a]	−0.61[c]	0.35[c]	0.40[c]	−0.11[a]	0.02[a]	−0.24[c]	−0.14[b]	−0.12[a]	−0.09[a]	0.05[a]
Latitude	0.15[b]	0.67[c]	−0.12[a]	−0.26[c]	0.11[a]	0.00[a]	0.23[c]	0.18[c]	0.26[c]	0.18[c]	0.14[b]

[a] Not significant.
[b] Significant at the 95% probability level.
[c] Significant at the 99% probability level.

Table 7.   Significant correlation degree (Pearson's *r* test) between the variation in defoliation from 1995 to 1997 and the variation in foliage element concentrations and macronutrient ratios in the same period

	ΔCu	ΔK	ΔMn	ΔN	ΔNa	ΔCa : Mg	ΔK : Mg	ΔN : Mg
ΔDef	−0.21[a]	−0.21[a]	0.24[a]	−0.23[a]	−0.20[a]	−0.26[a]	−0.25[a]	−0.20[a]

[a] Significant at the 95% probability level.

rotation gives five groups: Na–Zn–Fe; S–N–P; Mn–Ca; K–Cu; and Mg alone (Table 5).

Finally, correlation degrees between geographic coordinates of sampled stands and element concentrations were also calculated. The similar distribution patterns shown by Na, Zn, and Fe concentrations, which decrease from NW to SE, are the most interesting results obtained (Table 6).

Results corresponding to each sampling year were used separately, in order to calculate correlation degrees using the variations of every parameter for each tree during the interval between samplings.

Weak correlation degrees between the variation in defoliation from 1995 to 1997 (ΔDef) and the variation in some element concentrations and macronutrient ratios in the same period were also obtained: a positive correlation degree between ΔDef and ΔMn was found, and negative correlation degrees between ΔDef and ΔCu, ΔK, ΔN, ΔNa, ΔCa/Mg, ΔK/Mg, ΔN/Mg were found (Table 7). Using factor analysis, ΔDef is situated in the same group as ΔCu, ΔNa and ΔN (Table 8).

Interesting, although weak, correlations between the variation in the amount of precipitation from 1995 to 1997 and the variation in P foliage concentration, N/P ratio, and defoliation in the same period were obtained (Table 9).

Table 8. Groups of variations of different parameters obtained by Principal Components Analysis

	Component		
	1	2	3
ΔN/Mg	0.886		
ΔK/Mg	0.847		0.412
ΔCa/Mg	0.784		
ΔDef		−0.669	
ΔCu		0.591	
ΔNa		0.579	
ΔRainfall		0.534	
ΔN		0.481	
ΔK			0.806
ΔMn			0.613

Table 9. Significant correlation degree (Pearson's *r* test) between the variation in yearly rainfall from 1995 to 1997 and the variation in different parameters in the same period

	ΔP	ΔN:P	ΔDefoliation
ΔRainfall	−0.32[b]	0.26[a]	−0.30[b]

[a]Significant at the 95% probability level.
[b]Significant at the 99% probability level.

4. Discussion

As mentioned, most of the values for Ca, Mg, and S are well above the referenced data. Nevertheless, macronutrient concentrations are most probably determined by ecological and geological conditions (soil, bedrock, etc.). The possibility of an atmospheric contribution to these concentrations must be discounted as, for example, the lowest S values have been found in the stands of Goizueta and Uitzi, which are close to a paper mill. In all Navarre, SO_2 concentration in the troposphere is exceeded only in the vicinity of this paper mill.

Calcium concentrations clearly exceed referenced values, even in the acidophilic stands, which are characterized by Si-rich oligotrophic soils.

As far as microelement concentrations are concerned, the same explanation given for the distribution of macronutrients is valid for Mn concentration.

The excess of Na found in all the stands is due to the proximity of the Atlantic Ocean, as the correlation degrees between Na concentration and the geographic coordinates (location of the stands) show. A decreasing gradient from NW to SE is evident. The low values found for Fe, Zn, and Cu concentrations show that the sampled territory is not subject to heavy anthropogenic pollution. However, taking into account the prevailing direction of the winds in the area (NW component), the Pearson's test and factor analysis results for Na, Zn, and Fe, together with their identical geographical behavior could indicate a long-range transport of Zn and Fe (in small quantities) from emitter points situated in the Bay of Biscay area, as has already been reported (Santamaría and Martín, 1997, 1998). Nevertheless, as the acidophilic stands are located northwest of the basophilic ones, the possibility that soil-related differences may be responsible for the geographical behavior shown by Fe concentrations cannot be ruled out. However, taking into account that differences in Zn foliage concentrations between vegetation series are not significant, the above explanation is unlikely to be the reason for such a gradient in this case.

With reference to macronutrient ratios, their values can show if there are any disturbances in the nutritional status of trees (Fürst, 1992; Stefan, 1995). Most of the stands exceed the referenced maximum for some of the main macronutrient ratios, so these stands may be suffering nutritional imbalances. According to Flückiger and Braun (1998, 1999), these imbalances increase the susceptibility of beech trees to pathogens and sucking insects, more on trees growing in poor acid soils than on trees belonging to basophilic forests.

Regarding differences between vegetation series, they are mainly due to the very different ecological conditions in which they grow. Lower Ca and K concentrations in acidophilic stands are explained by the leaching of base cations from the clay particles in acid soils, whereas higher Mn and Fe concentrations can be explained by the greater mobility of both elements in acid soils (and subsequent greater uptake by trees). Besides, Mn-rich soils are more frequent in the area occupied by the acidophilic forests (Alonso, 1998). The fact that acidophilic forests are nearer to the sea explains the higher Na concentrations found in these stands.

Some of the correlations obtained do not agree with the interactions between elements known for plants in general (Shuman, 1994): although P–Cu, Zn; Mn–Mg; Cu–P, Zn; Fe–Zn; and P–Zn are known to be antagonistic pairs, we have obtained positive correlation coefficients between these pairs of elements. Some of our results also contradict results obtained in other studies on beech foliage: Bidló and Kovács (1998) found a negative correlation between P–Cu, and positive correlations between Ca–Mn and S–N. Our results are just the opposite. The reason for such contradictory results could be that ours were unwashed samples, so that the measured concentrations are both internal and exogenous.

Several authors have stated that yearly rainfall is a predominant causal factor responsible for changes in beech crown density (tree vitality) between successive years (Neirynck and Roskams, 1999). The correlation between the variation in the amount of rainfall and the variation in defoliation agrees with such findings. The drier the year preceding crown condition assessment, the worse the tree vitality of beech forests is. On the other hand, results suggest that increasing amounts of rainfall lead to increasing N/P ratio. Taking into account that most stands in this study exceed referenced values for this ratio, wetter years could lead to further nutritional imbalances in beech forests, also increasing the susceptibility of trees to pest and pathogen attacks (Flückiger and Braun, 1998, 1999).

The negative correlations found between the evolution of Cu, K, N, and Na concentrations, Ca/Mg, K/Mg, and N/Mg ratios, and the evolution of defoliation indicate that changes in these element concentrations and ratios can cause visible changes in beech tree vitality. According to factor analysis results, of them, all the changes in Cu and N concentrations are best related to the variation in defoliation. It suggests that these two could be the key elements involved in the health of the studied beech stands. Moreover, as an increase in the foliage N concentration improves beech tree vitality, it also suggests that beech forests in Navarre are not N-saturated yet. Therefore, it may mean that the literature data reported for N/X (X = Mg, P, K) ratios are not very useful as a reference in this study.

The relationship between the variation in Na concentration and in defoliation is probably indirect and due to the close association between the changes in the amount of rainfall and the changes in foliage Na concentration (most of the rainfall events in the studied area come from the Atlantic Ocean).

5. Conclusions

At present, we do not know if the deficits and excesses found for several element concentrations and ratios are causing nutritional imbalances or are simply typical values for beech forests in the studied area. The active and exclusive ability of beech trees to fit to particular ecological conditions is already well known (Heinze, 1998). Consequently, the evolution of these parameters over the next few years should be studied in order to obtain more conclusive information.

Even though the studied area is not polluted, there are signs of a possible long-range transport of anthropogenic pollutants to this area that indicate the need for follow-up action in the future.

References

Alonso, J.I., 1998. Estudio del contenido y distribución de los metales pesados en suelos de Navarra: cadmio, cobre, manganeso, niquel, plomo y cinc. Doctoral thesis, Universidad de Navarra.

Anonymous, 1994–1997. Coyuntura agraria. Departamento de Agricultura, Ganadería y Alimentación, Government of Navarre.

Anonymous, 1997. Results of large-scale foliar chemistry surveys (survey 1995 and data from previous years). In: Forest Foliar Condition in Europe. UN/ECE, Austrian Federal Forest Research Centre.

Bidló, A., Kovács, G., 1998. Investigations on nutrient content in beech (*Fagus sylvatica* L.) seedlings of various provenances. Agrokémia És Talajtan 47, 317–328.

Brumme, R., Leimcke, U., Matzner, E., 1992. Interception and uptake of NH_4 and NO_3 from wet deposition by above-ground parts of young beech (*Fagus sylvatica* L.) trees. Plant and Soil 142, 273–279.

Bussotti, F., Ferretti, M., Cenni, E., Grossoni, P., 1995. Monitoring of mineral nutrients and trace elements in broadleaves: a survey in Tuscany. In: Nutrient Uptake and Cycling in Forest Ecosystems. European Commission, Ecosystem Research Report 21, pp. 99–106.

DeVries, W., Leeters, E.E.J.M., Hendriks, C.M.A., van Dobben, H., van den Burg, J., Boumans, L.J.M., 1995. Large scale impacts of acid deposition on forests and forest soils in the Netherlands. In: Heij, G.J., Erisman, J.W. (Eds.), Acid Rain Research: Do We Have Enough Answers?, pp. 261–277.

Fischer, B., Schweingruber, F.H., Keller, T., 1993. Impact of emmissions from a garbage incinerator in Switzerland on the radial increment of beech trees. Dendrochronologia 11, 153–158.

Flückiger, W., Braun, S., 1998. Nitrogen deposition in Swiss forests and its possible relevance for leaf nutrient status, parasite attacks and soil acidification. Environ. Pollut. 102, 69–76.

Flückiger, W., Braun, S., 1999. Nitrogen and its effect on growth, nutrient status and parasite attacks in beech and Norway spruce. Water Air Soil Pollut. 116, 99–110.

Fürst, A., 1992. Die Badeutung des Schwefel/Stickstoffverhältnisses für die Beurteilung des Ernährungszustandes von Fichten. FBVA Berichte 71, 51–54.

Heinze, M., 1998. Nutrition of forest trees on gypsum sites. Forstwiss. Centralbl. 117 (5), 267–276.

Innes, J.L., 1993. Forest health: its assessment and status. Wallingford, CAB International.

Maňkovská, B., 1997. Concentrations of nutritional and trace elements in spruce and beech foliage as an environmental indicator in Slovakia. Lesnictví–Forestry 43, 117–124.

Maňkovská, B., 1998. The chemical composition of spruce and beech foliage as an environmental indicator in Slovakia. Chemosphere 36, 949–953.

Meiwes, K.J., Merino, A., Beese, F.O., 1998. Chemical composition of throughfall, soil water, leaves and leaf litter in a beech forest receiving long term application of ammonium sulphate. Plant Soil 201, 217–230.

Neirynck, J., Roskams, P., 1999. Relationships between crown condition of beech (*Fagus sylvatica* L.) and throughfall chemistry. Water Air Soil Pollut. 116, 389–394.

Rivas-Martínez, S., Báscones, J.C., Fernández-González, F., Loidi, J., 1991. Vegetación del Pirineo Occidental y Navarra. Itinera Geobot. 5, 1–26.

Santamaría, J.M., Martín, A., 1997. Moss bags as biomonitors of heavy metal deposition in Navarre, Spain. Toxicol. Environ. Chem. 60, 65–73.

Santamaría, J.M., Martín, A., 1998. Influence of air pollution on the nutritional status of Navarra's forests, Spain. Chemosphere 36 (4–5), 943–948.

Shuman, L.M., 1994. Mineral nutrition. In: Wilkinson, R.E. (Ed.), Plant–Environment Interactions. Marcel Dekker, New York, pp. 149–182.

Stefan, K., 1995. Schwefel- und Nährstoffversorgung der Fichtennadeln mi Gleinalmgebiet Mitt Der Forstl. Bundesversuchsanstalt Wien 163/5, 53–126.

Szarek, G., Braniewski, S., Chrzanowska, E., Rieger, R., Rutkowska, L., 1993. Nutrients and pollutants in forest vegetation of the Ratanica watershed (Carpathian foothills, Southern Poland). Ekologia Polska 41, 375–392.

UN/ECE, 1997. Manual on Methodologies and Criteria for Harmonised Sampling, Assessment, Monitoring and Analysis of the Effects of Air Pollution on Forests. 4th Edition, Hamburg.

Air Pollution, Global Change and Forests in the New Millennium
D.F. Karnosky et al., editors
© 2003 Elsevier Ltd. All rights reserved.

Chapter 17

Atmospheric contamination of a national forest near a copper smelter in Northern Michigan

W. Beer[1], E. Jepsen*, J. Roth[2]

Wisconsin Department of Natural Resources, Bureau of Air Management (AM/7), Box 7921, Madison, WI 53707-7921, USA

Abstract

The Copper Range Smelter in White Pine, Michigan, was a significant source of copper for the Ottawa National Forest. Air dispersion modeling (ISC3 model) accurately predicted the pattern of peak, half-peak, and near-background smelter particulate deposition. However, the model significantly underestimated total copper loading at the medium and high deposition sites due to incomplete emissions information. Excessive copper accumulation occurred under both eastern hemlock (*Tsuga canadensis* (L.) Carr.) and sugar maple (*Acer saccharum* Marsh.) forest types in the litter layer and to a depth of 0–10 cm in mineral soils at the peak and half-peak deposition sites. Mean copper concentrations at the peak deposition sites ranged from 2260–3050 mg/kg in the forest litter and 360–875 mg/kg in the 0–10 cm mineral soil layer. The retention of copper in the organically enriched upper horizons minimized movement of copper to the 10–20 cm mineral soil layer. Copper concentrations (range 17–47 mg/kg) in the 10–20-cm soil samples were similar to background values for Great Lakes region northern hardwood forests.

1. Introduction

The Upper Peninsula of Michigan has been an important source of copper for centuries. The indigenous peoples of the Great Lakes region traded the native copper extensively with other tribes of central North America. After European settlement in the 1830s, the region developed into a significant copper mining district. The Copper Range Company constructed the largest mine-smelter facility in this mining district near the village of White Pine, Michigan, in the

---

*Corresponding author.
[1] Water Beer, Ph.D., Visiting Scientist, University of Freiburg.
[2] John Roth, same contact info as corresponding author.

DOI:10.1016/S1474-8177(03)03017-1

mid 1950s. Smelter emissions exceeded air permit limits and contributed significant amounts of copper-rich particulates to adjacent forest ecosystems. The mine and smelter ceased operation in 1995.

Numerous studies from Europe and North America have documented significant, adverse impacts in local terrestrial ecosystems from smelting activities. Notable examples include Sudbury, Ontario (Freedman and Hutchinson, 1980a, 1980b, 1981; Cox and Hutchinson, 1981; Hutchinson and Whitby, 1974, 1977); Gusum, Sweden (Bengtsson and Rundgren, 1984; Folkeson, 1984; Folkeson and Anderson-Brinkmark, 1988; Ruehling et al., 1984; Tyler, 1984); Harjavalta, Finland (McEnroe and Helmisaari, 2001); and Arizona (Dawson and Nash, 1980; Wood and Nash, 1976). The effects noted in these studies included acute toxicity to flora and fauna, soil acidification, heavy metal accumulation in soil, and loss of species diversity. Much of the gross contamination at these sites occurred before modern mining regulations were implemented and before air pollution control devices were made mandatory.

This study was undertaken in the fall of 1996 and spring of 1997, approximately 18 months after the smelter ceased operation. An air-quality dispersion model was used to predict particulate deposition patterns and guide the soil contamination assessment efforts. Sampling efforts focused on the predicted peak, half-peak, and low (near-background) deposition sites under mature coniferous and deciduous cover types. Study objectives were threefold:

- Assess the accuracy of the predicted deposition pattern—i.e., peak, half-peak, and near-background deposition;
- Determine whether copper deposition or accumulation/retention rates in forest litter and mineral soils were influenced by forest type:
  - coniferous (eastern hemlock—*Tsuga canadensis* (L.) Carr.) and
  - deciduous (sugar maple—*Acer saccharum* Marsh.);
- Evaluate copper concentrations and loadings in the forest floor, 0–10 cm and 10–20 cm soils.

2. Methods

2.1. Air dispersion and particulate deposition modeling

The air dispersion/deposition model used was the Industrial Source Complex 3 or ISC3 (Environmental Protection Agency, 1995b). Because of a lack of information on smelter particulate density and chemical composition, four main modeling assumptions were made to calculate particle deposition. First, smelter particles less than 10 microns in diameter were assumed to have a density of 1.83 g/cm^3 and particles larger than 10 microns were assumed to have a density of 6.56 g/cm^3. Second, the particle size distribution provided in a 1992 smelter stack test report (TRC, 1992) was assumed to have occurred throughout the life of the project. Third, each particle, regardless of particle diameter and density, was assumed to have the same percent copper. Fourth, predicted deposition depends upon the stack discharge height ($\sim$ 167 meters above ground level), plume characteristics (e.g., temperature and velocity), height of the terrain, and the roughness of the surface (i.e., topography and vegetation cover).

The isopleths of air particulate concentrations were initially derived assuming all receptor sites were at the same elevation as the base of the stack. These initial modeling results were superimposed on United States Geological Survey quad maps to identify potential impact areas. The actual elevations for each of the sample areas were then used to recalculate deposition rates.

2.2. Study area

The smelter was located in the western Upper Peninsula of Michigan (46° 45′ N, 89°35′ W), approximately 10 km south of Lake Superior at an elevation of 305 meters. The mean annual temperature is 5.4 °C, with an average annual precipitation is 85.9 cm.

The sample sites were located on a gently sloping, well-dissected glacial lake plain within the Porcupine Mountains Wilderness State Park on the west and the Ottawa National Forest to the southwest and south of the smelter, respectively. The forest type is northern hardwoods dominated by second-growth sugar maple-hemlock forests, with most of the trees less than 60 years old. The most common tree species are sugar maple, eastern hemlock, birch (*Betula* spp.), aspen (*Populus* spp.), balsam fir (*Abies balsamifera* L. Mill.), ash (*Fraxinius americana* Marsh.), elm (*Ulmus* spp.) and cedar (*Thuja occidentalis*). The soils in the area have not been mapped, but isolated descriptions suggest a Spodosol classification with loamy to sandy loam textures in the upper pedons. Spodosols are common under the northern hardwood forests in the Great Lakes region (Frelich et al., 1993; Hix and Barnes, 1984; MacDonald et al., 1991). The depth of forest litter beneath sugar maple and hemlock in the region is often similar (Hix and Barnes, 1984). Bedrock geology is dominated by Precambrian sandstones, shales, and conglomerates.

2.3. Field sampling protocol

Sampling sites were selected based on forest type, age, and lack of disturbance. All sites were located on public lands: near-background deposition near Union Springs in the Porcupine Mountain Wilderness State Park, and the peak and half-peak deposition sites were located on US Department of Agriculture— Forest Service lands in the Ottawa National Forest.

Table 1. Field sampling design

	Low deposition						Half-peak deposition						Peak deposition					
	Maple			Hemlock			Maple			Hemlock			Maple			Hemlock		
	FI	S1	S2	FI	S1	S2	FI	S1	S2	FI	S1	S2	FI	S1	S2	FI	S1	S2
#	5	5	5	5	5	5	5	5	5	5	5	5	5	5	5	5	5	5
$\sum$		15			15			15			15			15			15	
$\sum$			30							30						30		
$\sum$									90									

$\sum$ = sum of samples for FI = forest litter; S1 = soil 0–10 cm; S2 = soil 10–20 cm; # = number of sample points.

Three types of samples were collected at each sample point: forest litter and mineral soil from depths of 0–10 cm and 10–20 cm (Table 1). Five samples of each type were collected beneath the drip line of mature sugar maple and eastern hemlock at each of the three deposition regimes. Samples were stored in plastic bags and kept in a chilled cooler in the field. All forest litter material within a 30×30^2 cm plot was collected and processed at the lab. The mineral soil samples were collected directly beneath the forest floor sampling area.

All five samples within a given cover type for a specific deposition regime were collected within an area of about 300 m^2 to minimize spatial variability. The topographic surface at the sample sites was approximately level to minimize erosional and depositional processes.

All coarse materials were removed from the forest floor samples in the laboratory and weighed separately. Only the leaves, needles, and fine twigs (less than 2 mm diameter) were dried and divided into two subsamples; one for physical and chemical analysis and the other for archival purposes. At the lab, all samples were stored at 2 °C until processing and analysis.

2.4. Physical and chemical analysis

All samples were analyzed at the University of Wisconsin Soil and Plant Analysis Lab in Madison, Wisconsin within 2 weeks of collection using methods described in UW Extension Publication (Schulte et al., 1987). All samples were dried at 55 °C for 7 days. After drying, the samples were ground to pass through a 12-mesh screen. Total copper concentrations were determined by ICP.

2.5. Statistics

Systat Windows (version 7.0) was used to calculate all regressions, pair-wise comparisons and ANOVAs. The probabilities for the comparisons were computed using Fisher's LSD test.

3. Results

3.1. Deposition modeling

The modeling predicted a striking southwest to northeast pattern that was strongly influenced by the unique meteorological conditions created by the Porcupine Mountains and the cold waters of Lake Superior. A substantial majority of the airborne particulates were deposited northeast of the smelter on company lands or in Lake Superior (Fig. 1). Significant deposition was also predicted southwest of the smelter in the Ottawa National Forest. The modeling predicted a doubling of the deposition between the low and half-peak sites and a fourfold increase between the low and peak deposition regimes (Table 2) over public lands.

Using copper concentrations from literature values and low deposition site forest floor and soil physical characteristics, a background copper load of approximately 29 kg/ha was calculated. The measured accumulation was significantly greater than the modeled deposition plus background, indicating enrichment at all sites. The estimated background copper is about half the measured load at the low deposition sites, fourfold less than the half-peak sites and thirteen-fold less than the peak sites. Potential reasons for the discrepancies between the modeled and measured loads are described in the Discussion section.

3.2. Forest type

Tables 3, 4 and 5 summarize the copper concentrations, loadings, and statistical comparisons for the soils within and between forest types and deposition regimes, respectively. The copper concentrations and loads in the forest floors

Table 2. Comparison of estimated background plus predicted copper deposition with measured soil load[a]

Low deposition		Half-peak deposition		Peak deposition	
Soil background plus deposition	Soil load kg/ha	Soil background plus deposition	Soil load kg/ha	Soil background plus deposition	Soil load kg/ha
$29 + 6.6 = 35.6$	63	$29 + 12 = 41$	178	$29 + 27 = 56$	759

[a] Average of both cover types.
Low (near-background) deposition = 0.164 kg Cu ha^{-1} yr^{-1} of copper $\times$ 40 yr = 6.6 kg Cu ha^{-1}. Half-peak deposition = 0.296 kg Cu ha^{-1} yr^{-1} of copper $\times$ 40 yr = 12 kg Cu ha^{-1}. Peak deposition = 0.664 kg Cu ha^{-1} yr^{-1} of copper $\times$ 40 yr = 27 kg Cu ha^{-1}. Estimated soil background Cu load of 29 kg/ha (assuming 20 mg/kg Cu in forest floor, 15 mg/kg Cu in the mineral soils and low deposition forest floor and soil bulk densities and thickness represented background conditions).

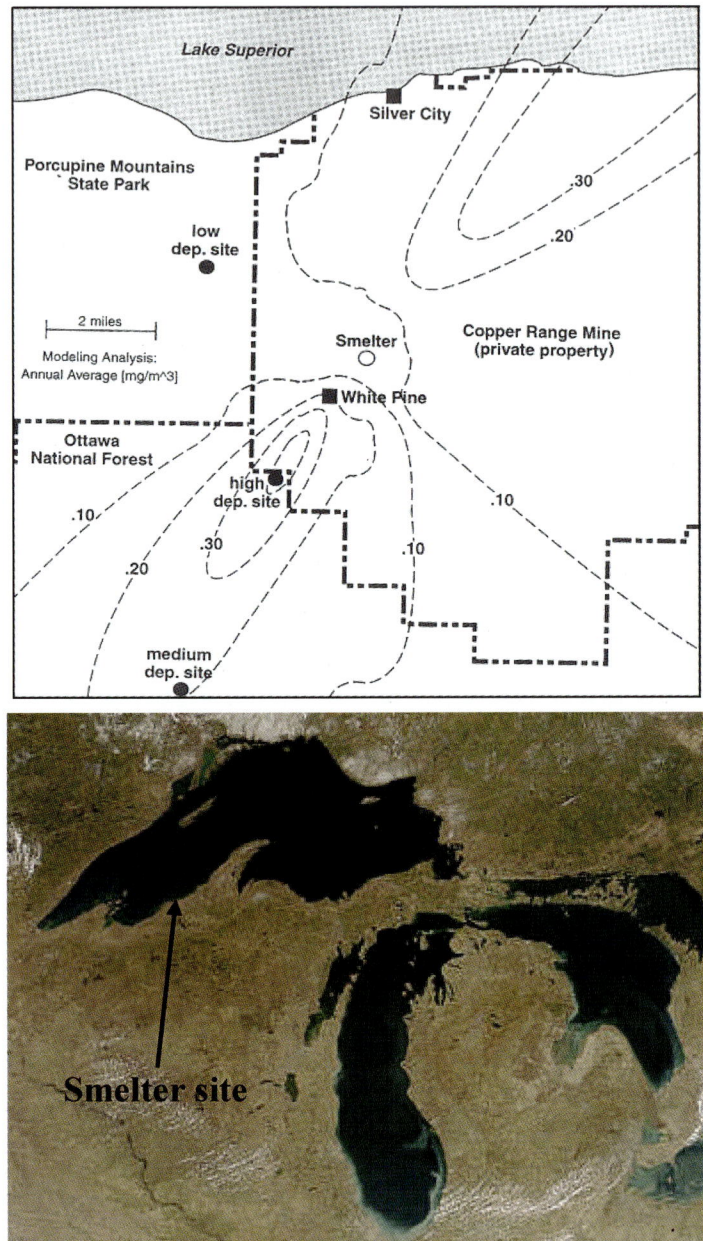

Figure 1. Predicted annual particulate deposition isopleths (mg/m³) for the Copper Range smelter.

Table 3.   Copper concentration forest floor and mineral soil layers mean (SE)

	Sugar maple			Eastern hemlock		
	Forest floor [mg/kg]	Soil: 0–10 cm [mg/kg]	Soil: 10–20 cm [mg/kg]	Forest floor [mg/kg]	Soil: 0–10 cm [mg/kg]	Soil: 10–20 cm [mg/kg]
Low	91 (9.8)	47 (2.6)	17 (1.5)	122 (7.1)	54 (3.6)	20 (2.7)
Half-peak	297 (49)	149 (29)	18 (2.2)	412 (48)	199 (42)	17 (3.8)
Peak	2260 (254)	875 (438)	47 (17)	3050 (55)	360 (89)	20 (5.9)

Table 4.   Total copper accumulation in leaf litter and soil layers mean (SE)

	Sugar maple			Hemlock			Grand total
	Forest floor [kg/ha]	Soil: 0–10 cm [kg/ha]	Soil: 10–20 cm [kg/ha]	Forest floor [kg/ha]	Soil: 0–10 cm [kg/ha]	Soil: 10–20 cm [kg/ha]	
Low	0.21 (0.1)	33 (2)	21 (2)	0.72 (0.1)	49 (3)	23 (3)	127
Half-peak	0.75 (0.1)	136 (27)	26 (3)	2.2 (0.4)	169 (35)	21 (5)	355
Peak	29 (4.5)	876 (483)	76 (28)	66 (8)	449 (111)	23 (7)	1519
Total	30	1045	123	69	667	67	

Note: Forest floor loading calculations only include leaf and fine twig litter mass. Coarse branches, bark, and seed material excluded.

Table 5.   ANOVA probabilities for differences in copper loading between forest types by soil depth

		Leaf litter	Soil: 0–10 cm	Soil: 10–20 cm
Low deposition	n	10	10	10
	p	0.001	0.003	0.522
Half-peak deposition	n	10	10	10
	p	0.004	0.470	0.403
Peak deposition	n	10	10	10
	p	0.003	0.372	0.104

n = number of samples; p = probabilities.

under hemlock were significantly greater than sugar maple in all deposition regimes. However, the hemlock forest floor (0.037 g/cm^3) was significantly ($p = 0.05$) more dense than that of the sugar maple (0.014 g/cm^3) and, adjusting for this difference, it indicates copper accumulation within the forest types was roughly similar. Copper concentrations and loads in the 0–10 and 10–20 cm mineral soils also increased with increasing deposition, but the standard

errors increased substantially too. This increased variability resulted in no significant interaction between copper accumulation and/or atmosphere–canopy
linkage.

3.3. Soil copper distribution

Highly significant differences ($p = 0.000$) in copper concentrations and loads
existed between deposition regimes and soil layers. Copper accumulations varied greatly between the forest litter and the 0- to 10-cm soil layers. The 0–
10 cm mineral soil layer contained the vast majority of the copper relative to
either the forest floor or the 10–20 cm soil layer. The loading analysis results
were similar to the concentration data trends with low < half peak ≪ peak
deposition for the forest floor and 0–10 cm soils.

The forest floor at the half-peak and peak sites had significantly greater copper concentrations than the mineral soils. The high concentrations were offset
by the lower bulk density of the forest floors and their thinness (Table 6). There
were significant increases in the soil organic matter at the half peak and peak
deposition sites (Table 7) in the 0–10 cm soils. Mineral soil bulk densities increased with depth at all sites (e.g., 0.70–1.25 g/cm^3 for the 0–10 cm and
1.17–1.6 g/cm^3 for the 10–20 cm soils). There were no significant differences
in bulk density for the mineral soil samples in the sugar maple and hemlock
stands. As a result, the overall contribution of the forest floor to the total copper
loading was substantially diminished.

The 10–20 cm soil layers had similar copper loads regardless of deposition
regime. Only the peak deposition sugar maple site displayed a significant relative increase in both copper concentration and load. However, this increase
was within the range reported for other non-polluted sites.

Table 6. Forest floor thickness (cm) mean (SE)

	Low	Half-peak	Peak
Hemlock	1.5 (0.16)	1.7 (0.2)	5.0 (0.27)
Sugar maple	1.7 (0.2)	2.0 (0.16)	6.0(0.7)

Table 7. Soil organic matter (kg/m^2) mean (SE)

	Sugar maple 0–10 cm	Sugar maple 10–20 cm	Hemlock 0–10 cm	Hemlock 10–20 cm
Low	8 (0.9)	3.8 (0.6)	9 (0.5)	4.5 (0.1)
Half-peak	23 (1.2)	4.1 (0.3)	24 (4.1)	4.2 (0.6)
Peak	37 (2.2)	6.5 (1.9)	30 (9.7)	4.7(1.0)

4. Discussion

4.1. Deposition modeling

The ISC3 dispersion model accurately predicted the location of peak, half-peak and low (near-background) deposition areas near the smelter. The copper concentration and loading results from the field sampling strongly confirmed the pattern predicted by the model. However, the model significantly underestimated copper loading at the half-peak and peak sites. The differences between the deposition estimates and the measured amounts can be attributed to two factors.

- Modeling results could have been improved if the physical and chemical characteristics of the particulate matter had been better defined (e.g., particle size distribution and the chemical composition and density of each size class).
- Particulate information used in the model was based on stack emissions test results from the early 1990s. The smelter operated for a number of years without any form of air pollution control in the 1950s to the mid-1960s. Deposition can be considered to have occurred in two phases—one before control and another after an electrostatic precipitator was placed in operation. Significantly greater emissions and deposition probably occurred during the pre-precipitator phase.

4.2. Forest type

Deciduous and coniferous species have distinctly different crown architectures and leaf characteristics. Factors such as longevity of the leaf-on season, leaf shape and orientation, and leaf chemistry can affect tree crown–atmosphere interactions, thus influencing particle deposition phenomena and long-term retention of copper in the forest ecosystem. According to Burton et al. (1991), Fassnacht (1995, 1996), and Miller and Lin (1985) sugar maple stands in the northern hardwood forests typically have leaf area index values (LAIs) between 6–8. Tucker et al. (1993) reported LAIs of 5–6 for dominant sugar maple trees with heights between 15–20 m. LAI values obtained from University of Wisconsin-Madison Forestry staff indicate LAIs of 5–8 for hemlock would be reasonable. The similarities in LAI values between these species suggest LAI would not have been a significant variable influencing deposition.

Comparing the substantial copper loadings in the 0–10 cm soil layers relative to the forest litter and the 10–20 cm soils does not support the contention that forest type significantly influenced deposition phenomena. The differences in copper concentrations between forest types did not significantly influence loading estimates after bulk density and organic matter factors were taken into account. The hemlock stands with year-round foliage and similar growing season LAIs were not more efficient at capturing smelter particulates than sugar maple. Particle aerodynamic and density characteristics were more influential than tree crown variables on deposition rates at these sites.

4.3. Soil contamination

Table 8 provides estimates of relatively non-polluted copper concentrations for northern hardwood forests in the Great Lakes region. Copper values were approximately 15 mg/kg (range 3–60 mg/kg) in the forest floor and 20 mg/kg (range 5–100 mg/kg) in mineral soils. The United States Environmental Protection Agency has established a maximum acceptable copper concentration in soils of 769 mg/kg for land disposal of sewage sludge on cropland (Environmental Protection Agency, 1995a). Canadian soil guidelines for residential/park land recommend mineral soil copper concentrations not exceed 63 mg/kg (CCME, 1997). These values suggest slight enrichment of the forest floors, and perhaps the 0–10 cm soils, may have occurred at the low-deposition sites. The 10–20 cm soils at the high-deposition maple site had a statistically greater accumulation of copper than the other deposition regimes, but it also had a statistically greater organic matter content. Based on these comparisons, the low-deposition site values and the 10–20 cm soils under all deposition regimes were similar to literature estimates of background copper concentrations and loadings.

Copper concentrations and loadings in the forest floor and 0–10 cm soils at the peak-deposition sites were grossly in excess of those at the low-deposition sites the background data references and most samples exceeded soil protection guidelines. Forest litter depths and soil organic matter accumulation at the peak sites were significantly greater than the low-deposition sites. At the half-peak deposition sites, copper loadings in the forest litter and 0–10 cm soils

Table 8. Background soil concentrations of copper

Freedman & Hutchinson (1)	Freedman & Hutchinson (2)	Hutchinson & Whitby (3)	Grigal & Ohmann (4)	Jepsen (5)
mean 20 mg/kg range 2–100 mg/kg	60 mg/kg	26 mg/kg	mean 11 mg/kg range 3–27 mg/kg	FF 5–20 mg/kg (0–10): 3–18 mg/kg

(1): Mineral soils; no information about specific horizons or depth. (2): Data from 1981 Sudbury study; samples of forest litter 76 km away from the stack. (3): Data from 1974 Sudbury study; soil samples (0–10 cm) 50 km away from the stack. (4): Data from a mid-1980s survey of selected Great Lake region FIA plots; forest floor samples. (5): Data from soil survey of trembling aspen plots in 1993. FF = forest floor; (0–10) = mineral soil.

were elevated and the soil standards were also exceeded. Litter depth at the half-peak sites was similar to the low-deposition sites, while organic matter in the 0–10 cm soil was significantly higher.

Copper forms strong bonds with chelate complexes (Alloway, 1995; Friedland et al., 1986). This can result in copper accumulation in the forest floor and the organically enriched upper soil horizons, with rapid decreases in concentrations to near-background levels in the underlying soil horizons. The accumulation of organic matter near smelters may be related to elevated copper concentrations adversely affecting litter decomposition rates (Freedman and Hutchinson, 1980b; McEnroe and Helmisaari, 2001). Both cover types had significantly thicker forest floors at the peak-deposition sites compared with the other deposition regimes. Soil organic matter content in the 0–10 cm soil layer increased with increasing copper deposition, regardless of cover type. The patterns of anthropogenic enrichment of copper in this study were similar to those reported in the literature.

5. Conclusions

The ISC3 air dispersion model proved to be an extremely useful tool for assessing deposition patterns and focusing sampling efforts. A lack of significant disturbance in the forested areas and the magnitude of the smelter emissions relative to other sources contributed to the success of the terrestrial verification of the model predictions. The spatial distribution predictions were exceptionally accurate, but total deposition and accumulation were significantly underestimated at the medium- and high-deposition sites. The underestimation was believed to be related to the particle density and chemistry assumptions, as well as to changes in air pollution control equipment, rather than an inherent flaw in the ISC3 model.

Excessive copper concentrations occurred under both hemlock and sugar maple in the forest litter and 0–10 cm soil layer at the peak and half-peak deposition sites in the Ottawa National Forest. Several of the sites exceeded soil guidelines established to protect plants and animals. However, the excessive accumulation was restricted to the upper soil layers, with copper concentrations in the 10–20 cm soils at these sites near background levels regardless of deposition regime or forest type. The copper concentrations for all soil layers sampled in the low (i.e., near-background) deposition stands were within acceptable soil guidelines.

In general, there were no significant differences in overall copper accumulation between hemlock and sugar maple stands. Forest type did not appear to have a significant effect on deposition phenomena or on copper movement beyond the 0–10 cm soil layer. The excessive copper in the forest floor and

0–10 cm soil layer at the higher deposition sites appears to adversely affect organic matter decomposition processes.

References

Alloway, B.J. (Ed.), 1995. Heavy Metals in Soils, 2nd Edition. Wiley, New York.

Bengtsson, G., Rundgren, S., 1984. Ground living invertebrates in metal-polluted forest soils. Ambio 13 (1), 29–33.

Burton, A.J., Pregitzer, K.S., Reed, D.D., 1991. Leaf area and foliar biomass relationships in northern hardwood forests located along an 800 km acid deposition gradient. Forest Sci. 37, 1041–1059.

CCME Subcommittee on Environmental Quality, 1997. Canadian Soil Quality for Copper: Environmental and Human Health. Canadian Council of Ministers of the Environment. Winnipeg, Manitoba, pp. 99.

Cox, R.M., Hutchinson, T.C., 1981. Environmental factors influencing the fate of spread of the grass *Deschampsia cespitosa* invading areas around the Sudbury nickel–copper smelters. Water Air Soil Pollut. 16, 83–106.

Dawson, J.L., Nash, T.N., 1980. Effects of air pollution from copper smelters on a desert grassland community. Environ. Exp. Bot. 20, 61–72.

Environmental Protection Agency, 1995a. A Guide to the Biosolids Risk Assessments for the EPA Part 503 Rule. Office of Wastewater Management. EPA832-B-93-005. September 1995.

Environmental Protection Agency, 1995b. User's Guide for the Industrial Source Complex (ISC3) Dispersion Models. Office of Air and Radiation. EPA 454/B-95-003b, 1995.

Fassnacht, K.S., 1995. Estimating the Leaf Area Index of Northern Central Wisconsin Forests using the Landsat Thematic Mapper. M.S. Thesis. University of Wisconsin, Madison, WI.

Fassnacht, K.S., 1996. Characterization of the Structure and Function of Upland Forest Ecosystems in Northern Central Wisconsin. Ph.D. Thesis. University of Wisconsin, Madison, WI.

Folkeson, L., 1984. Deterioration of the moss and lichen vegetation in a forest polluted by heavy metals. Ambio 13 (1), 37–39.

Folkeson, L., Anderson-Brinkmark, E., 1988. Impoverishment of vegetation in a coniferous forest polluted by copper and zinc. Can. J. Bot. 66, 417–428.

Freedman, B., Hutchinson, T.C., 1980a. Pollutant inputs from the atmosphere and accumulation in soils and vegetation near a nickel–copper smelter at Sudbury, Ontario, Canada. Can. J. Bot. 58, 108–132.

Freedman, B., Hutchinson, T.C., 1980b. Effect of smelter pollutants on forest leaf litter decomposition near a nickel–copper smelter at Sudbury, Ontario. Can. J. Bot. 58, 1722–1736.

Freedman, B., Hutchinson, T.C., 1981. Sources of metal and elemental contamination of terrestrial environments. In: Lepp, N.W. (Ed.), Metals in the Environment. In: Effect of Heavy Metal Pollution on Plants, Vol. 2, pp. 35–94.

Frelich, L.E., Calcote, R.R., Davis, M.B., Pastor, J., 1993. Patch formation and maintenance in an old-growth hemlock-hardwood forest. Ecology 74 (2), 513–527.

Friedland, A.J., Johnson, A.H., Siccama, T.G., 1986. Zinc, Cu, Ni and Cd in the forest floor in the northeastern United States. Water Air Soil Pollut. 29, 233–243.

Hix, D.M., Barnes, B.V., 1984. Effects of clear-cutting on the vegetation of an eastern hemlock dominated ecosystem, Western Upper Michigan. Can. J. Forest Res. 14 (6), 914–923.

Hutchinson, T.C., Whitby, L.M., 1974. Heavy metal pollution in the Sudbury mining and smelting region of Canada I Soil and vegetation contamination by nickel, copper and other metals. Environ. Conserv. 1, 123–132.

Hutchinson, T.C., Whitby, L.M., 1977. The effect of acid rainfall and heavy metal particulates on a boreal forest ecosystem near the Sudbury smelting region of Canada. Water Air Soil Pollut. 7, 421–438.

MacDonald, N.W., Burton, A.J., Jurgensen, M.F., McLaughlin, J.W., Mroz, G.D., 1991. Variation in forest soil properties along a Great Lakes air pollution gradient. Soil Sci. Soc. Am. J. 55, 1709–1715.

McEnroe, N.A., Helmisaari, H.-S., 2001. Decomposition of coniferous forest litter along a heavy metal pollution gradient, southwest Finland. Environ. Pollut. 113, 11–18.

Miller, D.R., Lin, J.D., 1985. Canopy architecture of a red maple edge stand measured by a point drop method. In: Hutchinson, B.A., Hicks, B.B. (Eds.), The Forest–Atmosphere Interaction, pp. 59–70.

Ruehling, A., Baath, E., Nordgren, A., Soedenstroem, B., 1984. Fungi in metal-contaminated soil near the Gusum Brass Mill, Sweden. Ambio 13 (1), 34–36.

Schulte, E.E., Peters, J.B., Hodgson, P.R., 1987. Wisconsin Procedures for Soil Testing, Plant Analysis and Feed & Forage Analysis. Department of Soil Science, University of Wisconsin, Madison, WI.

Tucker, G.F., Lassoie, J.P., Fakey, T.J., 1993. Crown architecture of stand-grown sugar maple (*Acer saccharum* Marsh.) in the Adirondack Mountains. Tree Physiol. 13, 297–310.

Tyler, G., 1984. The impact of heavy metal pollution on forests: A case study of Gusum, Sweden. Ambio 13 (1), 18–24.

TRC Environmental Cooperation, 1992. Emission Characterization Program. Copper Range Company, White Pine, MI.

Wood, C.W., Nash, T.N., 1976. Copper smelter effluent effects on Sonoran Desert vegetation. Ecology 57, 1311–1316.

Air Pollution, Global Change and Forests in the New Millennium
D.F. Karnosky et al., editors
© 2003 Elsevier Ltd. All rights reserved.

Chapter 18

Mycorrhizal community structure of Scots pine trees influenced by emissions from aluminum smelter

M. Rudawska*, B. Kieliszewska-Rokicka, T. Leski

Institute of Dendrology, Polish Academy of Sciences, 5 Parkowa Str., 62-035 Kórnik, Poland

T. Staszewski, P. Kubiesa

Institute for Ecology of Industrial Areas, 6 Kossutha Str., 40-832 Katowice, Poland

Abstract

Ectomycorrhizas of Scots pine were studied in three forest stands in the vicinity of an aluminum smelter: one young stand (0.4 km from the smelter) and two mature stands (2.5–3.0 km from the smelter). The aim of the study was to evaluate a possible influence of pollutants emitted from the smelter on the abundance and diversity of mycorrhizas in the upper soil layer (0–5 cm). Seventeen mycorrhizal morphotypes were distinguished in the root samples from the three study sites (10–13 at one site). The same dominant morphotype was observed at each of the sites. Biomass of fine roots and total number of mycorrhizas were significantly higher at the young forest stand than at the two mature stands, however the fungal biomass in the fine roots, measured as ergosterol content, was lowest at the young stand. The results suggest a negative, indirect effect of pollution on fine root production, and thus fewer short roots available for mycorrhizal colonization in mature forest stands, rather than a direct effect through the soil.

1. Introduction

Aluminum smelters emit considerable amounts of air pollutants that are known to influence the growth of woody plants. The main chemicals that contaminate both the air and the soil in the vicinity of aluminum smelters are sulfur dioxide (SO_2), fluorides, and oxides of nitrogen (NO_x). Although injury from SO_2 and NO_x is more common, fluorides are the most phytotoxic air pollutants and may injure plants at much lower concentrations (10–1000 times lower) than the other pollutants (Weinstein and Alscher-Herman, 1982). Gaseous and particulate fluorides can be taken up and accumulated by leaves

---

* Corresponding author.

DOI:10.1016/S1474-8177(03)03018-3

of coniferous and broad-leaved species and may reduce total photosynthe-sis, mainly because they may precipitate loss of leaves (Keller, 1977). Scots pine is considered one of the most sensitive tree species to these air pollu-tants.

Completed in 1966, the Konin aluminum smelter is located in a forested area in the middle of the Polish lowlands. Since 1966, it has emitted toxic pol-lutants into the atmosphere. In 1969, emission of fluorine exceeded 600 tons per year, but since then, it has slowly decreased due to technological mod-ernization. Currently, emissions are 37 tons per year. Annual sulfur emissions decreased from more than 200 tons per year in the 1980s to 60 tons in 1999. In the late 1970s, a protective area with four sub-zones indicating hazardous concentrations of fluorine was established in the vicinity of the smelter. Mature and young forests in the protective zone were composed mostly of Scots pine (*Pinus sylvestris* L.).

Ectomycorrhizal symbiosis is an integral part of the root system of Scots pine and is widely regarded as a key component for water and nutrient up-take and healthy growth. Because ectomycorrhizas are sensitive to changes in the environment (Smith and Read, 1997; Cairney and Meharg, 1999), we in-vestigated the mycorrhizal coenosis (i.e., composition and abundance of mor-photypes) of Scots pine trees growing near the aluminum smelter. Roots were sampled from 50-year-old trees that had been exposed to emissions from the smelter for 30 years and from 20-year-old trees that were planted after emis-sions were reduced by technological modernization. The concentration of er-gosterol, the main fungal sterol and an important component of membranes of the mycelia of mycorrhizal fungi (Weete, 1989), was analyzed in fine root sam-ples. Ergosterol is considered to be a sensitive, rapid, and convenient measure of metabolically active fungal biomass (Nylund and Wallander, 1992).

2. Materials and methods

2.1. Study sites

Three sites represent Scots pine forests located in the vicinity of the Konin alu-minum smelter (Fig. 1). Site 1 (Sulanki) was situated 0.4 km from the smelter and represents 20-year-old Scots pine trees planted after emission controls had been incorporated. Sites 2 and 3 (Anielew and Rudzica, respectively) are com-posed of mature Scots pine trees (50 years old) located 2.5–3.0 km from the smelter. These sites were exposed to emissions for more than 30 years. The lev-els of emission of fluorine and sulfur from 1969 to 1999 are shown in Fig. 2. Currently, the annual input of fluorine in throughfall is 5 kg ha^{-1} in Sulanki and about 2 kg ha^{-1} in Anielew and Rudzica. The mean concentration of fluorine in 1993, based on 24-hour measurement, was about 1.4 μg m^{-3} at a distance of

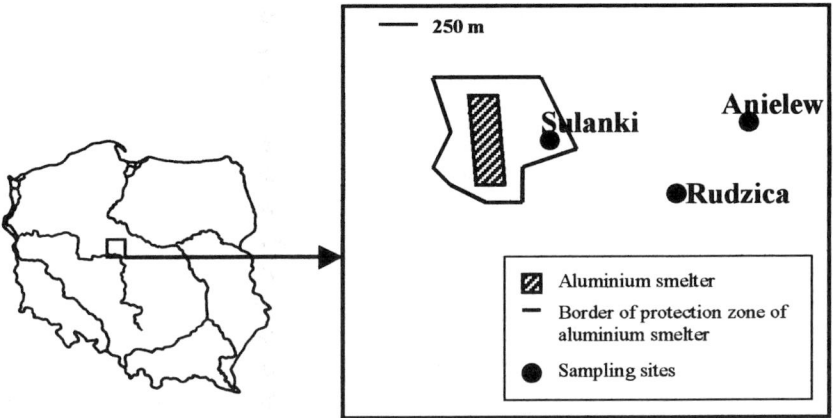

Figure 1. Monitoring sites.

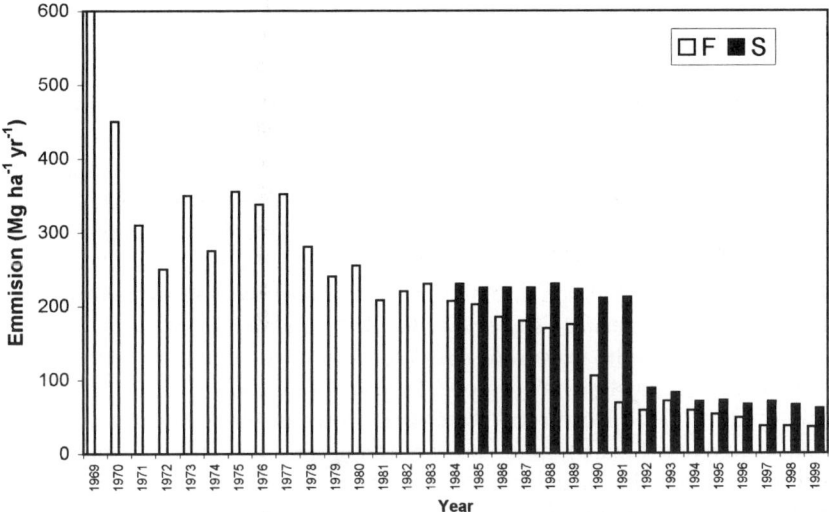

Figure 2. Emission of fluorine and SO_2 from the Konin aluminum smelter.

0.5 km from the smelter and below 1 $\mu g\,m^{-3}$ at 2.5–3.0 km from the emission source. The annual input of sulfur was similar at the three study sites, ranging from 22 to 27 $kg\,ha^{-1}$. Dust pollutants were also deposited in the area. In 1994, the annual dust deposit in Sulanki was 128 $tons\,km^{-2}$, in Rudzica 412 $tons\,km^{-2}$, and in Anielew 59 $tons\,km^{-2}$. Total concentrations of selected trace metals in soil in 1994 are shown in Table 1. The soils are dystric cam-

Table 1. Total concentration of selected trace metals in soil (mg kg^{-1}) in the vicinity of the study sites in 1994 (Pogodski, 1995)

Element	Study site		
	Sulanki	Rudzica	Anielew
Pb	8.9	13.7	53.5
Cd	14.6	2.2	5.3
Zn	3568	3712	352
Mn	194	798	6612
Ni	6.5	30.2	17.7
Cu	14.1	25.1	15.1

bisols at Sulanki, eutric cambisols at Anielew, and haplic podzols at Rudzica (FAO, 1988). Mean annual temperature in this region for the last decade was 9.1 °C and annual rainfall was 472 mm (Pogodski, 1995). Further characteristics of the investigation sites are summarized in Table 2 and are presented elsewhere (Staszewski et al., 1998).

2.2. Root samples

The soil core samples were collected in May and August 1999, using a soil core sampler (diameter = 2 cm), from the organic layer (F/H horizon) to a depth of 5 cm after removal of the litter layer. Eighteen cores, randomly selected, were taken at each study site twice a year (May and August). All cores were taken 0.7–1 m from tree base. Intact soil cores were sealed in plastic bags and kept frozen for ergosterol analysis or stored for up to 3 months at 4 °C until used for morphotyping. Fine roots (< 2 mm) were separated from soil and organic matter using a sieve placed under a stream of cold water, and were then excised from the main root. Final separation and counting of roots were conducted under a stereomicroscope. Live and dead roots were determined by color and firmness. Dead roots were dark brown or black in color and shriveled. Ectomycorrhizal morphotypes were determined according to morphological traits (color, shape, and surface texture) and expressed as the total number of mycorrhizal root tips per 100 cm^3. Identifications were based on published descriptions (Ingleby et al., 1990; Agerer, 1987–1995; Danielson and Visser, 1989; Bradbury, 1998). Each Scots pine root tip colonized by an ectomycorrhizal fungus was counted. Dichotomous and coralloid ectomycorrhizas were counted as one morphotype.

2.3. Ergosterol analyses

Mycorrhizal roots for ergosterol analyses were separated from soil and organic matter within 2 days following sampling. They were excised from the main root and cleaned with running water to remove adhering organic matter. Healthy-looking ectomycorrhizas (approximately 50–70 mg fresh weight) were chosen for subsequent ergosterol extraction, frozen in liquid nitrogen and kept at −80 °C until used. Freeze-dried root samples were ground with mortar and pestle. Ergosterol was extracted with 80% ethanol according to the procedure of Beguiristain and Lapeyrie (1997), separated by reverse phase HPLC on a Waters system, and detected using a UV detector at 280 nm, Waters Nova-Pak C18 column (150 × 4 mm) and 100% methanol (Baker) as a solvent, according to Nylund and Wallander (1992). Commercial ergosterol (5,7,22-Ergostarien-3ß-ol, Sigma) was used as a standard.

2.4. Chemical analyses

The total concentration of metals in the soil profile was extracted from replicate samples and quantified by atomic absorption spectrophotometry (Varian Spectra AA 300). Ions were determined by ion chromatograph Dionex X-100 with column AS4A. Organic C and total N were determined according to standard research methods (Ostrowska et al., 1991). Total fluorine concentration in soil, after combustion, was determined by the fluorine ion selective electrode (Reusmann and Westphalen, 1969). Soluble fluorine content in soil was determined according to Kaniewski and Kaniewski (1985).

2.5. Soil acidity

The pH of the soil layer from which roots were removed was measured using a glass electrode in soil-water and soil-salt (0.01 M KCl) suspensions (1 : 2 soil : solvent) immediately following sampling.

2.6. Statistical analysis

One-way ANOVA followed by the Tukey's t-test (StatSoft Inc. 1997, Statistica for Windows, 5.1, Tulsa, OK) was used to assess the effect of site and time of sampling on the different variables.

3. Results

Soil properties of all three stands are summarized in Table 2. The pH_{salt} of the soil at Sulanki was neutral (6.08) and medium acidic at Rudzica and Anielew (4.37 and 4.9, respectively). Concentrations of Ca, Mg, K, and Na in the soil and the value of cation exchange capacity (CEC) were relatively high at all three sites. The soils had medium concentrations of nitrogen, ranging from

Table 2. Parameters of soil (0–5 cm depth) at the three study sites

Soil parameters	Study site		
	Sulanki	Rudzica	Anielew
pH_{water}	6.66	6.01	6.06
pH_{salt}	6.08	4.37	4.9
Ca $(cmol_c \, kg^{-1})$	5.36	6.8	2.67
Mg $(cmol_c \, kg^{-1})$	0.75	1.19	0.52
K $(cmol_c \, kg^{-1})$	0.94	3.22	1.15
Na $(cmol_c \, kg^{-1})$	0.12	0.68	0.49
CEC $= cmol_c \, kg^{-1})$ (Ca + Mg + K + Na)	7.2	11.9	4.8
N_{total} (%)	0.091	0.171	0.14
$C_{organic}$ (%)	0.91	3.16	0.89
C/N	10	18.5	6.35
F_{total} $(mg \, kg^{-1})$	48.5	51.8	26.5
$F_{soluble}$ $(mg \, kg^{-1})$	11.2	< 1	3.8

0.09% in Sulanki to 0.17% in Rudzica. The soil C : N ratio ranged from 6.3 in Anielew to 18.5 in Rudzica. Total concentrations of fluorine in soil were similar in Sulanki and Rudzica and lower in Anielew; however, the highest concentration of soluble fluorine was found in the stand situated closest to the smelter (Table 2).

An Anova with site as a single factor yields a significant effect on fine root biomass of tested trees ($F = 5.38$, $P = 0.010045$). Screening of the roots in May and August 1998 showed no difference in the fine root biomass between the time of sampling ($F = 0.32$, $P > 0.5$). Tukey's t-test revealed a significant difference between the young and the mature stands ($P < 0.05$) but there was no significant difference between the two mature stands ($P = 0.95$) (Fig. 3(A)). However, in both mature stands (Rudzica and Anielew), there was a tendency for a lower fine root biomass to occur in August than in May. Almost all Scots pine fine root tips were mycorrhizal at all three sites, but the total number of mycorrhizas per 100 cm^3 was significantly lower (Tukey's t-test, $P < 0.05$) in the two mature stands (Rudzica and Anielew) than in the young stand (Sulanki) (Fig. 3(B)). The fewest number of mycorrhizal tips was recorded for the mature Scots pine at Rudzica. The relationship between live and dead mycorrhizas was statistically not significant irrespective of site ($F = 0.23$, $P > 0.7$) and time of sampling ($F = 0.55$, $P > 0.4$); however, slightly more dead mycorrhizas were observed in May than in August (Fig. 3(C)).

Approximately 6 400 ectomycorrhizal tips were harvested and examined in the course of this study. Based on the gross morphology, 17 mycorrhizal morphotypes were distinguished (Table 3). Patterns of abundance of mycorrhizal morphotypes associated with young and mature pine demonstrated that each

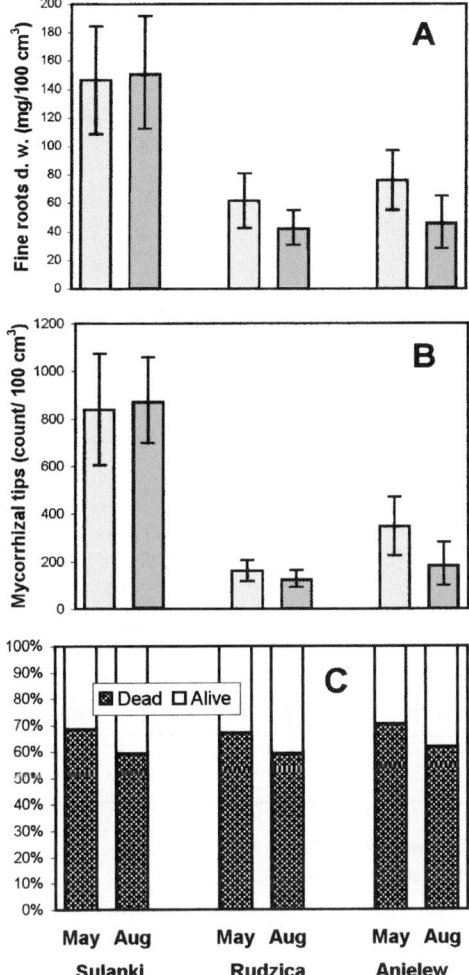

Figure 3. (A) Fine root biomass taken from 0–5 cm depth; (B) Total number of mycorrhizal tips in soil samples (per 100 cm^3); (C) Frequency of dead and alive mycorrhizas at three Scots pine sites (one young stand in Sulanki and two mature stands in Anielew and Rudzica) in the vicinity of the Konin aluminum smelter. Bars indicate standard deviation ($n = 18$).

site was characterized by one or two dominant types and a greater number of rare morphotypes (Fig. 4). The richness of mycorrhizal morphotypes was quite similar among the three pine stands. The young, 20-year-old Scots pine at Sulanki averaged 13 morphotypes and the 50-year-old trees at Rudzica and Anielew were characterized by 12 and 10 morphotypes, respectively. Three of the 17 ectomycorrhizal morphotypes (Nos. 4, 5, 7) occurred at only one study

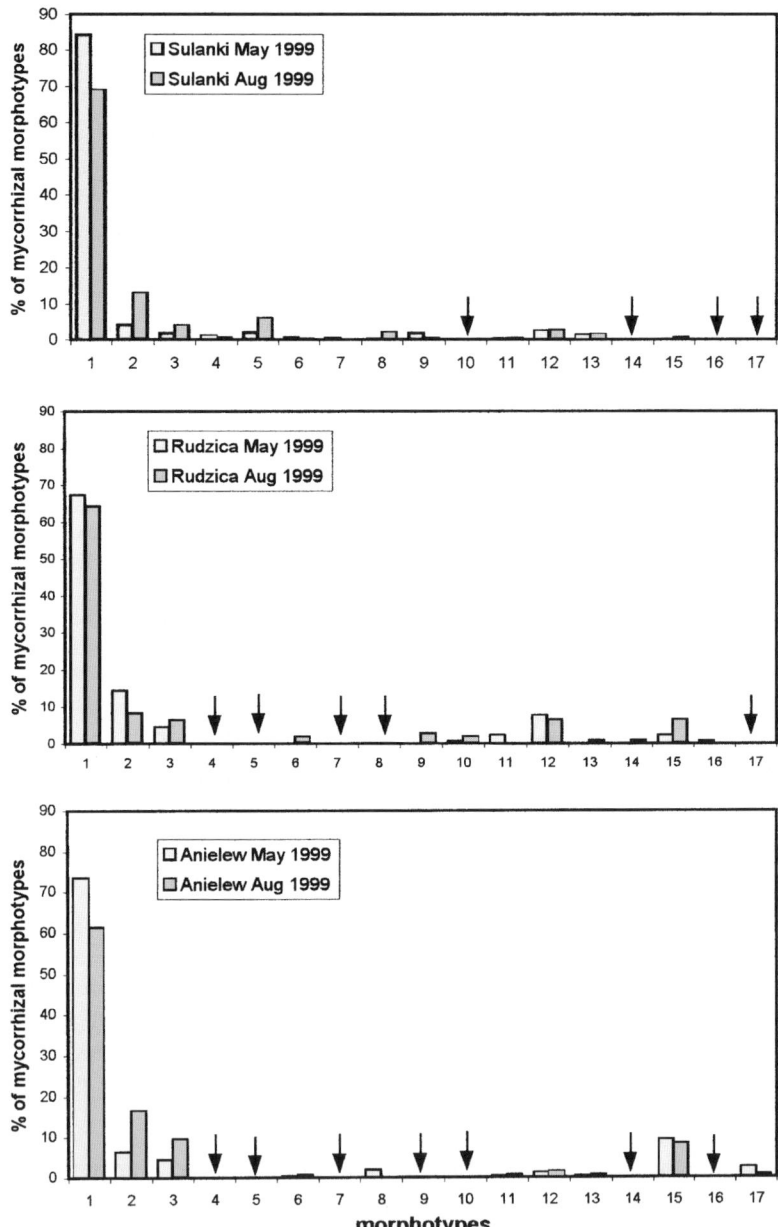

Figure 4. Patterns of abundance of the ectomycorrhizal morphotypes at three Scots pine sites, demonstrating that one mycorrhizal morphotype dominated on each site (↓ morphotype lacking at given site).

Table 3. Classsification of ectomycorrhizal (ECM) morphotypes found on the roots of Scots pine trees in the vicinity of the Konin aluminum smelter

ECM morphotype	Description
Type 1	Dark brick to brown, with a pale tip, fairly straight and infrequently dichotomously branched, discontinuous mantle reticulate and shiny, ectendomycorrhizal character
Type 2	Orange to light brown, mostly unbranched, extended, often segmented with thick, smooth, and shiny mantle
Type 3	Light brown, fairly long, simple to branched with dense, abundant brown hyphae surrounding the mantle
Type 4	Brown, fairly long, simple to branched with loose gray hyphae surrounding the mantle
Type 5	Light brown, simple and dichotomously branched, tips smooth and shiny, lower part covered by white-silver hyphae, compact strands emerging from the mantle
Type 6	Light brown, thick, mostly unbranched and or dichotomizing at the tip, smooth and matte
Type 7	Black to dark maroon, unbranched or dichotomously branched, smooth and shiny
Type 8	Light brown, unbranched or infrequently dichotomously branched, dense silver-white hyphae surrounding the mantle, *Hebeloma*-like
Type 9	Beige, dichotomous to coralloid with a thick mantle overgrowing branching short roots, abundant rhizomorphs with beige and pinkish hues
Type 10	White, dichotomously branched with thick, smooth greyish-white mycelium, thin white strands
Type 11	Whitish to buff, dichotomous, thick, short and stubby, smooth mantle
Type 12	Beige to brown, coralloid, smooth mantle
Type 13	Beige to brown, coralloid with abundant light brown hyphae surrounding the mantle
Type 14	Beige to brown, coralloid with abundant greyish brown hyphae surrounding the mantle
Type 15	Black, mantle angular stellate arrangement, with smooth, stiff hyphae radiating from mantle, *Cenococum*-like
Type 16	Brown, coralloid to subtuberculate and clusters, loose brown strands, *Suillus*-like
Type 17	Pallid to black, subfloccose, simple and very infrequently branched, abundant black hyphae loosely surround the base of the mycorrhiza, *Mycelium radicis atrovirens*-like

site (Site 1 in Sulanki, the young stand of Scots pine situated closest to the aluminum smelter). Several morphotypes were found on only a very few ectomycorrhizal root tips, sometimes only in one survey, May or August. At all three sites, the dominant ectomycorrhizal morphotype was Type 1, accounting for 85% of the morphotypes at Sulanki (young stand) and about 70% at Rudzica and Anielew. This dominant morphotype was brown, smooth, and rarely branched, showing evidence of ectendomycorrhizal penetration. Other mor-

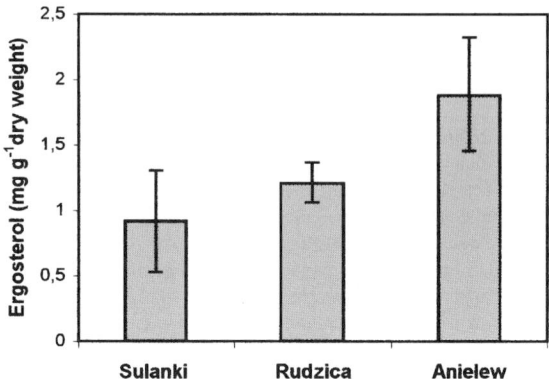

Figure 5. Average ergosterol concentration in mycorrhizas of Scots pine, collected in August 1999 at one young (Sulanki) and two mature (Anielew, Rudzica) forest sites in the vicinity of the Konin aluminum smelter. Bars indicate standard deviation ($n = 18$).

photypes accounted for only a small percentage of the mycorrhizas colonizing the roots of Scots pine.

Ergosterol, as a marker of living fungal mycelium, was analyzed in mycorrhizas collected in August. Ergosterol concentration ranged between 0.9 and 1.85 mg g^{-1} dry weight and was significantly influenced by site ($F = 21.39$, $P = 1 \times 10^{-6}$). Mycorrhizas from the mature forest site at Anielew contained significantly more ergosterol than mycorrhizal samples from Sulanki or Rudzica (Tukey's t-test, $P < 0.005$) (Fig. 5).

4. Discussion

Soil acidity (medium acidic at Rudzica and Anielew and neutral at Sulanki) was satisfactory for Scots pine growth (Puchalski and Prusinkiewicz, 1975) and mycorrhizal development (Smith and Read, 1997). All study sites were characterized by a high saturation of the soil sorption complex by base cations (Ca, Mg, K, and Na) and by rather high effective cation exchange capacity. However, soils in the vicinity of the Konin aluminum smelter were polluted with phytotoxic trace metals (Cd, Zn) that exceeded "normal" levels (Cd: 0.01–7 mg kg^{-1}; Zn: 10–300 mg kg^{-1} (Kabata-Pendias and Pendias, 1993)). Availability of the metals was probably low because of the rather high soil pH (Tables 1, 2). In the present study, the C : N ratio indicated a high nitrogen soil content, especially in Sulanki and Anielew (Uggla and Uggla, 1979).

All three Scots pine stands had similar numbers of ectomycorrhizal morphotypes, which may be considered an index of richness of fungal species. The sites were characterized by one very common morphotype and greater numbers of several rare morphotypes. This pattern of mycorrhizal diversity has been observed in other studies using sporocarp counting (Ohtonen and Markkola, 1989; Bradbury et al., 1998; Jonsson et al., 1999) and direct mycorrhizal morphotyping using morphological and molecular techniques (Visser, 1995; Gardens and Bruns, 1996; Kåren and Nylund, 1997; Gehring et al., 1998; Jonsson et al., 1999). The dominant morphotype at all three sites was brown, smooth, thin-mantle ectendomycorrhiza (Type 1), which occupied 60–85% of the root tips, depending on the site and time of the sampling (Fig. 4). Ectendomycorrhizas are noticed by other authors as typical for highly fertilized nurseries and disturbed sites (Mikola, 1965, 1988; Danielson, 1991). A similar morphotype was also found in several studies at highly contaminated sites in Poland (Kowalski, 1987; Kowalski et al., 1989; Kieliszewska-Rokicka et al., 1997). Ectendomycorrhizas can develop under conditions of very low irradiance and presumably low photosynthate production (Mikola, 1965), which may suggest higher competitiveness of this kind of symbiosis over other ectomycorrhizal fungi (Yu et al., 2001).

Ectomycorrhizal richness observed in this study was slightly lower than that found by other authors (Gehring et al., 1998; Jonsson et al., 1999; Saikkonen et al., 1999). Some authors determined that ectomycorrhizal morphotypes were not highly correlated with attributes of ecosystem productivity (i.e., nutrients and moisture) (Gehring et al., 1998) or environmental stress (Rygiewicz et al., 2000), suggesting a degree of stability in the overall structure of the community (Kåren and Nylund, 1997; Goodman and Trofymow, 1998). Similarly, others have found declining changes in the frequency of individual morphotypes due to environmental stresses (Jones et al., 1997; Wöllecke et al., 1999; Peter et al., 2001). In our study, environmental pollution seems to influence ectomycorrhizal richness and total number of mycorrhizal tips, especially in the case of mature stands of Scots pine. The total number of fine root biomass and mycorrhizal tips produced by young Scots pine trees at the forest site (Sulanki) located 0.5 km from the aluminum smelter was similar to that found in earlier studies in Scots pine stands of comparable age (Rudawska et al., 1995, 1996; Kieliszewska-Rokicka et al., 1997). Mature Scots pine trees located 2.5–3 km from the smelter (Rudzica and Anielew) developed significantly fewer mycorrhizas and fine root biomass than Scots pine trees of comparable age at a moderately polluted forest stand in our previous study (Kieliszewska-Rokicka et al., 1997). The number of mycorrhizas was similar to that occurring on mature Scots pine trees growing in a heavily polluted forest stand exposed for more than 40 years to gaseous and dust pollutants emitted by the adjacent urban-industrial area of Kraków and by the distant industrial area of the Upper Silesian Industrial District, one of the most polluted areas of central Europe (Kieliszewska-Rokicka et al., 1997). This indicated that the old stands growing in the vicinity of the aluminum smelter are significantly affected by the emitted pollutants which negatively influence the mycorrhizal status of tested Scots pine. Observations from polluted forests across Europe (Jansen, 1988) suggest that the mycorrhizal fungi occurring in late succession, characteristic of older stands, are particularly sensitive to atmospheric pollutants and that the decreasing abundance of fungal species may be a factor contributing to forest decline.

The range of ergosterol concentrations in mycorrhizas of Scots pine found in this study (Fig. 5) was similar to that reported by Wallander et al. (1997) in different morphotypes of a mature Scots pine forest in Sweden. Variation in ergosterol concentrations between the mycorrhizal samples may reflect differences in fungal biomass (Sung et al., 1995) or in the extent of fungal membranes (Ruzicka et al., 2000) in the mycorrhizal root tips. Van Praag et al. (1994) reported lower ergosterol contents in rootlets of trees in a spruce stand showing early symptoms of decline. Acid rain treatment resulted in lower ergosterol content in the *Lactarius* mycorrhiza of *Betula pubescens* (Zobel et al., 1994); however, no significant differences in ergosterol concentrations were found in fine roots of Scots pine influenced by various SO_2 and NO_x emissions (Markkola et al., 1995). Ergosterol concentrations can vary between morphotypes of Scots pine ectomycorrhizas (Wallander et al., 1997; Rudawska et al., 2000). Low ergosterol content was reported for ectendomycorrhizas and simple ectomycorrhizal morphotypes with smooth, thin fungal mantles, and high ergosterol levels were reported for the coralloid morphotypes with thick mantles.

Surface characteristics of needles collected at all study sites (Sulanki, Anielew, Rudzica) shown by Staszewski et al. (1998) using SEM analysis were typical of this area, which has been influenced by industrial activities: high fluorine concentrations in the needles, low wettability (determined by the contact angle values), and wax structure degradation. These symptoms may be related to lower mycorrhizal colonization of roots in the mature stands. It has been also suggested that mature Scots pine forests have a greater sensitivity to environmental pollution than young stands (Shaw et al., 1992; Termorshuizen, 1993; Kieliszewska-Rokicka et al., 1997).

Air pollution is often only considered injurious to forest ecosystems if trees are visibly damaged. However, many physiological reactions may be affected even if the trees do not show any clear symptoms of injury. Negative effects of some environmental pollutants are often visible first in roots and their mycorrhizas. As far as we are aware, there is little or no information about the direct effect of fluorides on mycorrhizal fungi and mycorrhizas. However, it is unlikely that roots are exposed to high concentrations of fluorine. Except for a few mycorrhizal roots or hyphae occurring in the upper litter layers, it is unlikely that gaseous pollutants will have a direct effect on the physiology of mycorrhiza.

Trees and their mycorrhizas may be indirectly affected by pollutants. Scots pine is a fluoride-sensitive species and is characterized by a low fluoride accumulation coefficient ($0.24 \text{ m}^3 \text{ g}^{-1}$ dry weight day^{-1}), compared with the median value ($0.8 \text{ m}^3 \text{ g}^{-1}$ dry weight day^{-1}) estimated for more than 50 other species (Horntvedt, 1997). At background levels of fluoride, Scots pine may show slight damage (indicated by leaf necroses), leaving the cause of the damage in doubt. But indirect effects, due to reduced photosynthesis and belowground carbon allocation, may significantly influence the ectomycorrhizal community of tested trees.

In conclusion, this study has demonstrated the toxicity of the environment surrounding the aluminum smelter to ectomycorrhizas and plant symbionts, which has manifested itself in reductions in root biomass and ectomycorrhizal colonization. A molecular study is planned that will enable us to investigate in greater detail the effect of emissions from the aluminum smelter on the fungal community.

Acknowledgements

The authors wish to thank the Polish Scientific Council (KBN) for financial support (project No. 5PO6H 01016). We are grateful to Ms. Halina Narożna for the laborious fine root sorting, to Ms. Małgorzata Łuczak for help with the ergosterol analyses, and to Mikołaj Rudawski for his help in soil determination. We also appreciate the comments of two anonymous reviewers on an earlier version of the manuscript.

References

Agerer, R., 1987–1995. Colour atlas of ectomycorrhizae. Einhorn-Verlag Eduard Dietenberger, Schwäbisch Gmünd, Germany.

Beguiristain, T., Lapeyrie, F., 1997. Host plant stimulates hypaphorine accumulation in Pisolithus tinctorius hyphae during ectomycorrhizal infection while excreted fungal hypaphorine controls root hair development. New Phytol. 136, 525–532.

Bradbury, S.M., 1998. Ectomycorrhizas of lodgepole pine (*Pinus contorta*) seedlings originating from seed in southwestern Alberta cut blocks. Can. J. Bot. 76, 213–217.

Bradbury, S.M., Danielson, R.M., Visser, S., 1998. Ectomycorrhizas of regenerating stands of lodgepole pine (*Pinus contorta*). Can. J. Bot. 76, 218–227.

Cairney, J.W.G., Meharg, A.A., 1999. Influences of anthropogenic pollution on mycorrhizal fungal communities. Environ. Pollut. 106, 169–182.

Danielson, R.M., 1991. Temporal changes and effects of amendments on the occurrence of sheathing (ecto-) mycorrhizas of conifers growing in oil sands tailings and coal spoil. Agric. Ecosyst. Environ. 35, 261–281.

Danielson, R.M., Visser, S., 1989. Effects of forest soil acidification on ectomycorrhizal and vesicular–arbuscular mycorrhizal development. New Phytol. 112, 41–47.

FAO, 1988. FAO/UNESCO soil map of the world. Revised legend 1989. Reprint of the World Soil Resources Report 60. FAO, Rome.

Gardens, M., Bruns, T.D., 1996. Community structure of ectomycorrhizal fungi in a *Pinus muricata* forest: above- and below-ground views. Can. J. Bot. 74, 1572–1583.

Gehring, C., Theimer, T.C., Whitham, T.G., Keim, P., 1998. Ectomycorrhizal fungal community structure on pinyon pines growing in two environmental extremes. Ecology 79, 1562–1572.

Goodman, D.M., Trofymow, J.A., 1998. Comparison of communities of ectomycorrhizal fungi in old-growth and mature stands of Douglas-fir at two sites of southern Vancouver Island. Can. J. Forest Res. 28, 574–581.

Horntvedt, R., 1997. Accumulation of airborne fluorides in forest trees and vegetation. Eur. J. Forest Pathol. 27, 73–82.

Ingleby, K., Mason, P.A., Last, F.T., Fleming, L.V., 1990. Identification of ectomycorrhizas. Institute for Terrestrial Ecology, Natural Environmental Research Council, London, UK Res. Publ. No. 5.

Jansen, A.E., 1988. The influence of acid rain on mycorrhizal fungi and mycorrhizas of Douglas fir (*Pseudotsuga menziesii*) in The Netherlands. In: Mathy, P. (Ed.), Air Pollution and Ecosystems, Proceedings of International Symposium, 18–22 May 1987, Grenoble, France. Reidel, Dordrecht, pp. 859–863.

Jones, M.D., Durall, D.M., Harniman, S.M.K., Classen, D.C., Simard, S.W., 1997. Ectomycorrhizal diversity on *Betula papyrifera* and *Pseudotsuga menziesii* seedlings growth in the greenhouse or outplanted in single-species and mixed plots in southern British Columbia. Can. J. Forest Res. 27, 1872–1889.

Jonsson, T., Kokalj, S., Finlay, R., Erland, S., 1999. Ectomycorrhizal community structure in a limed spruce forest. Mycol. Res. 103, 501–508.

Kabata-Pendias, A., Pendias, H., 1993. Biogeochemistry of trace elements. PWN, Warszawa.

Kaniewski, W., Kaniewski, A., 1985. Comparision of two different methods of fluorine estimation in plants, soil, water and air. Arch. Ochrony Środowiska 3–4, 95–101.

Kåren, O., Nylund, J.E., 1997. Effects of ammonium sulphate on the community structure and biomass of ectomycorrhizal fungi in a Norway spruce stand in southwestern Sweden. Can. J. Bot. 75, 1628–1642.

Keller, T., 1977. Der Einfluss von Fluorimmissionen auf die Nettoassimilation von Waldbaumarten. Mitteilungen Eidgenossische Anstatt für das Forstliche Versuchswesen 53, 161–198.

Kieliszewska-Rokicka, B., Rudawska, M., Leski, T., 1997. Ectomycorrhizae of young and mature Scots pine trees in industrial regions in Poland. Environ. Pollut. 98, 315–324.

Kowalski, S., 1987. Mycotrophy of trees in converted stands remaining under strong pressure of industrial pollution. Angew. Botanik 61, 65–83.

Kowalski, S., Wojewoda, W., Bartnik, C., Rupik, A., 1989. Mycorrhizal species composition and infection patterns in forest plantations exposed to different levels of industrial pollution. Agric. Ecosyst. Environ. 28, 249–255.

Markkola, A.M., Ohtonen, R., Tarvainen, O., Ahonen-Jonnarth, U., 1995. Estimates of fungal biomass in Scots pine stands on an urban pollution gradient. New Phytol. 131, 139–147.

Mikola, P., 1965. Studies on the ectendotrophic mycorrhiza of pine. Acta Forestalia Fennica 79, 1–56.

Mikola, P., 1988. Ectendomycorrhiza of conifers. Silva Fennica 22, 19–27.

Nylund, J.-E., Wallander, H., 1992. Ergosterol analysis as a mean of quantifying mycorrhizal biomass. Methods Microbiol. 24, 77–88.

Ohtonen, R., Markkola, A.M., 1989. Effect of local air pollution on the sporophore production of mycorrhizal fungi, mycorrhizae and microbial activity in Scots pine forests. Meddelelser Norsk Institutt for Skogforskning 42, 121–132.

Ostrowska, A., Gawliński, S., Strzubiałka, Ż., 1991. Methods of soil and plant analysis and evaluation. Instytut Ochrony Środowiska, Warszawa, Poland, pp. 67–77.

Peter, M., Ayer, F., Egli, S., 2001. Nitrogen addition in a Norway spruce stand altered macromycete sporocarp production and below-ground ectomycorrhizal species composition. New Phytol. 149, 311–325.

Pogodski, T., 1995. Library of Environmental Monitoring, Konin, Poland.

Puchalski, T., Prusinkiewicz, Z., 1975. Ecological Basis of Forest Habitat Knowledge. PWRIL, Warszawa, Poland.

Reusmann, J., Westphalen, J., 1969. Ein elektrometrisches Verfahren zur Bestimmung des Fluorgehaltes in Pflanzenmaterial. Staub-Rainhalt Luft 29, 10.

Rudawska, M., Kieliszewska-Rokicka, B., Leski, T., Oleksyn, J., 1995. Mycorrhizal status of a Scots pine (*Pinus sylvestris* L.) plantation affected by pollution from a phosphate fertilizer plant. Water Air Soil Pollut. 85, 1281–1286.

Rudawska, M., Kieliszewska-Rokicka, B., Leski, T., 1996. Mycorrhizal status as an indicator of stress of Scots pine *Pinus sylvestris* L. grown under influence of industrial pollutants. In: Siwecki, R. (Ed.), Biological Reactions of Trees to Industrial Pollution. Material from the 3rd National Symposium, Poznan-Kórnik, 23–26.05, 1994, pp. 611–623 (in Polish).

Rudawska, M., Leski, T., Kieliszewska-Rokicka, B., 2000. Characterization and identification of ectomycorrhizal fungi and ectomycorrhizas using morphological, biochemical and molecular criteria. In: Lisiewska, M., Ławrynowicz, M. (Eds.), Monitoring of Fungi. Bogucki Wydawnictwo Naukowe, Poznan-Lodz, pp. 109–121.

Ruzicka, S., Edgerton, D., Norman, M., Hill, T., 2000. The utility of ergosterol as a bioindicator of fungi in temperate soils. Soil Biol. Biochem. 32, 989–1005.

Rygiewicz, P.T., Kendall, J.M., Tuininga, A.R., 2000. Morphotype community structure of ectomycorrhizas on Douglas-fir (*Pseudotsuga menziesii* Mirb. Franco) seedlings grown under elevated atmospheric CO_2 and temperature. Oecologia 124, 299–308.

Saikkonen, K., Ahonen-Jonnarth, U., Markkola, A.M., Helender, M., Tuomi, J., Roitto, M., Ranta, H., 1999. Defoliation and mycorrhizal symbiosis: a functional balance between carbon sources and below-ground sinks. Ecol. Lett. 2, 19–26.

Shaw, P.J.A., Dighton, J., Poskitt, J., McLeod, A.R., 1992. The effects of sulphur dioxide and ozone on the mycorrhizas of Scots pine and Norway spruce in a field fumigation system. Mycol. Res. 96, 785–791.

Smith, S.E., Read, D.J., 1997. Mycorrhizal symbiosis, 2nd Edition. Academic Press.

Staszewski, T., Uziębło, A.K., Szdzuj, J., 1998. Characteristics of pine needles from trees of different age growing in the protective zone of Konin aluminum smelter. Chemosphere 36 (4–5), 1013–1018.

Sung, S.-J.S., White, L.M., Marx, D.H., Otrosina, W.J., 1995. Seasonal ectomycorrhizal fungal biomass development on loblolly (*Pinus taeda* L.) seedlings. Mycorrhiza 5, 439–447.

Termorshuizen, A.J., 1993. The influence of nitrogen fertilization on ectomycorrhizas and their fungal carpophores in young stands of *Pinus sylvestris*. Forest Ecol. Manag. 57, 179–189.

Uggla, H., Uggla, Z., 1979. Forest Pedology. PWRiL, Warszawa, Poland.

van Praag, H.J., Lognay, G., Carletti, G., Weissen, F., Severin, M., 1994. Temporal and spatial variations of root tip density and ergosterol content of mycorrhizal roots of *Picea abies* Karst. and *Fagus sylvatica* L. Soil Biol. Biochem. 7, 833–840.

Visser, S., 1995. Ectomycorrhizal fungal succession in jack pine stands following wildfire. New Phytol. 129, 389–401.

Wallander, H., Massicotte, H.B., Nylund, J.-E., 1997. Seasonal variation in protein, ergosterol and chitin in five morphotypes of *Pinus sylvestris* L. ectomycorrhizae in a mature Swedish forest. Soil Biol. Biochem. 29, 45–53.

Weete, J.D., 1989. Structure and function of sterols in fungi. Adv. Lipid Res. 23, 115–167.

Weinstein, L.H., Alscher-Herman, R., 1982. Physiological responses of plants to fluoride. In: Unsworth, M.H., Omrmrod, D.P. (Eds.), Effects of Gaseous Air Pollution on Agriculture and Horticulture. Butterworth, London, pp. 139–167.

Wöllecke, J., Münzenberger, B., Huttl, F., 1999. Some effects of N on ectomycorrhizal diversity of Scots pine (*Pinus sylvestris* L.) in northern Germany. Water Air Soil Pollut. 116, 135–140.

Yu, T.E., Egger, J.-C., Peterson, K.N., 2001. Ectendomycorrhizal associations—characteristics and functions. Mycorrhiza 11, 167–177.

Zobel, M., Sarv, O., Komulainen, M., 1994. Mycorrhizal root colonization and ergosterol content in an experimentally polluted subarctic birch–pine forest. In: Helmisaari, H.-S., Smolander, A., Suokas, A. (Eds.), Role of Roots, Mycorrhizas and Rhizosphere Microbes in Carbon Cycling in Forest Soil. In: Research Papers, Vol. 537. The Finnish Forest Research Institute, Helsinki, pp. 147–150.

Air Pollution, Global Change and Forests in the New Millennium
D.F. Karnosky et al., editors
© 2003 Elsevier Ltd. All rights reserved.

345

Chapter 19

Responses of forests in the eastern US to air pollution and climate change

R.A. Mickler*

*ManTech Environmental Technology, Inc., Southern Global Change Program,
920 Main Campus Drive, Venture Center II, Suite 300, Raleigh, NC 27606, USA*

S.G. McNulty

*USDA Forest Service, Southern Global Change Program, 920 Main Campus Drive,
Venture Center II, Suite 300, Raleigh, NC 27606, USA*

R.A. Birdsey, J. Hom

*USDA Forest Service, Northern Global Change Program, 11 Campus Boulevard, Suite 200,
Newtown Square, PA 19073, USA*

Abstract

The interactions of elevated atmospheric carbon dioxide (CO_2) with physical (i.e., precipitation, light, and temperature) and chemical (i.e., ozone (O_3), nitrogen and sulfur deposition, and nutrients) environmental factors that affect plant growth have been demonstrated in experiments that simulate managed and natural forest ecosystems in the eastern United States. Elevated atmospheric CO_2 has been shown to substantially enhance photosynthesis and carbon gain. The response of a southern tree species, loblolly pine (*Pinus taeda* L.), to a doubling of ambient CO_2 was a 50% increase (to 130%) in the rate of net photosynthesis and a 20% reduction in dark respiration, depending on the study and treatment conditions. Volume change showed similar trends with increases in stem wood volume growth of 52% to 152%. Carbon gain for northern tree species with similar experimental treatments showed a 37% increase in dry weight biomass for trembling aspen (*Populus tremuloides* Michx.) and a 73% increase in dry weight biomass for yellow poplar (*Lirodendron tulipifera* L.). The impact of a doubling of atmospheric CO_2 on forest net primary productivity at the regional scale indicates a potential increase of 49% in the southeastern United States and an increase of 30–37% in the northeastern United States.

---

*Corresponding author.

DOI:10.1016/S1474-8177(03)03019-5

1. Introduction

Terrestrial vegetation initially evolved in a carbon dioxide (CO_2) atmosphere that saturated photosynthesis and enhanced the growth of C_3 plants. Estimated levels of atmospheric CO_2 for 420 MYA suggest that the first terrestrial plants grew in CO_2 concentrations 16 times greater than those of the present day (Yapp and Poths, 1992). In contrast, over the last 160 000 years, atmospheric CO_2 concentrations have been atypically low, ranging from 190 to 280 $\mu l\,l^{-1}$ (Barnola et al., 1994), until stabilizing at about 280 $\mu l\,l^{-1}$ CO_2 following the last glacial period. Beginning in the 19th century, CO_2 concentration began to rise in a logarithmic manner until it reached the 1999 annual mean value of 368 $\mu l\,l^{-1}$. Global CO_2 has been increasing annually by 1.43 $\mu l\,l^{-1}$ (Conway et al., 1994), but the 1997–1998 increase of 2.9 $\mu l\,l^{-1}$ represents the largest annual increase on record since measurements began in 1958. Trees from long-lived species in natural forest ecosystems will encounter a more than 100% increase in atmospheric CO_2 during their life span into the middle of this century, when atmospheric CO_2 is projected to be 530 to 600 $\mu l\,l^{-1}$ (Trabalka et al., 1986; Watson et al., 1990).

Terrestrial ecosystems are experiencing changes in their chemical and physical environments at unprecedented rates. One compilation of literature studying the effects of a doubling of atmospheric CO_2 found that the average growth stimulation among 156 plant species was 41% for C_3 plants, 22% for C_4 plants, and 15% for CAM species (Poorter, 1993). An analysis of the response of 39 tree species reported by Gunderson and Wullschleger (1994) found that trees grown at elevated atmospheric CO_2 had an average photosynthetic enhancement of 44%. Findings from these and additional forest ecosystem studies indicate the importance of rising atmospheric CO_2 concentrations and other environmental resources and ambient levels of pollutants in modifying the response of forests and associated plant communities.

The interaction of elevated CO_2 with other factors that affect plant growth has already been demonstrated to occur in agricultural crops and managed and natural forest ecosystems. Several forest research reviews by Eamus and Jarvis (1989), Bazzaz (1990), Musselman and Fox (1991), Strain and Thomas (1992), Rogers and Runion (1994), Gunderson and Wullschleger (1994), Ceulemans and Mousseau (1994), Idso and Idso (1994), Mickler and Fox (1998), and Mickler et al. (2000) have all shown that rising CO_2 will alter the competitive interaction that influences forest ecosystems by direct effects on plant growth and development.

Environmental stresses and limiting resources are frequently identified as potential factors that currently limit growth in forest ecosystems and that may reduce or eliminate any promotion of growth by elevated CO_2. Several studies have reported that the full potential of forest ecosystems to increase net primary productivity (NPP) in a rising CO_2 environment will not be achieved because of nutrient and water limitations (Kramer, 1981; Allen et al., 1990; Thomas et al., 1994; Joyce et al., 2000). However, the benefits of increased photosynthetic water-use and nutrient-use efficiency observed in research studies (Dougherty et al., 1998; Teskey, 1998; Hennessey and Harinath, 1998; Isebrands et al., 2000) with trees growing in a rising CO_2 environment, when other environmental parameters may be limiting, have important implications for long-term forest ecosystem productivity and sustainability.

Although experimental manipulations are the best way to conclusively establish cause-and-effect relationships, extrapolating from relatively simple experiments to effects on complex ecosystems should be done cautiously. Typical experiments on tree responses to environmental stresses address one or a few factors, for a short period of time, using seedlings or small trees in artificial growth chambers. Only recently have experimental methods evolved to allow treatment of forest stands in natural conditions and, because of the expense, such experiments are few. Nevertheless, through judicious analysis of experimental evidence coupled with increasingly sophisticated model representations of complex systems, the scientific community is achieving a measure of capability to assess consequences of environmental change on large spatial and temporal scales. On a regional, national, and worldwide scale, estimates of the direction and magnitude of changes to NPP and the possible enhancement of temperate forest ecosystem carbon accumulation are being made with increasing confidence.

The studies described in this paper represent some of the research conducted in the first 10 years of the USDA Forest Service Global Change Research Program throughout the eastern United States, and are designed to provide a sound scientific basis for making regional and national management and policy decisions regarding forest ecosystems in the context of global change challenges. Scientific findings from related published studies are discussed as part of the general effects of global change on eastern US forest growth and physiology.

2. Physiological and growth responses of trees to environmental change

2.1. Effects of elevated CO_2

The enhancement of net photosynthesis due to elevated CO_2 concentrations is one consistent finding among studies in the southern, northeast, and north-central regions of the United States. In experiments conducted in the southern United States, when the ambient CO_2 concentration was doubled, rates of net photosynthesis in loblolly pine (*Pinus taeda* L.) increased by 50% to 130%, depending on the study and the treatment conditions. This increase was consistent for all age classes (i.e., seedlings, saplings, and trees). The average increase in net photosynthesis under twice-ambient CO_2 concentration, relative to the rates in the current ambient concentration, was about 90–100%, irrespective of growing conditions or age of the trees. There was a positive linear relationship between net photosynthesis and atmospheric CO_2 concentration for the CO_2 range of 350 to 700 $\mu l\, l^{-1}$ (Teskey, 1995). In this range, a change in CO_2 concentration of 10 $\mu l\, l^{-1}$ produces a positive change in the rate of net photosynthesis in loblolly pine of about 3% and, more importantly, indicates that the process is more limited by the availability of the substrate, i.e., CO_2, than by nutrients or other environmental factors.

Studies on other pine species generally have reported similar responses to elevated CO_2. Large increases in rates of net photosynthesis have been reported in Scots pine (*Pinus sylvestris* L.) (Wang et al., 1995) and Monterey pine (*Pinus radiata* D. Don) (Conroy et al., 1990). Exceptions to this have occurred in studies where photosynthetic compensation, or down regulation, has been reported in pot studies using ponderosa pine seedlings (*Pinus ponderosa* P. Laws. ex C. Laws) (Callaway et al., 1994; Grulke et al., 1993).

In the northeast and north central United States, tree responses to elevated CO_2 were evaluated for trembling aspen (*Populus tremuloides* Michx.), yellow poplar (*Liriodendron tulipifera* L.) and eastern white pine (*Pinus strobus* L.). In general, increasing atmospheric CO_2 concentrations increased photosynthetic rates, leaf production, height growth, and dry weight for these species. For example, rates of net photosynthesis in yellow poplar increased from 12% to 144% across studies with potted seedlings and field-grown saplings given various fertilization and irrigation treatments. Most of the enriched CO_2 responses observed for yellow poplar in Ohio (Isebrands et al., 2000) were similar to those reported in the southeastern US (Norby et al., 1992; Wullschleger et al., 1992; Gunderson et al., 1993). These findings suggest that field-planted yellow poplar, when exposed to enriched CO_2 and grown with limited nutrients, limited soil moisture, and ambient or elevated O_3, will display enhanced growth and photosynthetic assimilation.

For other tree species, the effect of elevated CO_2 concentrations on net photosynthesis is almost always positive, but the magnitude of the response appears to be quite variable, depending on species and growth conditions. Photosynthetic responses in the range found for loblolly pine and yellow poplar have been reported in deciduous hardwood species, such as European beech (*Fagus sylvatica* L.) (El Kohen et al., 1993), white oak (*Quercus alba* L.) (Gunderson et al., 1993), *Populus* hybrids (Ceulemans et al. cited in Ceulemans and Mousseau, 1994), chestnut oak (*Quercus montana* Willd.) (Bunce, 1992), as well as other conifers, e.g., Sitka spruce (*Picea sitchensis* (Bong.) Carr.) (Townend, 1993).

In addition to the direct effect of CO_2 concentration on net photosynthesis, elevated CO_2 concentrations directly affected dark respiration. Rates of dark respiration reported in two studies found reductions in the rate of respiration when foliage was exposed to higher concentrations of CO_2 (Teskey, 1995, 1998). The observed 20% reduction is significant, but the cause of this effect remains unknown. The apparent reduction in dark respiration appears to be further evidence that carbon gain will be enhanced in elevated CO_2 conditions. However, the relative contribution of increasing atmospheric CO_2 to overall carbon balance will be less than the stimulatory effect of CO_2 on rates of net photosynthesis.

2.2. Effects of elevated temperature

The effect of elevated air temperature reported in studies that manipulated air and soil temperature, provides some indication of the magnitude of the effect we can expect from air temperature on carbon gain. In the southern United States, a 2 °C increase in air temperature resulted in a less than 10% decrease in net photosynthesis in loblolly pine (Teskey, 1998). In comparison, the effect of elevated CO_2 on net carbon assimilation at twice ambient CO_2 concentrations was more than 100% greater than that at ambient CO_2 concentrations. Branch growth and leaf area development were slightly lower in the higher air temperature treatment, but again the effect was much smaller than the effect of elevated CO_2 concentration. There was no apparent effect of the 2 °C elevation in air temperature on the timing of budburst or the duration of the growing season. However, elevated air temperature in another study caused earlier pollen release and the initiation of female strobili development in loblolly pine trees (Connor et al., 1998).

Air temperature is expected to increase as the concentration of greenhouse gases increases in the atmosphere. A future +2 °C increase in air temperature in the southeastern US will have little negative impact on carbon gain or phenological growth patterns in loblolly pine, particularly as it will occur under conditions of much higher CO_2 concentrations (Burkett et al., 2000).

Increases in air temperature in the eastern United States have the potential to affect the belowground portions of forest ecosystems. These effects may be observed through at least three major pathways: (1) decreases in soil moisture resulting from increased evapotranspiration, (2) increased rates of root respiration and perhaps of root growth, and (3) increased rates of soil organic matter decomposition and accompanying impacts upon nutrient availability, especially nitrogen (Joslin and Johnson, 1998). Reductions in soil moisture will undoubtedly result if precipitation, solar radiation, humidity, and cloud cover remain constant while average temperature increases.

Lockaby et al. (1998) noted that small increases in soil temperature (1–3 °C) increased soil decomposition rates and nitrogen immobilization at two sites in the southeastern US. The study demonstrates the sensitivity of loblolly pine litter early in the decomposition processes, especially nitrogen immobilization, to small shifts in the temperature of the forest floor. Differences in litter decomposition among sites suggest that the positive effect of temperature on early litter decomposition may not be manifested on forest sites with poor litter quality or reduced precipitation. Litter, senescent roots, decaying macrofauna, and the organic component in the A horizon of southern forest soils account for up to 90% of the soils' cation exchange capacity and nearly all the mineralizable nitrogen and phosphorus. In the northeastern United States, the responses of red maple (*Acer rubrum* L.), red spruce (*Picea rubens* Sarg.), and American beech (*Fagus grandifolia* Ehrh.) litter to a 5 °C increase in soil temperature showed significant losses of foliage mass and nutrients (Rustad et al., 2000). Red maple litter lost 27% more mass and 33% more carbon than control plots during the first 6 months of decay. After 30 months, red spruce lost 19% more mass and carbon than control plots. American beech lost 19% more mass and 16% more carbon after 1 year of decay than did litter on control plots.

Experimental evidence supports the hypothesis that an increase in soil temperature of 1.0–5.0 °C may have significant effects on belowground C and N cycling in eastern United States forests (Lockaby et al., 1998; Rustad et al., 2000). Soil C and N cycling are important to eastern forest ecosystems because of potential feedbacks to the atmosphere that could affect climate change, the relationship of these cycles to forest productivity and health, and the potential for nutrient export from watersheds to sensitive downstream wetlands and coastal water bodies. Oxidation of methane (CH_4), nitrogen oxides NO_x flux, and litter decomposition showed variable responses that depended on litter quality, nitrogen availability, and soil moisture. Because of this complexity, it is not possible to definitively state whether eastern United States forest soils will be a net source or sink of carbon as a consequence of atmospheric warming. However, the balance of experimental evidence and observations suggests that increased soil respiration and litter decomposition, together with decreasing soil organic matter with increasing air temperature, will result in a net efflux of C from the soil to the atmosphere. Increases in air temperature and the subsequent release of nutrients and organic constituents in the soil solution may strongly affect the cycling of C and N within forest ecosystems.

2.3. Effects of water availability

In addition to changes in atmospheric CO_2 and temperature, precipitation may also change, but the magnitude and direction of change across the eastern United States is uncertain. Water availability and water-use efficiency will remain important factors contributing to the actual level of productivity of eastern United States forests. Several studies have reported that drought can significantly decrease the stimulatory effect of CO_2 enrichment in tree species (Guehl et al., 1994; Tschaplinski et al., 1995), yet it has also been shown that CO_2 enrichment can decrease whole-plant water use by increasing water-use efficiency (Overdieck and Forstreuter, 1994). Dougherty et al. (1998) used irrigated and non-irrigated treatments under elevated CO_2 concentrations to study stomatal conductance and leaf area development. Irrigation increased annual volume growth by 30% in drought years, and had less effect in years with average or above-average precipitation. Although a 30% increase in growth is appreciable, it was far less than the 108% average yearly increase over 4 years in fertilized plots, compared with growth in the unfertilized plots. These results illustrate the relative importance of nutrients in determining productivity in the southern United States.

The effect of elevated CO_2 on stomatal conductance was not consistent among studies involving loblolly pine. Most of the studies found no significant changes in stomatal conductance to water vapor for foliage measured in both ambient and elevated CO_2 concentrations, including all the studies using mature forest trees and saplings (Groninger et al., 1998; Sword, 1998; Teskey, 1998). However, the studies in open-top chambers reported a reduction in stomatal conductance in the range of 15% due to elevated CO_2 (Alemayehu et al., 1998). The differences in the results among the studies suggest that stomatal conductance in loblolly pine trees may be relatively insensitive to changes in CO_2 concentrations, at least in the 350 to 700 $\mu mol\,mol^{-1}$ range, but also indicate that the stage of development and the growth conditions may alter the sensitivity of the stomata to CO_2 concentration.

2.4. Effects of ozone and interactions

Ozone is considered one of the most significant pollutants affecting forest growth and health in the eastern United States. The effects of O_3 on eastern forest-tree species growth and physiological processes have been studied in the eastern United States for more than 20 years. The response of trembling aspen to seasonal O_3 doses comparable to O_3 concentrations found across the lower Great Lakes region was 43, 21 and 33% decrease in leaf, stem, and root biomass, respectively (Isebrands et al., 2000). Work by Karnosky (1981) suggests that ambient concentrations of O_3 in Wisconsin have contributed to mortality of sensitive genotypes of eastern white pine and that in the northeast and north central US, natural selection has already contributed to the alteration of genetic diversity of forest stands in high O_3 areas.

In the southern United States, loblolly pine (McLaughlin and Downing, 1998) and shortleaf pine (Flagler et al., 1998) experienced significant physiological effects from ambient O_3 concentrations. Flagler et al. (1998) found that ambient O_3 concentrations reduced foliage biomass and leaf area compared with a charcoal-filtered control treatment. Additionally, there was no interaction between O_3 and soil water availability so ozone uptake continued even when the plants were under moderate and severe water stress. McLaughlin and Downing (1998) also reported that the current O_3 concentrations in the region were correlated with reductions in tree growth.

Atmospheric CO_2 and O_3 have substantial impacts on forest growth and are increasing in the atmosphere as a consequence of human activity. Tree responses to CO_2 and O_3 are complex, and become even more difficult to interpret in the presence of other known stressors (i.e., nitrogen limitation or saturation, increasing temperature, precipitation extremes, and insects and diseases). In general, increasing atmospheric CO_2 increases photosynthetic rates, height growth, and biomass production. Increasing atmospheric O_3 decreases photosynthetic rates and biomass production, and increases leaf senescence. The amount and sometimes the direction of growth and physiological parameter changes are dependent on plant factors, such as tree age and genotype. Higher CO_2 concentrations may compensate for some other environmental stresses (see reviews by Mickler and Fox, 1998; Mickler et al., 2000). For example, most studies show that CO_2 enrichment increases growth even though light and/or nutrients are limiting. It is becoming evident that increasing CO_2, O_3, temperature, and drought impact fundamental plant processes, which then affect susceptibility to plant-feeding insects, such as southern pine beetle (McNulty, 1998; Wilkens, 1998) and gypsy moth (Williams et al., 2000).

Studies on trembling aspen show that O_3 usually decreases growth although the effect varies significantly with genotype (Isebrands et al., 2000). Similar results have been reported by Flagler et al. (1998) in shortleaf pine (*Pinus echinata* Mill.) and in other southern pine species. Root growth appears particularly sensitive to O_3. In contrast, substantial increases in relative belowground C allocation were found in response to elevated CO_2. Trembling aspen experiments with both elevated CO_2 and O_3 suggest that elevated CO_2 does not compensate for reduced growth caused by elevated O_3. When N limits growth, there appears to be no response to elevated CO_2. Because CO_2 and O_3 change the chemical composition of the foliage, resistance to insect attack and nutritional value of foliage are altered. Elevated O_3 appears to increase insect growth and elevated CO_2 decreases insect growth. Under field conditions, changes in insect physiology may offset increases or decreases in biomass production that are associated with a changing atmosphere.

Consistent growth responses of yellow poplar to O_3 have not been reported, even though the species shows visual foliar symptoms of exposure. After 2 years of exposure to $2 \times CO_2$ and $2 \times O_3$ treatment in an open-top chamber experiment, yellow poplar had 6% increased leaf biomass, 14% increased stem biomass, and 20% increased root biomass compared with other treatments (Rebbeck, 1996). Elevated CO_2 appears to increase yellow poplar growth regardless of level of exposure to O_3 (Isebrands et al., 2000). In general, research on yellow poplar suggests that, under field conditions, this species will increase biomass production in elevated atmospheric CO_2 even when nutrients and moisture are limited and in the presence of O_3.

After 4 years of CO_2 and O_3 treatments on white pine, no significant growth effects were detected despite some annual stimulatory effects in height growth, increased stem dry weight and total plant dry weight, and increased needle retention (Rebbeck, 1996).

Although the findings from these studies are not conclusive, they add to the existing body of information (Eagar and Adams, 1992; Fox and Mickler, 1996; Olson et al., 1992) that has clearly demonstrated that concentrations of O_3 that are higher than the current ambient concentration are detrimental to the growth and physiological activity of forest throughout the US. However, the balance between O_3 and CO_2 concentrations, as well as other regional pollutants (e.g., nitrogen deposition), will be important in determining the absolute growth response of pine and hardwood forests in the eastern US.

3. Discussion

A positive response of carbon gain to elevated CO_2 concentrations lasting up to 3 years has been demonstrated for some eastern forest species. A key question still remaining is whether the effect will last for extensive periods of time, and whether it will result in a dramatic long-term stimulation of growth for forest species in the eastern United States. The answer to whether long-term increases in productivity are possible in an enriched CO_2 environment appears to be determined by whether or not the plant growth is constrained by important physiological or growth parameters. A species such as loblolly pine has a greater potential for utilizing elevated CO_2. Loblolly pine has a long growing season, including multiple flushes of foliage growth and it is not a high nutrient-demanding species. Loblolly pine growth characteristics allow it to incorporate carbohydrates into growing sinks for long periods during the year when photosynthetic rates are highest and the potential for accumulating carbohydrates is greatest. But the growth potential may be constrained by the availability of nutrients for growth. In the northeast and north central US, 61% of the timberland is dominated by maple–beech–birch and oak–hickory forest type groups. The shorter growing season, single flush of foliar growth, and higher nutrient demands of these mixed deciduous forest species may present a lesser opportunity for increased growth in the short term. Additionally, the effect of nitrogen saturation and calcium depletion on some forest soils may constrain future long-term potential productivity in the region (Fenn et al., 1998). These issues still remain unresolved but are critical to determining the long-term growth potential that could result from increased atmospheric CO_2 concentrations.

4. Conclusions

Predicting the current and future impacts of global change on eastern United States forests is predicated on experimental manipulations of individual tree species or a small number of naturally co-occurring tree species. The experimental methods employed with forest trees have a major impact on how results from these studies should be interpreted at the stand and regional scales. Significant greenhouse, open-top chamber, and pollutant-delivery system effects are common in tree experiments, limiting extrapolation of many experimental results to natural field conditions. Tree-level experiments are typically short lived (1–5 years) yet trees are long-lived perennial species, so there is no direct evidence showing how changing atmospheric chemistry and environmental conditions would affect plant processes over the tree's life span or the longer term. Extrapolations of experimental results to natural ecosystems using process models have concluded that the long-term effects of increasing atmospheric CO_2 are likely to decline in magnitude over time (Luo et al., 1999). However, field exposures of loblolly pine seedlings (Alemayehu et al., 1998), saplings (Dougherty et al., 1998; Hennessey and Harinath, 1998), mature loblolly pine (Teskey, 1998), and yellow poplar (Isebrands et al., 2000) to elevated CO_2 have not demonstrated any acclimation. Open-air exposure experiments and field physiological studies conducted over long time periods will eventually increase our understanding of individual species and forest community responses to elevated CO_2 and global change stressors, as will improvements to physiological process models.

Acknowledgements

The research reported in this paper was supported by the USDA Forest Service Southern and Northern Global Change Research Programs.

References

Alemayehu, M., et al., 1998. Effects of elevated carbon dioxide on the growth and physiology of loblolly pine. In: Mickler, R.A., Fox, S. (Eds.), The Productivity and Sustainability of Southern

Forest Ecosystems in a Changing Environment. In: Ecological Studies, Vol. 128. Springer-Verlag, New York, pp. 93–101.

Allen, H.L., Dougherty, P.M., Cambell, R.G., 1990. Manipulation of water and nutrients—practice and opportunity in Southern US pine forests. Forest Ecol. Manag. 30, 437–453.

Barnola, J.M., et al., 1994. Historical CO_2 record from the Vostok ice core. In: Boden, T.A., Kaiser, D.P., Sepanski, R.J., Stoss, R.W. (Eds.), Trends '93: A Compendium of Data on Global Change. ORNL/CDIAC-65. Carbon Dioxide Information Analysis Center, Oak Ridge National Laboratory, Oak Ridge, TN, pp. 7–10.

Bazzaz, F.A., 1990. The response of natural ecosystems to the rising global CO_2 levels. Annu. Rev. Ecol. Syst. 21, 167–196.

Bunce, J.A., 1992. Stomatal conductance, photosynthesis and respiration of temperate deciduous tree seedlings grown outdoors at elevated concentration of carbon dioxide. Plant Cell Environ. 15, 541–549.

Burkett, V.R., et al., 2000. Potential Consequences of Climate Variability and Change for the Southeastern United States. In: United States National Assessment of Climate Change Impacts. Foundation Document. US Government Printing Office. Washington, DC.

Callaway, R.M., et al., 1994. Compensatory responses of CO_2 exchange and their effects on the relative growth rate of ponderosa pine in different CO_2 and temperature regimes. Oecologia 98, 159–166.

Ceulemans, R., Mousseau, M., 1994. Effects of elevated atmospheric CO_2 on woody plants. New Phytol. 127, 425–446.

Connor, K.F., et al., 1998. Environmental stresses and reproductive biology of loblolly pine (*Pinus taeda* L.) and flowering dogwood (*Cornus florida* L.). In: Mickler, R.A., Fox, S. (Eds.), The Productivity and Sustainability of Southern Forest Ecosystems in a Changing Environment. In: Ecological Studies, Vol. 128. Springer-Verlag, New York, pp. 103–116.

Conroy, J.P., et al., 1990. Influence of phosphorus deficiency on the growth response of four families of *Pinus radiata* seedlings to CO_2 enriched atmospheres. Forest Ecol. Manag. 30, 175–188.

Conway, T.J., Tans, P.P., Waterman, L.S., 1994. Atmospheric CO_2 record from sites in the NOAA/CMDL air sampling network. In: Boden, T.A., Kaiser, D.P., Sepanski, R.J., Stoss, F.W. (Eds.), Trends '93: A Compendium of Data on Global Change. ORNL/CDIAC-65. Carbon Dioxide Information Analysis Center, Oak Ridge National Laboratory, Oak Ridge, TN, pp. 41–119.

Dougherty, P.M., et al., 1998. An investigation of the impacts of elevated carbon dioxide, irrigation, and fertilization on the physiology, and growth of loblolly pine. In: Mickler, R.A., Fox, S. (Eds.), The Productivity and Sustainability of Southern Forest Ecosystems in a Changing Environment. In: Ecological Studies, Vol. 128. Springer-Verlag, New York, pp. 149–168.

Eagar, C., Adams, M.B., 1992. Ecology and Decline of Red Spruce in the Eastern United States. In: Ecological Studies, Vol. 96. Springer-Verlag, New York.

Eamus, D., Jarvis, P.G., 1989. The direct effects of increase in the global atmospheric CO_2 concentration on natural and commercial temperate trees and forests. Adv. Ecol. Res. 19, 1–55.

El Kohen, A., Vener, L., Mousseau, M., 1993. Growth and photosynthesis of two deciduous forest tree species exposed to elevated carbon dioxide. Funct. Ecol. 7, 480–486.

Fenn, M.E., et al., 1998. Nitrogen excess in North American ecosystems: Predisposing factors, ecosystem responses, and management strategies. Ecol. Applic. 8, 706–733.

Flagler, R.B., Brissette, J.C., Barnett, J.P., 1998. Influence of drought stress on the response of shortleaf pine to ozone. In: Mickler, R.A., Fox, S. (Eds.), The Productivity and Sustainability of Southern Forest Ecosystems in a Changing Environment. In: Ecological Studies, Vol. 128. Springer-Verlag, New York, pp. 73–92.

Fox, S., Mickler, R.A. (Eds.), 1996. The Impacts of Air Pollution on Southern Pine Forests. In: Ecological Studies, Vol. 118. Springer-Verlag, New York.

Groninger, J.W., et al., 1998. Interactions of elevated carbon dioxide, nutrient stress, and water stress on physiological processes and competition interactions among three forest tree species. In: Mickler, R.A., Fox, S. (Eds.), The Productivity and Sustainability of Southern Forest Ecosystems in a Changing Environment. In: Ecological Studies, Vol. 128. Springer-Verlag, New York, pp. 117–130.

Grulke, N.E., Hom, J.L., Roberts, S.W., 1993. Physiological adjustment of two full-sib families of ponderosa pine to elevated CO_2. Tree Physiol. 12, 391–401.

Guehl, J.M., et al., 1994. Interactive effects of elevated CO_2 and soil drought on growth and transpiration efficiency and its determinants in two European forest tree species. Tree Physiol. 14, 707–724.

Gunderson, C.A., Norby, R.J., Wullschleger, S.D., 1993. Foliar gas exchange responses of two deciduous hardwoods during 3 threes of growth in elevated CO_2: no loss of photosynthetic enhancement. Plant Cell Environ. 16, 797–807.

Gunderson, C.A., Wullschleger, S.D., 1994. Photosynthetic acclimation in trees to rising atmospheric CO_2: A broader perspective. Photosynth. Res. 39, 369–388.

Hennessey, T.C., Harinath, V.K., 1998. Effects of elevated carbon dioxide, water, and nutrients on photosynthesis, stomatal conductance, and total chlorophyll content of young loblolly pine (*Pinus taeda* L.) trees. In: Mickler, R.A., Fox, S. (Eds.), The Productivity and Sustainability of Southern Forest Ecosystems in a Changing Environment. In: Ecological Studies, Vol. 128. Springer-Verlag, New York, pp. 169–184.

Idso, K.E., Idso, S.B., 1994. Plant responses to atmospheric CO_2 enrichment in the face of environmental constraints: a review of the past 10 years' research. Agric. For. Meteor. 69, 153–203.

Joslin, J.D., Johnson, D.W., 1998. Effects of soil warming, atmospheric deposition, and elevated carbon dioxide on forest soils in the southeastern United States. In: Mickler, R.A., Fox, S. (Eds.), The Productivity and Sustainability of Southern Forest Ecosystems in a Changing Environment. In: Ecological Studies, Vol. 128. Springer-Verlag, New York, pp. 571–587.

Joyce, L.J., et al., 2000. Potential consequences of climate variability and change for the forests of the United States. In United States National Assessment of Climate Change Impacts. Executive Summary. US Government Printing Office. Washington, DC.

Isebrands, J.G., et al., 2000. Interacting effects of multiple stresses on growth and physiological processes in northern forest trees. In: Mickler, R.A., Birdsey, R.A., Hom, J. (Eds.), Responses of Northern US Forests to Environmental Change. In: Ecological Studies, Vol. 139. Springer-Verlag, New York, pp. 149–180.

Karnosky, D.F., 1981. Changes in eastern white pine stands related to air pollution stress. Mitt Forst Bundes Wien 137, 41–45.

Kramer, P.J., 1981. Carbon dioxide concentration, photosynthesis, and dry matter production. BioScience 31, 29–33.

Lockaby, B.G., et al., 1998. Influence of microclimate on short-term litter decomposition in loblolly pine ecosystems. In: Mickler, R.A., Fox, S. (Eds.), The Productivity and Sustainability of Southern Forest Ecosystems in a Changing Environment. In: Ecological Studies, Vol. 128. Springer-Verlag, New York, pp. 525–542.

Luo, Y., et al., 1999. Validity of extrapolating field CO_2 experiments to predict carbon sequestration in natural ecosystems. Ecology 80 (5), 1568–1583.

McLaughlin, S.B., Downing, D.L., 1998. Dynamic responses of mature forest trees to changes in physical and chemical climate. In: Mickler, R.A., Fox, S. (Eds.), The Productivity and Sustainability of Southern Forest Ecosystems in a Changing Environment. In: Ecological Studies, Vol. 128. Springer-Verlag, New York, pp. 207–230.

McNulty, S.G., 1998. Predictions of southern pine beetle populations using a forest ecosystem model. In: Mickler, R.A., Fox, S. (Eds.), The Productivity and Sustainability of Southern Forest

Ecosystems in a Changing Environment. In: Ecological Studies, Vol. 128. Springer-Verlag, New York, pp. 617–634.

Mickler, R.A., Fox, S. (Eds.), 1998. The Productivity and Sustainability of Southern Forest Ecosystems in a Changing Environment. In: Ecological Studies, Vol. 128. Springer-Verlag, New York.

Mickler, R.A., Birdsey, R.A., Hom, J. (Eds.), 2000. Responses of Northern US Forests to Environmental Change. In: Ecological Studies, Vol. 139. Springer-Verlag, New York.

Musselman, R.C., Fox, D.G., 1991. A review of the role of temperate forests in the global CO_2 balance. J. Air Waste Manag. Assoc. 41 (8), 798–807.

Norby, R.J., et al., 1992. Productivity and compensatory responses of yellow-poplar trees in elevated CO_2. Nature 357, 322–324.

Olson, R.K., Binkley, D., Bohm, M. (Eds.), 1992. The Response of Western Forest to Air Pollution. In: Ecological Studies, Vol. 97. Springer-Verlag, New York.

Overdieck, D., Forstreuter, M., 1994. Evapotranspiration of beech stands and transpiration of beech leaves subject to atmospheric CO_2 enrichment. Tree Physiol. 14, 997–1003.

Poorter, H., 1993. Interspecific variation in the growth response of plants to an elevated ambient CO_2 concentration. Vegetatio 104/105, 77–97.

Rebbeck, J., 1996. The chronic response of yellow-poplar and eastern white pine to ozone and elevated carbon dioxide: three-year summary. In: Hom, J., Birdsey, R., O'Brian, K. (Eds.), Proceedings of 1995 Meeting of the Northern Global Change Research Program, 14–16 March 1995, Pittsburgh, PA, UDSA Forest Service, General Technical Report NE-214, Radnor, PA, pp. 23–30.

Rogers, H.H., Runion, G.B., 1994. Plant responses to atmospheric CO_2 enrichment with emphasis on roots and their rhizosphere. Environ. Pollut. 83, 155–189.

Rustad, L.E., et al., 2000. Effects of soil warming on carbon and nitrogen cycling. In: Mickler, R.A., Birdsey, R.A., Hom, J. (Eds.), Responses of Northern US Forests to Environmental Change. In: Ecological Studies, Vol. 139. Springer-Verlag, New York, pp. 357–381.

Strain, B.R., Thomas, R.B., 1992. Field measurements of CO_2 enhancement and climate change in natural vegetation. Water Air Soil Pollut. 64, 45–60.

Sword, M.A., 1998. Ecophysiological response of managed loblolly pine to changes in stand environment. In: Mickler, R.A., Fox, S. (Eds.), The Productivity and Sustainability of Southern Forest Ecosystems in a Changing Environment. In: Ecological Studies, Vol. 128. Springer-Verlag, New York, pp. 185–206.

Teskey, R.O., 1995. A field study of the effects of elevated CO_2 on carbon assimilation, stomatal conductance and leaf and branch growth of *Pinus taeda* trees. Plant Cell Environ. 18, 565–573.

Teskey, R.O., 1998. Effects of elevated carbon dioxide levels and air temperature on carbon assimilation of loblolly pine. In: Mickler, R.A., Fox, S. (Eds.), The Productivity and Sustainability of Southern Forest Ecosystems in a Changing Environment. In: Ecological Studies, Vol. 128. Springer-Verlag, New York, pp. 131–148.

Thomas, R.B., Lewis, J.D., Strain, B.R., 1994. Effects of leaf nutrient status on photosynthesis capacity in loblolly pine (*Pinus taeda* L.) seedlings grown in elevated atmospheric CO_2. Tree Physiol. 14, 947–960.

Townend, J., 1993. Effects of elevated carbon dioxide and drought on the growth and physiology of clonal Sitka spruce plants (*Picea sitchensis* (Bong.) Carr.). Tree Physiol. 13, 389–399.

Trabalka, J.R., et al., 1986. Atmospheric CO_2 projection with globally averaged carbon cycle models. In: Trabalka, J.R., Reichle, D.E. (Eds.), The Changing Carbon Cycle: A Global Analysis. Springer-Verlag, New York, pp. 534–560.

Tschaplinski, T.J., et al., 1995. Interactions between drought and elevated CO_2 on growth and gas exchange of seedlings of three deciduous tree species. New Phytol. 129, 63–71.

Wang, K., Kellomäki, S., Laitinen, K., 1995. Effects of needle age, long-term temperature and CO_2 treatments on the photosynthesis of Scots pine. Tree Physiol. 15, 211–218.

Watson, R.T., Rohde, H., Oeschger, H., Siegenthaler, U., 1990. Greenhouse gases and aerosols. In: Houghton, J.T., Jenkins, G.J., Ephraum, J.J. (Eds.), Climate Change: the IPCC Scientific Assessment. Cambridge Univ. Press, Cambridge, UK, pp. 1–40.

Williams, D.W., et al., 2000. Effects of climate change on forest insect and disease outbreaks. In: Mickler, R.A., Birdsey, R.A., Hom, J. (Eds.), Responses of Northern US Forests to Environmental Change. In: Ecological Studies, Vol. 139. Springer-Verlag, New York, pp. 455–494.

Wilkens, R.T., 1998. Environmental effects on pine tree carbon budgets and resistance to bark beetles. In: Mickler, R.A., Fox, S. (Eds.), The Productivity and Sustainability of Southern Forest Ecosystems in a Changing Environment. In: Ecological Studies, Vol. 128. Springer-Verlag, New York, pp. 591–616.

Wullschleger, S.D., Norby, R.J., Hendrix, D.L., 1992. Carbon exchange rates, chlorophyll content, and carbohydrate status of two forest tree species exposed to carbon dioxide. Tree Physiol. 10, 21–31.

Yapp, C.J., Poths, H., 1992. Ancient atmospheric CO_2 pressures inferred from natural goethites. Nature 355, 342–344.

Air Pollution, Global Change and Forests in the New Millennium
D.F. Karnosky et al., editors
© 2003 Elsevier Ltd. All rights reserved.

359

Chapter 20

An intensive monitoring study of air pollution stress in a beech forest in Spain

J.M. Santamaría*, G. Amores, J. Garrigó

Department of Chemistry and Soil Science, University of Navarre, Irunlarrea s/n, 31080 Pamplona, Spain

B.S. Gimeno

Department CIEMAT-IMA 3b, Avda. Complutense 22, 28040 Madrid, Spain

L. Luchetta

Ecole Nationale Supérieure de Chimie de Toulouse, 118 Route de Narbonne, 31077 Toulouse, France

N. Madotz, R. Cavero, A. Ederra

Department of Botany, University of Navarre, Irunlarrea s/n, 31080 Pamplona, Spain

Abstract

The main objectives of this study were to gain insight into relationships between crown condition, foliar composition, and soil chemistry on the one hand and various environmental factors, including atmospheric deposition and air pollution, on the other. Defoliation of beech (*Fagus sylvatica* L.) trees was very similar across the different surveys, reaching a maximum in 1996, a year following a period of drought. Biotic effects also influenced forest health, especially *Rhynchaenus fagi* L., which was present during all surveys. Visible ozone-injury-like symptoms observed on beech and *Vaccinium myrtillus* L. leaves suggested that ozone and/or other air pollutants could be additional damaging factors in this area. The high ozone concentrations recorded in the field, and the detection of several anthropogenic hydrocarbons, supported this hypothesis. Foliar analysis of beech trees revealed relatively low contents of Cu and Fe and high levels of Ca and Na. Beech trees showed a good nutritional balance for other elements. Soils were desaturated in basic cations and had low nutrient contents. Several acid episodes (pH < 5.6) were detected, especially in winter, when industrial activity increases. Annual throughfall fluxes of elements were significantly higher than in Northern Europe but similar to those observed in the Mediterranean area. Finally, the vegetation assessment carried out in the Level II plot did not allow

---

*Corresponding author.

DOI:10.1016/S1474-8177(03)03020-1

us to draw any conclusion about the vegetation dynamics. However, this is the first step in a long-term study that will provide information about changes in ecosystem variables in future inventories.

1. Introduction

The Pan-European Programme for Intensive and Continuous Monitoring of Forest Ecosystems was implemented in 1994 in order to gain a better understanding of the effects of air pollution and other stress factors on forests. At present, 864 permanent observation plots have been established in Europe, 513 in the European Union, and 351 in several non-EU countries.

This Programme is based on both the European Scheme on the Protection of Forests against Atmospheric Pollution (Council Regulation EEC No. 2528/86) and the International Cooperative Programme on Assessment and Monitoring of Air Pollution Effects on Forests (ICP Forests) under the Convention on Long-Range Transboundary Air Pollution (CLRTAP, UN/ECE).

Spain has participated in the Level II Monitoring Programme since 1994, establishing a total of 53 intensive monitoring plots throughout the country by the end of 1995. On all these plots, several surveys are carried out, including crown condition, foliage and soil chemistry, and forest growth. At 11 of these plots, assessments of atmospheric deposition, meteorology and ground vegetation are also made.

The forests of the province of Navarre, in the north of Spain, represent a great environmental, cultural, and productive resource, as well as a scenic landscape. For these reasons, the regional government has shown special interest in monitoring the state of the forest environment and in 1994, joined the Pan-European Programme for Intensive and Continuous Monitoring of Forest Ecosystems. Since then, the University of Navarre has responsibility for a Level II plot made up of a beech (*Fagus sylvatica* L.) forest, the most characteristic species of trees growing in this region.

This paper describes the results of the Level II surveys carried out in Navarre for the period 1994–1999. Besides the mandatory activities, several pilot studies have also been carried out, including the determination of biogenic and anthropogenic emissions of volatile organic compounds (VOCs), a short-term study to monitor ozone concentrations, and a phytosociological assessment.

This research will provide a better insight into crown, soil, and foliar condition of beech forests, focusing specifically on relationships between stand and site characteristics and environmental factors, such as meteorology and air pollution.

2. Materials and methods

2.1. Site

The permanent observation plot under study is a pure beech (*Fagus sylvatica*) stand that lies in the Cantabrian-Atlantic Province within the Eurosiberian region. This beech forest consists of a mosaic of two different communities: *Saxifrago hirsutae–Fagetum sylvaticae* subassociation *luzuletosum pilosae* and *Carici sylvaticae–Fagetum sylvaticae* subassociation *isopyretosum thalictroidis*. It is a 90-year-old homogeneous beech stand, located 963 m above sea level, with a basal area of 19 $m^3 ha^{-1}$. It is subject to exploitation (2 $m^3 ha^{-1}$) and has a silvicultural rotation of 150 years.

The Level II plot (0.25 ha, $43° 00' 03'' N$, $1° 20' 41'' W$) is located in the Plain of Auritz-Burguete. This plain, carved of soft materials and partially filled with alluvial material, is formed by an east-west band limited by high areas like Tiratun (1217 m) and the Corona (1387 m) mountains. Geologically, the area belongs to the Quaternary period, made up from a mixture of colluvial apron and terrace containing semiangular pebbles, gravels, sands, limes and clays (Íñiguez et al., 1992).

The climate is hyperhumid, with a historic annual mean precipitation of 2183 mm (173 days of rain per year) and a thermal index of 123 according to Rivas-Martínez (1995). Consequently, the area belongs to the mountain floor, where climate conditions are characterized by cold winters with mean minimum temperatures of between 0 and $-4 °C$. From September to June, frosts are statistically probable, especially at higher altitudes.

The soils of this beech forest (profile: Ao–Eg–Bt–BC–C–2Crg) are developed on Quaternary colluviums over Paleogen marls (Tertiary). According to spatial location, these colluviums have different thicknesses and compositions. On convex topographic locations, colluvium is approximately 1 m in depth and is dominated by marl residuum; on debris cones, it is several meters thick and it is composed of early Devonian (Paleozoic) sediments that are rich in schist, limes, and dolomites. All these materials are subjected to a very wet climate (> 2000 mm yr^{-1}) with an udic soil moisture regime. Because of the environmental conditions, the soil is subject to very intense leaching, and even the marls lose all the limes before they lose their lithic structure.

Leaching is also the main pedogenic process characterizing these soils. The soils are very acid ($pH_{KCl} = 3.7–3.9$) and weathered to more than 1 m in depth. The residuum is a very fine material with a large proportion of clay and fine silt. On stable surfaces, there may be slight clay illuviation processes but it is not of taxonomic importance.

These incipient argilic horizons favor the appearance of slight redoxomorphic features, reddish mottles of pseudogley at the bottom of the eluvial horizons (Eg). The weathering faces of the marl (Crg) also display some black coloration due to the presence of manganese oxides. Despite these acid and oxyaquic conditions, the short dry periods and surplus of nutrients provided by the beech trees mean there is good biological activity and a high rate of decomposition and removal of soil organic matter. The type of humus in these soils is an acid mull with a good incorporation of organic compound.

2.2. *Field sampling and chemical analysis*

We assessed crown condition by measuring defoliation (5% classes), discoloration, and the presence and intensity of insects and fungi (UN/ECE, 1998) in 30 sample trees. Assessments were carried out in July–August by forestry staff (3 people) previously trained in a 1-week calibration course aimed at harmonizing assessment criteria.

Deposition samples were collected from the forest stand and an adjacent open area throughout 1997–1999. The stand throughfall samples were collected at 2-week intervals using six systematically located precipitation collectors (1.5 m above the soil). The corresponding number of collectors for the adjacent open area was four. Precipitation amount was measured using four Helmann collectors (two under the forest canopy and two in an open field).

Additional meteorological parameters were solar radiation, precipitation, relative humidity, air temperature, wind speed, and wind direction, which were measured continuously in the open field at a fully automated meteorological station.

Deposition samples were analyzed for pH, conductivity, amount of precipitation, Cl, NO_3, SO_4, Na, K, Ca, Mg, Cu, Cd, Pb, Zn, Fe, and Mn. Anions were analyzed by ionic chromatography (Waters 432) and cations by plasma emission spectrometry (Jovin Yvon J-38 S) and atomic absorption (Perkin Elmer Analyst 800).

The chemical composition of the foliage was measured in eight predominant or dominant beech trees selected in the buffer area (5 m width around forest plot). Leaves were collected during the summers of 1995 and 1997 from the upper third of the crown using a long pruning device.

The unwashed samples were dried at 60 °C and pulverized in a mixer. The concentrations of Ca, Cd, Cu, Fe, K, Mg, Mn, Na, P, Pb, S, and Zn were determined, following wet digestion ($HNO_3 : HClO_4$), by plasma emission spectrometry (Jovin Yvon J-38 S) and atomic absorption (Perkin Elmer Analyst 800). Total nitrogen concentration was determined by the Kjeldahl method. The performance of the analysis was verified by analyzing certified reference material (CRM—*F. sylvatica*) and no method bias was detected. More details about the treatment and analysis of samples may be found in Santamaría and Martín (1998).

Besides the nutritional status, we also assessed the health of the sampled trees (*see Crown Condition*).

Soil samples (0–100 cm depth) were collected from the buffer zone of the beech forest. The soil was dried at 35 °C and gently ground to pass through a 2-mm sieve. In order to characterize the soil, the total content of each element studied in the exchangeable fraction was measured. Particle size was determined using a Robinson pipette, total N by Kjeldahl's method (semimicro Afora-Boet), and organic carbon according to Walkley-Blach. Metals were analyzed by atomic absorption (Perkin Elmer Analyst 800) and cation exchange capacity (CEC) was obtained by the ammonium acetate method. Finally, pH (1 : 2.5) was determined in water and KCl.

We determined biogenic emissions from *F. sylvatica* during a short-term study conducted in June 1998 through the bag enclosure method using Teflon cuvettes (volume = 21 l). The air circulating through the cuvettes was sampled and analyzed every hour by a cyclic procedure using a fully automated adsorption–desorption device. The measurements were conducted on June 15, 17, and 18. Sample analysis was done by thermal desorption (ATD 400, Perkin Elmer) gas chromatography (HP-5890)/mass spectrometry (HP-5972 MSD).

At the same time as VOC biogenic emissions were determined, the levels of anthropogenic hydrocarbons were also measured. Air was sampled during the course of the day (June 15 and 18) using Chrompack tubes filled with Carbotrap B and C.

Ozone concentrations were recorded in the Level II plot for a period of 8 days (13–20 June) using a continuous measuring device (Dasibi 1008 RS).

Finally, data related to vegetation were collected from five small sampling units (10 × 10 m), representative of the forest. Plots were visited monthly throughout 1999 (twice a month in the spring) and phytosociological inventories were carried out following the Braun-Blanquet method (Braun-Blanquet, 1979).

We collected data for each plant, including the abundance-dominance index, the phenological stage, and the life form. After completing 1 year of study, we deduced the degree of vitality of each species: (1) well-developed plants that complete their life cycle regularly; (2) poorly-developed plants that complete their life cycle or well-developed plants that have not completed their life cycle. In addition to this, a checklist from all the forest was recorded as a tool to explain future changes in the sampling units' vegetation.

3. Results and discussion

Excluding the 1994 survey, defoliation of sampled trees remained almost constant over the years, reaching a maximum in 1996, when mean defoliation

Table 1.   Variation in defoliation of sampled trees, 1994–1999

Year	Mean	Minimum	Maximum	Std. deviation
1994	10.0	10	10	0.0
1995	18.5	10	25	4.3
1996	24.1	15	45	7.6
1997	18.0	10	25	3.6
1998	19.0	15	30	4.0
1999	18.7	15	25	3.5

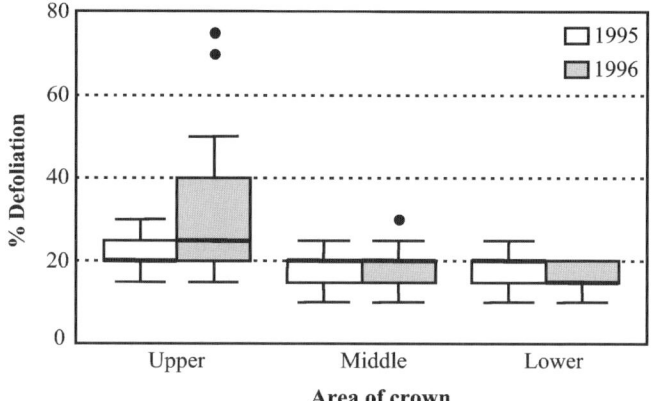

Figure 1.   Variation of defoliation by crown thirds.

amounted to 24.6% (Table 1). The correlation analysis (Pearson's *r* test) carried out using all data showed a significant correlation between defoliation of years 1997–1998 ($r = 0.39$; $P < 0.01$) and 1998–1999 ($r = 0.58$; $P < 0.01$). These results suggest that the process of defoliation does not occur randomly among stand trees but mainly affects those individuals that are more predisposed and weakened by other circumstances. In the case of the 1996 survey, no correlation was found.

A high proportion of damaged trees (defoliation > 25%) were detected in 1996, when 33.3% of all trees showed defoliation > 25%. This is probably related to the drought of the previous year, when the annual mean precipitation (721 mm) was much lower than the historic mean precipitation (2.183 mm) of this area.

During the 1995 and 1996 surveys, canopy defoliation was estimated by thirds (Fig. 1). In both years, the upper third of the crown was the most defoliated, a fact that underlines the greater vulnerability of this part of the crown, which is more exposed to stress factors (biotic and abiotic ones). This was es-

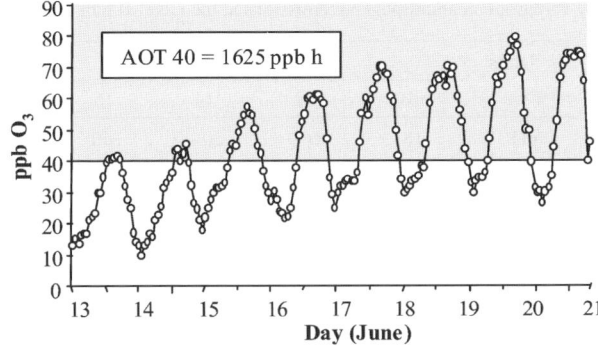

Figure 2. Ozone diurnal fluctuation by sampling day and calculation of AOT 40.

pecially true in 1996, when the increase in mean defoliation of stand trees was mainly related to the worsening of the upper leaves of the crown.

In general, the most frequently assessed damage causes were abiotic agents, mainly explained by drought and late frosts. In addition, classic damaging agents must be mentioned, especially the insect *Rhynchaenus fagi* L., which was present in all surveys.

Results support the hypothesis that the major cause of damage in forests is due to the adverse effects of stresses such as drought or pest infestations (De Vries et al., 2000). However, other stress factors (especially air pollutants) must not be overlooked. A small number of young beech trees (at the end of summer 1998) had visible injury symptoms attributable to ozone (dark pigmented stipple on interveinal tissue of upper leaf surface). Similar symptoms were also recorded in some individuals of *Vaccinium myrtillus* L. These observations are consistent with the ozone levels recorded in the forest stand (Fig. 2) during the short-term field study carried out in 1998 and they underscore the usefulness of both species as ozone bioindicators (Nygaard, 1994; Braun and Flückiger, 1995).

The AOT 40 calculated for the 8 sampling days was 1625 ppb h (Fig. 2), a value abnormally high for a remote area located far away from any pollution source. Moreover, the toxicity threshold fixed at 65 $\mu g\,m^{-3}$ for a 24-h period was exceeded from June 15–20, which may have negatively affected vegetation. With respect to ozone variations, the fact that the mean profile showed minimum concentrations close to 30 ppb and no lower suggests that the measured ozone comes from long- or medium-range transport rather than from local production due to photochemical activity.

This hypothesis is also confirmed by the 15 anthropogenic hydrocarbon compounds detected during the sampling period (Table 2), probably related to road transport. The concentrations recorded were similar to those detected in

Table 2.   Hourly variations in anthropogenic emissions at the Level II plot

Emission type	Units	Time (h)					
		10	12	14	16	18	20
Butene	$pl\,l^{-1}$	181	163	204	146	178	200
Pentene	$pl\,l^{-1}$	58	39	156	71	59	88
Hexene	$pl\,l^{-1}$	26	27	43	39	43	78
Heptene	$pl\,l^{-1}$	5	7	4	7	8	10
Octene	$pl\,l^{-1}$	7	42	12	9	7	5
Butane	$pl\,l^{-1}$	16	9	10	11	14	54
Pentane	$pl\,l^{-1}$	111	28	77	51	34	70
Hexane	$pl\,l^{-1}$	28	25	30	41	32	36
Heptane	$pl\,l^{-1}$	42	49	42	80	84	49
Octane	$pl\,l^{-1}$	7	13	38	33	15	25
Benzene	$pl\,l^{-1}$	128	125	105	195	126	208
Toluene	$pl\,l^{-1}$	156	160	271	292	161	253
Ethylbenzene	$pl\,l^{-1}$	10	9	11	13	14	18
M + p-Xylene	$pl\,l^{-1}$	28	28	25	52	27	29
o-Xylene	$pl\,l^{-1}$	8	7	5	30	8	9

Table 3.   Hourly variations in biogenic emission rates, PAR, and foliar temperature (*Fagus sylvatica* L.)

Time (h)	Emission rate ($\mu g\,g^{-1}\,h^{-1}$)	PAR ($\mu mol\,m^{-2}\,s\,g^{-1}$)	Foliar temp. (K)
7	0.02	46	277.8
9	9.56	698	292.4
11	18.41	1146	302.6
14	12.63	828	296.4
16	7.24	335	297.0
18	0.53	68	294.3

the forests of Italy and Holland (Gelencsér et al., 1994; Kalakobas et al., 1997). The meteorological conditions recorded on these days (prevailing winds from N and NE) support the notion that all these contaminants come from the south of France.

Aside from the anthropogenic sources, several biogenic compounds emitted from vegetation (mainly isoprene and monoterpenes) are also involved in the production of photooxidants such as ozone (Guenther et al., 1993). Analysis of biogenic emissions showed that the main monoterpene emitted by *F. sylvatica* L. was isoprene, which accounted for 95% of all emitted compounds (Table 3). These results are similar to those obtained by several authors (König et al., 1994; Tollsten and Müller, 1996; Schuh et al., 1997).

Table 4. Differences in defoliation and foliar element contents of sampled trees between 1995, 1997, and the foliar reference values for *Fagus sylvatica* L.

Element	Units	Foliar concentrations			Reference
		1995	1997	Variation	values
Ca	mg/g	10.6	9.4	1.2[a]	4.0–8.0
K	mg/g	7.7	7.5	0.2	5.0–10.0
Mg	mg/g	1.9	1.4	0.5	1.0–1.5
N	mg/g	21	22	−1	18.0–25.0
P	mg/g	1.0	1.3	−0.3	1.0–1.7
S	mg/g	2.6	1.7	0.9[a]	1.3–2.0
Cd	µg/g	0.05	0.04	0.01	< 0.5
Cu	µg/g	5.7	5.6	0.1	6–14
Fe	µg/g	117	101	16[a]	200–2000
Mn	µg/g	1022	1017	5	1000
Na	µg/g	306	233	73[b]	< 100
Pb	µg/g	1.2	1.3	−0.1	2–6
Zn	µg/g	23	25	−2	20–80
Def.	%	18.4	17.6	0.8	

[a]Significant at the 95% probability level.
[b]Significant at the 99% probability level.

The statistical analysis of data (Pearson's r test) showed that the mean diurnal variation profile of the monoterpene emission was strongly correlated to light intensity ($r = 0.98$), foliar temperature ($r = 0.86$), transpiration ($r = 0.95$), and photosynthetic activity ($r = 0.97$). Maximum emission rates were recorded around midday.

In order to extrapolate these results, we calculated the standard emission rate of beech trees using Guenther's G2 algorithm. This index suits beech well as it depends simultaneously on radiation and temperature. The standard emission rate calculated for beech was 21.7 $\mu g\, g^{-1}\, h^{-1}$, indicating that this species is a strong emitting isoprene, an organic compound that may play an important role in ozone production.

The chemical composition of the foliage of forest trees is another important indicator for the functioning of trees, especially with respect to nutrition and the influence of air pollutants (Innes, 1993). Our evaluation of beech tree nutrition conditions (Table 4) revealed relatively low contents of Cu and Fe and high levels of Ca and Na (this last due to the proximity of the study stand to the Bay of Biscay). Beech trees showed a good nutritional status for the remaining elements, exhibiting no deficiency symptoms. Although no correlation between defoliation and nutrient contents of leaves has been found, the disturbances detected for several nutrients could be inducing a negative effect on the health of beech trees.

Table 5. Main nutrient ratios of sampled trees and reference values for *Fagus sylvatica* L.

Ratio	Mean values		Reference values
	1995	1997	(range)
S/N	0.12	0.08	0.05–0.11
N/P	20.8	16.9	10.6–25.0
N/K	2.7	2.9	1.8–5.0
N/Ca	2.0	2.3	2.3–6.3
N/Mg	11.0	16.1	12.0–25.0
K/Ca	0.7	0.8	0.6–2.5
K/Mg	4.0	5.4	3.3–10.0
Ca/Mg	5.6	6.7	3.7–8.0

The chemical composition of leaves was very similar in both sampling years. The paired-*t* test showed a significant increase in foliar N contents in 1997, and Ca, S, Fe, and Na concentrations were higher in 1995. Other elements remained constant throughout the two studies. With respect to heavy metals, all analyzed elements were well below toxicity levels.

As the absolute concentration of a particular element is sometimes of little value as an index of nutrition, the relationships between several elements were examined (Table 5). In all cases, ratios were within the reference values cited in the literature (Anonymous, 1997; Mankovska, 1998).

Other factors, such as soil acidification or soil nutrient imbalances, cannot be excluded as contributing factors to forest decline either. The Level II stand soil has the following profile: Ao–Eg–Bt–BC–C–2Crg. Analysis (Table 6) showed high acidity in this soil, even in the weathered marl (2Crg). Except on the superficial horizon (0–8 cm), the soil is pale brown when dry, and brown or yellowish brown when moist.

The soil particle size is very fine, with a clay increase between 20 and 47 cm depth. The soil is associated with clay skins but the illuviation index is much too low to consider that an argilic endopedon (SSSA, 1999) or argic horizon (FAO, ISRIC and ISSS, 1998) exists. The soil is strongly desaturated (less than 30%) despite the contribution of beech leaves and parent material to the bases. According to analytical results, this soil is an *Inceptisol* of the *Udept*, *Dystrudept* great group and *oxyaquix* subgroup because it has mottles due to precipitation of iron and manganese oxides. For the same reasons, it is classified as *Cambisol stagnic dystric* according to WRB.

The main characteristic of Burguete soils is their very high leaching rate because of the large amount of precipitation they receive. Consequently, soils are desaturated in basic cations and have low nutrient contents. As well, temporal anoxic conditions are developed. This fact, together with the high acidity levels of this soil, induces mobilization of Al and Mn, elements that can cause

Table 6.   Variations in chemical and physical parameters according to depth of sampled soil

Parameters	Depth (cm)					
	0–8	8–20	20–47	47–65	65–100	> 100
pH H$_2$O	4.1	4.3	4.6	4.8	4.6	4.3
pH KCl	3.7	3.8	3.9	3.9	3.9	3.8
Organic C (%)	12.1	2.7	1.6	1.3	0.9	0.6
Organic matter (%)	20.9	4.6	2.7	2.2	1.6	1.0
N (%)	0.76	0.23	0.15	0.14	0.13	0.09
C/N	16	12	11	9	7	6
CEC (cmol(+)/kg)	53.9	25.2	23.4	21.7	20.1	21.1
Ca (cmol(+)/kg)	10.4	5.3	5.8	5.1	4.1	3.8
Mg (cmol(+)/kg)	2.1	1.1	1.1	1.0	0.8	0.7
K (cmol(+)/kg)	1.3	0.5	0.5	0.4	0.3	0.2
Na (cmol(+)/kg)	0.8	0.3	0.2	0.2	0.2	0.1
V (%)	27.1	28.5	32.7	31.0	26.4	22.8
Particle size (%)						
Coarse sand	1.9	2.1	2.5	2.8	2.7	2.1
Fine sand	5.7	6.8	6.5	6.0	5.9	2.5
Coarse silt	9.6	11.4	7.4	10.7	10.1	6.5
Fine silt	38.9	44.3	44.1	42.1	42.1	46.8
Clay	43.9	35.4	39.5	38.4	39.2	42.1

adverse effects on vegetation, especially young trees, that have less developed root systems.

Aside from all the above-mentioned factors, climate and atmospheric deposition also influence the soil solution chemistry and the nutritional status of forests (De Vries, 1996).

The amount of precipitation collected throughout the different sampling years was lower than the historic mean recorded for this area. This was especially true in 1995, when the total amount of precipitation collected did not exceed 721 mm.

This drought period influenced the phytosanitary state of beech trees, an important contributing factor being the high defoliation levels recorded in the 1996 survey. Winter and spring were the seasons with the highest precipitation amounts, summer being the driest period.

Several acid episodes (pH < 5.6) were detected, although acid input was not very high (Fig. 3). Generally, the pH of throughfall was higher than bulk deposition, a fact mainly related to the alkaline effect of Ca within the forest. The most acidic rains were collected in winter, when industrial activity is increased. Otherwise, the highest pH values corresponded to the summer season, when rain is often enriched with airborne alkaline particles.

Whether the input of S and N compounds from the atmosphere causes acidification is strongly influenced by the deposition of accompanying base

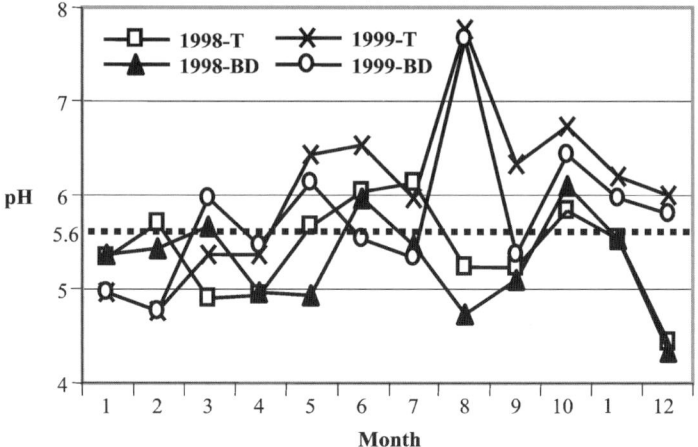

Figure 3.   Seasonal variation of pH (throughfall and bulk deposition).

Table 7.   Base cation to S + N ratios and acid input in throughfall and bulk deposition

Year	Throughfall		Bulk deposition	
	S + N ratio	Acid input	S + N ratio	Acid input
1997	4.7	37.9	0.2	−7.0
1998	1.1	−1.6	−0.3	−16.6
1999	2.1	11.5	0.5	−9.7
Mean	2.4	13.7	0.2	−11.5

cations. In this way, the relationship between Cl corrected base cation deposition (Ca + Mg + Na + K − Cl) and the sum of N + S deposition may be used
as an indicator of the potential acid input. As displayed in Table 7, the input
of corrected base cations in bulk deposition was lower than the S and N input,
resulting in a positive acid input. In the case of throughfall, the results were the
opposite, that is mainly because of the higher loads of K and Ca in throughfall.

These results partially explain the low pH values of soils found at the beech
study site. Within a forest, Ca is the most important base cation, neutralizing
the potential acid input from the atmosphere.

As expected, annual throughfall fluxes of elements were higher than in bulk
deposition (Table 8), reflecting the effect of canopy interaction. The resulting
deposition rates were significantly higher than in northern Europe but similar
to those observed in the Mediterranean area. The N input received throughout
the sampling years was below 15 kg ha^{-1} yr^{-1}, the threshold at which species
diversity may be at risk. However, below this value, tree growth may be inhibited. Sulfur inputs were also below critical levels.

Table 8. Annual average throughfall and bulk deposition fluxes ($kg\,ha^{-1}\,yr^{-1}$)

| | Throughfall | | | | | | | Bulk deposition | | | | | | |
Year	Cl^-	NO_3^-	SO_4^{2-}	Na	K	Ca	Mg	Cl^-	NO_3^-	SO_4^{2-}	Na	K	Ca	Mg
1997	17.7	19.9	22.8	13.4	18.9	34.1	2.9	17.9	22.0	21.8	9.8	3.3	9.9	2.0
1998	43.1	29.6	28.0	19.2	17.3	20.3	3.5	37.0	21.6	25.1	17.1	6.3	9.1	2.7
1999	35.0	29.4	26.8	17.0	24.9	16.7	3.6	32.9	26.0	26.0	15.8	3.2	15.6	2.8
Mean	33.1	26.8	26.1	16.9	20.6	22.4	3.4	30.3	23.3	24.5	14.8	4.4	11.7	2.6

Finally, the vegetation assessment within the framework of the intensive monitoring activities at Level II is important for two important reasons. Firstly, vegetation plays a direct role in water or nutrient cycling, and interacts strongly with other biotic components. Secondly, vegetation is a good bioindicator of environmental changes. Thus, the current knowledge of the ecological niche of numerous plant species allows us to deduce changes in underlying environmental factors from vegetation changes.

Results of the flora inventory (Table 9) and the soil pH values of the Level II plot located in Burguete show that this beech forest consists of a mosaic of two different communities: *Saxifrago hirsutae–Fagetum sylvaticae* subassociation *luzuletosum pilosae* and *Carici sylvaticae–Fagetum sylvaticae* subassociation *isopyretosum thalictroidis*. Future monitoring of vegetation in this beech forest must take this into account.

The greatest flora richness was observed in May or June, when most of the species were in flower or fruiting, and identification was easier. Once the basal flora composition of the beech forest is known, long-term study of vegetation dynamics will provide information about changes due to natural or anthropogenic environmental factors. The best time to conduct the inventories is May or June and, if species vitality is to be studied, sampling frequency should be monthly (twice a month in spring).

4. Conclusions

The main aim of this study was to gain insight into relationships between crown condition, foliar composition, and soil chemistry and various environmental factors, including atmospheric deposition and air pollution.

Beech tree defoliation remained almost constant throughout the various surveys, reaching a maximum in 1996, a year that was preceded by a drought period. Aside from abiotic factors, biotic damaging effects were also noted, especially *R. fagi* L., which was present during all surveys, even before the appearance of leaves. Visible ozone-injury-like symptoms were observed on *F. sylvatica* L. and *V. myrtillus* L. leaves, suggesting that ozone and/or other air

Table 9. Abundance-dominance, phenological stages, degree of vitality and life forms in sampling unit 1

Date of inventory	Oct.	Nov.	Dec.	Mar.	Apr.	May	May	June	July	Aug.	Sep.	Vit.	Life form
Number of inventory	1	2	3	4	5	6	7	8	9	10	11	12	
Tree layer													
Fagus sylvatica	5^b	5^e	5^e	5^e	5^e	5^b	5^c	5^c	5^a	5^a	5^a	1	Phanerophyte
Shrub layer													
Fagus sylvatica	2^d	1^e	1^e	1^e	1^e	1^d	2^d	2^d	1^d	1^d	1^d	1	Phanerophyte
Rubus sp.	1^b	1^b	1^b	2^b	2^b	2^b	2^b	2^c	1^c	1^a	1^a	1	Nanophanerophyte
Herbaceous layer													
Agrostis capillaris	$+^b$	$+^b$	$+^b$	$+^b$	$+^b$	$+^b$	1^b	1^b	1^b	1^b	$+^b$	2	Hemicryptophyte
Ajuga reptans	$+^b$	$+^b$	1^b	1^b	1^b	1^b	1^c	1^a	1^b	1^b	1^b	1	Hemicryptophyte
Cardamine pratensis	$+^b$	1^b	1^b	2^b	2^c	2^c	2^c	2^a	1^b	$+^b$	$+^b$	1	Hemicryptophyte
Carex sylvatica	1^b	1^b	1^b	2^b	2^b	2^c	2^c	2^a	1^a	1^b	1^b	1	Hemicryptophyte
Circaea lutetiana						$+^b$	$+^b$	$+^b$	$+^c$	$+^a$	$+^b$	1	Hemicryptophyte
Crocus nudiflorus	$+^c$			1^b	1^b	1^b	1^b					1	Geophyte
Deschampsia flexuosa			$+^b$	$+^b$	$+^b$	1^b	1^b	1^c	$+^c$	$+^a$	$+^b$	1	Hemicryptophyte
Euphorbia amygdaloides	$+^b$	$+^b$	$+^b$	1^b	1^c	1^c	1^c	1^c	1^a	1^b	$+^b$	1	Camephyte
Euphorbia angulata				$+^b$	1^b	1^c	1^c	1^a	1^b	1^b	$+^b$	1	Geophyte
Galium odoratum	3^b	3^b	3^b	3^b	3^b	3^c	3^c	3^a	3^a	3^b	3^b	1	Geophyte
Geranium robertianum	$+^b$	$+^b$	$+^b$	$+^b$	$+^b$	$+^b$	$+^b$	$+^b$	$+^a$	$+^b$	$+^b$	1	Hemicryptophyte
Helleborus viridis				$+^b$	1^b	1^b	1^b	1^b	$+^b$	$+^b$	$+^b$	2	Geophyte
Isopyrum thalictroides				1^b	1^b	2^b	1^b	$+^b$				2	Geophyte
Labiate	1^b	1^b	1^b	1^b	1^b	1^b	1^b	1^b	1^b	1^b	1^b	2	Hemicryptophyte
Potentilla sterilis			$+^b$	$+^b$	$+^b$	1^c	1^a	1^b	$+^b$	$+^b$	$+^b$	1	Hemicryptophyte
Scrophularia alpestris					$+^b$	$+^b$	$+^c$	1^c	$+^a$	$+^a$		1	Hemicryptophyte
Veronica montana	1^b	1^b	1^b	1^b	1^b	2^c	2^c	2^a	2^a	2^a	2^b	1	Hemicryptophyte
Veronica officinalis										$+^b$	$+^b$	2	Hemicryptophyte
Viola riviniana	2^b	1^b	1^b	1^b	2^c	2^c	2^a	2^a	2^a	2^b	2^b	1	Hemicryptophyte
Mosses layer													
Atrichum undulatum	+				+	+	+	+	+	+	+		
Dicranum scoparium	+	+	+	+	+		+	+	+	+	+		
Fissidens exilis			+										
Frullania tamarisci			+				+						
Hypnum cupressiforme	+	+			+	+	+	+	+	+	+		
Isothecium myurum									+	+			
Plagiochila asplenioides						+		+	+	+	+		
Polytrichum formosum	1	1	1	1	1	1	1	1	1	1	1		
Thuidium tamariscinum					+	+	+	+	+	+	+		

[a] Fruit.
[b] Vegetative.
[c] Flower.
[d] Sterile.
[e] Leaf buds.

pollutants may be an additional damaging factor in this area. The high ozone concentrations recorded during the field work and the detection of several anthropogenic hydrocarbons support this hypothesis.

Chemical composition of beech leaves is clearly related to soil composition. Thus, the high leaching of basic cations of these soils is correlated with the low foliar nutrient contents. Likewise, the high acidity levels of the soil, related to the positive acid input, induce mobilization of Al and Mn, elements that can have adverse effects on vegetation.

We cannot draw any conclusions yet about vegetation dynamics from the vegetation assessment carried out in the Level II plot. However, this is the first step in a long-term study that will provide information about changes in ecosystem variables in future inventories.

References

Anonymous, 1997. Forest foliar condition in Europe. Results of large-scale foliar chemistry surveys. EC, UN/ECE, Austrian Federal Forest Research Centre.

Braun-Blanquet, J., 1979. Fitosociología. Bases Para el Estudio de las Comunidades Vegetales. Blume Ediciones.

Braun, S., Flückiger, W., 1995. Effects of ambient ozone on seedlings of *Fagus sylvatica* L. and *Picea abies* (L.) Karst. New Phytol. 129, 33–44.

De Vries, W., 1996. Critical loads for acidity and nitrogen for Dutch forests on a 1 km × 1 km grid. Wageningen, The Netherlands, DLO Winand Staring Centre for Integrated Land, Soil and Water Research. Report 113.

De Vries, W., Reinds, G.J., van Kerkvoorde, M.S., Hendriks, C.M,A,, Leeters, F.E.J.M., Gross, C.P., Voogd, J.C.H., Vel, E.M., 2000. Intensive monitoring of forest ecosystems in Europe. Technical Report 2000. The Netherlands.

FAO, ISRIC and ISSS, 1998. World reference base for soil resources. 84 World Soil Resources Reports FAO. Rome.

Gelencsér, A., Kiss, G., Hlavay, J., Hafkenscheid, T.L., Peters, R.J.B., De Leer, E.W.B., 1994. The evaluation of the Tenax GR diffusive sampler for the determination of benzene and other volatile aromatics in outdoor air. Talanta 41 (7), 1095–1100.

Guenther, A.B., Zimmerman, P.R., Harley, P.C., 1993. Isoprene and monoterpene emission rate variability: Model evaluations and sensitivity analyses. J. Geophys. Res. 28, 1197–1210.

Íñiguez, J., Sánchez-Carpintero, I., Val, R.M., Vitoria, G., Peralta, F.J., 1992. Mapa de suelos de Navarra, E: 1:50.000. Hoja 116. Garralda.

Innes, J.L., 1993. Forest Health: Its Assessment and Status. CAB International, Wallingford.

Kalakobas, P., Bartzis, J.G., Bomboi, T., Ciccioli, P., Cieslik, S., Dlugi, R., Foster, P., Kotzias, D., Steinbrecher, R., 1997. Ambient atmospheric trace gas concentrations and meteorological parameters during the first BEMA measuring campaign on May 1994 at Castelporziano, Italy. Atmos. Environ. 23 (5), 921–927.

König, G., Brunda, M., Puxbaum, H., Hewitt, C.N., Duckham, C., Rudolph, J., 1994. Relative contribution of oxygenated hydrocarbons to the total biogenic VOC emissions of selected mild-European agricultural and natural plant species. Atmos. Environ. 29, 861–874.

Mankovska, B., 1998. The chemical composition of spruce and beech foliage as an environmental indicator in Slovakia. Chemosphere 36, 949–953.

Nygaard, P.H., 1994. Effects of ozone on Vaccinium myrtillus, Hylocomium splendens, Pleurozium schrebery and Dicranum polysetum. Naturens Talegrenser. Fagrapport No. 61.

Rivas-Martínez, S., 1995. Clasificación bioclimática de la Tierra. Folia Botanica Matritensis 16, 1–25.

Santamaría, J.M., Martín, A., 1998. Influence of air pollution on the nutritional status of Navarre's forests, Spain. Chemosphere 36, 943–948.

Schuh, G., Heiden, A.C., Hoffmann, T.H., Kahl, J., Rockel, P., Rudolph, J., Wildt, J., 1997. Emissions of volatile organic compounds from sunflower and beech: dependence on temperature and light intensity. J. Atmos. Chem. 27, 291–318.

SSSA, 1999. Soil Taxonomy. A Basic System of Soil Classification for Making and Interpreting Soil Surveys. 2nd Edition, USDA-NRCS Agricultural Handbook No. 436. Washington.

Tollsten, L., Müller, P.M., 1996. Volatile organic compounds emitted from beech leaves. Phytochemistry 43, 759–762.

UN/ECE, 1998. Manual on Methodologies and Criteria for Harmonised Sampling, Assessment, Monitoring and Analysis of the Effects of Air Pollution on Forests. 4th Edition, Hamburg, Geneva.

Air Pollution, Global Change and Forests in the New Millennium
D.F. Karnosky et al., editors

Chapter 21

Effects of elevated carbon dioxide and acidic rain on the growth of holm oak

E. Paoletti*

IPAF-CNR, P. Cascine 28, I-50144 Florence, Italy

F. Manes

Dipartimento di Biologia Vegetale, Università La Sapienza, P. Moro 5, I-00185 Rome, Italy

Abstract

In order to study the interactive effects of elevated atmospheric CO_2 and acid rain on an evergreen oak, holm oak (*Quercus ilex* L.) (which is the keystone tree species in the Mediterranean environment) potted seedlings were grown for 90 days at two CO_2 concentrations, ambient and $+400\ \mu mol\,mol^{-1}$. Half of each group was sprayed once a week with deionized (pH 5.6) or acidified (pH 2.5) water. Elevated CO_2 enhanced growth: shoot and leaf sprouting, shoot length, total leaf area, total and individual leaf mass, stem, and fine root mass were increased. Acid rain increased leaf and shoot turnover by stimulating both abscission and sprouting and, because abscission was more common than sprouting, the result was growth inhibition. The trend of these results was visible after 30–40 days. Total leaf area was increased by the increase in leaf number at elevated CO_2, and decreased by the reduction in leaf size in the acid treatments. Elevated CO_2 favored biomass partitioning to stems, whereas acid rain did not modify allocation. Root-to-shoot ratio was not significantly affected.

Despite the elevated CO_2-enhanced growth of acid-stressed holm oak seedlings, significant $CO_2 \times pH$ interactions were recorded only in cases where the effect of acidity was null (biomass allocation to stem) or positive (shoot and leaf sprouting). Thus, we conclude that short-term CO_2 enrichment did not alleviate the negative effect of acid rain on holm oak growth and that a strong acid load may inhibit the CO_2-promoted biomass partitioning to stems.

1. Introduction

Atmospheric carbon dioxide (CO_2) concentration is increasing at a rate of about $1.5\ \mu mol\,mol^{-1}$ yearly (Watson et al., 1990) and is predicted to reach

---

*Corresponding author.

DOI:10.1016/S1474-8177(03)03021-3

600–$800\,\mu mol\,mol^{-1}$ by the end of the 21st century (Conway et al., 1988). This global phenomenon may interact locally with the deposition of acidifying substances, whose critical loads (the levels of deposition above which long-term harmful effects can be expected) are exceeded in about 10% of the European area (European Environment Agency, 1998).

Growth enhancement under elevated CO_2 is a general response for young trees (Ceulemans and Mousseau, 1994). Simulated acid rain may induce either an increase or a reduction in growth or may have no effect on growth (Lee and Weber, 1979; Jacobsen et al., 1990; Neufeld et al., 1985). Possible synergistic effects of acid rain and elevated CO_2 on aboveground tree growth have not yet been investigated.

We explored their short-term effects on growth, anatomy, and biomass partitioning of holm oak (*Quercus ilex* L.) seedlings. Holm oak is a drought- and shade-tolerant evergreen broadleaf tree, which forms climax forest communities over large areas of the Mediterranean basin. Holm oak biomass and metabolism have been shown to be stimulated in CO_2-enriched environments (Chaves et al., 1995; Hättenschwiler et al., 1997a, 1997b; Tognetti et al., 1998; Tognetti and Johnson, 1999). Holm oak root and mycorrhizal response to elevated CO_2 and acid rain has been the object of a previous study (Puppi et al., 1992). Other responses to acid depositions in holm oak are unknown. The object of this study was to determine if a strong acid load may inhibit the growth stimulation induced in holm oak by doubling atmospheric CO_2.

2. Material and methods

One hundred and twenty 2-year-old holm oak seedlings, grown in pots (2.5 l, peat : vermiculite : nursery soil $= 70 : 15 : 15$) in the same nursery and all the same size, were placed in two controlled chambers (LABCO Mod. CT15, 25/18 °C day/night temperature, 10-hour photoperiod, $250\,\mu E\,m^{-2}\,s^{-1}$ PPFD, 60% RH, $0.5\,m\,s^{-1}$ air flow), irrigated daily with water and once a week with Hoagland solution to attain field capacity, and rotated once a week to avoid position effects. The growing conditions were intended to simulate a Mediterranean spring under a closed canopy, as holm oak is a shade-tolerant species (Valladares et al., 2000).

Carbon dioxide was maintained at two constant concentrations, ambient ($\approx 360\,\mu mol\,mol^{-1}$) and ambient $+\,400\,\mu mol\,mol^{-1}$, by an electronic flowmeter and a gas analyzer (CIM-1 and EGM-1, PP Systems, UK). In each controlled chamber, half of each CO_2 group (30 seedlings each) was sprayed with deionized (pH 5.6) or acidified (pH 2.5, $H_2SO_4 : HNO_3 = 2 : 1$ in equivalents) water through stainless steel nozzles at a $3\,mm\,min^{-1}$ rate for $15\,min\,week^{-1}$

in the same day each week, until dripping point. The treatments (control, acid, elevated CO_2, combined) lasted 90 days.

After labelling, the number and length of shoots and number of leaves were recorded every 4–9 days. At harvest, leaf area was measured by LI-COR 3100, and dry weights of leaves, stems, and roots were recorded after reaching a constant value in an oven. The root system was divided into coarse (diameter > 2 mm) and fine (diameter < 2 mm) roots. New leaf area/dry mass (specific leaf area, SLA), new leaf area/new shoot dry mass (leaf area ratio, LAR), ratios of new leaf (leaf weight ratio, LWR), new stem (stem weight ratio, SWR), and fine root (root shoot ratio, R/S) dry mass to new shoot dry mass were calculated. Cross-sections of leaves and stems sprouted after 55–75 days of treatment, were sampled from the central part of mature, fully expanded leaves, and from stem segments 1.5 cm below the shoot apex. There was one sample per seedling, ten samples per treatment. At harvest, ten 1-cm long segments were removed 2 cm below the fine root apex from each of three seedlings per treatment. After fixation in formalin–acetic acid–alcohol and embedding in butyl and methyl methacrylate (7 : 3), samples were sectioned into 2 μm segments by ultramicrotome Reichert om U3 and stained with periodic acid-Schiff's reaction or Lugol (Jensen, 1962) for starch grains and with toluidine blue O (Trump et al., 1961) for anatomical measurements. Because the pith in the stem was in the shape of a five-pointed star, the mean diameter for each sample was calculated as mean length of the lines through the star points. Six cross-sectional measurements were taken for each leaf and each root segment. Spongy cell area and size, and starch grain area, with respect to cell area were quantified by the WinDias image analysis system (1.5 Delta-T Devices) on one photograph per cross-section.

Data were analyzed by two-way ANOVA ($CO_2 \times$ pH). The statistical unit was the individual pot ($n = 30$, except for fine root anatomy where $n = 3$). Data expressed in percentages (variation in shoot and leaf final number with respect to the initial one, spongy cell area with respect to the spongy mesophyll area, starch area with respect to the cell area) were previously arc sine transformed.

3. Results

3.1. Growth

Elevated CO_2 stimulated shoot and leaf sprouting (Fig. 1(A), (D)), new shoot elongation (Fig. 1(B)), and leaf sprouting in each shoot (Fig. 1(C)). Acid rain stimulated shoot and leaf sprouting (Fig. 1(A), (D)) and inhibited shoot length and leaf sprouting in each shoot (Fig. 1(B), (C)). A significant increase

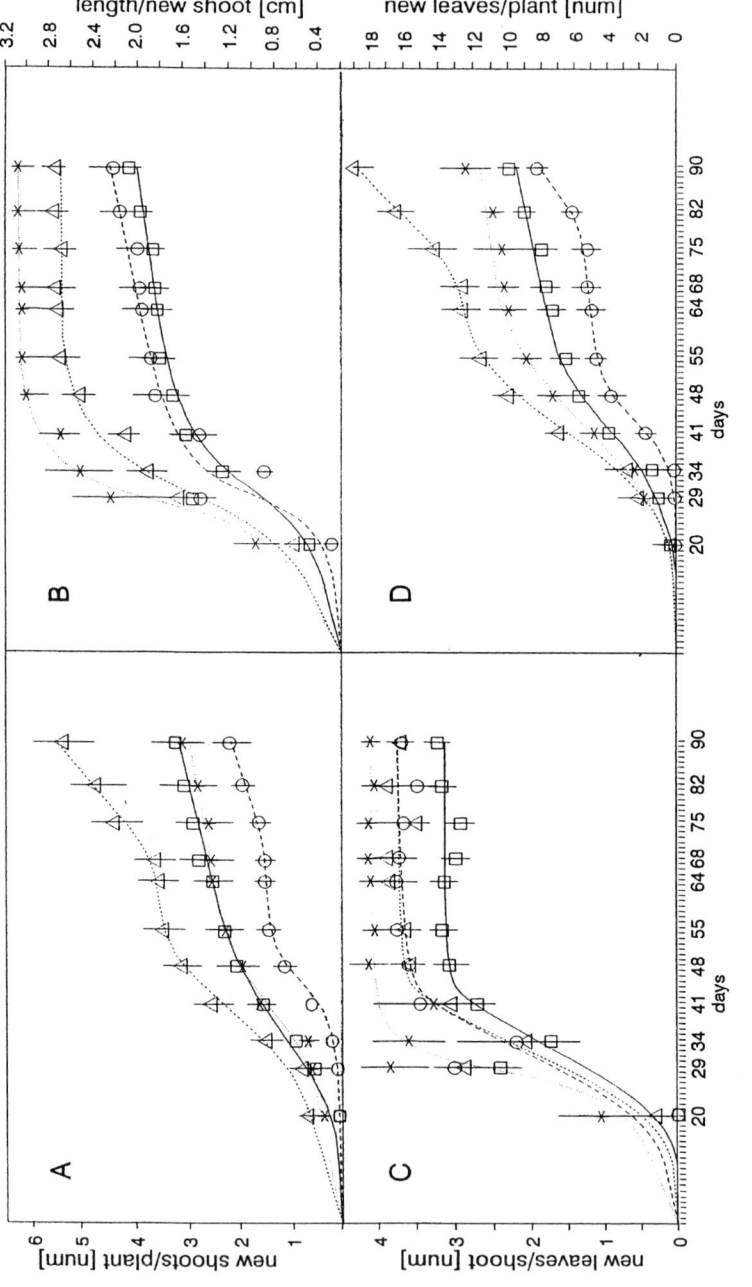

Figure 1. Shoot number (A) and length (B), leaf number per shoot (C) and per seedling (D) during the 90-day test period for the control (O, ambient CO_2 + pH 5.6 rain), acid ($\square$, ambient CO_2 + pH 2.5 rain), elevated CO_2 ($\times$, +400 μmol mol^{-1} CO_2 + pH 5.6 rain), and combined ($\triangle$, +400 μmol mol^{-1} CO_2 + pH 2.5 rain) treatment. Two-way analysis of variance was applied to the final data, in the box: * $p \leq 0.05$, o $p \leq 0.1$, ns: $p > 0.1$.

	ANOVA		
	CO_2	pH	$CO_2 \times$ pH
A	o	o	o
B	*	o	ns
C	o	o	ns
D	o	o	o

Figure 1. (Continued)

Table 1. Morphobiometric characteristics at the harvest. Increase/decrease in shoots and leaves were calculated with respect to the initial number

	Treatments				ANOVA[a]		
	Control	Acid	Elevated CO_2	Combined	CO_2	pH	$CO_2 \times$ pH
Increase/decrease in shoots [%]	8.5	−1.9	25.3	0.4	*	*	ns
Fallen shoots per plant [num]	1.7	3.4	2.3	5.5	o	*	ns
Increase/decrease in leaves [%]	−2.5	−22.2	17.2	8.8	*	*	ns
Abscissed leaves per plant [num]	9.8	24.5	6.5	17.6	o	*	ns
Leaf area per plant [cm^2]	81.2	60.8	105.6	100.0	*	o	ns
Leaf area per leaf [cm^2]	9.9	6.0	8.5	5.2	o	o	ns

[a] $* \, p \leqslant 0.05$, o $p \leqslant 0.1$, ns: $p > 0.1$.

in shoot and leaf sprouting was recorded in the $CO_2 \times$ pH treatment. Such results were significant ($p < 0.05$) as early as 30–40 days after beginning the treatment (Fig. 1). After 75 days, seedlings began a second growth flush (Fig. 1(A), (D)).

Elevated CO_2 increased final shoot and leaf number despite increased abscission, and increased leaf area despite decreasing leaf size (Table 1). Acid rain favored both shoot and leaf number and abscission, and also inhibited leaf area and size (Table 1). Leaf area showed no significant $CO_2 \times$ pH interaction.

3.2. Biomass

Elevated CO_2 increased dry weight of the total leaves, single leaf, stems, new shoots, and fine roots (Table 2). Coarse root biomass was not influenced by any

Table 2. Dry weight (dw) for roots and shoots sprouted during the test period [mg]

	Treatments				ANOVA[a]		
	Control	Acid	Elevated CO_2	Combined	CO_2	pH	$CO_2 \times$ pH
Leaf dw per plant	565	440	889	915	*	ns	ns
Leaf dw per leaf	69	44	72	48	o	o	ns
Stem dw per plant	121	100	297	222	*	o	ns
Shoot dw per plant	696	544	1189	1142	*	o	ns
Fine root dw per plant	338	520	774	740	o	ns	ns
Coarse root dw per plant	4520	4115	3596	4772	ns	ns	ns

[a] $* \ p \leqslant 0.05$, o $p \leqslant 0.1$, ns: $p > 0.1$.

Table 3. Allocation parameters for fine roots and for shoots sprouted during the test period. SLA: Specific Leaf Area; LAR: Leaf Area Ratio; LWR: Leaf Weight Ratio; SWR: Stem Weight Ratio; R/S: fine Root/new Shoot dry weight

	Treatments				ANOVA[a]		
	Control	Acid	Elevated CO_2	Combined	CO_2	pH	$CO_2 \times$ pH
SLA [$cm^2\,mg^{-1}$]	0.142	0.138	0.119	0.109	o	ns	ns
LAR [$cm^2\,mg^{-1}$]	0.118	0.113	0.090	0.088	o	ns	ns
LWR [$mg\,mg^{-1}$]	0.823	0.816	0.758	0.814	ns	ns	ns
SWR [$mg\,mg^{-1}$]	0.174	0.184	0.250	0.194	o	ns	o
R/S [$mg\,mg^{-1}$]	0.486	0.956	0.651	0.648	ns	ns	ns

[a] o $p \leqslant 0.1$, ns: $p > 0.1$.

of the treatments. Acid rain reduced the dry weight of the single leaf, stems, and new shoots. Interactions between CO_2 and pH were not significant.

Specific leaf area and LAR were reduced and SWR was increased at elevated CO_2 (Table 3). Leaf weight ratio and R/S did not vary significantly. Acid rain did not change biomass allocation parameters. A significant $CO_2 \times$ pH interaction was found for SWR.

3.3. Anatomy

Leaf tissue thickness did not vary (Table 4). The palisade was consistently composed of at least one layer, and more frequently two layers, of cells. At elevated CO_2, the shape of spongy cells was irregularly elongated, but did not vary in size, and a reduction in intercellular spaces was observed (Table 4). These results suggest that the number of spongy cells increased (data not shown). Starch was more abundant at elevated CO_2, particularly in the palisade cells (Table 4).

Table 4. Thickness of leaf tissues [µm], cell area in the spongy mesophyll [%], spongy cell size [µm^2], and area of starch grains in the palisade and spongy cells [%]

	Treatments				ANOVA[a]		
	Control	Acid	Elevated CO_2	Combined	CO_2	pH	$CO_2 \times$ pH
Upper epidermis	14	14	15	15	ns	ns	ns
Palisade	75	75	80	66	ns	ns	ns
Spongy	85	100	85	85	ns	ns	ns
Lower epidermis	10	11	10	11	ns	ns	ns
Total thickness	184	200	190	176	ns	ns	ns
Cell area in the spongy mesophyll	58	59	66	65	*	ns	ns
Spongy cell size	201	200	254	227	ns	ns	ns
Starch area in the palisade	2.4	2.6	56.4	49.5	**	ns	ns
Starch area in the spongy mesophyll	1.5	1.4	2.7	2.7	*	ns	ns

[a] ** $p \leqslant 0.01$, * $p \leqslant 0.05$, ns: $p > 0.1$.

Table 5. Thickness of stem tissues [µm] and area of starch grains in the cortex and pith cells [%]

	Treatments				ANOVA[a]		
	Control	Acid	Elevated CO_2	Combined	CO_2	pH	$CO_2 \times$ pH
Periderm	36	39	41	37	ns	ns	ns
Cortex	135	181	159	210	ns	ns	ns
Sclerenchyma	35	35	36	43	ns	ns	ns
Phloem	91	62	103	105	o	ns	ns
Xylem	106	127	211	228	*	ns	ns
Pith	492	614	594	612	ns	ns	ns
Total ray	919	1065	1150	1227	*	ns	ns
Starch area in the cortex	0	0	0.5	0.6	*	ns	ns
Starch area in the pith	0.14	0.15	11.3	10.9	**	ns	ns

[a] ** $p \leqslant 0.01$, * $p \leqslant 0.05$, o $p \leqslant 0.1$, ns: $p > 0.1$.

Stem thickness was increased at elevated CO_2 because of the increase in phloem and xylem production (Table 5). Stems at elevated CO_2 showed a starch surplus in the cortex and especially in the pith (Table 5).

Fine root thickness and starch content did not vary significantly (data not shown). Effects of acid treatments and $CO_2 \times$ pH interactions were not significant.

4. Discussion

4.1. Overall growth

Elevated CO_2 promoted shoot and leaf initiation, and shoot length. Growth enhancement under elevated CO_2 is a general response for young trees (Ceulemans and Mousseau, 1994), although no effect on height has been reported in *Picea glauca* or *Pinus radiata* (Brown and Higginbotham, 1986; Conroy et al., 1986) and a shoot growth reduction has been recorded for *Castanea sativa* (Mousseau and Enoch, 1989). A weakening in apical dominance and a greater number of side shoots under CO_2 enrichment have also been recorded on *Liquidambar styraciflua*, *Citrus aurantium*, and two *Populus* clones (Sionit et al., 1985; Idso et al., 1991; Ceulemans et al., 1995).

Acid rain promoted shoot initiation and inhibited leaf initiation—as shown by the variation in the number of new leaves per plant and per shoot—and shoot length. Simulated acid rain has been found to induce both an increase and a reduction in shoot elongation as well as a lack of variation in many conifers and deciduous broadleaves (Lee and Weber, 1979; Jacobsen et al., 1990; Neufeld et al., 1985). Nonetheless, stimulation of shoot and leaf sprouting by acid spray on epigeous parts confirmed the results of Winner et al. (1985), supporting the hypothesis that photosynthates were partitioned above all into portions under stress. However, acidity also promoted abscission, so that leaf and shoot turnover increased. Shoot abscission in the *Quercus* genus is a common phenomenon (Millington and Chaney, 1973), although it occurs mainly in autumn. Serious and unusual environmental stresses also promote the loss of plant parts (Kozlowski, 1973). Controls showed slight shoot and leaf abscission, even though the second growth flush indicated that chamber conditions were optimal for growth. Oak seedlings are known to show recurrent cyclic flushes which, if not synchronized, can induce excessive variability in samples (Norby and O'Neill, 1989). Periodic non-destructive measurements allowed us to compare the growth models of each plant, thus avoiding sampling errors. Furthermore, our treatments did not modify the growth flush synchrony (Fig. 1). Significant $CO_2 \times pH$ interactions were recorded for total leaf and shoot sprouting when both elevated CO_2 and acid rain individually stimulated initiation. Thus, no inhibiting effect of acidity on CO_2-enhanced growth was detected.

4.2. Leaf area

Elevated CO_2 increased total leaf area and decreased individual leaf area, i.e., leaf size. A previous investigation on holm oak response to CO_2 enrichment also promoted total leaf area, but leaf size did not change (Tognetti and Johnson, 1999). An increase in total leaf area has often been found under elevated

CO_2 concentrations in both conifers and deciduous broadleaves (Ceulemans and Mousseau, 1994; Janssens et al., 2000). The findings of the present study show that the increase in total leaf area in the elevated CO_2 treatment was associated with an increase in leaf number, indicating a strong morphogenic effect of CO_2 on leaf initiation, as found in *Liquidambar styraciflua* (Tolley and Strain, 1984), *Quercus petraea* (Guehl et al., 1994) and two *Populus* clones (Ceulemans et al., 1995). Leaf size may respond to an elevated CO_2 supply by increasing (Conroy et al., 1986; Koch et al., 1986; Gaudillère and Mousseau, 1989), not changing (Tolley and Strain, 1984; Radoglou and Jarvis, 1990) or decreasing (Mousseau and Enoch, 1989). A CO_2-induced decrease in leaf size contrasts with the enhancement in leaf cell expansion reported for some grassland herbs (Ferris and Taylor, 1994) and *Populus* clones (Radoglou and Jarvis, 1990; Gardner et al., 1995), and supports the hypothesis that leaf ontogenesis response to CO_2 is species dependent.

Acid rain reduced both total and individual leaf area. The reduction in total leaf area, also reported in *Brassica oleracea* (Caporn and Hutchinson, 1986), was related to the decreasing leaf size as the total number of leaves increased, demonstrating that acidity significantly influenced both shoot initiation and leaf ontogenesis.

Despite the ability of both CO_2 and acid rain to influence leaf initiation and ontogenesis in holm oak, no significant alleviating or inhibiting interaction was recorded for leaf area.

4.3. Leaf biomass and anatomy

The increase in total leaf mass at elevated CO_2 was associated with a greater number of leaves and, in contrast to the above-mentioned decreasing leaf size, with an increase in individual leaf mass. As leaf thickness did not vary, the higher starch content and smaller intercellular spaces in the spongy mesophyll may have contributed to this increase. A greater starch content in leaves grown at elevated CO_2 has frequently been reported (Patterson and Flint, 1980; Cave et al., 1981; Wulff and Strain, 1982; Yelle et al., 1989), even for holm oak (Tognetti and Johnson, 1999). The increase in individual leaf mass was the result of increased cell division, which was sensitive to CO_2, whereas cell enlargement apparently was not. In *Populus* clones and *Phaseolus vulgaris*, more extensive and slightly decreasing intercellular air spaces were recorded (Radoglou and Jarvis, 1990, 1992, respectively), again suggesting a species-dependent response of leaf ontogenesis to CO_2. In the same species, elevated CO_2 increased leaf thickness, mainly as a result of larger mesophyll cells (Radoglou and Jarvis, 1990, 1992). In C_3 plants at high CO_2, Thomas and Harvey (1983) observed an increase in leaf thickness that was associated with a third layer of palisade cells. In our study, similarly to Hofstra and Hesketh

(1975) for *Glycine max* and Radoglou and Jarvis (1993) for *Vicia faba*, the number of palisade layers did not change with variation in CO_2.

An increase in total leaf mass has been found in all high CO_2 studies (e.g., Norby and O'Neill, 1989; O'Neill et al., 1987; Koch et al., 1986; Sionit et al., 1985), but exposure to simulated acid depositions has been found to result in no variation (Neufeld et al., 1985) or reduction (Evans and Lewin, 1980). The acid-induced decrease in total leaf mass observed in the present experiment was not significant as it was offset by the increase in leaf number. The reduction in individual leaf mass in the acid treatments was due to the decrease in individual leaf area, as no significant anatomical modification was found.

Interactions between CO_2 and pH were not significant, with acid rain reducing individual leaf mass to the same extent in seedlings raised at ambient and elevated CO_2. Acid rain did not change total leaf mass in either ambient or elevated CO_2 treatments.

4.4. Stem biomass and anatomy

Stem mass at elevated CO_2 increased, similarly to Rogers et al. (1983), Higginbotham et al. (1985), Koch et al. (1986), O'Neill et al. (1987), Norby and O'Neill (1989), as a result of the stimulation in shoot elongation and phloem and xylem production, thus suggesting that CO_2 promoted activity in both apical meristems and vascular cambium. The starch surplus found in the cortex and pith at elevated CO_2 may have contributed to the increase in stem dry weight.

In accordance with Neufeld et al. (1985), acid rain decreased stem mass, as a result of the inhibition in shoot elongation rather than of anatomical modifications, thus suggesting that acidity depressed apical meristem activity and did not influence the vascular cambium.

Despite the influence of both CO_2 and acid rain on stem meristematic tissues, in the combined treatment there was no evidence that CO_2 enrichment afforded additional protection against stem biomass decrease induced by acidity.

4.5. Root biomass and anatomy

Elevated CO_2 increased fine root biomass, as reported elsewhere for holm oak seedlings (Tognetti and Johnson, 1999) and in agreement with many studies recording substantial root growth at elevated CO_2 (Dahlman, 1993). No variation was found in coarse root dry weight, as expected, given that coarse roots developed during the 2 years preceding the experiment.

No significant differences in starch distribution and tissue thickness were observed, probably because of sampling artifacts due to the inability to perform direct observations on root growth and to collect roots at the same developmental stage.

Although Neufeld et al. (1985) and Lee et al. (1981) found that simulated acid depositions applied to foliage reduced root dry weight, fine root biomass in this experiment did not change. The increase in fine root biomass induced by elevated CO_2 under acid rain was similar to that induced by CO_2 under pH 5.6 rain.

4.6. Biomass partitioning

Most experiments cited in Ceulemans and Mousseau (1994) and Janssens et al. (2000) reported an increase in root/shoot ratio, suggesting that CO_2 enrichment preferentially induces extra root storage. The lack of response observed in our experiment could be attributable to the small pot size (Arp, 1991) or to other environmental variables (Janssens et al., 2000). Another study using holm oak seedlings grown in one-quarter smaller pot volumes has found evidence of a preferential shift of biomass to belowground tissues in response to CO_2 enrichment (Tognetti and Johnson, 1999).

The decrease in LAR at elevated CO_2 observed in our experiment was considered a result of the reduction in SLA, as reported elsewhere for holm oak seedlings (Tognetti and Johnson, 1999). This suggests that foliage mass increased sufficiently to compensate for the reduction in leaf size. According to Tolley and Strain (1984), the decrease in SLA in plants grown at high CO_2 could be due to an increase in leaf thickness and/or quantity of starch. In our experiment, starch accumulation (in agreement with Tognetti and Johnson, 1999) was associated with greater cell density in the spongy cells, while leaf thickness remained constant.

The increase in SWR at high CO_2 and the lack of variation in LWR indicate that biomass allocation tended to favor the stems, as confirmed by the increase in stem thickness and shoot elongation.

Leaf sprouting was likely to be the most important carbon sink in the acid-treated seedlings. Abscission was also promoted, so that acid rain reduced shoot mass without apparent shifts in allocation. Studies on several deciduous broadleaves and conifers have also found no influence of acid depositions on dry matter partitioning (Tolley and Strain, 1984; Neufeld et al., 1985; Norby and O'Neill, 1989; Deans et al., 1990).

Acid rain reduced the CO_2-induced increase in SWR in our combined treatment, probably because of the reduction in stem length, given that no anatomical difference was detected. This was the only case in which the acid treatment inhibited a positive effect induced by the elevated CO_2.

5. Conclusions

Both elevated CO_2 and acid rain induced morphogenetic responses in holm oak, the effects of which are likely to be important in predicting stand productivity. Elevated CO_2 promoted the growth of holm oak seedlings, despite the fact that holm oak, like all sclerophyllous species, shows relatively low CO_2 exchange rates (Mooney, 1986). Acid mist accelerated leaf and shoot turnover by stimulating both abscission and sprouting, and inhibited the overall growth, as abscission was more common than sprouting.

Trees in natural stands may morphologically adjust to increasing CO_2 and reduce CO_2-induced initial growth stimulation. In fact, contrary to our results, mature holm oaks in naturally CO_2-enriched sites showed decreased branching and lower total leaf area (Hättenschwiler et al., 1997a). They also showed an increase in stem mass that was largely due to responses when trees were young (Hättenschwiler et al., 1997b) and confirms the increased biomass partitioning to stems reported here.

In contrast, acid rain reduced leaf and stem mass without apparent significant shifts in allocation. Together with stem and fine roots, leaf sprouting was one of the most important carbon sinks in the CO_2-treated seedlings, and was likely the most important one in the acid-treated seedlings, even if the increase in abscission concealed it.

Our study shows that increasing CO_2 concentrations enhanced the growth of acid-stressed holm oak seedlings. However, significant $CO_2 \times pH$ interactions were recorded only for the final number of shoots and leaves, and for biomass allocation to stems. In both cases, the effect of acidity was null or positive, and the effect of CO_2 was positive. Thus, no ameliorating effect of short-term CO_2 enrichment on negative responses induced by acid rain was detected, although an inhibiting effect of acidity on the CO_2-increased biomass partitioning to stems was noted.

Acknowledgements

This study is dedicated to the memory of Prof. Romano Gellini, who suggested we undertake a study on this subject.

References

Arp, W.J., 1991. Effects of source–sink relations on photosynthetic acclimation to elevated CO_2. Plant Cell Environ. 14, 869–875.

Brown, K., Higginbotham, K.O., 1986. Effects of carbon dioxide enrichment and nitrogen supply on growth of boreal tree seedlings. Tree Physiol. 2, 223–232.

Caporn, S.J.M., Hutchinson, T.C., 1986. The contrasting response to simulated acid rain of leaves and cotyledons of cabbage (*Brassica oleracea* L.). New Phytol. 103, 311–324.

Cave III, G.H., Tolley, L.C., Strain, B.R., 1981. Effect of carbon dioxide enrichment on chlorophyll content, starch content and starch grain structure in *Trifolium subterraneum* leaves. Physiol. Plant. 51, 171–174.

Ceulemans, R., Mousseau, M., 1994. Effects of elevated atmospheric CO_2 on woody plants. New Phytol. 127, 425–446.

Ceulemans, R., Jiang, X.N., Shao, B.Y., 1995. Effects of elevated atmospheric CO_2 on growth, biomass production and nitrogen allocation in two *Populus* clones. J. Biogeogr. 22, 261–268.

Chaves, M.M., Pereira, J.S., Cerasoli, S., Clifton-Brown, J., Miglietta, F., Raschi, A., 1995. Leaf metabolism during summer drought in *Quercus ilex* trees with lifetime exposure to elevated CO_2. J. Biogeogr. 22, 255–259.

Conroy, J., Barlow, E.W.R., Bevege, D.I., 1986. Response of *Pinus radiata* seedlings to carbon dioxide enrichment at different levels of water and phosphorous: growth, morphology and anatomy. Ann. Bot. 57, 165–177.

Conway, T.J., Tans, P., Waterman, L.S., Thoning, K.W., Masarie, K.A., Gammon, R.M., 1988. Atmospheric carbon dioxide measurements in the remote global troposphere, 1981–1984. Tellus 40B, 81–115.

Dahlman, R.C., 1993. CO_2 and plants: revisited. Vegetatio 104/105, 339–355.

Deans, J.D., Leith, I.D., Sheppard, L.J., Cape, J.N., Fowler, D., Murray, M.B., Mason, P.A., 1990. The influence of acid mists on growth, dry matter partitioning, nutrient concentrations and mycorrhizal fruiting bodies in red spruce seedlings. New Phytol. 115, 459–464.

European Environment Agency, 1998. Europe's Environment: The Second Assessment. Elsevier Science, London, UK. p. 295.

Evans, L.S., Lewin, K.F., 1980. Growth, development and yield responses of pinto beans and soybeans to hydrogen ion concentrations of simulated acid rain. Environ. Exp. Bot. 21, 103–113.

Ferris, R., Taylor, G., 1994. Stomatal characteristics of four native herbs following exposure to elevated CO_2. Ann. Bot. 73, 447–453.

Gardner, S.D.L., Taylor, G., Bosac, C., 1995. Leaf growth of hybrid poplar following exposure to elevated CO_2. New Phytol. 131, 81–90.

Gaudillère, J.P., Mousseau, M., 1989. Short term effect of CO_2 enrichment on leaf development and gas exchange of young poplars (*Populus euramericana* cv 1214). Acta Oecolog.: Oecolog. Plant. 10, 95–105.

Guehl, J.M., Picon, C., Aussenac, G., Gross, P., 1994. Interactive effects of elevated CO_2 and soil drought on growth and transpiration efficiency and its determinants in two European forest tree species. Tree Physiol. 14, 707–724.

Hättenschwiler, S., Miglietta, F., Raschi, A., Körner, C., 1997a. Morphological adjustments of mature *Quercus ilex* to elevated CO_2. Acta Oecolog. 18, 361–365.

Hättenschwiler, S., Miglietta, F., Raschi, A., Körner, C., 1997b. Thirty years of *in situ* tree growth under elevated CO_2: a model for future forest responses? Global Change Biol. 3, 463–471.

Higginbotham, K.O., Mayo, J.M., L'Hirondelle, S., Krystofiak, D.K., 1985. Physiological ecology of lodgepole (*Pinus contorta*) in an enriched CO_2 environment. Can. J. Forest Res. 15, 417–421.

Hofstra, G., Hesketh, J.D., 1975. The effects of temperature and CO_2 enrichment on photosynthesis in soybean. In: Marcelle, R. (Ed.), Environmental and Biological Control in Photosynthesis. J. Junk, The Hague, pp. 71–80.

Idso, S.B., Kimball, B.A., Allen, S.G., 1991. CO_2 enrichment of sour orange trees: 2.5 years into a long-term experiment. Plant Cell Environ. 14, 351–353.

Jacobsen, J.S., Heller, L.I., Yamada, K.E., Osmeloski, J.F., Bethard, T., Lassoie, J.P., 1990. Foliar injury and growth response of red spruce to sulfate and nitrate acidic mist. Can. J. Forest Res. 20, 58–65.

Janssens, I.A., Mousseau, M., Ceulemans, R., 2000. Crop ecosystem responses to climatic change: tree crops. In: Reddy, K.R., Hodges, H.F. (Eds.), Climate Change and Global Crop Productivity. CAB International, pp. 245–270.

Jensen, W.A., 1962. Botanical histochemistry. Freeman, San Francisco, CA.

Koch, K.E., Jones, P.H., Avigne, W.T., Allen Jr., L.H., 1986. Growth, dry matter partitioning, and diurnal activities of RuBP carboxylase in citrus seedlings maintained at two levels of CO_2. Physiol. Plant. 67, 477–484.

Kozlowski, T.T., 1973. Extent and significance of shedding of plant parts. In: Kozlowski, T.T. (Ed.), Shedding of Plant Parts. Academic Press, New York, pp. 1–44.

Lee, J.J., Neely, G.E., Perrigan, S.C., Grothaus, L.S., 1981. Effects of simulated acid rain on yield, growth and foliar injury of several crops. Environ. Exp. Bot. 21, 171–185.

Lee, J.J., Weber, D.E., 1979. The effect of simulated acid rain on seedling emergence and growth of eleven woody species. Forest Sci. 25, 393–398.

Millington, W.F., Chaney, W.R., 1973. Shedding of shoots and branching. In: Kozlowski, T.T. (Ed.), Shedding of Plant Parts. Academic Press, New York, pp. 149–204.

Mooney, H.A., 1986. Photosynthesis. In: Crawley, M.J. (Ed.), Plant Ecology. Blackwell, pp. 345–373.

Mousseau, M., Enoch, H.Z., 1989. Carbon dioxide enrichment reduces shoot growth in sweet chestnut seedlings (*Castanea sativa* Mill.). Plant Cell Environ. 12, 927–934.

Neufeld, H.S., Jernstedt, J.A., Haines, B.L., 1985. Direct foliar effects of simulated acid rain. I. Damage, growth and gas exchange. New Phytol. 99, 389–405.

Norby, R.J., O'Neill, E.G., 1989. Growth dynamics and water use in seedlings of *Quercus alba* L. in CO_2-enriched atmosphere. New Phytol. 111, 491–500.

O'Neill, E.G., Luxmoore, R.J., Norby, R.J., 1987. Elevated atmospheric CO_2 effects on seedlings growth, nutrient uptake and rhizosphere bacterial populations of *Liriodendron tulipifera* L. Plant Soil 104, 3–11.

Patterson, D.T., Flint, E.P., 1980. Potential effects of global atmospheric CO_2 enrichment on the growth and competitiveness of C_3 and C_4 weed and crop plants. Weed Sci. 28, 71–75.

Puppi, G., Paoletti, E., Manes, F., 1992. Effect of high CO_2 concentration and simulated acid rain on root parameters and mycorrhizal status of *Quercus ilex* seedlings. In: Kutschera, L., Hübl, E., Lichtenegger, E., Persson, H., Sobotik, M. (Eds.), Root Ecology and its Practical Application, 3. ISRR Symp. Wien, Univ. Bodenkultur, 1991, pp. 262–264.

Radoglou, K.M., Jarvis, P.G., 1990. Effects of CO_2 enrichment on four poplar clones. I. Growth and leaf anatomy. Ann. Bot. 65, 617–626.

Radoglou, K.M., Jarvis, P.G., 1992. The effects of CO_2 enrichment and nutrient supply on growth morphology and anatomy of *Phaseolus vulgaris* L. seedlings. Ann. Bot. 70, 245–256.

Radoglou, K.M., Jarvis, P.G., 1993. Effects of atmospheric CO_2 enrichment on early growth of *Vicia faba*, a plant with large cotyledons. Plant Cell Environ. 16, 93–98.

Rogers, H.H., Bingham, G.E., Cure, J.D., Smith, J.M., Surano, K.A., 1983. Responses of selected plant species to elevated carbon dioxide in the field. J. Environ. Qual. 12, 569–574.

Sionit, N., Strain, B.R., Hellmers, H., Reichers, G.H., Jaeger, C.H., 1985. Long-term atmospheric CO_2 enrichment affects the growth and development of *Liquidambar styraciflua* L. and *Pinus taeda* L. seedlings. Can. J. Forest Res. 15, 468–471.

Thomas, J.F., Harvey, C.N., 1983. Leaf anatomy of four species grown under continuous CO_2 enrichment. Bot. Gazzette 144, 303–309.

Tognetti, R., Johnson, J.D., 1999. Responses to elevated atmospheric CO_2 concentration and nitrogen supply of *Quercus ilex* L. seedlings from a coppice stand growing at a natural CO_2 spring. Ann. Forest Sci. 56, 549–561.

Tognetti, R., Johnson, J.D., Michelozzi, M., Raschi, A., 1998. Response of foliar metabolism in mature trees of *Quercus pubescens* and *Quercus ilex* to long-term elevated CO_2. Environ. Exp. Bot. 39, 233–245.

Tolley, L.C., Strain, B.R., 1984. Effects of CO_2 enrichment on growth of *Liquidambar styraciflua* and *Pinus taeda* seedlings under different irradiance levels. Can. J. Forest Res. 14, 343–350.

Trump, B.F., Smuckler, E.A., Benditt, E.P., 1961. A method for staining epoxy sections for light microscopy. J. Ultrastruct. Res. 5, 343–348.

Valladares, F., Martinez-Ferri, E., Balaguer, L., Perez-Corona, E., Manrique, E., 2000. Low leaf-level response to light and nutrients in Mediterranean evergreen oaks: a conservative resource-use strategy? New Phytol. 148, 79–91.

Watson, R.T., Rodhe, H., Oeschger, H., Siegenthaler, U., 1990. Greenhouse gases and aerosols. In: Houghton, J.T., Jenkins, G.T., Ephramus, J.J. (Eds.), Climate Change. The IPCC Scientific Assessment. Cambridge Univ. Press, Cambridge, pp. 1–40.

Winner, W.E., Mooney, H.A., Williams, K., von Lammerer, S., 1985. Measuring and assessing SO_2 effects on photosynthesis and plant growth. In: Winner, W.E., Mooney, H.A., Goldstein, R.A. (Eds.), Sulfur Dioxide and Vegetation. Stanford Univ. Press, Stanford, CA, pp. 118–132.

Wulff, R.D., Strain, B.R., 1982. Effects of CO_2 enrichment on growth and photosynthesis in *Desmodium paniculatum*. Can. J. Bot. 60, 1084–1091.

Yelle, S., Beeson Jr., R.C., Trudel, M.J., Gosselin, A., 1989. Acclimation of two tomato species to high atmospheric CO_2. I. Sugar and starch concentrations. Plant Physiol. 90, 1465–1472.

Air Pollution, Global Change and Forests in the New Millennium
D.F. Karnosky et al., editors

Chapter 22

Effects of elevated CO_2 and O_3 on aspen clones of varying O_3 sensitivity

B.A. Wustman

*Amicus Therapeutics Inc., Commercialization Center for Innovative Technologies,
675 US Highway One, North Brunswick, NJ 08902, USA*

E. Oksanen

*University of Kuopio, Department of Ecology and Environmental Science,
POB 1627, 70211 Kuopio, Finland*

D.F. Karnosky

*Michigan Technological University, School of Forest Resources and Environmental Science,
1400 Townsend Drive, Houghton, MI 49931-1295, USA*

A. Noormets

*The University of Toledo, Department of EEES, LEES Lab, Mail Stop 604,
Toledo, OH 43606, USA*

J.G. Isebrands

*Environmental Forestry Consultants, LLC, P.O. Box 54, E7323 Hwy 54,
New London, WI 54501, USA*

K.S. Pregitzer

*Michigan Technological University, School of Forest Resources and Environmental Science,
1400 Townsend Drive, Houghton, MI 49931-1295, USA*

G.R. Hendrey

*Brookhaven National Lab, Division of Environmental Biology and Instrumentation,
1 South Technology Street, Upton, NY 11973, USA*

J. Sober

*Michigan Technological University, School of Forest Resources and Environmental Science,
1400 Townsend Drive, Houghton, MI 49931-1295, USA*

G.K. Podila* *

*University of Alabama, Department of Biological Sciences, 301 Sparkman Drive,
WH142, Huntsville, AL 35899, USA*

---

* Corresponding author.

DOI:10.1016/S1474-8177(03)03022-5

Abstract

To determine whether elevated CO_2 reduces or exacerbates the detrimental effects of O_3 on aspen (*Populus tremuloides* Michx.). Aspen clones 216 and 271 (O_3 tolerant), and 259 (O_3 sensitive) were exposed to ambient levels of CO_2 and O_3 or elevated levels of CO_2, O_3, or $CO_2 + O_3$ in the FACTS II (Aspen FACE) experiment, and physiological and molecular responses were measured and compared. Clone 259, the most O_3-sensitive clone, showed the greatest amount of visible foliar symptoms as well as significant decreases in chlorophyll, carotenoid, starch, and ribulose-1,5-bisphosphate carboxylase/oxygenase (Rubisco) concentrations and transcription levels for the Rubisco small subunit. Generally, the constitutive (basic) transcript levels for phenylalanine ammonia-lyase (Pal) and chalcone synthase (Chs) and the average antioxidant activities were lower for the ozone sensitive clone 259 as compared to the more tolerant 216 and 271 clones. A significant decrease in chlorophyll a, b and total (a + b) concentrations in CO_2, O_3, and $CO_2 + O_3$ plants was observed for all clones. Carotenoid concentrations were also significantly lower in all clones; however, Chs transcript levels were not significantly affected, suggesting a possible degradation of carotenoid pigments in O_3-stressed plants. Antioxidant activities and Pal and 1-aminocyclopropane-1-carboxylic acid (ACC)-oxidase transcript levels showed a general increase in all O_3 treated clones, while remaining low in CO_2 and $CO_2 + O_3$ plants (although not all differences were significant). Our results suggest that the ascorbate-glutathione and phenylpropanoid pathways were activated under ozone stress and suppressed during exposure to elevated CO_2. Although $CO_2 + O_3$ treatment resulted in a slight reduction of O_3-induced leaf injury, it did not appear to ameliorate all of the harmful affects of O_3 and, in fact, may have contributed to an increase in chloroplast damage in all three aspen clones.

1. Introduction

Atmospheric CO_2 is increasing rapidly and is expected to double by the end of the next century (Barnola et al., 1995). Tropospheric ozone (O_3) is also increasing globally at a rate of 1–2% per year (Chameides et al., 1995) and it is likely that these two gases will have significant impacts in the future on forest tree species and ecosystems (Matyssek and Innes, 1999; Reilly et al., 1999). However, it is not clear at this point whether or not elevated CO_2 reduces or exacerbates the detrimental effects of O_3. Moreover, leaf age can greatly affect the direction of the $CO_2 + O_3$ response. McKee et al. (1995) showed no protective effect of elevated CO_2 against O_3 on emerging wheat flag leaves, but strong effects in mature leaves. Experimental results provide evidence for both hypotheses, but neither has been tested under open field conditions, and both could be strongly influenced by the lower boundary layer conductances that characterize enclosures (Polle and Pell, 1999). There is clearly a need to test this in an open-air facility, such as FACE, where artifacts are minimized. The view that elevated CO_2 should reduce the detrimental effects of O_3 is based primarily on the expected decrease in stomatal conductance in elevated CO$_2$, which reduces O$_3$ flux received by the plant. The increase in substrate availability for repair and detoxification at elevated CO$_2$ may also reduce the effect of elevated O$_3$ (Rao et al., 1995).

The alternate hypothesis, that elevated CO$_2$ will exacerbate the detrimental effects of O$_3$, is based primarily on the prediction that reduced photorespiration, decreased Rubisco content, increased availability of CO$_2$ for photosynthesis, and a decrease in non-photochemical energy dissipation will reduce the need for cellular detoxification of reactive oxygen species (Polle et al., 1993). Thus, the lower antioxidant levels result in reduced tolerance to O$_3$. In aspen, there is some evidence of antioxidants playing a key role in O$_3$ tolerance (Sheng et al., 1997; Noormets et al., 2000) and an apparent increase in O$_3$ sensitivity with elevated CO$_2$ (Kull et al., 1996). In aspen softwood cuttings grown in open-top chambers, CO$_2$ did not compensate for the deleterious effects of elevated O$_3$ and, in some cases, photosynthetic capacity decreased more than with O$_3$ alone (Kull et al., 1996). This was particularly true for the O$_3$-sensitive clone. Moreover, the tolerant clones sometimes became more sensitive to O$_3$ with CO$_2$ enrichment (Kull et al., 1996; Karnosky et al., 1998). Data from various tree species is conflicting on the interacting effects of O$_3$ and CO$_2$. These findings fall into two categories, (1) the elevated CO$_2$ has some positive effects in ameliorating the negative effects of O$_3$ (Mortensen, 1995; Kellomaki and Wang, 1997; Dickson et al., 1998; Sehmer et al., 1998; Grams et al., 1999; Utrainen et al., 2000; Broadmeadow et al., 1999) or (2) it increases the negative effects of O$_3$ (Kull et al., 1996; Karnosky et al., 1998; Niewiadomska et al., 1999).

To better understand the interactive affects of increased CO$_2$ and O$_3$ levels, we have measured various leaf antioxidant activities and gene expression patterns of several stress-response pathways for aspen trees growing in a Free Air CO$_2$ and/or O$_3$ Enrichment (FACE) experiment. Chlorophyll, carotenoid, and Rubisco contents have also been measured to help determine how the photosynthetic capacities are affected for clones of differing O$_3$ sensitivity. Results from this open air experiment indicate that elevated CO$_2$ levels increase O$_3$ sensitivity in all three clones by suppressing ascorbate-glutathione and phenylpropanoid pathways. An increase in AOS, consequently, may be responsible for the observed decreases in chlorophyll and carotenoid pigments.

2. Materials and methods

2.1. Plant material and fumigation protocols

The research facility is located in the USDA Forest Service Harshaw farm site in Oneida County, Wisconsin (Karnosky et al., 1999; Dickson et al., 2000).

Three replicate FACE rings were established for each treatment (control (= ambient), elevated O_3, elevated CO_2, and elevated $O_3 + CO_2$). In elevated-O_3 rings, O_3 concentrations were maintained approximately 1.5 times higher than the ambient concentrations. The mean O_3 concentrations during the 1998 growing season were 36 ppb for ambient air and 56 ppb for elevated O_3 exposures. The monthly AOT40 values ranged from 5.2–9.2 ppm-h (mean 7.0 ppm-h) for elevated O_3 treatments and ranged from 25.9–30.6 ppm-h for the seasonal dose (Table 1). The mean concentrations and total cumulative O_3 exposures over a threshold of 0 ppb (AOT00) and of 40 ppb (AOT40) calculated for daylight hours for each exposure ring are given in Table 1. Carbon dioxide was dispersed during daytime hours at 560 ppm, approximately 200 ppm above our ambient. The plants were established at 1 m × 1 m spacing across each ring to simulate an aggrading aspen forest. Details regarding the aspen clones are presented in Karnosky et al. (1998) and Sheng et al. (1997) and the planting and handling of the aspen clones is described in Karnosky et al. (1999).

The same plants were used for biomaterial measurements, determinations of visible leaf injury, antioxidant activities, chlorophyll, carotenoid, rubisco and starch concentrations, and gene expression levels (Rubisco small subunit and defense-related proteins), and therefore the data from different analyses are comparable. In addition, chlorophyll, carotenoid and rubisco concentrations were determined for the same leaf extraction samples.

2.2. Samples

Foliar samples from aspen were collected in late July of 1998 from each treatment ring as close as possible to mid-day (noon). Samples were collected from 6–10 trees in each treatment from each of the three replicates. The samples were immediately frozen in liquid nitrogen and stored at $-80\,^{\circ}C$ until use. Leaf samples were collected from the developing leaf zone (LPI 3 to 5 (Larson and Isebrands, 1971)) and care was taken to collect same age leaf samples from each tree.

2.3. Sample extraction and analysis

Frozen leaves (0.25 g) were ground with liquid nitrogen and extracted briefly with 6 ml of cold 100 mM potassium phosphate buffer (pH 7.0) containing 0.1 mM EDTA and 1% (w/v) insoluble polyvinylpyrrolidone. The homogenate was centrifuged at $3000g$ ($4\,^{\circ}C$) for 2 min, 1 ml of supernatant was added to a 1 ml column of P2 Biogel (Bio Rad) and centrifuged at $600g$ for 1 min. Biogel purified samples were immediately assayed for SOD, APX and CAT activity. Samples assayed for GR activity were extracted as described above,

Table 1. Mean O$_3$ concentrations (ppb), total cumulative O$_3$ exposures over a threshold of 0 ppb (AOT00, ppm-h) and over a threshold of 40 ppb (AOT40, ppm-h) calculated for daylight hours during the 1998 growing season. Ambient O$_3$ was calculated for three points (North, East and South) within the exposure area

Month	O$_3$	Ambient			Elevated O$_3$			Elevated O$_3$ + CO$_2$		
		North	East	South	1.3	2.3	3.3	1.4	2.4	3.4
May	Mean ppb	44	40	40	52	54	53	52	53	50
	AOT00 ppm-h	12.6	11.5	11.5	17.9	18.6	18.1	17.2	18.3	17.5
	AOT40 ppm-h	0.3	0.3	0.3	5.7	6.6	6.2	6.1	6.3	5.2
June	Mean ppb	35	31	32	55	57	55	55	57	55
	AOT00 ppm-h	12.5	11.3	11.7	19.7	20.5	19.7	19.9	20.3	19.7
	AOT40 ppm-h	0	0	0	6.7	7.9	6.9	7.3	7.9	6.7
July	Mean ppb	32	27	31	57	58	56	55	57	57
	AOT00 ppm-h	11.9	10	11.5	21.2	21.6	20.8	20.5	21.2	21.2
	AOT40 ppm-h	0	0	0	7.6	8.0	7.1	7.2	7.9	7.5
August	Mean ppb	44	37	41	59	58	59	64	62	57
	AOT00 ppm-h	13.6	11.4	12.6	18.4	18	18.3	19.8	19.1	17.8
	AOT40 ppm-h	0	0	0	7.0	8.1	7.0	9.2	8.4	6.5
May–August	Mean ppb	39	34	36	56	57	56	57	57	55
	AOT00 ppm-h	50.6	44.2	47.3	77.2	78.7	76.9	77.4	78.9	76.2
	AOT40 ppm-h	0.3	0.3	0.3	27.0	30.6	27.2	29.8	30.5	25.9

using 40 mM potassium phosphate buffer (pH 7.4) containing 1 mM EDTA, 3% (w/v) insoluble polyvinylpyrrolidone, and 40 mM 3-amino-1, 2,4-triazole. Homogenate was centrifuged at 6000g for 10 min (4 °C) and the supernatant was assayed for GR activity.

2.3.1. Antioxidant enzymes

Aspen clones (from samples collected in July) were analyzed to determine the levels of antioxidant enzymes. Biochemical analyses included enzyme assays for SOD (Dhindsa et al., 1981), ascorbate peroxidase (Chen and Asada, 1989), GR activity (Price et al., 1990), and catalase (Kato and Shimizu, 1987). All the biochemical analyses were normalized for age of the plants, weight of the sampled foliar material and for amount of total proteins or cell lysates used for each analysis. Assays were repeated on each extracted sample at least 3 times.

2.3.2. Rubisco, soluble protein, starch and pigments

For Rubisco and soluble protein assays, frozen leaf samples were weighed, and a crude extract was prepared using 2 ml of extraction buffer (Gezelius and Hallen, 1980). The amount of total Rubisco protein was determined by PAGE as described by Rintamäki et al. (1988), using purified Rubisco protein (Sigma Chemical Co.) as a standard. The areas and intensities of Rubisco bands were determined by scanning the gels with the Adobe PhotoShop Program (Version 5.0), and the Rubisco concentrations were calculated on a dry weight basis. An aliquot for soluble proteins was quantified as described by Pääkkönen et al. (1998). Chlorophyll analysis was done as per the protocol of Porra et al. (1989) and carotenoid measurements were done as per the protocol of Wellburn and Lichtenthaler (1984). In order to avoid variability between samples for Rubisco and protein analysis and chlorophyll and associated pigments, samples for pigments were taken from the crude extract before centrifugation and subjected to proper extraction protocol for either chlorophyll or carotenoid pigments. Leaf samples for starch determination were freeze-dried, milled, and analyzed by standard enzymatic techniques (Boehringer Kit for Food analysis).

2.3.3. Gene expression

Leaf material (1–2 g) was ground in liquid nitrogen, and added to the pre-warmed extraction buffer (100 mM Tris-HCl pH 8.0, 20 mM EDTA, 0.5 M NaCl, 2% PVP, 0.5% SDS, 0.5% β-mercaptoethanol). After three extractions with chloroform: IAA (24 : 1), 1/5 × volume of 10 M LiCl was added and the RNA was precipitated one hour on ice. The pellet was dissolved into DEPC-TE, extracted once more with chloroform:IAA, and reprecipitated in ethanol at $-80\,°C$ overnight. The RNA pellet was purified with 70% ethanol, dried and resuspended in DEPC-treated water. RNA quality was checked by gel electrophoresis, and concentration determined by GeneQuant II RNA/DNA Calculator (Pharmacia Biotech). Single-stranded digoxigenin (DIG)-labeled DNA-probes were generated by PCR using DynaZyme DNA-polymerase (Finnzymes, Finland) and DIG-DNA labeling mixture (Boehringer Mannheim, Germany). Birch RbcS, Pal, Chs, and AccOx probes (see Pääkkönen et al., 1998) were used after careful optimization of hybridization conditions for aspen samples. RNA hybridizations were performed according to the DIG System User's Guide for Filter Hybridization (Boehringer Mannheim, Germany). Total RNA, 10 µg each per lane, was separated in formaldehyde containing 1% agarose gel. Equivalent loading of RNA was verified by SYBR-Green staining (Molecular Probes Inc., OR). After gel electrophoresis, RNA was transferred to nylon membranes overnight at room temperature with $10 \times SSC$ (1.5 M NaCl, 150 mM sodium citrate pH 7.0). RNA was fixed to the filter by UV-cross-linking, followed by washing in $2 \times SSC$ for 10 minutes. The filters were prehybridized for $1\frac{1}{2}$ h at $48\,°C$ in DIG Easy Hyb (Boehringer Mannheim, Germany) and hybridized overnight at $48\,°C$ with denatured DIG-labeled purified DNA-probes. Filters were washed in $2 \times SSC$, 0.1% SDS at room temperature (2×5 min) and in $0.1 \times SSC$, 0.1% SDS at $68\,°C$ (3×15 min). Hybridization signals were visualized using the DIG Luminescent Detection Kit (Boehringer Mannheim, Germany). Membranes were incubated in blocking reagent and treated with anti-DIG-alkaline phosphatase-conjugate. The chemiluminescent substrate was pipetted onto the filter and the light signal was recorded on X-ray films (Fig. 1). The relative intensities of stress protein inductions were determined by scanning the X-ray films and measuring the band densities with Adobe PhotoShop program (Version 5.0). The relative band densities were compared only within the film and the band intensities were indicated as percentage of the maximal intensity on the film.

2.3.4. Growth analysis

Fifteen plants per clone per treatment were measured between 27 and 31 July 1998 for base diameter (mm), number of leaves per branch (3–6th branch from the top), mean leaf size (cm^2; 9–11th leaf from the top) and proportion of visibly injured leaves (%; determined for the 3–6th branch from the top) as described (Karnosky and Steiner, 1980; Karnosky et al., 1996). Visible injuries were necrotic or dark-brown pigmented dots and flecks appearing on the upper leaf surface.

B.A. Wustman et al.

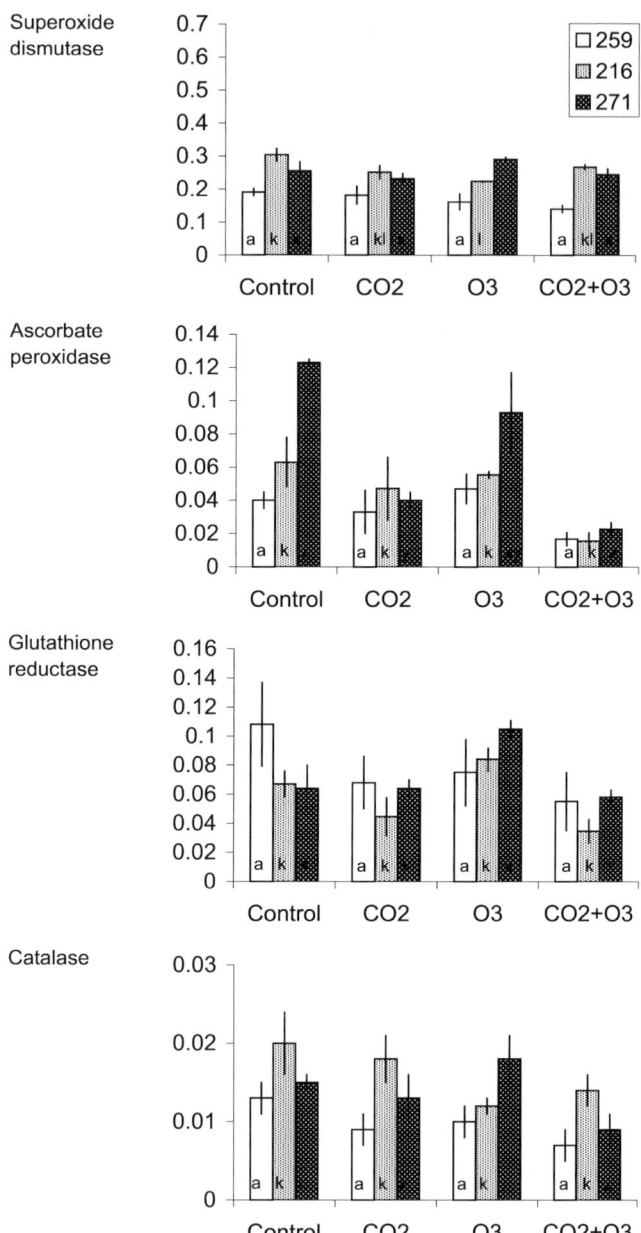

Figure 1. Effects of elevated O_3 and CO_2 alone or in combination on SOD, APX, GR and CAT activities in aspen clones 259, 216 and 271. Data for three replicate FACE rings/treatment were pooled and expressed as standard deviations.

Table 2. Effects of elevated O_3 and CO_2 alone and in combination on growth parameters in aspen (*Populus tremuloides*) clones. Significant block effects were not found, and therefore the data for three replicate FACE rings/treatment were pooled for further analyses. Anova, Tukey's test. $P < 0.05$, $n = 15$. Values are means ± S.E. Significant differences between the treatments are indicated by different letters

Response	Clone	Control	CO_2	O_3	$CO_2 + O_3$
Base diameter, mm	259	18.8 ± 0.4 bc	20.3 ± 0.4 c	16.9 ± 0.5 a	20.1 ± 0.7 c
	216	22.2 ± 0.4 b	24.4 ± 0.4 c	18.9 ± 0.3 a	22.2 ± 0.4 b
	271	22.4 ± 0.5 b	22.6 ± 0.4 b	19.5 ± 0.5 a	21.9 ± 0.4 b
Number of leaves/branch	259	37.5 ± 1.0 b	36.5 ± 0.9 b	23.6 ± 0.6 a	23.2 ± 0.7 a
	216	34.8 ± 0.6 b	41.2 ± 1.2 c	23.9 ± 0.7 a	24.4 ± 0.7 a
	271	39.8 ± 0.9 b	39.9 ± 1.5 b	26.4 ± 0.7 a	25.5 ± 0.7 a
Mean leaf size, cm^2	259	80.9 ± 4.7 a	67.8 ± 2.8 a	70.8 ± 5.3 a	70.1 ± 4.4 a
	216	67.4 ± 2.2 b	61.7 ± 3.4 ab	57.6 ± 2.1 ab	56.4 ± 2.9 a
	271	66.6 ± 3.0 b	63.1 ± 3.6 ab	59.8 ± 1.9 ab	54.3 ± 2.1 a
Injured leaves/branch, %	259	0.0 ± 0.0 a	0.0 ± 0.0 a	10.76 ± 0.65 b	8.11 ± 0.65 b
	216	0.0 ± 0.0 a	0.0 ± 0.0 a	8.42 ± 0.72 b	4.26 ± 0.54 b
	271	0.0 ± 0.0 a	0.0 ± 0.0 a	7.51 ± 0.71 b	3.30 ± 0.46 b

3. Results

3.1. Leaf growth and visible injury

Ozone treated plants of all three clones had significantly smaller base diameters, average number of leaves per branch, and mean leaf sizes when compared to their respective controls, and similar results were obtained for average leaf sizes and leaves per branch of $CO_2 + O_3$ treated plants (Table 2). Average leaf sizes for clones 216 and 271 were most significantly affected by $CO_2 + O_3$, indicating an increased sensitivity to O_3 in the presence of elevated CO_2 (Table 2). However, visible leaf injury (average number of injured leaves per branch as well as injury index per leaf measured as described by Karnosky and Steiner, 1980; Karnosky et al., 1996) was significantly higher for O_3 treated plants when compared to $O_3 + CO_2$ treated plants, and clone 259 showed the greatest amount of leaf injury (Table 2).

3.2. Antioxidant activities

With the exception of samples from clone 259, SOD activities varied little, while APX, GR and CAT activities were generally highest for O_3 treated plants and lowest for CO_2 and $O_3 + CO_2$ treated plants for all three clones when compared to controls (Fig. 1).

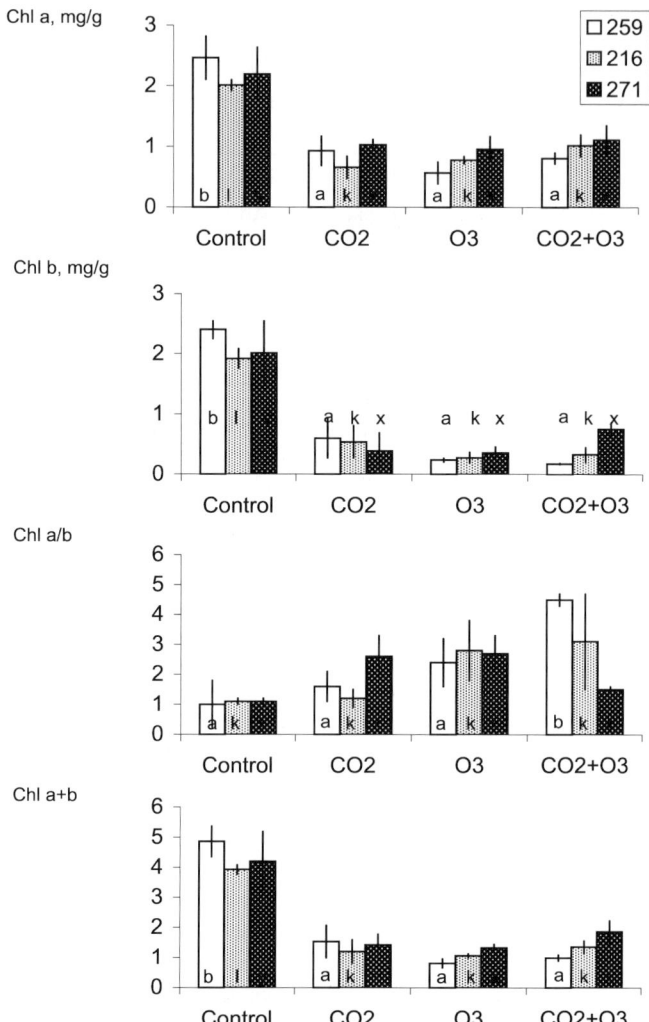

Figure 2. Effects of elevated O_3 and CO_2 alone or in combination on chl a and b, total chl (mg/g dwt), carotenoid (µg/g dwt), total Rubisco (mg/g dwt), and starch (mg/g dwt) concentrations, and chl a/b ratio in aspen clones 259, 216 and 271. Anova, Tukey's test. $P < 0.05$, $n = 15$. Values are means ± S.E. Significant differences between the treatments are indicated by different letters. (*Continued on next page*).

3.3. Pigment, rubisco, and starch contents

Ozone, CO_2, and $O_3 + CO_2$ treatments had significant effects on Chl a and Chl b content, whereas Rubisco content was not significantly affected (Fig. 2).

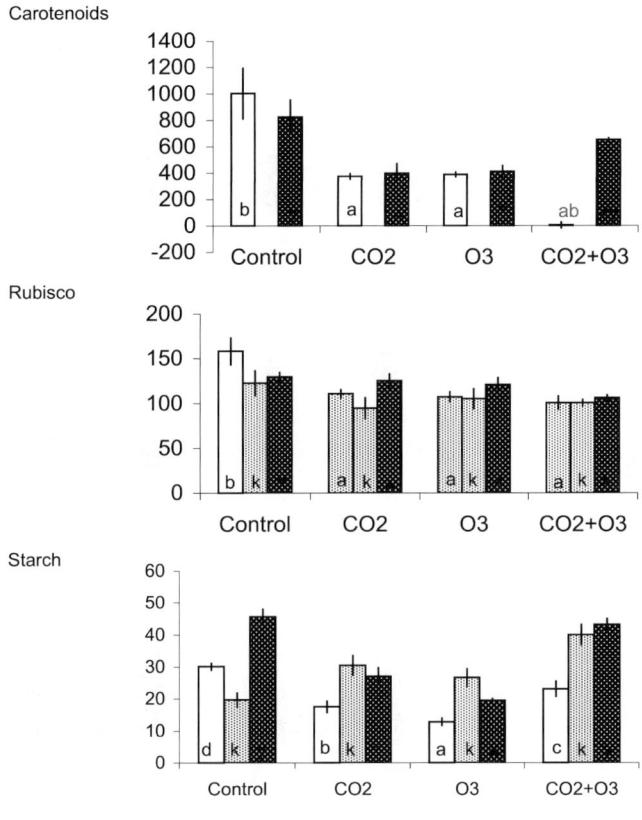

Figure 2. (*Continued*)

Chl a, chl b, and total chl (a + b) contents significantly decreased for all CO_2, O_3, and $CO_2 + O_3$ treated clones (Fig. 2). Chl a/b ratios only varied significantly for clone 259, with an increase (low chl b content) (Fig. 2). Total carotenoid and Rubisco concentrations generally decreased in CO_2, O_3, and $CO_2 + O_3$ treated plants. Starch concentrations for clones exposed to all three treatments, exhibited some trends, with CO_2, O_3, and $CO_2 + O_3$ plants generally having higher starch concentrations in clone 216 and lower concentrations in clones 259 and 271 when compared to controls (Fig. 2).

3.4. Transcription levels of RbcS, Chs, Pal, AccOx

Expression levels of RbcS, for all clones and all treatments, generally decreased compared to controls with the exception of O_3 treated clone 271 plants, which showed a significant increase (Figs. 3, 4). These results cor-

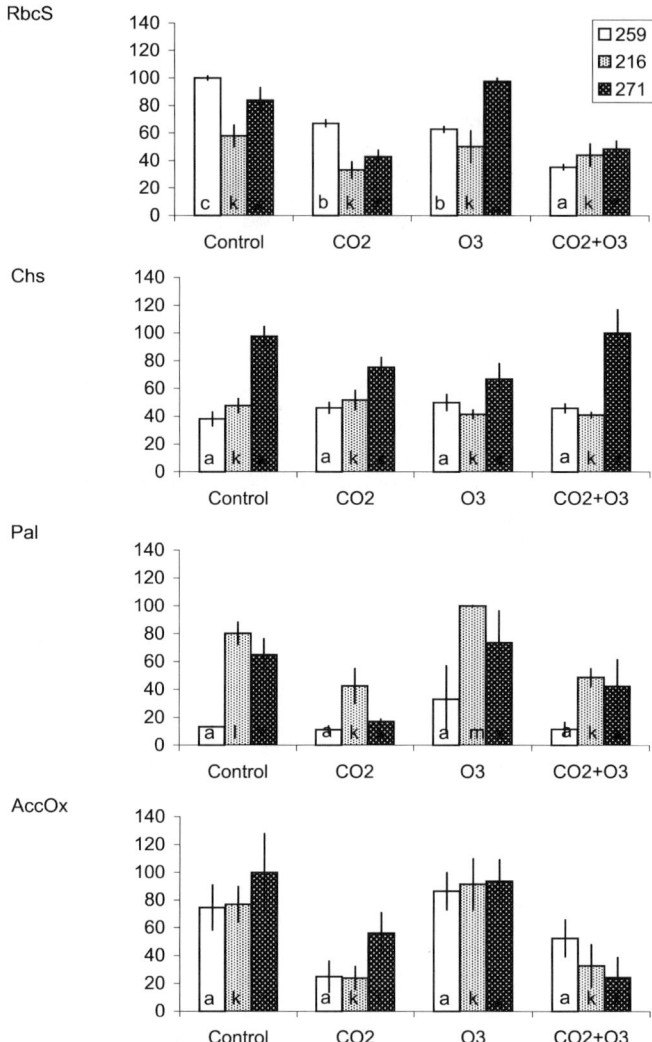

Figure 3. Transcript levels of Rubisco small subunit (RbcS), chalcone synthase (Chs), pheny-lalanine ammonia-lyase (Pal) and AccOx (means ± S.E.) in leaves of aspen clones 259, 216 and 271. Maximum density of the band in each gel = 100. Kruskall–Wallis test. Significant differences between the treatments are indicated by different letters.

relate, most significantly for clone 259, with the Rubisco contents shown in Fig. 2, however, transcription levels do not appear to correlate with starch contents (Figs. 2, 3). Chs expression levels did not vary significantly, except for

Clone	259	216	271

Treatment

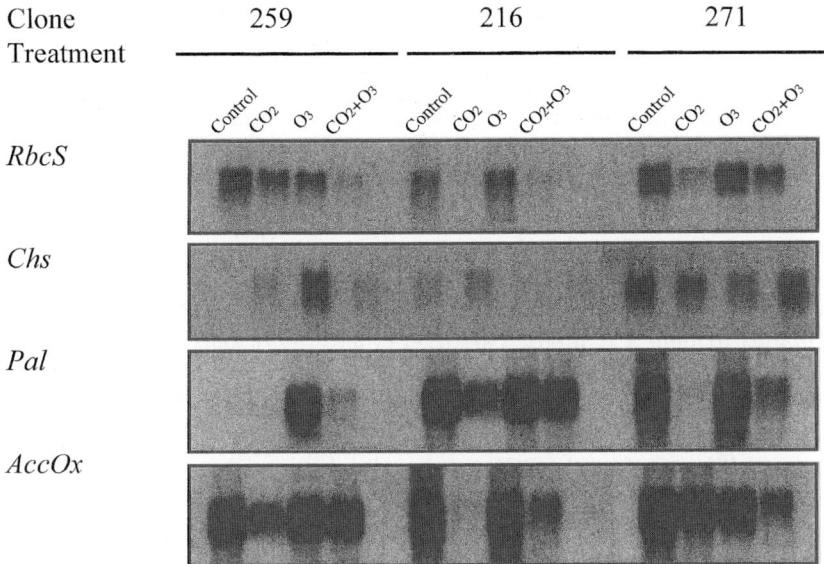

| | Control | CO$_2$ | O$_3$ | CO$_2$+O$_3$ | Control | CO$_2$ | O$_3$ | CO$_2$+O$_3$ | Control | CO$_2$ | O$_3$ | CO$_2$+O$_3$ |

RbcS

Chs

Pal

AccOx

Figure 4. Northern analysis of RbcS, Pal, Chs, and AccOx in aspen clones. Total RNA (10 μg each) from aspen clones 259, 216, and 271 was probed with single-stranded digoxigenin (DIG)-labeled Birch RbcS, Pal, Chs, and AccOx DNA probes.

O$_3$ treated clone 271 plants, which showed a significant increase compared to controls. Both clones 259 and 271 showed a significant decrease in total carotenoid contents under CO$_2$ and O$_3$ treatments (Fig. 2). The combination of CO$_2$ + O$_3$ had a rather dramatic effect in the decrease of carotenoid component on 259, but not on gene expression of Chs. Pal expression levels of all three clones were significantly increased by O$_3$ treatments and significantly decreased by CO$_2$ and CO$_2$ + O$_3$ treatments. Although the differences are not significant, changes in AccOx expression levels correlate with changes in Pal mRNA levels.

4. Discussion

Polle et al. (1997) provided evidence that oxidative stress in beech leaves (*Fagus sylvatica* L.) increases when N and/or C assimilation was limited. Podila et al. (1998) found that clones grown in soils with the lowest average NH$_4^+$ and C:N ratios demonstrated a slight decrease in antioxidant activity. However, trends in the effects of elevated CO$_2$ on antioxidant activities could still be seen in these aspen clones. Furthermore, Coleman et al. (1998) and Sober et al. (2001), found that N levels were linearly related to photosynthesis.

It has been reported that elevated CO_2 may result in decreased production of antioxidant metabolites (e.g., glutathione and ascorbate) and in antioxidant enzymes (Polle et al., 1993; Badiani et al., 1993; Schwanz et al., 1996; Polle et al., 1997; Polle and Pell, 1999) including aspen (Karnosky et al., 1998), while O_3 generally increases antioxidant activities. Furthermore, several reports have demonstrated that overexpression of SOD increases resistance to oxidative stress (Jansen et al., 1989; Sen Gupta et al., 1993a, 1993b), and that APX activities also increased in the leaves of transformed plants. In aspen (*Populus tremula* x *P. alba*), Tyystjärvi et al. (1999) found that an 5–8 fold increase in FeSOD expression was unable to protect the plants from photoinhibition, while tobacco plants, with a similar increase in GR expression levels, had a lower level of photoinhibition. Thus, the ascorbate-glutathione pathway may not be limited by the SOD catalyzed reduction of superoxide to H_2O_2, but instead by the conversion of H_2O_2 to H_2O by APX (and catalases) and the reduction of oxidized dehydroascorbate and glutathione by dehydroascorbate reductase and glutathione reductase. In our experiments we found that APX, CAT and GR activities for all three clones generally increased as a result of exposure to elevated O_3, and decreased due to elevated CO_2 and $CO_2 + O_3$ levels, suggesting a suppression of the ascorbate-glutathione pathway by CO_2.

A decrease in antioxidant activities would decrease the plants primary defense against AOS, and thus, an increase of oxyradicals within the chloroplast may account for the significant decreases observed for chl a and b and total carotenoid contents under elevated O_3, especially since transcription levels of Chs were not significantly affected. This may also account for the decrease in Rubisco content. However, RbcS expression levels were also significantly lower in plants treated with CO_2, O_3, and $CO_2 + O_3$. While it appears that more damage within the chloroplast occurs due to elevated CO_2 and O_3 levels, it was found in our work and in previous reports (Karnosky et al., 1999; Tjoelker et al., 1993; Volin et al., 1993) that CO_2 can reduce the amount of visible leaf injury. It has also been shown that CO_2 increases the light saturated photosynthetic rate for aspen (Karnosky et al., 1999; Volin and Reich, 1996; Volin et al., 1998); thus, it is possible that, in the presence of CO_2, an increased rate of photosynthesis increases the concentration of available NADPH in the chloroplast, which can then be used by GR to reduce the glutathione pool and increase the flux through the ascorbate-glutathione cycle. However, leaf size was most significantly reduced for $CO_2 + O_3$ treated plants, suggesting that photosynthesis may be inhibited. Thus, it is also possible that an increase in AOS, due to suppression of antioxidants as discussed above, may be offset by a decrease in rate of photosynthesis which may only occur in $CO_2 + O_3$ treated plants, resulting in a decrease in both visible leaf injury and leaf growth.

Ozone sensitivity in *Populus* may be related to a lack of defense gene activation (Koch et al., 1998, 2000). Salicylic acid (SA) and jasmonic-acid-mediated (JA) signaling pathways, known mediators of pathogen and wound responses, may also be involved in responses to oxidative stress (Örvar et al., 1997; Koch et al., 1998, 2000; Rao et al., 1997, 2000). Our results indicate that Pal expression may be suppressed by CO$_2$ since Pal mRNA levels were lowest for CO$_2$ and CO$_2$ + O$_3$ treated plants (all clones). Suppression of the phenylpropanoid pathway, and a resulting decrease in various pathogen defense compounds, could increase the plants sensitivity to secondary pathogens especially when exposed to O$_3$ in combination with elevated CO$_2$. This may explain the increased rate of infection by a sooty mold fungus, (*Alternaria* spp.) observed for all three clones grown at field sites with intermediate and high O$_3$ levels (Karnosky et al., 1999) and a 3- to 5-fold increase in *Melampsora* leaf rust in aspen exposed to O$_3$ or O$_3$ + CO$_2$ (Karnosky et al., 2001). Just as Karnosky et al. (1999) found clone 259 to be most dramatically affected by the secondary infection, we found that Pal expression was most significantly suppressed in clone 259. However, increased infection levels for clone 259 correlated well with increased visible leaf damage (Karnosky et al., 1999), suggesting that O$_3$-induced foliar injury may result in a high susceptibility to secondary infections. It is also probable that clone 259 may be inherently less sensitive to SA-mediated signaling which may explain for its susceptibility to both O$_3$ and pathogens (Koch et al., 2000; Overmyer et al., 2000). On the other hand, clones 216 and 271 may be more sensitive to SA-mediated signaling resulting in launching a more effective defense against O$_3$-mediated damage and/or pathogens.

The observation that AccOx levels were reduced under elevated CO$_2$ conditions for all three clones suggests that under elevated CO$_2$ conditions the production of ethylene may be reduced, leading to less visible injury due to oxidative damage as well as damage by pathogens. Recently, it has been shown that in *Arabidopsis* ethylene promotes SA-mediated (ozone-induced) lesion formation and the ozone-induced AccOx transcript levels match with ozone-induced ethylene production (Overmyer et al., 2000). Similarly, the reduction in AccOx levels under CO$_2$ + O$_3$ may have a role in reducing ethylene induced SA-mediated or O$_3$-induced damage in clones 216 and 271. However, the low level of reduction for AccOx in clone 259 may yet lead to increased visible damage under CO$_2$ + O$_3$ conditions.

5. Conclusions

Although responses to elevated CO$_2$, O$_3$, and CO$_2$ + O$_3$ were similar in all clones, clone 259 showed the highest sensitivity to all treatments. Our results suggest that elevated CO_2 levels may exacerbate the harmful effects of O_3 by suppressing the ascorbate-glutathione and Pal pathways, resulting in increased cellular damage as a result of increasing AOS levels and a reduced resistance to secondary infections. Degradation due to an increase in AOS, resulting from a decrease in antioxidant activities, was most likely responsible for the observed decrease in chlorophyll and carotenoid contents, while decreased expression levels of RbcS correlated well with decreased Rubisco contents.

Acknowledgements

This research was partially supported by the USDA Forest Service Global Change Program, the US Department of Energy, Office of Biological and Environmental Research (BER) (DE-FG02-95ER62125 and DE-FG02-93ER6166), and the Brookhaven National Labs/US Department of Energy (725079), the National Science Foundation (DBI-9601942, DBI-9601942, and IBN-9652675), the National Council of Air and Stream Improvement of the Pulp and Paper Industry (NCASI), and the Academy of Finland.

References

Badiani, M., D'Annibale, A., Paolacci, A.R., Miglietta, F., Raschi, A., 1993. The antioxidant status of soybean (*Glycine max*) leaves grown under natural CO_2 enrichment in the field. Aust. J. Plant Physiol. 20, 275–284.

Barnola, J.M., Anklin, M., Porheron, J., Raynaud, D., Schwander, J., Stauffer, B.T.I., 1995. CO_2 evolution during the last millennium as recorded by Antartic and Greenland ice. Tellus B 47, 264–272.

Broadmeadow, M.S.J., Heath, J., Randle, T.J., 1999. Environment limitations to O_3 uptake—some key results from young trees growing at elevated CO_2 concentrations. In: Sheppard, L.J., Cape, J.N. (Eds.), Forest Growth Responses to the Pollution Climate of the 21st Century. Water Air Soil Pollut., pp. 299–310.

Chameides, W.L., Kasibhatla, P.S., Yienger, J., Levy II, H., 1995. Growth of continental-scale metro-agro-plexes, regional ozone pollution, and world food production. Science 264, 74–77.

Chen, G.S., Asada, K., 1989. Ascorbate peroxidase in the tea leaves: Occurrence of two isozymes and the differences in their enzymatic and molecular properties. Plant Cell Physiol. 30, 987–998.

Coleman, M.D., Dickson, R.E., Isebrands, J.G., 1998. Growth and physiology of aspen supplied with different fertilizer addition rates. Physiol. Plant. 103, 513–526.

Dickson, R.E., Coleman, M.D., Riemenschneider, D.E., Isebrands, J.G., Hogan, G.D., Karnosky, D.F., 1998. Growth of five hybrid poplar genotypes exposed to interacting elevated CO_2 and O_3. Can. J. Forest Res. 28, 1706–1716.

Dickson, R.E., Lewin, K.F., Isebrands, J.G., Coleman, M.D., Heilman, W.E., Riemenschneider, D.E., Sober, J., Host, G.E., Zak, D.F., Hendrey, G.R., Pregitzer, K.S., Karnosky, D.F., 2000. Forest atmosphere carbon transfer storage-II (FACTS II)—The aspen free-air CO_2 and O_3 enrichment (FACE) project in an overview. USDA Forest Service North Central Research Station. General Tech. Rep. NC-214.

Dhindsa, R.S., Plumb-Dhindsa, P., Thorpe, T.A., 1981. Leaf senescence correlated with increased levels of membrane permeability and lipid peroxidation, and decreased levels of superoxide dismutase and catalase. J. Exp. Bot. 32, 93–101.

Gezelius, K., Hallen, M., 1980. Seasonal variation in ribulose bisphosphate carboxylase activity in *Pinus sylvestris*. Physiol. Plant. 48, 88–98.

Grams, T.E.E., Anegg, S., Haberle, K.H., Langebartels, C., Maytessek, R., 1999. Interactions of chronic exposure to elevated CO_2 and O_3 levels in the photosynthetic light and dark reactions of European beech (*Fagus sylvatica*). New Phytol. 144, 95–107.

Jansen, A.K.J., Shaaltiel, Y., Kazzes, D., Canaani, O., Malkin, S., Gessel, J., 1989. Increased tolerance to photoinhibitory light in paraquat-resistant *Conyza bonariensis* measured by photoacoustic spectroscopy and $^{14}CO_2$-fixation. Plant Physiol. 91, 1174–1178.

Karnosky, D.F., Steiner, K.C., 1980. Provenance and family variation in response of *Fraxinus americana* and *F. pennsylvanica* in response to ozone and sulfur dioxide. Phytopathology 71, 804–807.

Karnosky, D.F., Gagnon, Z.E., Dickson, R.E., Coleman, M.D., Lee, H., Isebrands, J.G., 1996. Changes in growth, leaf abscission and biomass associated with seasonal tropospheric ozone exposures of *Populus tremuloides* clones and seedlings. Can. J. Forest Res. 26, 23–37.

Karnosky, D.F., Podila, G.K., Gagnon, Z., Pechter, P., Akkapeddi, A., Sheng, Y., Riemenschneider, D.E., Coleman, M., Dickson, R.E., Isebrands, J.G., 1998. Genetic control of responses to interacting tropospheric ozone and CO_2 in *Populus tremuloides*. Chemosphere 36, 807–812.

Karnosky, D.F., Mankovska, B., Percy, K., Dickson, R.E., Podila, G.K., Hendrey, G., Coleman, M.D., Kubiske, M., Pregitzer, K.S., Sober, J., Isebrands, J.G., 1999. Effects of tropospheric O_3 and interaction with CO_2: Comparison of open-top chamber, field and FACE results with trembling aspen genotypes differing in O_3 sensitivity. J. Air Water Soil Pollut. 116, 311–322.

Karnosky, D.F., Percy, K.E., Xiang, B., Callan, B., Noormets, A., Mankovska, B., Hopkin, A., Sober, J., Jones, W., Dickson, R.E., Isebrands, J.G., 2001. Interacting CO_2-tropospheric O_3 and predisposition of aspen (*Populus tremuloides* Michx.) to infection by *Melampsora medusae* rust. Global Change Biol. 8, 329–338.

Kato, M., Shimizu, S., 1987. Chlorophyll metabolism in higher plants VII. Chlorophyll degradation in senescing tobacco leaves phenolic-dependent peroxidative degradation. Can. J. Bot. 65, 729–735.

Kellomaki, S., Wang, K.Y., 1997. Effects of elevated O_3 and CO_2 concentrations on photosynthesis and stomatal conductance in Scots pine. Plant Cell Environ. 20, 995–1006.

Koch, R.K., Scherzer, A.J., Eshita, S.M., Davis, K.R., 1998. Ozone sensitivity in hybrid poplar is correlated with a lack of defense-gene activation. Plant Physiol. 118, 1243–1252.

Koch, J.R., Creelman, R.A., Eshita, S.M., Seskar, M., Mullet, J.E., Davis, K.R., 2000. Ozone sensitivity in hybrid poplar correlates with insensitivity to both salicylic acid and jasmonic acid. The role of programmed cell death in lesion formation. Plant Physiol. 123, 487–496.

Kull, O., Sober, A., Coleman, M.D., Dickson, R.E., Isebrands, J.G., Gagnon, Z., Karnosky, D.F., 1996. Photosynthetic response of aspen clones to simultaneous exposures of ozone and CO_2. Can. J. Forest Res. 16, 639–648.

Larson, P.R., Isebrands, J.G., 1971. The plastochron index as applied to developmental studies of cottonwood. Can. J. Forest Res. 1, 1–11.

Matyssek, R., Innes, J.L., 1999. Ozone—a risk factor for trees and forests in Europe? J. Water Air Soil Pollut. 116, 199–226.

McKee, I.F., Farage, P.K., Long, S.P., 1995. The interactive effects of elevated CO_2 and O_3 concentration on photosynthesis in spring wheat. Photosynth. Res. 45, 111–119.

Mortensen, L.M., 1995. Effect of carbon dioxide concentration on biomass production and partitioning in *Betula pubescens* Ehrh. Seedlings at different ozone and temperature regimes. Environ. Pollut. 87, 337–343.

Niewiadomska, E., Gaucher-Velleux, C., Chevrier, N., Maufette, Y., Dizengremel, P., 1999. Elevated CO_2 does not provide protection against ozone considering activity of several antioxidant enzymes in the leaves of sugar maple. J. Plant Physiol. 155, 70–77.

Noormets, A., Podila, G.K., Karnosky, D.F., 2000. Rapid response of antioxidant enzymes to O_3-induced oxidative stress in *Populus tremuloides* clones varying in O_3 tolerance. Forest Genetics 7, 339–342.

Örvar, B.J., McPherson, J., Ellis, B.E., 1997. Pre-activating wounding response in tobacco prior to high-level ozone exposure prevents necrotic injury. Plant J. 11, 203–212.

Overmyer, K., Tuominena, H., Kettunena, R., Betzb, C., Langebartels, C., Sandermann Jr., H., Kangasjärvi, J., 2000. Ozone-sensitive arabidopsis *rcd1* mutant reveals opposite roles for ethylene and jasmonate signaling pathways in regulating superoxide-dependent cell death. Plant Cell 12, 1849–1862.

Pääkkönen, E., Vahala, J., Pohjola, M., Holopainen, T., Kärenlampi, L., 1998. Physiological, stomatal and ultrastructural ozone responses in birch (*Betula pendula* Roth.) are modified by water stress. Plant Cell Environ. 21, 671–684.

Podila, G.K., Wustman, B.A., Wang, Y., Zak, D.R., Sober, J., Hendrey, G.R., Isebrands, J.G., Pregitzer, K., Karnosky, D.F., 1998. Effects of elevated CO_2 on the antioxidant potential of aspen clones varying in O_3 sensitivity: Results from a free air CO_2 exposure (FACE) project. In: Proc. 18th International Meeting for Specialists in Air Pollution Effects on Forest Ecosystems, Edinburgh, Scotland (Abstract, p. 97).

Polle, A., Pfirrmann, T., Chakrabarti, S., Rennenberg, H., 1993. The effects of enhanced ozone and enhanced carbon dioxide concentrations on biomass, pigments and antioxidative enzymes in spruce needs (*Picea abies* L.). Plant Cell Environ. 16, 311–316.

Polle, A., Eiblmeier, M., Sheppard, L., Murray, M., 1997. Responses of antioxidant enzymes to elevated CO_2 in leaves of beech (*Fagus sylvatica* L.) seedlings grown under a range of nutrient regimes. Plant Cell Environ. 20, 1317–1321.

Polle, A., Pell, E., 1999. Role of carbon dioxide in modifying the plant response to ozone. In: Luo, Y., Mooney, H.A. (Eds.), Carbon Dioxide and Environmental Stress. Academic Press, New York, pp. 193–213.

Porra, R.J., Thompson, W.A., Kriedemann, P.E., 1989. Determination of accurate extinction coefficients and simultaneous equations for assaying chlorophylls a and b extracted with four different solvents: verification of the concentration of chlorophyll standards by atomic absorption spectroscopy. Biochem. Biophys. Acta 975, 384–394.

Price, A., Lucas, P.W., Lea, P.J., 1990. Age-dependent damage and glutathione metabolism in ozone-fumigated barley: A leaf section approach. J. Exp. Bot. 41, 1309–1317.

Rao, M.V., Hale, B.A., Ormrod, D.P., 1995. Amelioration of ozone-induced oxidative damage in wheat plants grown under high carbon dioxide. Plant Physiol. 109, 421–432.

Rao, M.V., Lee, H., Creelman, R.A., Mullet, J.E., Davis, K.R., 2000. Jasmonic acid signaling modulates ozone-induced hypersensitive cell death. Plant Cell 12, 1633–1646.

Rao, M.V., Paliyath, G., Ormrod, D.P., Murr, D.P., Watkins, C.B., 1997. Influence of salicylic acid on H_2O_2 production, oxidative stress, and H_2O_2-metabolizing enzymes. Plant Physiol. 115, 137–149.

Reilly, J., Prinn, R., Harnisch, J., Fitzmaurice, J., Jacoby, H., Kicklighter, D., Mellilo, J., Stone, P., Sokolov, A., Wang, C., 1999. Multi-gas assessment of the Kyoto Protocol. Nature 401, 549–555.

Rintamäki, E., Keys, A.J., Parry, M.A.J., 1988. Comparison of the specific activity of ribulose-1,5-bisphosphate carboxylase-oxygenase from some C_3 and C_4 plants. Physiol. Plant. 74, 326–331.

Schwanz, P., Kimball, B.A., Idso, S.B., Hendrix, D.L., Polle, A., 1996. Antioxidants in sun and shade leaves of sour orange trees (*Citrus aurantium*) after long-term acclimation to elevated CO_2. J. Exp. Bot. 47, 1941–1950.

Sehmer, L., Fontaine, V., Antoni, F., Dizengremel, P., 1998. Effects of ozone ad elevated atmospheric carbon dioxide on carbohydrate metabolism of spruce needles. Catabolic and detoxification pathways. Physiol. Plant. 102, 605–611.

Sen Gupta, A., Heinen, J.L., Holaday, A.S., Burke, J.J., Allen, R.D., 1993a. Increased resistance to oxidative stress in transgenic plants that overexpess chloroplastic Cu/Zn superoxide dismutase. Proc. Nat. Acad. Sci. USA 90, 1629–1633.

Sen Gupta, A., Webb, R.P., Holaday, A.S., Allen, R.D., 1993b. Overexpression of superoxide dismutase protects plants from oxidative stress: Induction of ascorbate peroxidase in super oxide dismutase-overexpressing plants. Plant Physiol. 103, 1067–1073.

Sheng, Y., Podila, G.K., Karnosky, D.F., 1997. Differences in O_3-induced SOD and glutathione antioxidant expression in tolerant and sensitive aspen (*Populus tremuloides* Michx.) clones. Forest Genetics 4, 25–33.

Sober, A., Noormets, A., Kull, O., Isebrands, J.G., Dickson, R.E., Sober, J., Karnosky, D.F., 2001. Photosynthetic parameters in aspen grown with interacting elevated CO_2 and tropospheric ozone concentrations as affected by leaf nitrogen. Tree Physiol. (in press).

Tjoelker, M.G., Volin, J.C., Oleksyn, J., Reich, P., 1993. Light environment alters response to ozone stress seedlings of *Acer saccharum* Marsh. and hybrid *Populus* L. I. *In situ* net photosynthesis, dark respiration and growth. New Phytol. 124, 627–636.

Tyystjärvi, E., Riikonen, M., Arisi, A.-C.M., Kettunen, R., Jouanin, L., Foyer, C.H., 1999. Photoinhibition of photosystem II in tobacco plants overexpressing glutathione reductase and poplars overexpressing superoxide dismutase. Physiol. Plant. 105, 409–416.

Utrainen, J., Janhunen, S., Helmisaari, H.S., Holopainen, T., 2000. Biomass allocation, needle structural characteristics and nutrient composition in Scots pine seedlings exposed to elevated CO_2 and O_3 concentrations. Trees 14, 475–484.

Volin, J.C., Tjoelker, M.G., Oleksyn, J., Reich, P.B., 1993. Light environment alters response to ozone stress in seedlings of *Acer saccharum* Marsh. and hybrid *Populus* L. II. Diagnostic gas exchange and leaf chemistry. New Phytol. 124, 637–646.

Volin, J.C., Reich, P.B., 1996. Interaction of elevated CO_2 and O_3 on growth, photosynthesis and respiration of three perennial species grown in low and high nitrogen. Physiol. Plant. 97, 674–684.

Volin, J.C., Reich, P.B., Givinsh, T.J., 1998. Elevated carbon dioxide ameliorates the effects of ozone on photosynthesis and growth: species respond similarly regardless of photosynthetic pathway or plant functional group. New Phytol. 138, 315–325.

Wellburn, A.R., Lichtenthaler, H., 1984. Formulae and program to determine total carotenoids and chlorophylls a and b of leaf extracts in different solvents. In: Sybesma, C. (Ed.), Adv. Photosynth. Res., Vol. 2, pp. 9–12.

Air Pollution, Global Change and Forests in the New Millennium
D.F. Karnosky et al., editors
© 2003 Elsevier Ltd. All rights reserved.

Chapter 23

Growth responses of aspen clones to elevated carbon dioxide and ozone

J.G. Isebrands*

Environmental Forestry Consultants, LLC, P.O. Box 54, E7323 Hwy 54, New London, WI 54961, USA

E.P. McDonald

USDA Forest Service, Forestry Sciences Laboratory, Rhinelander, WI 54501, USA

E. Kruger

University of Wisconsin-Madison, Department of Forest Ecology, 1630 Linden Dr., Madison, WI 53706, USA

G.R. Hendrey

Brookhaven National Laboratory, Upton, NY 11973, USA

K.E. Percy

Canadian Forest Service, Fredericton, NB E3B5P7, Canada

K.S. Pregitzer, J. Sober, D.F. Karnosky

School of Forest Resources and Environmental Science, Michigan Technological University, Houghton, MI 49931, USA

Abstract

The International Panel of Climate Change (IPCC) has concluded that the greenhouse gases, carbon dioxide (CO_2) and tropospheric ozone (O_3), are increasing concomitantly globally. Little is known about the effect of these interacting gases on growth, survival, and productivity of forest ecosystems. In this study we assess the effects of three successive years of exposure to combinations of elevated CO_2 and O_3 on growth responses in a 5 trembling aspen (*Populus tremuloides*) clonal mixture in a regenerating stand. The experiment is located in Rhinelander, Wisconsin, USA (45° N 89° W) and employs free air carbon dioxide and ozone enrichment (FACE) technology. The aspen stand was exposed to a factorial combination of 4 treatments consisting of elevated CO_2 (560 ppm), elevated O_3 (episodic exposure—90 $\mu l \, l^{-1} \, h^{-1}$), a combination of elevated CO_2 and O_3, and ambient control in 30 m treatment rings with 3 replications.

---

*Corresponding author.

DOI:10.1016/S1474-8177(03)03023-7

Our overall results showed that our 3 growth parameters including height, diameter and volume were increased by elevated CO_2, decreased by elevated O_3, and were not significantly different from the ambient control under elevated CO_2 + O_3. However, there were significant clonal differences in the responses; all 5 clones exhibited increased growth with elevated CO_2, one clone showed an increase with elevated O_3, and two clones showed an increase over the control with elevated CO_2 + O_3, two clones showed a decrease, and one was not significantly different from the control. Notably, there was a significant increase in current terminal shoot dieback with elevated CO_2 during the 1999–2000 dormant seasons. Dieback was especially prominent in two of the five clones, and was attributed to those clones growing longer into the autumnal season where they were subject to frost. In summary, our results show that elevated O_3 negates expected positive growth effects of elevated CO_2 in *Populus tremuloides* in the field. Our results suggest that future climate model predictions should take into account the offsetting effects of elevated O_3 on CO_2 enrichment when estimating future growth of trembling aspen stands.

1. Introduction

There are increasing concerns about the effects of global climate change on the growth and sustainability of forest ecosystems. In their 2001 assessment report, the Intergovernmental Panel on Climate Change (IPCC) comprised of hundreds of scientists from many countries concluded that, "there is new and stronger evidence that human activities are affecting our climate." Moreover, they cite the increase of the greenhouse gases carbon dioxide (CO_2) and tropospheric ozone (O_3) as important external factors affecting climate change (IPCC, 2001). This report also confirms the previous reports that CO_2 (Keeling et al., 1995) and O_3 (Marenco et al., 1994; Taylor et al., 1994; Stevenson et al., 1998; Fowler et al., 1999; Ryerson et al., 2001) are increasing steadily in the atmosphere.

Hundreds of research studies have been done on the single factor effects of elevated CO_2 and O_3 on plants, and there are numerous review articles available on these subjects. Ceulemans and Mousseau (1994), Ceulemans et al. (1999), and Saxe et al. (1998) reviewed the effects of elevated CO_2 on plants, and Jones and Curtis (1999) compiled an on-line bibliography of CO_2 effects on vegetation and ecosystems. In general, the effects of elevated CO_2 on most plants are increased growth through increased photosynthetic rates, individual leaf area, leaf area duration, and water use efficiency. The effects of O_3 on plants have been reviewed by Adams et al. (1989), and on forests by Taylor et al. (1994), Hogsett et al. (1997) and Bortier et al. (2000). In general, O_3 has profound negative effects on growth, development and productivity of many plants. The phytotoxic effects of O_3 on plants are generally due to decreased photosynthetic rates, decreases in leaf surface area, premature leaf abscission, and weakened branch and root growth.

Little is known about interacting effects of elevated CO_2 and O_3 on plants, however (Allen, 1990; Isebrands et al., 2000). This lack of knowledge is problematic because both gases are concurrently increasing in the atmosphere, as mentioned above. This paucity of information is also related to the difficult logistical problems and expense of multiple factor experiments. The research on interacting effects of CO_2 and O_3 on plants has shown contradictory results. In some studies, elevated CO_2 offsets decreases in photosynthesis and growth caused by elevated O_3 (Barnes and Wellburn, 1998; Volin et al., 1998; Donnelly et al., 2000). In other studies, elevated CO_2 did not offset the effects of elevated O_3 on plants (Barnes et al., 1995; Barnes and Wellburn, 1998; Bortier et al., 2000). Most studies, however, did not evaluate the genetic responses to elevated CO_2 and O_3. We found the elevated CO_2 did not ameliorate the effects of elevated O_3 on photosynthesis parameters in trembling aspen clones (Kull et al., 1996). In hybrid poplar clones, the effect of elevated CO_2 on negating detrimental O_3 effects on growth varied by clone (Dickson et al., 1998).

Trembling aspen (*Populus tremuloides*), hereafter aspen, is an important ecological and economic tree species in the Great Lakes region of the USA and Canada. According to the International Poplar Commission, the aspen forest types make up more than 8.8 Mha in the US and 17.8 Mha in Canada. Moreover, other aspen species are significant in China, Russia and Scandinavia. It is known that aspen is sensitive to abiotic stressors (Kozlowski and Constantinidou, 1986). Berrang et al. (1989) collected aspen clones from across its natural range in the US to study the effects of O_3 on their growth and physiological processes. This collection led to a series of studies by D.F. Karnosky and coworkers on aspen clone vulnerability to O_3 and CO_2. In those studies, aspen clones were also shown to vary in their sensitivity to CO_2 (Kubiske et al., 1998; Zak et al., 2000), O_3 (Coleman et al., 1995a, 1995b; Karnosky et al., 1996), and to interacting effects of elevated CO_2 and O_3 (Karnosky et al., 1998; Kull et al., 1996; Isebrands et al., 2000). We also found that in certain aspen clones elevated CO_2 did not offset the detrimental effects of O_3 on growth and physiological processes (Isebrands et al., 2000).

Previous studies conducted in growth chambers and open-top chambers showed that the size and scale of the chamber studies were too small and there was a significant chamber effect that often confounded the results and conclusions of those experiments (Karnosky et al., 1999a). Therefore, we initiated a free air carbon dioxide and ozone enrichment experiment (Aspen FACE) near Rhinelander, Wisconsin, USA (Dickson et al., 2000) to study the effects of combinations of elevated CO_2 and O_3 and their interactive effects on regenerating northern hardwood stands under field conditions.

Other research groups are using FACE technology to expose forest ecosystems to elevated CO_2 (Hendrey et al., 1999). Results of those studies have shown that CO_2 increased growth in: (1) a loblolly pine plantation when nitrogen was not limiting (DeLucia et al., 1999), and (2) in sweetgum plantations (Norby et al., 1999), and a desert shrub ecosystem, when water was not limiting (Smith et al., 2000). Our FACE experiment is unique in that we are studying the combined effects of elevated CO_2 and O_3 predicted for the year 2100 on a regenerating trembling aspen stand beginning with establishment (Dickson et al., 2000).

Our objective was to determine the interacting effects of elevated CO_2 and O_3 alone and in combination, on trembling aspen growth, survival and productivity over time. Our results based upon 3 years of exposure in the field confirm our open-top chamber study findings that showed aspen clones differ in their sensitivity to elevated CO_2 and O_3, and that in some cases elevated CO_2 did not offset the detrimental effects of elevated O_3 on growth.

2. Materials and methods

2.1. Plant material

Five trembling aspen clones were selected from the Great Lakes Region for study based upon results of our previous open-top chamber research. Three of the clones (216, 259, 271) were selected for their differing sensitivity to O_3 (Karnosky et al., 1996) and two clones were selected for early (42E) and late (8L) leaf phenology and their differing response to elevated CO_2 (Kubiske et al., 1998). The clones were propagated from green wood cuttings, and grown in a greenhouse, then hardened in a shade frame before outplanting (Dickson et al., 2000).

2.2. FACE experimental design

The Aspen FACE study is located within a fenced 32 ha field of sandy loam soil on the Harshaw Experimental Farm near Rhinelander, Wisconsin, USA, (45 °N 89 °W). The study consists of 12 individual treatment rings (30 m diameter), spaced 100 m apart to minimize drift of the CO_2 and O_3. The treatments are arranged in a full factorial design with 3 replications. Treatments include 3 control rings (ambient air), 3 elevated CO_2 (560 ppm), 3 elevated O_3 rings (episodic total seasonal exposure 90 $\mu l\, l^{-1}\, h^{-1}$), and 3 rings of combination elevated CO_2 and O_3. Exposure was with a FACE delivery and control system designed by Brookhaven National Laboratory (Hendrey et al., 1999) consisting of a high volume blower, plenum pipe, vertical vent pipes for emitting CO_2

and O_3, and a computer control system linked to micrometeorology stations at the center of each ring. The gas delivery was controlled by valves that were opened in the upwind direction upon computer instruction. The FACE equipment was tested and calibrated in 1997, which was considered an establishment period and a full complement of treatments began on May 1, 1998. Details of exposures are outlined in Dickson et al. (2000).

The aspen clones were planted in June 1997, as randomized pairs of individual clones with a hand held 10 cm diameter auger in the east half of each treatment ring at 1-m spacing (Fig. 1). Each tree was given a unique study identification number and each ring was unique (Dickson et al., 2000, Fig. 8). The aspen clonal mix in each ring consisted of a 100-tree study core surrounded on all sides by six buffer rows. All growth measurements were made on the core trees only. See Dickson et al. (2000) for all details.

2.3. Plant growth measurements

Growth parameters were measured at the end of each growing season after leaf fall. A telescoping height pole was used for all tree height measurements, and a digital caliper was used to measure basal diameter at 3 cm above the soil surface. To minimize the potential for injury, a single basal diameter measurement was made along an N–S axis for consistency. By 1999 and 2000, the size of the trees allowed for caliper readings taken at N–S and E–W axes. The tree sizes at the end of the 1997 establishment period are considered the initial size of individuals for the subsequent growth analysis. The basal diameter^2 × height, or D^2H, is an exceptionally robust, non-destructive measure that is linearly related to aboveground tree biomass (Crow, 1988). Destructive biomass estimates are not yet available.

Each spring after bud-break, trees were visually inspected for damage or dieback on the current terminal. In May 2000, a high incidence of dieback was observed. We measured the length of dieback using the height pole, and the amount of dieback was expressed as a percentage of the total length of the 1999 current shoot length. Mortality at the end of 2000 ranged from 2–4% and was not significant among treatments.

2.4. Statistical analyses

The details of the Aspen-FACE design and statistical considerations are described thoroughly by Dickson et al. (2000). For analysis, the experiment was considered a randomized complete block design, with 3 replications of atmospheric treatments at the whole-plot level, and between clone effects and interactions of clone and treatments evaluated as sub-plot factors. The treatments represent ambient atmospheric conditions and a single elevated target

Ring 3.4

	15	16	17	18	19	20	21	22	23	24	25	26	27	28	29	
AD	271	271	259	42E												AD
AC	271	271	259	42E	42E	259	259									AC
AB	8L	42E	216	42E	271	271	216	42E	8L							AB
AA	8L	42E	216	42E	271	271	216	42E	8L	271						AA
A	259	259	259	42E	8L	259	42E	271	271	216	42E					A
B	259	259	259	216	8L	259	42E	271	271	216	42E	8L				B
C	259	259	42E	216	8L	216	8L	259	271	216	8L	8L				C
D	259	259	42E	216	8L	216	8L	259	271	8L	8L	216	271			D
E	216	8L	259	216	216	271	216	271	271	8L	42E	216	271	8L		E
F	216	8L	259	42E	216	271	216	271	271	8L	42E	271	42E	8L		F
G	216	271	8L	42E	42E	8L	271	259	259	8L	259	259	42E	8L		G
H	216	271	8L	42E	42E	8L	271	259	259	216	259	42E	8L	8L	8L	H
I	271	8L	259	42E	259	271	271	8L	216	216	8L	42E	8L	8L	216	I
J	271	8L	259	216	259	271	271	8L	216	259	8L	259	271	271	216	J
K	42E	42E	42E	216	216	8L	42E	42E	216	259	216	259	271	271	216	K
L	42E	42E	42E	42E	216	8L	42E	42E	216	216	216	42E	8L	8L	216	L
M	259	259	42E	42E	216	259	8L	216	42E	216	8L	42E	8L	8L		M
N	259	259	42E	8L	216	259	8L	216	42E	8L	8L	259	259	42E		N
O	259	216	8L	8L	216	271	271	259	8L	8L	216	259	259	42E		O
P	259	216	8L	259	216	271	271	259	8L	271	216	271	8L	271		P
Q	259	8L	271	259	259	259	271	42E	259	271	271	271	8L	271		Q
R	259	8L	271	259	259	259	271	42E	259	271	216	259	271			R
S	42E	8L	216	259	259	216	271	271	271	271	271	259	271			S
T	42E	8L	216	42E	259	216	271	271	271	42E	271	271				T
U	8L	42E	216	42E	42E	42E	42E	8L	259	42E	271					U
V	8L	42E	216	216	42E	42E	42E	8L	259	216						V
W	216	271	216	216	42E	8L	216	216	216							W
X	216	271	216	8L	42E	8L	216	216								X
Y	42E	271	42E	8L	259	259										Y
Z	42E	271	42E													Z
	15	16	17	18	19	20	21	22	23	24	25	26	27	28	29	

Figure 1. Example treatment ring map showing paired arrangement of aspen clones and buffer rows. The dark line indicates the target exposure core. Ring 3.4 refers to replication 3, treatment 4 (elevated $CO_2 + O_3$) of the FACE array (Dickson et al., 2000).

level of the treatment gases, CO_2 and O_3. Because the aspen clones were chosen based on previous performance and we wished to continue long-term study of these genotypes, the treatment and clone effects were evaluated as fixed effects within analysis of variance. Replication effects and replication × treatment effects are considered random within this design. Thus, a mixed model analysis was required, and we used the PROC Mixed component of SAS software (SAS Institute Inc., 2000) for all analyses (Table 1). For the analysis of main effects of CO_2 and O_3, the replication × treatment error (a) terms can be pooled or partitioned, depending on the patterns of variation. To determine whether replication × CO_2 or replication × O_3 effects were significant partitioning was required, we tested the difference in Restricted Maximum Like-

Table 1. Analysis of variance table for growth measurements from 1998, 1999, and 2000 growing season

Source of variation	Degrees of freedom	Expected mean squares
Whole-plots		
Replications	2	$\sigma^2_{e(b)} + nc\sigma^2_{e(a)} + ntc\sigma_r$
Treatments	3	$\sigma^2_{e(b)} + nc\sigma^2_{e(a)} + nrcQ_t$
CO_2	1	
O_3	1	
$CO_2 + O_3$	1	
Error (a)	6	$\sigma^2_{e(b)} + nc\sigma^2_{e(a)}$
Sub-plots		
Clones	4	$\sigma^2_{e(b)} + nrtQ_c$
Clones × Treatments	12	$\sigma^2_{e(b)} + nrQ_{tc}$
Clones × CO_2	4	
Clones × O_3	4	
Clones × $CO_2 + O_3$	4	
Error (b)	$60(n-1)$	$\sigma^2_{e(b)}$
Total	$60n - 1$	

Analysis of variance (PROC MIXED SAS Institute Inc., 2000) for this split-plot design having n values per replication × treatment × clone combination, with the whole plot consisting of replications (r), treatments (t), and replications × treatments pooled for *Error* (a) (n varied from 100 to 110 for the treatments). Clones (c) and clones × treatments are the sub-plot factors, with *Error* (b) being a pooled term containing variation due to replication × subplot effect interactions and variation due to subsampling (individual trees). The specific errors associated with the Q terms are designated with subscripts. Replications, replications × treatments, *Error* (a) and *Error* (b) were assumed to be random; all other effects were assumed fixed. Variation due to CO_2, O_3, and $CO_2 + O_3$ effects was estimated by linear contrasts. F-ratios for subplot effects and *Error* (a) all used *Error* (b) as the denominator. Degrees of freedom for F-ratios are corrected for sampling imbalance using Satterthwaite approximations. F-ratios for whole-plot treatment combinations (single degree of freedom orthogonal linear contrasts) use the same denominator as the F-ratio for treatments.

lihood indices between pooled and partitioned models as described by Littell et al. (1996). In the majority of analyses, pooled error terms were appropriate.

Statistical analyses included only data from approximately 100 trees per FACE ring. To account for effects of initial size on growth, the height, diameter and D^2H values from 1997 were transformed to give linear relationships with their respective response variables in years 1998 through 2000. Once the linear relationships were established, the transformed initial size parameters were included as covariates in the mixed SAS models used in analyzing growth responses. The means and standard errors (based on $n = 3$) reported in tables and figures are for individual tree values adjusted for the covariate initial tree size by adding the residual from the expected value, based on initial size, to the overall grand mean for that parameter.

3. Results

3.1. Overall growth responses (1997–2000)

3.1.1. Height

During the study establishment year (1997), there were no elevated CO_2 or O_3 treatments administered during the season except for minor tests of equipment, so there were no observed treatment effects on growth. But there were clonal differences in height (cm) at the end of the first growing season related to establishment justifying the use of initial height as a covariate in the analyses. The height of the clones ranged from 35.6 cm for clone 8L to 68.0 cm for clone 271 (Table 2).

There were no overall treatment effects (i.e., CO_2, O_3, $CO_2 + O_3$) in subsequent year-end heights of the aspen stand (Table 3). In each of the three years, there was a highly significant effect of clone on year-end height. The $CO_2 \times$ clone interaction was not significant at the end of any of the years, but as expected from our previous research, the $O_3 \times$ clone interaction was significant each year. The $CO_2 + O_3 \times$ clone interaction was significant at the end of 1998 and 2000, but not in 1999 (Table 3).

3.1.2. Diameter

Aspen diameter growth in the elevated CO_2 treatment was not significant ($p < 0.05$) after the first treatment year (1998), but was near significant (0.06 level) at the end of the second and third treatment years (1999–2000). Similarly, the O_3 treatment did not affect diameter growth in 1998, but was near significant in 1999 and 2000. The $CO_2 + O_3$ treatment response for diameter

Table 2. Mean height (cm) $\pm$ SE of the 5 *Populus tremuloides* clones studied at the end of the establishment year (1997)

Clone	Mean height (cm)	SE
216	50.1	1.7
259	41.0	3.9
271	68.0	2.9
42E	48.0	2.5
8L	35.6	1.1

Table 3. Exact probability levels from SAS analyses of variance of the growth parameters—height (cm), diameter (cm), and diameter^2 × height (cm^3) for three growing seasons (1998–2000)

Effect	Height (cm)			Diameter (cm)			D^2H (cm^3)		
	1998	1999	2000	1998	1999	2000	1998	1999	2000
CO_2	0.519	0.491	0.115	0.326	0.056	0.064	0.962	0.015	0.040
O_3	0.819	0.410	0.343	0.522	0.054	0.063	0.497	0.075	0.021
$CO_2 + O_3$	0.903	0.653	0.927	0.522	0.043	< 0.001	0.983	0.029	0.360
Clone	< 0.001	< 0.001	< 0.001	< 0.001	< 0.001	< 0.001	< 0.001	< 0.001	< 0.001
CO_2 × Clone	0.195	0.691	0.277	0.058	0.001	0.057	0.420	0.289	0.022
O_3 × Clone	0.027	< 0.001	0.001	< 0.001	0.010	0.133	0.001	< 0.001	< 0.001
$CO_2 + O_3$ × Clone	0.036	0.161	0.003	0.065	0.002	< 0.001	0.337	0.031	0.078

growth was similar to the CO_2 effect, and the O_3 effect treatment for the first 2 years. However, at the end of the third treatment year it was highly significant (< 0.01). The clone effect for diameter at the end of each season was highly significant; and all treatment x clone interactions were significant or near significant for each year except at the end of 2000. The O_3 × clone interaction effect on diameter was highly significant at the end of 1998 and 1999, and the $CO_2 + O_3$ × clone interaction effect on diameter was highly significant at the end of the second and third treatment years (Table 3).

3.1.3. Volume estimate (D^2H)

There was no significant effect of treatments on overall aspen stand volume estimates (i.e., D^2H) after the first treatment year (Table 3). This finding was expected because it was the first year of exposure (1998) and volume estimate is a metric of height and diameter that had no treatment effects (above). However, volume growth was significantly affected by elevated CO_2 in the second and third treatment years (1999 and 2000). Elevated O_3 had a significant effect on volume growth by the end of the second ($p = 0.08$) and third treatment year ($p = 0.02$). Elevated $CO_2 + O_3$ had a significant effect on the overall stand volume in the second year, but not the third.

At the end of each of the treatment years, the clone effect was significant as was the case with height and diameter data. There was no significant $CO_2 \times$ clone interactions after the first and second year, but there was after three years. Moreover, there was a highly significant effect of $O_3 \times$ clone interaction at the end of each year as expected from our previous research. The elevated $CO_2 + O_3 \times$ clone interaction was not significant at the end of the first treatment, but was significant at the end of the second ($p = 0.03$) and near significant at the end of the third year ($p = 0.08$).

3.2. Yearly growth responses of clones

The overall mean growth data and their standard errors for each treatment and clone for each of the first three treatment years 1998, 1999, and 2000 are given in Table 4.

1998

In the first year, there were few treatment effects on the mean height of the 5 aspen clones. Assessment of the effect of treatment on height was complicated by the initial height differences at the end of the establishment year (Table 2). The only significant treatment effect on height of a clone was elevated O_3 for clone 8L. The mean heights of the clones in the control at the end of 1998 ranged from 161 cm for clone 259 to 212 cm for clone 271 (Table 4).

Mean diameters for the aspen clones growing in the control treatment ranged from 1.92 cm for clone 259 and 2.35 cm for clone 271. Thus, a pattern of growth emerged in 1998 whereby clone 259 was the smallest clone and clone 271 the largest that continued throughout the three years. Mean diameter of individual clones was not significantly affected by elevated CO_2 in 1998, but was significantly decreased by elevated O_3 for clone 216 and 259, and increased for clone 8L (Table 4).

In 1998, the mean estimated volume increment (D^2H) of the clones in the control treatment ranged from 824 cm^3 for clone 259 to 1451 cm^3 for clone 271. Elevated CO_2 did not affect volume increment of individual clones in 1998; however, in all clones but one, diameter growth was decreased significantly by elevated O_3. The exception was clone 8L that increased. In 1998, mean volume increment was significantly decreased for the elevated $CO_2 + O_3$ treatment for clones 216, 271, and 42E, and significantly increased for clone 8L, while clone 259 remained the same as the control (Table 4).

Table 4. Mean growth measurements (±SE) for each treatment and clone for each of the first three treatment years (1998–2000)

Clone	Height (cm)				Diameter (cm)				D^2H (cm³)			
	Control	CO_2	O_3	CO_2+O_3	Control	CO_2	O_3	CO_2+O_3	Control	CO_2	O_3	CO_2+O_3
1998												
216	205±7	192±24	194±5	202±1	2.26±0.09	2.26±0.10	1.95±0.06	2.07±0.06	1246±84	1246±284	984±67	1069±100
259	161±4	153±22	169±7	160±6	1.92±0.08	2.05±0.18	1.79±0.21	1.90±0.15	824±85	886±216	734±185	829±111
271	212±8	194±20	205±4	194±8	2.35±0.11	2.34±0.10	2.31±0.10	2.26±0.13	1451±219	1359±283	1236±149	1222±183
42E	175±7	165±16	162±8	164±8	2.09±0.06	2.17±0.12	2.00±0.09	2.06±0.14	998±30	977±128	837±67	833±123
8L	175±2	178±18	199±2	179±7	1.98±0.04	2.16±0.05	2.21±0.01	2.03±0.13	913±65	1089±84	1111±26	978±86
X̄±SE	194±4	180±20	184±3	180±5	2.18±0.08	2.23±0.10	2.03±0.10	2.08±0.12	1179±100	1150±211	963±99	987±109
1999												
216	302±12	318±33	294±7	309±11	3.58±0.11	3.86±0.08	3.09±0.11	3.33±0.11	4453±294	5691±822	3533±316	3966±204
259	248±10	260±28	252±10	250±8	2.83±0.09	3.25±0.11	2.46±0.25	2.69±0.19	2727±237	3526±459	2083±563	2669±457
271	346±11	356±23	321±6	328±11	3.76±0.09	3.98±0.04	3.68±0.15	3.45±0.17	5654±482	6833±364	4804±534	4533±545
42E	331±20	335±19	294±7	296±13	3.60±0.04	3.88±0.13	3.25±0.12	3.44±0.15	5040±318	5499±471	3749±173	4088±418
8L	291±14	342±20	319±6	308±14	3.30±0.10	3.76±0.09	3.61±0.02	3.14±0.21	3936±277	5172±230	4619±115	3972±370
X̄±SE	310±10	322±24	293±5	299±11	3.49±0.04	3.79±0.06	3.17±0.15	3.23±0.13	4584±199	5578±384	3644±337	3874±362
2000												
216	350±16	373±33	334±10	374±8	4.03±0.18	4.39±0.14	3.61±0.13	3.95±0.10	6542±734	8686±1359	5187±612	6750±189
259	287±9	307±28	294±14	296±11	3.07±0.12	3.57±0.10	2.68±0.27	2.89±0.18	3760±362	5105±509	2858±901	3760±778
271	411±3	419±25	373±16	415±15	4.39±0.07	4.81±0.02	4.21±0.21	4.15±0.15	8959±345	11706±940	7164±1202	8392±822
42E	433±30	455±18	377±12	411±26	4.38±0.15	4.66±0.15	3.86±0.15	4.14±0.25	9454±740	10637±989	6733±377	8098±1089
8L	364±15	436±12	397±5	393±16	3.98±0.10	4.61±0.06	4.35±0.04	3.76±0.28	7046±562	9275±505	8245±161	7112±883
X̄±SE	371±8	391±23	351±10	380±17	4.04±0.07	4.43±0.06	3.67±0.17	3.80±0.16	7291±199	9316±624	5793±668	6906±810

At the end of the second treatment year (1999), the mean height of the individual clones ranged from 248 cm for clone 259 to 346 cm for clone 271 for the control treatment (Table 4). Elevated CO_2 did not significantly affect height of clones, except for clone 8L which increased. Elevated O_3 did not affect height for clones 216 and 259, but decreased it for clones 271 and 42E, and increased it for clone 8L (Table 4).

Diameter at the end of 1999 ranged from 2.83 cm for clone 259 to 3.76 cm for clone 271 in the control. In all clones, elevated CO_2 increased diameter growth (Table 4). However, elevated O_3 decreased diameter in clones 216, 259, and 42E, increased it in clone 8L, and was not significantly different in clone 271. Notably, at the end of 1999, the diameter of all clones was decreased by the combination treatment of elevated CO_2 and O_3, except 8L which was not significantly different.

Volume estimate in the control treatment at the end of 1999 ranged from 2727 cm^3 for clone 259 to 5654 cm^3 for clone 271. A clear separation in volume increment became evident at the end of two treatment years (i.e., 3 years old). All 5 clones showed increased volume increment in the elevated CO_2 treatment, although clone 42E had the least (Fig. 3). Moreover, the clones showed decreased volume increment with elevated O_3, except for clone 8L. This ozonephilic clone after two years of treatment continued to show a pattern of increased growth under elevated O_3 (Table 4). Volume increment at the end of 1999 was not significantly affected by elevated CO_2 and O_3 in clones 259 and 8L, but was decreased in clones 216, 271, and 42E.

3.3. Dormant season dieback 1999–2000

After the 1999–2000 dormant season, we observed significant current terminal (CT) shoot length dieback throughout the experiment (Table 5). Dieback symptoms were dead buds and discolored stems. An analysis of the dieback showed that the percent of aspen trees affected varied by treatment and clone. The elevated CO_2 treatment had an average frequency of 19% of the trees with dieback, while the other treatments averaged from 2–7%. Clones 271 and 216 accounted for most of the dieback observed with 20% and 10% of the trees experiencing dieback, respectively. The other three clones had minimal dieback in the range of 2–4%.

The severity of the dieback as evidenced by the percent of the current terminal shoot length that died back did not vary by treatment, but was significantly different among the clones (Table 5). About 28% of the shoot length died back in clones 216 and 271, 34% (although variable) in clone 259, 19% in clone 8L, and 10% of 42E. The dieback appeared to be associated with the prolonged

Table 5. Mean percent of aspen trees during the 1999–2000 dormant season with dieback by treatment and clone, and the percent of the current terminal (CT) shoot length that died back ($\pm$SE)

	% of trees with dieback	SE
Clone		
216	10	0.75
259	4	1.40
271	20	3.38
42E	2	0.00
8L	2	0.03
Treatment		
Control	7	1.26
CO_2	19	4.65
O_3	6	1.81
$CO_2 + O_3$	2	0.55
	% of CT length with dieback	SE
Clone		
216	28	6.67
259	34	11.57
271	28	0.30
42E	10	0.00
8L	19	0.29
Treatment		
Control	25	4.79
CO_2	31	3.18
O_3	32	10.62
$CO_2 + O_3$	23	7.03

growing season that the region experienced in the autumn of 1999. Clones 271 and 8L grew well into October and did not completely lose leaves until approximately November 1. The first freeze was in mid October, which was much later than normal in Rhinelander.

2000

At the end of the third treatment season (2000), the average height of the clones ranged from 287 cm for clone 259 to 433 cm for clone 42E for the control. Note that clone 42E surpassed clone 271 in height for the first time in the experiment. There was a significant effect of elevated CO_2 on height or clone of 8L, but not the other clones. Elevated O_3 again significantly increased height of clone 8L, but decreased height in clones 271 and 42E. Elevated $CO_2 + O_3$

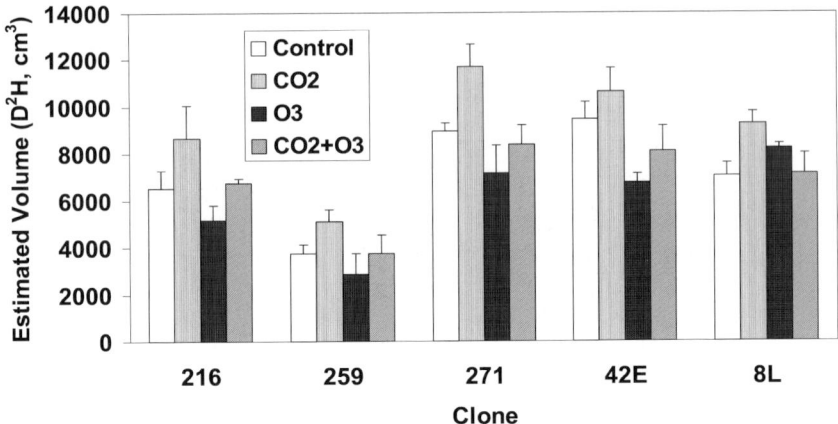

Figure 2. Summary of the effects of treatment for each clone on estimated volume increment (D^2H) at the end of the third treatment season (2000).

increased height in clones 216 and 8L, decreased height in 42E, and was not significantly different in clones 259 and 271 (Table 4).

Diameters at the end of 2000 ranged from 3.07 cm for clone 259 to 4.38 and 4.39 cm for clones 42E and 271, respectively, for the control treatments. In all clones elevated CO_2 increased diameter growth. Elevated O_3 decreased the diameter of all clones, except clone 8L, which increased. Elevated $CO_2 + O_3$ slightly decreased the diameter in clones 271 and 42E, and no significant effect on the other clones.

The mean volume increment (D^2H) in the control treatment for 2000 ranged from 3760 cm^3 for clone 259 to 9454 cm^3 for clone 42E (Fig. 2). Again, that clone 42E surpassed clone 271 in size by the end of the third treatment season. In all clones, elevated CO_2 increased the volume estimate by the end of 2000. Moreover, elevated O_3 decreased volume increment in all clones, except the ozonephilic clone 8L. The effect of elevated $CO_2 + O_3$ was not significantly different than the control, except for clones 271 and 42E (Table 4).

A summary of the effect of treatments on estimated volume increment for 2000 is given in Fig. 2. Note that for the most part, all clones show the same pattern for treatment effects on volume estimate even though the absolute values of the clone volume estimates were quite different. Elevated CO_2 increases volume estimate for all clones, elevated O_3 decreases it, except for clone 8L, and elevated $CO_2 + O_3$ is not significantly different from the control, except for clone 42E which was lower.

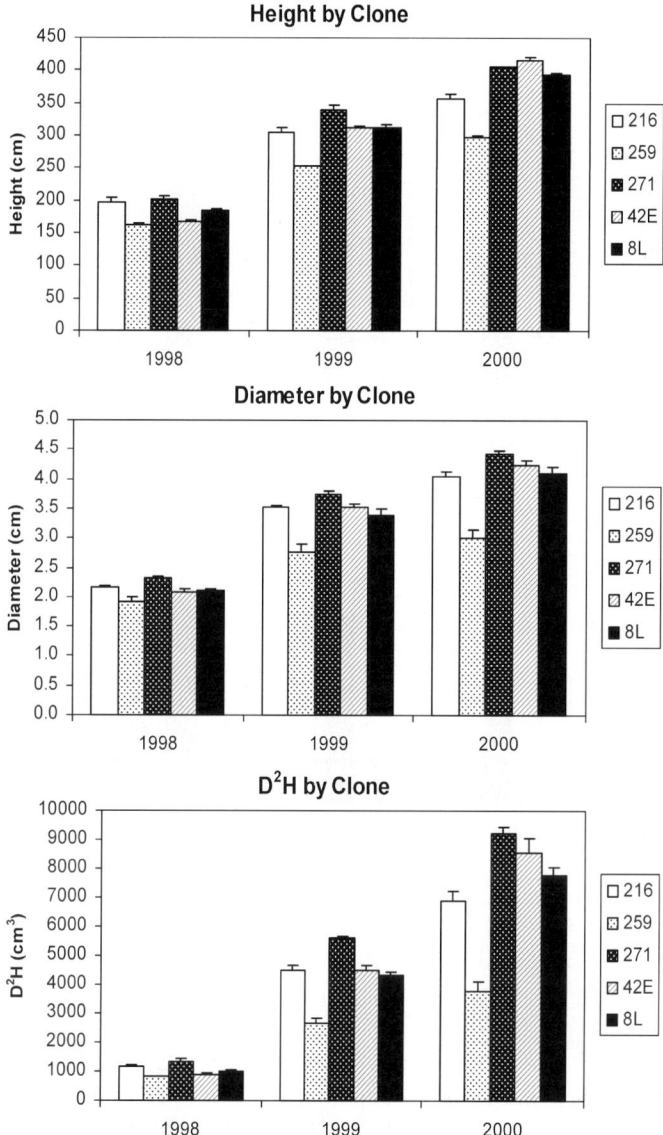

Figure 3. Overall growth response for each treatment and clone for three treatment years (1998–2000). (*Continued on next page*)

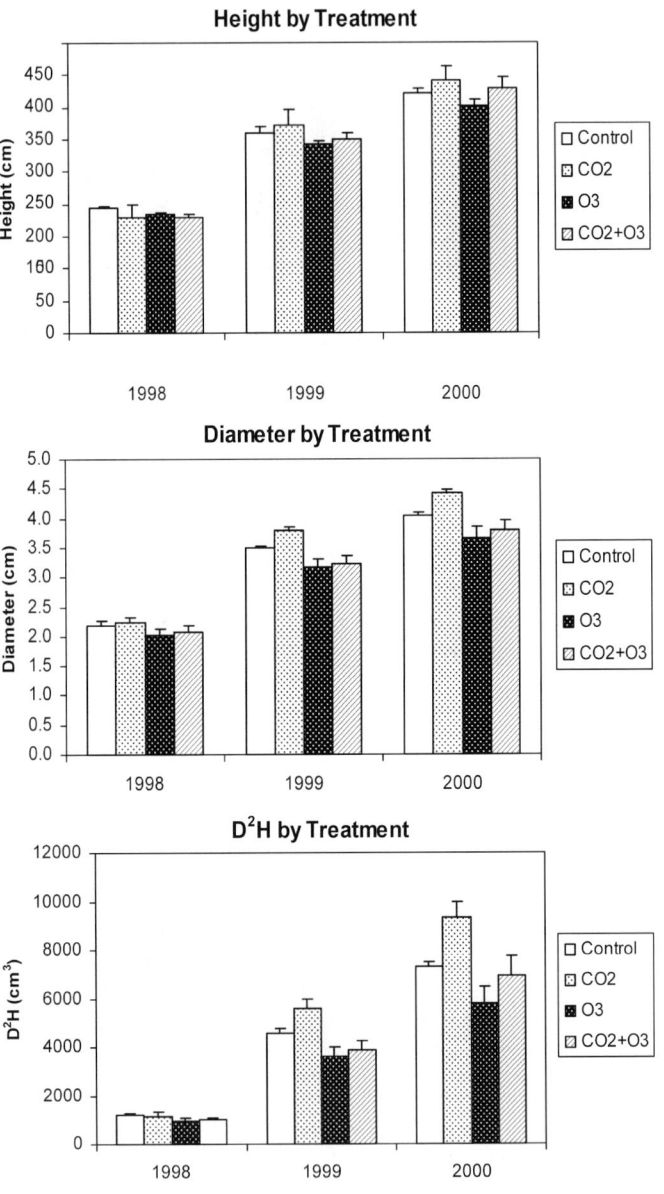

Figure 3.   (Continued)

3.4. Overall growth response summary for three years

The overall growth response for the three treatment years is shown in Fig. 3. The three panels on the left summarize the overall experiment clone averages for height, diameter and volume increment (D^2H) for the three years. Note that clone 271 was the largest clone throughout the first two years, but that clone 42E equaled its size in the third year. Clone 259 was the slow grower throughout the three years, and the other 2 clones were intermediate.

The overall growth response of all clones in the aspen stand to treatment is shown on the right three panels of Fig. 3. Note that there was no difference in tree height of the stand with elevated CO_2, and decreased height with elevated O_3 for the overall aspen stand at the end of 2000. Elevated CO_2 increased overall mean tree diameter for the stand, and elevated O_3 and elevated $CO_2 + O_3$ decreased mean tree diameter for the stand compared to the control.

The most significant treatment effects were an overall tree volume increment for the stand at the end of the second and third treatment year, compared to the control (Fig. 3). Elevated CO_2 alone increased tree volume by 22% by the end of 1999, and by 28% by the end of 2000. Elevated O_3 alone decreased tree volume by 26% by the end of 1999 and by 26% by the end of 2000. Elevated $CO_2 + O_3$ decreased tree volume compared to the control by 18% by the end of 1999, and by 6% by the end of 2000. The change in this decrease from 1999 to 2000 was due to the relative changes in individual clone responses to elevated $CO_2 + O_3$ (Fig. 4).

4. Discussion

What effect will atmospheric changes have on trembling aspen growth, survival and productivity by the year 2100? The Intergovernmental Panel on Climate Change (IPCC, 1998, 2001) predicts that atmospheric CO_2 will continue to rise by about 2% a year, so that the atmospheric CO_2 level in 2100 will approach 560 ppm. Moreover, tropospheric ozone (O_3) will concomitantly rise by 12–60% in certain regions, primarily due to rising atmospheric emission of several O_3 forming gases. Their report also predicts the freeze-free growing season will lengthen with increases in nighttime temperature of ca. 2 °C expected by 2100. The results of our experiment over the first three years of treatment, based upon a 2100 scenario, suggest that aspen growth, survival, and productivity will be greatly affected by the predicted 2001 climate. The extent of the effect will likely depend upon the genetic composition of the aspen stand, and the level of O_3 concentration in the region, as well as other interacting abiotic and biotic factors (Isebrands et al., 2000).

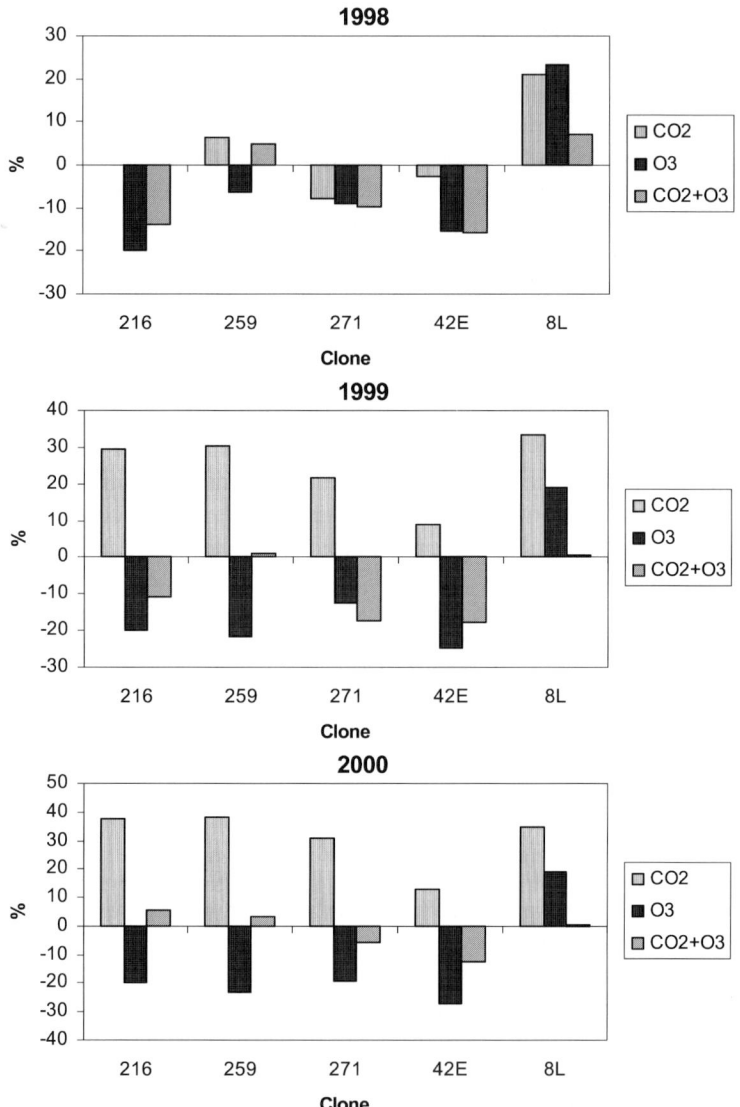

Figure 4. Volume increment (D^2H) for treatments, expressed as percent of control for five aspen clones over three treatment years (1998–2000).

If elevated CO_2 were to increase independently, as many researchers have suggested, our results suggest aspen volume growth would likely increase significantly (28%, Table 4), if all other factors such as soil fertility, water availability, temperature, and pests and diseases were equal. This result agrees with other research findings in open-top chambers with aspen (Curtis et al., 2000; Zak et al., 2000) and hybrid poplars (Curtis et al., 1995; Dickson et al., 1998; Wang et al., 2000), and is in agreement with the results of other field experiments conducted on loblolly pine (DeLucia et al., 1999) and desert ecosystems (Smith et al., 2000) using FACE technology. Height growth, on the other hand, would not be significantly affected as others have shown (Kubiske et al., 1998); however, diameter growth would increase dramatically with elevated CO_2, thereby contributing to increased volume growth.

If tropospheric ozone doses were to increase in aspen forests to the current level in many areas surrounding of our mid-western US cities, aspen volume growth would decrease significantly (26%, Table 4), if all related abiotic and biotic factors including CO_2 concentrations remained the same (an unlikely scenario). These results agree with many growth chamber and open-top chamber studies on the effects of elevated O_3 on aspen (Coleman et al., 1995a; Karnosky et al., 1996, 1998) and hybrid poplar (Dickson et al., 1998), and are in general agreement with reported elevated O_3 effects on woody plants (Bortier et al., 2000). Again, height growth was not affected by elevated O_3, while diameter growth decreased markedly as we found in our previous open-top chamber studies with aspen and hybrid poplar (Karnosky et al., 1996; Dickson et al., 1998).

If the predicted increases of atmospheric CO_2 and O_3 for the year 2100 occur concomitantly, our results suggest that elevated CO_2 will not ameliorate the negative effects of elevated O_3 on growth of aspen. In fact overall, the volume growth would be about 6% less than growth of the control. Again, these results assume that all interacting related abiotic and biotic factors would remain the same, which is unlikely. Our findings are counterintuitive with respect to results of most growth chamber and open-top chamber experiments with other plants (Olszyk and Wise, 1997; Barnes and Wellburn, 1998; Grams and Matyssek, 1999; Donnelly et al., 2000), and suggests that there is some doubt as to whether open-top experiments with aspen showing an amelioration effect (Volin and Reich, 1996, Volin et al., 1998) are valid under field conditions. No significant effect of elevated $CO_2 + O_3$ would be expected on height and/or diameter in aspen.

In our experiment, the growth response dynamics of aspen under elevated CO_2 and O_3 varied significantly by clone (Fig. 4). Moreover, the elevated CO_2 and O_3 effects in our FACE experiment developed over time. But, unlike our previous studies that showed clone 259 as an O_3 sensitive clone and clone 216 tolerant (Karnosky et al., 1996), we found that the five clones responded quite similarly to long-term field exposure to elevated CO_2 and O_3. It is notable that the faster-growing clones 271 and 42E responded negatively to the combination treatment of elevated $CO_2 + O_3$ over time. In addition, one of the clones, namely 8L, was ozonephilic throughout the three treatment years, growing greater than the control in all three years. Also, clone 8L exhibited a unique response to elevated $CO_2 + O_3$, in that it exhibited decreased volume growth negating the positive effect of either the elevated CO_2 and elevated O_3 treatment alone (1999 and 2000, Fig. 4). These dynamic and counterintuitive results illustrate the importance of long-term field experiments of climate change variables as opposed to short-term chamber experiments.

Another important prediction of the IPCC (2001) report was the extension of the freeze-free period in the autumn. In 1999, the growing season was extended by over 50 days. During that period, clones 271 and 216 grew late in the growing season in the elevated CO_2 treatments. They did not grow later in nearby ambient experimental plots (Karnosky et al., 1999a). The observed dormant season dieback that occurred in these clones under elevated CO_2 illustrates the potential negative effect that elevated CO_2 can have on plant growth. Apparently, certain aspen clones will be more vulnerable to freezing under an elevated CO_2 scenario, because over-wintering buds on the current terminal shoots do not adequately harden. This negative effect of elevated CO_2 prolonging the growing season has been observed by others (Lutze et al. 1998) and is yet another important ramification of global climate change.

There are many complex interacting factors in our experiment. One of the most important considerations in interpretation of the field results is soil nitrogen (N) availability. Many studies have shown the importance of soil N to understanding elevated CO_2 experiments (Oren et al., 2001). Kubiske et al. (1998) and Zak et al. (2000) showed that N levels greatly influence the results of elevated CO_2 with aspen and poplars. In their studies, lower N availability exhibited a lesser response to elevated CO_2. These findings have been found to be generally true of all plants (Jones and Curtis, 1999). Our experimental site had a history of agriculture prior to the establishment of our FACE study. As a result, the site remains high in soil N even after four growing seasons (Dickson et al., 2000). Moreover, Sober et al. (2003) have shown the importance of N nutrition to interpretation of our FACE study results. Leaf N content is highly correlated with photosynthetic rate responses under elevated CO_2 and O_3. In their studies, they found that leaf N is declining over the course of the three treatment years in our experiment. If soil N is quite high at our site, but is declining, and if soil N is often limiting in northern forests, one must ask the questions: (1) "How would aspen trees respond to elevated CO_2 when soil N is limiting?", and (2) "Is elevated CO_2 likely to draw down available soil N needed for sustainable growth of aspen in future climates?".

Another important consideration raised in our field study is "what effect will the differential growth response of clones to elevated CO_2 and O_3 have with stand development?" Clone 259 is rapidly being over-topped over time by the competing faster-growing clones. Moreover, clones 271 and 8L are, after three treatment years, assuming a more dominant role in the aspen clone mixture. Clonal mixtures are a common feature of aspen stands (Berrang et al., 1989) and our results suggest that certain aspen clones will not compete well with neighbors under future climate scenarios (McDonald et al., 2002). Such a response to elevated CO_2 and O_3 could lead to significant mortality in aspen stands in the future. In fact, we have observed substantial mortality in clone 259 in aspen clone mixtures growing in an elevated O_3 ambient environment near Kenosha, WI at present (Karnosky et al., 1999b).

Unfortunately, not all biotic stressors will remain constant with respect to any future climate scenario. We found in open-top chamber experiments that elevated CO_2 and O_3 and their combination affects chemical composition of aspen foliage, which in turn influences the incidence and severity of insect attack and disease occurrence (Isebrands et al., 2000). Herms et al. (1996) showed that insect feeding and larvae growth were affected by elevated CO_2 and O_3 with aspen. With elevated CO_2, the leaf feeding insects consumed more material because of the lower N content per unit area of leaves. Preliminary results at the Aspen FACE experiment suggest that both leaf feeding and wood boring insect populations, as well as their larval growth, will be affected by elevated CO_2 and O_3 (Mattson et al., unpublished).

Moreover, we have found in our Aspen FACE experiment that disease incidence and severity of the pathogens *Melampsora* and *Venturia* vary by clone and treatment. For example, the percent of trees infected by *Venturia* in 2000 was significantly different among clones and treatments, and the treatment x clone interactions were highly significant (Karnosky et al., 2002). These findings suggest that in an elevated CO_2 and O_3 world in 2100, pests and diseases may become very important interacting stressors for aspen stands. These stressors will likely have an impact on growth, survival and productivity of aspens in the future, and they should be taken into account by any models aimed at predictions of future climate effects on aspen forests.

The mechanisms of the counterintuitive effects of elevated CO_2 and O_3 on aspen clones that we observed are not yet clearly understood. However, we have some mounting anatomical, biochemical, and physiological evidence that may help explain our observed growth responses in aspen. Elevated O_3 and $CO_2 + O_3$ have direct effects on leaf surface chemical composition. For example, the epicuticular wax chemical composition varied by clone and treatment and may be implicated in the above-mentioned host pathogen/insect interactions (Percy et al., 2000). In addition, we find that elevated CO_2 did not ameliorate the negative effect of O_3 on biochemical and molecular responses (Karnosky et al., 1998). For example, significant decreases in chlorophyll, carotenoids, starch and Rubisco concentrations were observed in aspen under elevated CO_2 and O_3 in our experiment (Oksanen et al., unpublished). These results may explain why we found counterintuitive decreases in photosynthesis in aspen clones grown under elevated CO_2 and O_3 in open-top chambers (Kull et al., 1996) and in the field at Aspen FACE (Noormets et al., 2001).

In summary, the results of our research team at the Aspen FACE experiment suggest that for trembling aspen the favorable reports and model predictions of enhanced growth and productivity of seedlings grown in chambers under elevated CO_2 (Saxe et al., 1998) probably overestimate field performance (Norby et al., 1999). Many questions remain concerning the effects of interacting abiotic and biotic stressors on aspen growth. We found in our experiments that relatively low concentrations of tropospheric ozone (i.e., currently present in regions with aspen) dramatically decrease growth and productivity of an aspen stand. Moreover, the fact that soil N will probably decrease over time, will likely further decrease growth of aspen under elevated CO_2 and O_3. Add to that decrease the enhanced probability that dieback will occur in certain aspen clones with lengthened freeze-free periods by the year 2100, and the differential competition effects of elevated CO_2 and O_3 will have on growth of certain clones. In addition, there are higher probabilities that aspen damaging insects and diseases will increase under future climate scenarios. All in all based on these aforementioned considerations, the complex effects of elevated CO_2 and O_3, and climate change along with associated abiotic and biotic stressors are not likely, in our view, to enhance aspen growth, survival, and productivity in the future.

Acknowledgements

We gratefully acknowledge the support for this research by the US Department of Energy, Office of Biological and Environmental Research (BER) (DE-FG02-95ER62125), the USDA Forest Service Northern Global Change Program, the Canadian Forest Service, the USDA Forest Service, North Central Research Station and Michigan Technological University. We also thank Audra Kolbe for her valuable technical assistance and data analysis, and Penny Kluetz and Janet Pikkarainen for typing the manuscript.

References

Adams, R.M., Glyer, J.D., Johnsen, S.L., McCarl, B.A., 1989. A reassessment of the economic effects of ozone on US agriculture. J. Air Pollut. Control Assoc. 39, 960–968.

Allen Jr., L.H., 1990. Plant responses to rising carbon dioxide and potential interactions with air pollutants. J. Environ. Quality 19, 15–34.

Barnes, J.D., Pfirrman, T., Steiner, K., Lutz, C., Busch, U., Kuchenhoff, H., Payer, H.D., 1995. Effects of elevated CO_2, elevated O_3, and potassium defining on Norway spruce. Plant Cell Environ. 18, 1345–1357.

Barnes, J.D., Wellburn, A.R., 1998. Air pollutant combinations. In: De Kok, L.J., Stulen, I. (Eds.), Responses of Plant Metabolism to Air Pollution and Global Change. Backhuys Publishers, Leiden, The Netherlands, pp. 147–164.

Berrang, P.C., Karnosky, D.F., Bennett, J.P., 1989. Natural selection for ozone tolerance in *Populus tremuloides* II. Field verification. Can. J. Forest Res. 19, 519–522.

Bortier, K., Ceulemans, R., Temmerman, L., 2000. Effects of tropospheric ozone on woody plants. In: Agrawal, S.B., Agrawal, M. (Eds.), Environmental Pollution and Plant Responses. CRC Press, pp. 153–174.

Ceulemans, R., Mousseau, M., 1994. Effects of elevated atmospheric CO_2 on woody plants. New Phytol. 127, 425–446.

Ceulemans, R., Janssens, I.A., Jach, M.E., 1999. Effects of CO_2 enrichment on trees and forests: Lessons to be learned in view of future ecosystem studies. Ann. Bot. 84, 577–590.

Coleman, M.D., Dickson, R.E., Isebrands, J.G., Karnosky, D.F., 1995a. Carbon allocation and partitioning in aspen clones varying in sensitivity to tropospheric ozone. Tree Physiol. 15, 593–604.

Coleman, M.D., Isebrands, J.G., Dickson, R.E., Karnosky, D.F., 1995b. Photosynthetic productivity of aspen clones varying in sensitivity to tropospheric ozone. Tree Physiol. 15, 585–592.

Curtis, P.S., Vogel, C.S., Pregitzer, K.S., Zak, D.R., Teeri, J.A., 1995. Interacting effects of soil fertility and atmospheric CO_2 on leaf area growth and carbon gain physiology in *Populus* x *euramericana* (Dode) Guinier. New Phytol. 129, 253–263.

Curtis, P.S., Vogel, C.S., Wang, X., Pregitzer, K.S., Zak, D.R., Lussenhop, J., Kubiske, M., Teeri, J.A., 2000. Gas exchange, leaf nitrogen, and growth efficiency of *Populus tremuloides* in a CO_2-enriched atmosphere. Ecol. Applic. 10, 3–17.

Crow, T.R., 1988. A guide to using regression equations for estimating tree biomass. North. J. Appl. For. 5, 15–22.

DeLucia, E.H., Hamilton, J.G., Naidu, S.L., Thomas, R.B., Andrews, J.A., Finzi, A., Lavine, M., Matamala, R., Mohan, J.E., Hendrey, G.R., Schlesinger, W.H., 1999. Net primary production of a forest ecosystem with experimental CO_2 enrichment. Science 294, 1177–1179.

Dickson, R.E., Coleman, M.D., Riemenschneider, D.E., Isebrands, J.G., Hogan, G.D., Karnosky, D.F., 1998. Growth of five hybrid poplar genotypes exposed to interacting elevated CO_2 and O_3. Can. J. Forest Res. 28, 1706–1716.

Dickson, R.E., Lewin, K.F., Isebrands, J.G., Coleman, M.D., Heilman, W.E., Riemenschneider, D.E., Sober, J., Host, G.E., Hendrey, G.R., Pregitzer, K.S., Karnosky, D.F., 2000. Forest atmosphere carbon transfer storage (FACTS II)—The aspen free-air CO_2 and O_3 enrichment (FACE) project in an overview. USDA Forest Service, North Central Experiment Station, General Technical Report, NC-214.

Donnelly, A., Jones, M.B., Burke, J.I., Schnieders, B., 2000. Elevated CO_2 provides protection from O_3 induced photosynthetic damage and chlorophyll loss in flag leaves of spring wheat (*Triticum aestivum* L., cv 'Minaret'). Agric. Ecosyst. Environ. 80, 159–168.

Fowler, D., Cape, J.N., Coyle, M., Flechard, C., Kuylenstierna, J., Hicks, K., Derwent, D., Johnson, C., Stevenson, D., 1999. The global exposure of forests to air pollutants. Water Air Soil Pollut. 116, 5–32.

Grams, T.E.E., Matyssek, R., 1999. Elevated CO_2 counteracts the limitation by chronic ozone exposure on photosynthesis in *Fagus sylvatica* L.: Comparison between chlorophyll fluorescence and leaf gas exchange. Phyton (Austria) 39, 31–40.

Hendrey, G.R., Ellsworth, D.S., Lewin, K.F., Nagy, J., 1999. A free air enrichment system for exposing tall forest vegetation to elevated atmospheric CO_2. Global Change Biol. 5, 293–309.

Herms, D.A., Mattson, W.J., Karowe, D.N., Coleman, M.D., Trier, T.M., Birr, B.A., Isebrands, J.G., 1996. Variable performance of outbreak defoliators on aspen clones exposed to elevated CO_2 and O_3. In: Hom, J., Birdsey, R., O'Brian, K. (Eds.), Proceedings, 1995 Meeting of the

Northern Global Change Program, March 14–16, Pittsburgh, PA, Radnor PA. US Department of Agriculture, Forest Service, Northeastern Forest Experiment Station, General Technical Report NE-214, pp. 43–55.

Hogsett, W.E., Weber, J.E., Tingey, D., Herstrom, A., Lee, E.H., Laurence, J.A., 1997. Environmental auditing: an approach for characterizing tropospheric ozone risk to forests. Environ. Manag. 21, 105–120.

Intergovernmental Panel on Climate Change (IPCC), 1998. The regional impacts of climate change: An assessment of vulnerability. In: Watson, R.T., Zinyowera, M.C., Moss, R.H., Dokken, D.J. (Eds.), Special Report of IPCC Working Group II. Cambridge Univ. Press, New York, NY.

IPCC 2001. Third Assessment Report. Summary for policy makers. http://www.ipcc.ch/.

Isebrands, J.G., Dickson, R.E., Rebbeck, J., Karnosky, D.F., 2000. Interacting effects of multiple stresses on growth and physiological processes in northern forests. In: Mickler, R.E., Birdsey, R.A., Hom, J. (Eds.), Responses of Northern US Forests to Environmental Change. In: Ecological Studies, Vol. 139. Springer-Verlag, pp. 149–180.

Jones, M.H., Curtis, P.S., 1999. Bibliography on CO_2 Effects on Vegetation and Ecosystems: 1990–1999, The Ohio State University (OSU) and produced in cooperation with the Institute for Scientific Information. http://www.biosci.ohio-state.edu/~pcurtis/.

Karnosky, D.F., Gagnon, Z.E., Dickson, R.E., Coleman, M.D., Lee, E.H., Isebrands, J.G., 1996. Changes in growth, leaf abscission, and biomass associated with seasonal tropospheric ozone exposures of Populus tremuloides clones and seedlings. Can. J. Forest Res. 26, 23–37.

Karnosky, D.F., Podila, G.K., Gagnon, Z., Pechter, P., Akkapeddi, A., Coleman, M., Dickson, R.E., Isebrands, J.G., 1998. Genetic control of responses to interacting O_3 and CO_2 in Populus tremuloides. Chemosphere 36, 807–812.

Karnosky, D.F., Mankovska, B., Percy, K., Dickson, R.E., Podila, G.K., Sober, J., Noormets, A., Hendrey, G., Coleman, M.D., Kubiske, M., Pregitzer, K.S., Isebrands, J.G., 1999a. Effects of tropospheric O_3 on trembling aspen and interaction with CO_2: Results from an O_3-gradient and a FACE experiment. J. Water Air Soil Pollut. 116, 311–322.

Karnosky, D.F., Percy, K.E., Mankovska, B., Dickson, R.E., Isebrands, J.G., Podila, G.K., 1999b. Genetic implications for forest trees of increasing levels of greenhouse gases and UV-B radiation. In: Matyas, C. (Ed.), Forest Genetics and Sustainability. In: Forest Gen. Sust., Vol. 63, pp. 111–124.

Karnosky, D.F., Percy, K.E., Xiang, B., Callan, B., Hopkin, A., Dickson, R., Isebrands, J.G., 2002. Interacting elevated CO_2 and tropospheric ozone disposes aspen to infection by rust (Melampsora spp.). Global Change Biol. 8, 329–338.

Keeling, C.D., Whort, T.P., Wahlen, M., van der Plicht, J., 1995. Interannual extremes in the rate of rise of atmospheric carbon dioxide since 1980. Nature 375, 666–670.

Kozlowski, T.T., Constantinidou, H.A., 1986. Responses of woody plants to environmental pollution. Forestry Abstracts 47, 5–51.

Kubiske, M.E., Pregitzer, K.S., Zak, D.R., Mikan, C.J., 1998. Growth and C allocation of Populus tremuloides genotypes in response to atmospheric CO_2 and soil N availability. New Phytol. 140, 251–260.

Kull, O., Sober, A., Coleman, M.D., Dickson, R.E., Isebrands, J.G., Gagnon, Z., Karnosky, D.F., 1996. Photosynthetic responses of aspen clones to simultaneous exposures of ozone and CO_2. Can. J. Forest Res. 26, 639–648.

Littell, R.C., Milliken, G.A., Stroup, W.W., Wolfinger, R.D., 1996. SAS® System for Mixed Models. SAS Institute Inc, Cary, NC.

Marenco, A., Gouget, H., Nedelec, P., Pages, J.P., 1994. Evidence of a long term increase in tropospheric ozone from Pic du Midi series: consequences: Positive radiative forcing. J. Geophys. Res. 99 (D8), 16617–16632.

McDonald, E.P., Kruger, E.L., Riemenschneider, D.E., Isebrands, J.G., 2002. Competitive status intherences tree-growth responses to elevated CO_2 and O_3 in aggrading aspen stands. Functional Ecology 16, 792–801.

Noormets, A., McDonald, E.P., Kruger, E.L., Söber, A., Isebrands, J.G., Dickson, R.E., Karnosky, D.F., 2001. The effect of elevated carbon dioxide and ozone on leaf-and branch level photosynthesis and potential plant level carbon gain in aspen. Trees 15, 262–270.

Norby, R.J., Wullschleger, S.D., Gunderson, C.A., Johnson, D.W., Ceulemans, R., 1999. Tree responses to rising CO_2 in field experiments: implications for the future forest. Plant Cell Environ. 22, 683–714.

Olszyk, D.M., Wise, C., 1997. Interactive effects of elevated CO_2 and O_3 on rice and *flacca* tomato. Agricult. Ecosyst. Environ. 66, 1–10.

Oren, R., Ellsworth, D.S., Johnsen, K.H., Phillips, N., Ewers, B.E., Maier, C., Schäfer, K.V.R., McCarthy, H., Hendrey, G., McNulty, S.G., Katul, G.G., 2001. Soil fertility limits carbon sequestration by forest ecosystems in a CO_2-enriched atmosphere. Nature 411, 469–472.

Percy, K.E., Karnosky, D.F., Mankovska, B., Sober, J., Isebrands, J.G., Hendrey, G., Pregitzer, K., 2000. Interactive O_3 and CO_2 effects on leaf surface physiochemical characteristics in paper birch (*Betula papyrifera*) Marsh and three aspen (*Populus tremuloides* Michx.) clones: results from the aspen FACE project (FACTS II). In: Proc. Air Pollution, Global Change and Forests in the New Millennium, The 19th International Meeting for Specialists in Air Pollution Effects on Forest Ecosystems, Houghton, Michigan USA, p. 67, Abstract.

Ryerson, T.B., Trainer, M., Holloway, J.S., Parrish, D.D., Huey, L.G., Sueper, D.T., Frost, G.J., Donnelly, S.G., Schauffler, S., Atlas, E.L., Kuster, W.C., Goldan, P.D., Hubler, G., Meagher, J.F., Fehsenfeld, F.C., 2001. Observations of ozone formation in power plant plumes and implications for ozone control strategies. Science 292, 719–723.

SAS Institute Inc., 2000. Computer Software, SAS® System for Windows: Release 8.1. SAS Institute, Cary, NC.

Saxe, H., Ellsworth, D.S., Heath, J., 1998. Tree and forest functioning in an enriched CO_2 atmosphere Tansley Review No. 98. New Phytol. 139, 395–436.

Smith, S.D., Huxman, T.E., Zitzer, S.F., Charlet, T.N., Housman, D.C., Coleman, J.S., Fenstermaker, L.K., Seemann, J.R., Nowak, R.S., 2000. Elevated CO_2 increases productivity and invasive species success in an arid ecosystem. Nature 408, 79–82.

Sober, A., Noormets, A., Isebrands, J.G., Dickson, R.E., Sober, J., Karnosky, D.F., 2003. Photosynthetic parameters in aspen grown with interacting elevated CO_2 and tropospheric ozone concentrations as affected by leaf nitrogen. Tree Physiol. (in press).

Stevenson, D.S., Johnson, C.E., Collins, W.J., Derwent, R.G., Shine, K.P., Edwards, J.M., 1998. Evolution of tropospheric ozone radiative forcing. Geophys. Res. Lett. 25, 3819–3822.

Taylor Jr., G.E., Johnson, D.W., Andersen, C.P., 1994. Air pollution and forest ecosystems: a regional to global perspective. Ecol. Applic. 4, 662–689.

Volin, J.C., Reich, P.B., 1996. Interaction of elevated CO_2 and O_3 on growth, photosynthesis and respiration of three perennial species grown in low and high nitrogen. Physiol. Plant. 97, 674–684.

Volin, J.C., Reich, P.B., Givnish, T.J., 1998. Elevated carbon dioxide ameliorates the effects of ozone on photosynthesis and growth: species respond similarly regardless of photosynthetic pathway or plant function group. New Phytol. 138, 315–325.

Wang, X., Curtis, P.S., Pregitzer, K.S., Zak, D.R., 2000. Genotype variation in physiological and growth responses of *Populus tremuloides* to elevated atmospheric CO_2 concentration. Tree Physiol. 20, 1019–1028.

Zak, D.R., Pregitzer, K.S., Curtis, P.S., Vogel, C.S., Holmes, W.E., Lussenhop, J., 2000. Atmospheric CO_2 soil-N availability, and allocation of biomass and nitrogen by *Populus tremuloides*. Ecol. Applic. 10, 34–46.

Chapter 24

State of science and gaps in our knowledge in relation to air pollution

M. Ferretti

LINNAEA, Ambiente Srl., Via G. Sirtori, 371-50137 Firenze, Italy

J. Bucher

Swiss Federal Institute for Forest, Snow & Landscape Research, Zürcherstrasse 111, CH-8903 Birmensdorf, Switzerland

A. Bytnerowicz

USDA Forest Service, Pacific Southwest Research Station, 4955 Canyon Crest Drive, Riverside, CA 92507, USA

W. Prus-Glowacki

Adam Mickiewicz University, Faculty of Biology, Department of Genetics, Miedzychodzka 5, PL-60 371 Poznan, Poland

D.F. Karnosky

School of Forestry and Wood Products, Michigan Technological University, 1400 Townsend Drive, Houghton, MI 49931-1295, USA

K.E. Percy*

Natural Resources Canada, Canadian Forest Service-Atlantic Forestry Centre, P.O. Box 4000, Fredericton, New Brunswick E3B 5P7, Canada

Abstract

The IUFRO Research Group 7.04.00 "Impacts of Air Pollutants on Forest Ecosystems" in 1998 adopted, as one of its outputs from biennial international meetings, production of a series of general statements [Percy, et al., 1999, Water Air Soil Pollut. 116, 443–448] on the state of science from each of its Working Parties (WP). These discipline-oriented WP have continued to evolve in focus in tune with emerging issues and overall international research direction. The second IUFRO 7.04.00 report in this series emanating from concurrent sessions held during the 20th International Meeting for Specialists in Effects of Air Pollution on Forest Ecosystems "Air Pollution, Global Change and Forests in the New Millennium" follows.

* Corresponding author.

DOI:10.1016/S1474-8177(03)03024-9

1. Diagnosis, monitoring and evaluation

1.1. State of science

Ozone (O_3) continues to be of major concern in the field of air pollution effects on forest ecosystems. The need for adequate mapping of risks, identification and monitoring of symptoms, evaluation of effects and further research in this field were formerly mentioned as high priority areas (Percy et al., 1999) and were actually addressed in the meeting. Progress in mapping potential O_3 risks was presented, suggesting that effects-oriented monitoring programs could take advantage by preliminary identification of target areas. This is thought to also be a step ahead in orienting the research toward more politically under-standable end-points. Identification, confirmation and monitoring of O_3 symptoms were reported for both southern Europe (Lombardy, Northern Italy) and North America (Great Lakes States, USA). In southern Europe, the changes of species assemblage along with altitudinal gradients, prevent any clear relationship with O_3 exposure being identified. However, investigations carried out in the US provide convincing results that symptoms on native tree species correlate well with O_3 exposure indices.

These findings can obviously provide further advantages when used to validate risk maps. Besides symptoms, effects on tree performance are of obvious interest. In this respect, novel results were provided for beech and, to a lesser extent, Norway spruce growing in permanent monitoring plots in Switzerland. After removing the effects of concurrent factors, O_3 was reported to explain a considerable part of the variance in shoot growth for beech. Together with field surveys and investigations, the experimental approach remains important, especially to clarify what could be the effects of pollutants under specific conditions and/or to provide modellers with data that can be helpful when attempting to upscale the results of field measurements. In particular, competition was reported as an important factor to take into account when the response at the forest ecosystem level is to be investigated. Experimental results presented at the meeting actually indicate that competition is an important factor to take into account when attempting to evaluate the impact of O_3 on tree growth.

Together with O_3, exceedance of critical loads for acidity and nitrogen remains a high priority for today's air pollution researchers. Findings of increases in tree mortality and nitrogen content (both in humus layer and needles) and lowering of other nutrients in areas where critical loads are exceeded were reported for Norway spruce.

Additional investigations addressed different forests and air pollution issues. The effects of bauxite mining on different habitats were reported for India, where an important concern is placed on the need to address the social aspects related to the evaluation of air pollution effects on forests.

Following a monitoring approach widely adopted in Europe, a variety of measurements (deposition, air chemistry, foliage chemistry, tree condition, needle wax) was presented from Polish national parks.

1.2. Knowledge gaps

The role of scientists in existing monitoring activity, the value of critical levels/loads approach, the need for the research and the social aspects to be considered in, and outside, the developing countries were among the major discussion issues. Monitoring activity in the field of air pollution effects on the forest ecosystem is currently carried out through programs that, in some cases, are subjected to strong political control. Although policy is necessarily involved when setting priorities, identification of adequate methods of investigation and interpretation of results should be carried out by the scientific community. Unfortunately, this is not always the case, partly because communication and cooperation between scientists still need improvement and partly because the requirements that science may have in terms of quality of results and robustness of evidence may be problematic for politicians. At the end, however, there is a clear need to strengthen the role of scientists.

Aside from wide acceptance as operational tools, different settings of critical levels and loads are obvious between Europe and North America and this is a field where closer cooperation is needed. Here, a further problem is created by the need for definitions of exposure indices to be easily understood by many categories of users, within which politicians play a key role. Besides debating the value of a threshold against one another, modelers and ecologists should work together.

Stability and integrity of wilderness areas are of global interest and their monitoring is important. However, monitoring could be difficult in these areas because of a number of constraints, including costs and operational difficulties. Therefore, adequate investigation techniques should be proposed. Investigations into the effects of air pollution on forests can be difficult in developing countries, mostly because of financial constraints. In addition, there could be some reluctance to support studies about air pollution effects on forests in those places where air pollution is associated with some form of economical development. However, there is evidence that the contribution of developing countries to global air pollution will increase in the next decades (Fowler et al., 1999) and thus it is important to deal with the above problems, especially given that information gaps are much higher there.

From a different point of view, consideration about social aspects related to air pollution effects on forests is also relevant for developed countries. As was pointed out, demographic changes, population movements, increases in urban areas and mobility, and a comprehensive reshaping of needs and work access will ultimately result in both changes in air pollution and demand for environmental quality. There is certainly a considerable need to explore the role of concurrent factors in determining the effects of a given pollutant. However, experiments and investigations in this field are scarce, and much needed.

Although investigations concerning O_3 effects on forests are well underway, a number of gaps, needs and improvement areas were identified. The role of scientists in designing and implementing monitoring programs is often frustrated by the need for quick and sometimes prepackaged answers suggested by politicians. To strengthen the role of scientists in existing programs, communication and cooperative meetings should be encouraged. On the other hand, researchers should take more responsibility, be more understandable and more willing to draw conclusions when possible. Strengthening the role of scientists will be beneficial for the credibility of monitoring programs, and help avoid biased interpretation of the monitoring results. Despite steady improvements in communication tools over the past years, communication between scientists remains an area where substantial progress has to be made. Benefits accruing will include increased research opportunities worldwide and better harmonization of methods. Thus, communication between scientists continues to be an important area of development in many different respects.

The critical loads approach, coupled with adequate mapping of exposures, can be of considerable help in identifying areas where forests may be at risk due to air pollution. Adequate, preferably low-cost, and easy-to-run investigation techniques (e.g., passive samplers for O_3, bioindicators, ...) should be identified, tested and disseminated for applicability in wilderness areas. Together with investigation techniques, care should be devoted to select an appropriate design for the field studies. Upscaling is increasingly important as site-specific data often have little meaning for policy makers. However, upscaling needs adequate models and data. Although long-term monitoring programs can provide much basic information, the experimental approach is important to clarify specific questions about the response of trees. Monitoring and experiments, especially conducted at large scale, should not be alternatives, but complementary parts. Proper experiments should be carried out to have adequate evaluation of the effects of factors such as competition on the responsiveness of forests to pollutants.

Although there are many questions to be solved, studies about air pollution effects on forests are widely carried out in developed countries. At the same time, the need for such studies is dramatically increasing in many countries currently under rapid development, where the resources to be devoted to such studies are limited. As the development of these countries is of global interest, a major challenge for the future will be to try and find some way to support scientists from these countries. The easiest way to do this, once again, is through improved communication, ensuring they are well informed on the progress made in this field. Stronger support through cooperation is advisable and should be encouraged.

2. Mechanisms of action and indicator development

2.1. State of science

Presentations focused on four main topics: molecular mechanisms involved in O_3 injury, ecophysiological responses to long-term CO_2–O_3 interactions, modelling O_3 effects on gas exchange, and bioindication by N/P dynamics or wood anatomy. By using marker genes, it has been determined that O_3-, pathogen-, and wound-induced expression of both salicylic acid- and jasmonic acid-regulated defense genes, was attenuated in the O_3-sensitive hybrid clone. Cell death in the tolerant clone was caused by induction of a salicylic acid-dependent programmed cell-death pathway (a hypersensitive response), while cell death in the sensitive clone was likely to be caused by the lack of a sufficient level of antioxidant defences.

Gas exchange and water relations were the main parameters examined under elevated CO_2 and O_3. Carbon dioxide mitigated the negative effects induced by O_3, but there were indications suggesting some species and some genotypes within species may have their sensitivity changed by elevated CO_2 and by presence of aboveground competition. A new effect of CO_2 and O_3 on leaf area development was also identified; depending on the developmental stage of the leaf, the effect of both O_3 and CO_2 can either inhibit or stimulate the foliar distension rate.

Findings on nitrogen and phosphorus dynamics stressed once more the importance of climatic factors in determining plant response to elevated O_3 and CO_2. Wood production under elevated pollution was also examined, in terms of both quantity and quality.

Finally, two approaches for modelling O_3 uptake during different episodes were developed. In aspen, O_3-induced growth responses were simulated using a functional and structural tree growth model with built-in, process-based photosynthesis routines for sun and shade leaves. In another species and site-specific model it was possible to calculate stomatal O_3 uptake in ponderosa pine on the basis of episodic micrometeorological data only.

2.2. Knowledge gaps

Ozone degrades to Reactive Oxygen Species (ROS), such as superoxide, hydroxyl radicals and hydrogen peroxide in the plant cell wall, and elicits active ROS production in plant cells, similar to the oxidative burst in plant–pathogen interactions. The role of salicylic and jasmonic acids as mediators/modulators of ROS-induced plant responses has been elucidated by different research groups and by using different plant species (*Arabidopsis, Betula pendula*, and a hybrid poplar). Such results indicate a strong requirement for the extension of the molecular approach to other tree species, in order to provide information on the genetic and mechanistic basis of natural variation in O_3 sensitivity.

In terms of scaling up from molecules to ecosystems, gaps still remain with special reference to root physiology, plant competition, and progeny fitness. A new approach has to be developed to increase the scale of investigation. Good perspectives are implied in modelling, as a tool to coordinate and integrate experimental data, even if application might be difficult, as models are usually species- and site-specific. At any rate, models may help in defining the key processes and the key parameters to be investigated. Ozone flux entering the leaves appears to be more promising than an AOT40 approach, as it correlates better to physiological parameters. Further, it appears to be of special interest to determine how CO_2 can influence the detoxification processes in an increasing oxidizing environment.

On the more technical side, today there are different working tools. Among these, Free-Air Pollutant Exposure Systems (FACE) experiments are the most promising, but there was general agreement on the continued usefulness of conventional approaches such as OTC, field plots, or climate chambers to investigate specific processes.

The restructuring of this WP toward mechanisms of action and indicator development has opened new horizons and has provided new objectives for our biochemical and physiological research. This enables our group to adopt an explanatory and mechanistic approach, following the principles that all good research should pursue, and to meet the needs of modern research, combining ecophysiology, molecular biology, and modelling. In short, becoming more inter- and multidisciplinary will enable us to be more efficient in our research task of 'Mechanisms of action and indicator development'.

3. Atmospheric deposition, soils and biogeochemistry

3.1. State of science

The importance of dry deposition of gases and particles in the Mediterranean climate of southern California was emphasized. Dry deposition can provide as much as 90–95% of all atmospherically deposited nitrogen (N) to California forests and other ecosystems. Long-term elevated levels of N deposition change nutritional status of soils and affect forest and other ecosystems. For example, the coastal sage scrub community in southern California has experienced significant shifts in vegetation composition that may be attributed to these changes.

Nitrification is the key process leading to elevated N losses in the San Bernardino Mountains of southern California that experience long-term elevated levels of N deposition. As a result, streamwater is highly contaminated with nitrate at the highest levels for the undisturbed wildland watersheds in North America. At the high N deposition forest sites, after 3 years of N fertilization, growth of both pine and oak trees increased. It is suggested that due to the highly open nature of N cycling in the Mediterranean climate, stage 3 of the N saturation hypothesis (forest decline or reduced NPP) may be difficult to achieve in these forests.

A 4-year study on responses of ponderosa pine seedlings to elevated levels of CO_2 and O_3 is being performed in the outdoor exposure chambers. After two seasons of exposures, shoot growth of pines significantly increased at elevated levels of CO_2, especially at the low O_3 concentrations. When completed, this study will provide information on interactive effects of CO_2 and O_3 on biogeochemical cycles of pine/litter/soil mesocosm.

Ponderosa pine seedlings were grown with blue wild-rye grass to determine if the presence of natural competitors alters responses of pines to O_3. Grass presence significantly reduced total pine mass by nearly 50% after 3 years of O_3 exposure, but O_3 alone had no significant effects on pine growth. Competition for soil N, moisture, and light were the primary factors driving the responses of pines to grass competition.

Model ecosystems of beech and spruce seedlings were exposed to ambient and elevated levels of CO_2 and low and high wet N deposition for 4 years in open-top chambers. CO_2 effects on nutrient cycle and water relations were clearly species and soil dependent. Elevated CO_2 increased both growth and N-use efficiency of spruce on acidic loam and the calcareous sand, but in beech only on the nutrient-rich calcareous soils. This study suggested that elevated CO_2 increased N immobilization in soil, therefore, growth of forests may be restricted at the future CO_2-rich environment.

Short-term changes in dynamics of aspen and paper birch leaf decomposition due to elevated concentrations of CO_2 and O_3 were assessed in a litter bag experiment at the Aspen FACE study. This study revealed that effects of elevated CO_2 + O_3 levels were complex and could not be predicted from decay rates under elevated CO_2 or O_3 applied alone.

3.2. Knowledge gaps

Examples of research priorities stemming from discussions on knowledge gaps include the following: additional work on the effects of gaseous pollutants (O_3, CO_2 or HNO_3 vapor) on biogeochemical cycles and on the role of soil systems in responses of ecosystems to multiple stresses (air pollutants and other abiotic and biotic stressors) should be undertaken. Of course, effects of air pollution deposition on C sequestration and on the role of different forms of N deposition inputs on biodiversity changes as well as C sequestration need to be better understood.

The effects of deposition and biogeochemical changes in forests on water resources remain a priority in which relatively little work is concentrated. Development and evaluation of critical loads of N and S deposition for various ecosystems (taking into consideration complexities of systems in different geographical and ecological settings, effects of O_3 and CO_2, etc.) should be advanced.

Of major importance in the policy-setting framework is the development and use of models (linking atmospheric, plant and soil components) for evaluation of ecological risks from atmospheric deposition to plants, forest stands and landscapes. This must implicitly rely upon the development and use of new research approaches, such as large-scale use of passive samplers for air pollutants, air pollution gradient studies, etc.

4. Influence of air pollution and climate change on genetics, adaptation and succession

4.1. State of science

Presentations in this concurrent session focused on two main topics. The first topic centered on the genetic response of sensitive and tolerant clones of broadleaved trees (aspen, paper birch and silver birch) to elevated levels of O_3 and CO_2. Investigated responses were stomatal conductance, stomatal density, photosynthetic rate, rubisco activity, and epicuticular wax composition. Clear evidence of the impact of O_3 and of the interactive effect of O_3 and CO_2 was shown. The second main topic corresponded to the selection effects and differences in genetic structure of natural and semi-natural populations of Norway spruce, European silver fir, European beech and aspen exposed to O_3, heavy metals, or salts. Simulation of effects of global warming of climate on genetic structure of forest trees populations was also presented.

Individual case studies have shown clear evidence of large differences in response of various clones (genotypes) of silver birch, paper birch and aspen to elevated O_3 and CO_2 in terms of stomatal conductance, stomatal density, and also in the overall physiology of trees as measured by photosynthetic rate and rubisco activity. The influence of the combined effects of O_3 and CO_2 on epicuticular wax composition, structure, and relative amount of different wax compounds has also shown the negative impact of these gases on sensitive genotypes in O_3 gradient and FACE studies. The observation could have important implications for host-herbivore and pathogen interactions. Ozone impact on the competitive ability and fitness (mortality rate and volume relative growth) of clones of aspen sensitive and tolerant to O_3 is providing exciting new data demonstrating the selection for O_3 tolerance in aspen.

A case study on differences in genetic structure of groups of Norway spruce and European beech, both sensitive and tolerant, to industrial air pollution has demonstrated opposing trends for Norway spruce and European beech in terms of their isoenzyme variation. Simulation of climate warming and the influence of this phenomenon on genetic structure indicate that in populations of *Picea abies* trees from warmer conditions, a higher diversity occurs. However, different types of isoenzyme markers would be expected to produce different results. The influence of O_3 on genetic structure of the progeny from tolerant and sensitive *Picea abies* trees has shown clearly that selection processes are acting strongly on the embryos of sensitive trees.

The usefulness of new molecular markers (untranslated region of ribosomal protein gene) for heavy metal and salt stress in *Picea abies* was demonstrated.

4.2. Knowledge gaps

There is very little information on the dynamics and trends on changes connected to selection processes, and on the genetic structure of populations affected by air and soil pollution. There is an urgent need for more information about competition between genotypes in natural and artificial populations while exposed to various stress factors. There are also knowledge gaps with respect to adaptive strategies of particular forest tree species, as well as populations of these species, to different stress factors such as air and soil pollution. Such data are required in the case of elevated O_3, CO_2, and climate warming, to set policy for the protection of tree germplasm.

Little information is available on which isoenzymatic loci are diagnostic and could be used as good and reliable markers for selection/changes in genetic parameters for various tree species. We still need additional molecular markers, isoenzymatic, and DNA as indicators of genetic response to stress factors. We also need better linkage between the markers and physiology of plants under stress. For preservation of *in situ* and *ex situ* genetic diversity and genetic "richness" of forest tree populations as a necessary base for ability to adaptive processes, there is need for a monitoring study of tolerant stress factor populations and also "natural" ones.

In particular, there is insufficient information on the correlation between such phenotypic traits as growth, fertility, volume production, tolerance and sensitivity and molecular markers (isoenzymes and DNA). Interactions between different abiotic and biotic environmental stress factors also need to be further investigated.

References

Fowler, D., Cape, J.N., Coyle, M., Flechard, C., Kuylenstierna, J., Hicks, K., Derwent, D., Johnson, C., Stevenson, D., 1999. The global exposure of forests to air pollutants. Water Air Soil Pollut. 116, 5–32.

Percy, K., Bucher, J., Cape, J., Ferretti, M., Heath, R., Jones, H., Karnosky, D., Matyssek, R., Muller-Starck, G., Paoletti, E., Rosengren-Brinck, U., Sheppard, L., Skelly, J., Weetman, G., 1999. State of science and knowledge gaps with respect to air pollution impacts on forests reports from concurrent IUFRO 7.04.00 Working Party sessions. Water Air Soil Pollut. 116, 443–448.

Chapter 25

Air pollution and global change impacts on forest ecosystems: Monitoring and research needs

D.F. Karnosky*

School of Forest Resources and Environmental Science, Michigan Technological University, 101 U.J. Noblet Forestry Building, 1400 Townsend Drive, Houghton, MI 49931, USA

K.E. Percy

Natural Resources Canada, Canadian Forest Service-Atlantic Forestry Centre, P.O. Box 4000, Fredericton, New Brunswick E3B 5P7, Canada

A.H. Chappelka

Auburn University, School of Forestry & Wildlife Sciences, 206 M. White-Smith Hall, Auburn, AL 36849-5418, USA

S.V. Krupa

University of Minnesota, Plant Pathology Department, 495 Borlaug Hall, 1991 Upper Buford Circle, St. Paul, MN 55108, USA

Abstract

The start of the 21st century brings an increasing public awareness of environmental issues worldwide. In this book, we have attempted to present the current state of knowledge about CO_2 effects and related global warming on forest productivity and ecosystem function and to discuss research needs in that regard. In addition, we have discussed other air pollutants including O_3, nitrogen and sulfur compounds, and heavy metals. The status of those pollutants globally and some representative effects on forest trees and forest ecosystems have been presented. Certainly, there remains much to do in monitoring air pollutants, particularly in rural and forested areas. For example, very little is known about expanding pollutant loading in forest areas of developed countries or across both urban and forest areas in developing countries where the need to industrialize is generally outweighing the resources to control pollutant emissions. While great strides have been made in the past few decades to decrease acidic deposition in North America and Europe, there remain extensive areas of the world's forests being impacted by acidic deposition. Similarly, large areas (25%) of the world's forests are currently exposed to elevated levels of O_3 and this is projected to rise to fully 50% of the world's forests by the year 2100. Thus, research is needed to

---

* Corresponding author.

DOI:10.1016/S1474-8177(03)03025-0

document pollutant effects and to reduce uncertainties about forest productivity and forest ecosystem responses worldwide in the next century. In this chapter, we summarize the eight major research and monitoring needs for the investigation into air pollutant impacts on forests worldwide.

1. Introduction

Air pollution and climate change are two key factors threatening forest health and sustainability. Considerable scientific effort, mainly in northern hemisphere countries, has been devoted to the enhancement of our understanding of forest responses to global change at the process, organ, system, stand and ecosystem levels. Forest ecosystems around the world are being exposed to increasing levels of atmospheric CO_2 and changing climates. In addition continental and regional scale air pollutants, such as O_3 and acidic deposition, and numerous forms of S and N are impacting large sections of the world's forests.

Fowler et al. (1999) have calculated areas of forests where July peak surface O_3 concentrations have exceeded 60 ppb. In 1950, this area was largely restricted to the temperate latitude forests. By 1990, 25% of the forests were exposed to > 60 ppb July peak O_3. In 2100, fully 49.8% of the world's forests (17.0 million km^2) will likewise be exposed.

Fowler et al. (1999) have also estimated the area of global forests at risk from acidification (> 2 keq H^+ ha^{-1} yr^{-1} as S). They predicted a 21-fold increase in area of global forests at risk between 1985 (0.28 million km^2) and 2050 (5.9 million km^2), with the majority of the increase in sub-tropical and tropical forest regions.

In summary, forest trees and ecosystems are facing combinations of gradually increasing CO_2, warming temperatures, and changing seasonal phenology, often in concert with elevated air pollution (Houghton et al., 2001). Unless there is a strong downturn in global population growth and industrialization, forests will continue to be exposed to a deteriorating atmospheric environment. Areas of forests at risk from O_3, S, N and acidification are expanding under current economic and social trends. Modeling of future S and N scenarios for North America and Europe indicates that although the driver of acidification is changing (molar ratios in rain are now almost equal for N and S, whereas S in the past dominated for many decades), acidification potential in many areas remains high (http://nadp.sws.uiuc.edu).

The goal of this book has been to explore the impacts of air pollution and climate change on forests and forest ecosystems around the world. Trends in the major air pollutant types and their generalized effects on forests are summarized here in Table 1. In addition, we discuss the most outstanding monitoring and research questions still needing to be addressed by the international scientific community and list them in Table 2.

Table 1. The occurrence of major air pollutants are changing. In this table, we describe how these pollutants are changing and what is known about their impacts on forest ecosystems

Pollutant	Distribution and change	Forest effects	See discussion in chapters
CO_2	Increasing globally	• Short-term growth and productivity increase; long-term effects still undetermined • Implicated in global warming and predicted to cause massive shifts of species ranges	1–3, 19, 21–24
O_3	Global increases in O_3 and its precursors with largest increases coming in developing countries	• Growth and yield losses in sensitive species and impacts on relative fitness and on community dynamics • Predisposition of forest trees to insect and diseases	1, 2, 4–13, 19, 20, 22–24
Nitrogen (N) (NO_x, NH_3, etc.)	Global increases, particularly in developing countries (especially India and China)	• Stimulation of growth and productivity in N-poor soils • N-saturation in some forests causing decline and in many streams, ponds, and lakes causing fish species shifts and extinction • Contributes to increases in O_3	1, 2, 14, 15, 21, 24
Sulfur (SO_2, SO_x, H_2S, etc.)	Stable globally with steady decreases in the past few decades in developed countries but increasing in several developing countries (especially India and China)	• Acidified soils in many parts of the world are difficult to mitigate	1, 2, 14, 15, 21, 24
Heavy metals	Decreasing problem in developed countries, continued localized point-source problem in many developing countries and nations in transition	• Localized forest extinction and toxic soils which limit the ability to regenerate forests	1, 2, 17, 18, 24

2. Monitoring needs

1. Air pollutant concentrations are changing rapidly and these changes need to be documented (Table 2). For example, until recently, it was thought that daytime global CO_2 concentrations were similar. However, research has shown that while a daytime background level of ~ 360 ppm is relatively common around the world today, there are large metropolitan areas such as Phoenix

Table 2.   Important monitoring and research needs for the world's major air pollutants

Pollutant	Monitoring	Forest research[a]

CO_2	• Continued background CO_2 monitoring and expanded CO_2 monitoring in urban areas
	• Continued temperature measurements around the world
	• Continued phenological studies of forest trees
	• Examine trees along range margins where global warming responses are first likely to occur
		• Effects of elevated CO_2 and global warming on:
		  (a) Long-term growth and productivity
		  (b) Ecosystem-level responses such as biogeochemistry and cycling of nutrients, water balance, etc.
		  (c) Community dynamics and population changes (above- and below-ground)
		  (d) Effectiveness in carbon sequestration by various methods of afforestation, reforestation, and agroforestry
		  (e) Methods to increase carbon sequestration (amounts and duration of sequestration)
		• Interactions of elevated CO_2 with global warming, increased N deposition, or elevated O_3, etc.
		• Linking of FACE/FLUXNET networks to study C cycle

O_3	• Increased monitoring forest areas around the world • Increased monitoring in urban and rural areas of developing countries • Examine relationship of forest health to O_3 or other pollutants • Increased use of passive samplers for forest areas for monitoring • Link Forest Health Monitoring to O_3 monitoring • Link Forest Health monitoring to GIS, GPS, etc., to allow eventual satellite imagery use in forest health characterization	• Effects of elevated O_3 on stand, community, and ecosystem levels • Increased use of field techniques for ecosystem-level research such as FACE, O_3 gradients, and dendrochronology • Effects of elevated O_3, alone and in combination with other pollutants, on foliar symptoms and growth in developing countries • Increased understanding of why some trees or plants are sensitive to O_3 and some are not, facilitated by large scale gene expression studies • Effects of global change on volatile organic compound production • Effects of interacting O_3 and other pollutants

(*continued on next page*)

(Idso et al., 2000), Baltimore (Hom et al., 2001), and Chicago (Grimmond et al., 2002) where daytime CO_2 concentrations can be > 100 ppm higher. It is worth noting as well that night-time concentrations may be > 100 ppm higher in non-urban areas where plants are a source, rather than a sink at night (Legge and Krupa, 1990).

Table 2. (*Continued*)

Pollutant	Monitoring	Forest research[a]
N	• N deposition in developed countries and especially in developing countries with rapidly expanding automobile traffic and industry (i.e., India, China, Nepal, Chile, and Mexico) • Long-term monitoring of acidified streams, ponds, and lakes • Long-distance (intercontinental) N transport and the effects of this transported N on O_3 formation	• Effects of N additions in N-saturated or nearly N-saturated ecosystems • Effects of N additions to ecosystems experiencing other pollutants (i.e., CO_2, O_3, S, etc.) • Effectiveness of various N mitigation treatments on forest soils and waterways
S	• S deposition in countries in transition and in developing countries • Continued assessments of impacted forest ecosystems to ensure proper restoration	• Methods to mitigate long-term sulfur inputs into soils and to restore sustainable forest ecosystems • Effects of SO_2 and sulfur deposition on forest trees in developing countries, particularly tropical countries where little air pollution effects research has been done.
Heavy Metals	• Continued monitoring near point sources of heavy metals	• Mitigation and restoration of forest ecosystems devastated by heavy metal deposition

[a]While we highly endorse research to reduce emissions or the precursors of all of these major pollutants, we focus here on major knowledge gaps or research needs for forest ecosystems.

Other primary pollutants, such as nitrogen oxides, volatile organic compounds and sulfur oxides are transformed in the atmosphere, in some cases very rapidly and their products subjected to long-range transport. For those pollutants, the majority of effects may be much farther downwind of the source. For example, there are many Class I wilderness areas in the US and Canada that are being exposed to high levels of O_3 (NPS, 2002). In developing countries such as India, Malaysia, China, and also countries in transition in eastern Europe, with rapidly expanding economies, growing industrialization, rising populations of concern in some cases and increasing traffic volume, expanded air pollution monitoring is needed in and around major cities and in forested areas to document trends in air pollution occurrences (Gupta et al., 2002; Kimmel et al., 2002). Continued monitoring of restoration progress in previous seriously impacted forest ecosystems (as in the Black Triangle in central Europe (Fanta, 1994, 1997; Karnosky, 1997) is needed to determine if mitigation approaches and decreased emissions are helping those disturbed ecosystems.

2. With respect to the most pervasive air pollutant affecting forests, we know that ambient O_3 concentrations are highly variable in space and time and exhibit a dynamic flux to plant canopies (Krupa et al., 2003). Therefore, continued and expanded monitoring is needed in developed countries to better document pollutant deposition in forested areas (Bytnerowicz et al., 2002, Bytnerowicz et al., 2003; Zimmerman et al., 2003). The more widespread use of passive samplers has clearly extended the area for which O_3 data are now available, at least in the form of a single period, integrated concentration (Krupa and Legge, 2000). The subsequent development of a Weibull probability model and the most recent meteorology integrated statistical model to predict hourly ambient O_3 concentrations from single weekly passive sampler data (Krupa et al., 2001; Krupa and Nosal, 2001; Krupa et al., 2003) have enhanced the utility of passive samplers and their continued development and use should be expanded to decrease the uncertainty regarding O_3 concentrations in rural forested areas.

3. Forest health monitoring continues to be used in both the USA (Smith, 2002; USDA Forest Service, 2003) and Europe (DeVries et al., 2000; ICP, 2002) in an attempt to quantify visible symptoms and crown vigor and to link those conditions, among other causes, to ambient air pollution. Regrettably, there is no universally accepted definition of forest health (Percy, 2002). However, in the context of forest response to air pollution, the sustaining of ecosystem structure and function is especially important because process-oriented and pattern-oriented considerations have been shown to underpin any definition of forest health. McLaughlin and Percy (1999) have accordingly defined forest ecosystem health as the capacity to supply and allocate water, nutrients and energy in ways that increase or maintain productivity while maintaining resistance to biotic and abiotic stresses. This definition fits quite well within new forest health concepts built around issues such as long-term sustainability, resilience, maintenance of structure and functions and multiple benefits and products (Kolb et al., 1994). Thus, expanded visions are needed for forest health monitoring (Percy and Ferretti, 2003). For example, expanded use if GIS (Geographic Information System), GPS (Geographic Positioning System) and other satellite-based systems, linked to the ground through ground truthing, could be useful in Forest Health Monitoring in the future.

Approaches commonly used to assess forest health are generally inadequate for evaluation of trends, for detection of future change, and for elucidation of the roles of natural and anthropogenic stressors. Integrated approaches linking process-oriented empirical studies with pattern-oriented monitoring along defined spatial variations in pollution using clonal plantations (Karnosky et al., 1999), genetically-screened tree pairs (Muller-Starck et al., 2000), ecological analogues (Krupa and Legge, 1998), and ecosystem-based research on essential cycles (FACE or other non-exposure chamber based techniques) with better characterization of physical and chemical environments are needed. In turn, these will yield new approaches toward a statistically and conceptually sound monitoring system required if the interactive effects of global change (air pollution + climate change) on forest health and sustainability are to be understood in the 21st century. Long-term ecological research is essential to understand the status and trends in processes within forest ecosystems. Close linkages need to be made between atmospheric and ecological monitoring. For example, in the US, linking existing acidic deposition (NADP/NTN) and air quality (AIRs) monitoring networks with the Forest Health Monitoring program could yield some valuable insights into forest health.

4. Global warming, associated with greenhouse gas emission trapping of radiant energy near the earth's surface, has created concerns about species range shifts (Parmesan and Yohe, 2003; Alley et al., 2003; Bakkenes et al., 2002; Houghton et al., 2001; IPCC, 2001), extensive insect and disease outbreaks (Aber et al., 2001; Bale et al., 2002), forest fires and large-scale forest community changes. Extensive forest monitoring will need to be done to continue to document effects of elevated greenhouse gases and global warming on aboveground and belowground competitive interactions (McDonald et al., 2002; Poorter and Navas, 2003), canopy and soil community dynamics (Karnosky et al., 2001), and forest health (Percy and Ferretti, 2003). Continued monitoring of phenological gardens (Menzel and Fabian, 1999) and long-term studies along the edges of species ranges will be needed to document the rates and magnitude of climate change impacts on forest trees. Continued assessment of forest inventory plots in the US and other countries will also help document changes in forest productivity and community dynamics under climate change (Jenkins et al., 2003).

3. Research needs

1. Forest productivity continues to increase in the United States (USDA Forest Service, 2001) and Europe (Spiecker et al., 1996; Mäkinen et al., 2003). Whether this is due to rising CO_2 in the atmosphere, enhanced N deposition, or global warming is impossible to say (Aber et al., 2001). Long-term research projects are needed to address this question and to determine if these trends will continue. Little is yet known about whole ecosystem responses to either air pollution or global change (Aber et al., 2001; Karnosky et al., 2003). The current global networks of FACE (Hendrey et al., 1999; Karnosky et al., 2001) and Fluxnet (Buchmann and Schulze, 1999; Baldocchi et al., 2001; Baldocchi, 2003) offer opportunities to sort out trends in ecosystem productivity. Learning more about interactive effects of multiple stresses on realistic ecosystems will require experimental manipulations of ecosystems on a larger scale than yet

conducted (Parson et al., 2003). For example, FACE experiments with multiple, interacting stresses are needed to determine how forests will respond to combinations of air pollution and climate change (Karnosky et al., 2002; Karnosky, 2003). New, less costly and innovative designs are needed for exposing forest ecosystems simultaneously to elevated CO_2, warming environments and air pollutants. Effects of climate change and atmospheric pollution on forest soils and soil microorganisms remain largely uncertain (Zak et al., 2000), although impacts of these changes have been shown to affect soil biodiversity and are in need of additional research (Larson et al., 2002), given their crucial role in nutrient cycling and forest productivity.

2. To begin to understand forest responses to global change, greater effort needs to be directed to the multi-factor (e.g., moisture, nutrient availability) experimental approach and not just to CO_2 and/or O_3 (Shaw et al., 2002). Increasing basic understanding of forest tree responses to air pollution and climate change will help guide process-based modellers to more accurately project forest ecosystem responses to global change (Chappelka and Samuelson, 1998; Pitelka et al., 2001). Research on air pollutant and climate change effects should be done under realistic field conditions using FACE or other appropriate techniques (McLeod and Long, 1999), spatially differing pollutant exposures (Karnosky et al., 2003), and dendro-ecological techniques (McLaughlin et al., 2002). In addition, an increased understanding of stress physiology is important. Large-scale gene expression studies using microarrays laced with thousands of expressed sequence tags (ESTs) offer unprecedented opportunities to study complicated responses to stress and to determine functional patterns of gene expression. Thus, functional genomics will likely play an ever-increasing role in the future and will give physiologists increased understanding of forest responses to interacting stresses. Furthermore, this genomics research can facilitate additional studies on biodiversity. Little is known, for example, about the effects of air pollution and/or climate change on fitness and on genetic diversity.

3. Understanding feedbacks and interactions between forest ecosystems and the atmosphere is critical to understanding and mitigating various aspects of air pollution and climate change. For example, increasing atmospheric CO_2 is closely linked to carbon (C) sequestration in global forests and forest soils, and enhanced reforestation, afforestation and agroforestry have been proposed as methods to mitigate rising atmospheric CO_2. Impacts of such mitigation activities on global forest carbon budgets (Körner, 2003) and biodiversity (Schulze et al., 2002) remain uncertain. Furthermore, little is known about C sequestration under changing climate conditions (particularly changes in temperature and changes in organic matter decomposition), under elevated CO_2, or under interacting elevated CO_2 and elevated air pollutants. Improved forest management and silvicultural practices, advanced genetic selection and improvement,

and biotechnological approaches to improve C sequestration should be coupled with other approaches to mitigate the effects of global change.

4. Another example of a closely linked climate change/air pollution/forest canopy interaction is that of volatile organic compounds emitted by forest trees and involved as precursors in O_3 formation. Under warming conditions, VOC emissions are expected to rise as they have been shown by Sharkey and Singsaas (1995) to have a role in protecting leaves from short, high temperature events. Increased VOC emissions and warmer temperatures could contribute to higher O_3 production in the troposphere. How VOCs will respond to rising CO_2 and increasing air pollutants, such as O_3, has been listed as a critical research need both in the US (Fuentes et al., 2001) and Europe (Kellomäki et al., 2001).

4. Conclusions

Approximately 49% of forests of the world will be exposed to damaging concentrations of O_3 by 2100 and area at risk from S may reach 5.9 M km^2 by 2050, despite large reductions in SO_2 emissions in the developed countries. However, emissions of NO_x have changed little, or have increased. Coincidentally, shifts in precipitation and temperature patterns are occurring. Despite the fact that a number of reports suggest forests are being affected by air pollution, the extent remains uncertain. Routine monitoring systems provide many data, yet often they do not fit statistical requirements for detecting status and trends of forest health. There is a clear need for a new examination of monitoring concepts, designs and choice of ecological indicators especially as much of this information is often considered by decision makers.

Air pollution, climate change and increasing demands upon the forest resources are key factors comprising the global change threat to forest health and sustainability. Considerable scientific effort has been devoted to the enhancement of our understanding of forest responses to global change at the process, organ, system, stand and ecosystem levels. Much work, however, remains to be done, especially in the area of scaling up to landscape in the context of multiple stressors. Integrated approaches are required linking long-term process-oriented empirical studies with pattern-oriented monitoring along defined spatially differing pollutant exposures using ecosystem-based research. New approaches to monitoring will be required if the interactive effects global change on forest health and ecosystem function are to be understood in the 21st century. **Of paramount importance, air quality and climatology measurements must be coupled in time and space to effects analyses. Lack of such coordinated studies has been one of the single most important shortcomings to date.**

Acknowledgements

This research was partially supported by the US Department of Energy, Office of Science (BER) (DE-FG02-95ER-62125), the USDA Forest Service Northern Global Change Program, the USDA Forest Service North Central Research Station, and the Canadian Forest Service.

References

Aber, J., Neilson, R.P., McNulty, S., Lenihan, J.M., Bachelet, D., Drapek, R.J., 2001. Forest processes and global environmental change: predicting the effects of individual and multiple stressors. Bioscience 51, 735–751.

Alley, R.B., Marotzke, J., Nordhaus, W.D., Overpeck, J.T., Peteet, D.M., Pielke Jr., R.A., Pierre-humbert, R.T., Rhines, P.B., Stocker, T.F., Talley, L.D., Wallace, J.M., 2003. Abrupt climate change. Science 299, 2005–2010.

Bakkenes, M., Alkemade, J.R.M., Ihle, F., Leemans, R., LaTour, J.B., 2002. Assessing effects of forecasted climate change on the diversity and distribution of European higher plants for 2050. Global Change Biol. 8, 390–407.

Baldocchi, D.D., 2003. Assessing the eddy covariance technique for evaluating carbon dioxide exchange rates of ecosystems: past, present and future. Global Change Biol. 9, 479–492.

Baldocchi, D., Falge, E., Gu, L., Olson, R., Hollinger, D., Running, S., Anthoni, P., Bernhofer, Ch., Davis, K., Fuentes, J., Goldstein, A., Katul, G., Law, B., Lee, X., Malhi, Y., Meyers, T., Munger, J.W., Oechel, W., Pilegaard, K., Schmid, H.P., Valentini, R., Verma, S., Vesala, T., Wilson, K., Wofsy, S., 2001. FLUXNET: a new tool to study the temporal and spatial variability of ecosystem scale carbon dioxide, water vapor and energy flux densities. Bull. Amer. Met. Soc. 82, 2415–2434.

Bale, J.S., Masters, G.J., Hodkinson, I.D., Awmack, C., Bezemer, T.M., Brown, V.K., Butter-field, J., Buse, A., Coulson, J.C., Farrar, J., Good, J.E., Harrington, R., Hartley, S., Jones, T.H., Lindroth, R.L., Press, M.C., Symrnioudis, I., Watt, A.D., Whittaker, J.B., 2002. Herbivory in global climate change research: direct effects of rising temperature on insect herbivores. Global Change Biol. 8, 1–16.

Buchmann, N., Schulze, E.-D., 1999. Net CO_2 and H_2O fluxes of terrestrial ecosystems. Global Biogeochem. Cycles 13, 751–760.

Bytnerowicz, A., Tausz, M., Alonso, R., Jones, D., Johnson, R., Grulke, N., 2002. Summer-time distribution of air pollutants in Sequoia National Park, California. Environ. Pollut. 188, 187–203.

Bytnerowicz, A., Badea, O., Barbu, I., Fleischer, P., Fraczek, W., Gancz, V., Godzik, B., Grodzin-ska, K., Grodzki, W., Karnosky, D., Koren, M., Krywult, M., Krzan, Z., Longauer, R., Mankovska, B., Manning, W.J., McManus, M., Musselman, R.C., Novotny, J., Popescu, F., Postelnicu, D., Prus-Glowacki, W., Skawinski, P., Skiba, S., Szaro, R., Tamas, S., Vasile, C., 2003. New international long-term ecological research on air pollution effects on the Carpathian Mountain forests, Central Europe. Environ. Internat. 29 (2–3), 367–376.

Chappelka, A.H., Samuelson, L.J., 1998. Ambient ozone effects on forest trees of the eastern United States: A review. New Phytol. 139, 91–108.

DeVries, W., Reinds, G.J., Klap, J.M., van Leeuwen, E.P., Erisman, J.W., 2000. Effects of environmental stress on forest crown condition in Europe. Part III: Estimation of critical deposition and concentration levels and their exceedances. Water Air Soil Pollut. 119, 363–386.

Fanta, J., 1994. Forest ecosystem development on degraded and reclaimed sites. Ecol. Eng. 3, 1–3.

Fanta, J., 1997. Rehabilitating degraded forests in Central Europe into self-sustaining forest ecosystems. Ecol. Eng. 8, 289–297.

Fowler, D., Cape, J.N., Coyle, M., Flechard, C., Kuylenstierna, J., Hicks, K., Derwent, D., Johnson, C., Stevenson, D., 1999. The global exposure of forests to air pollutants. Water Air Soil Pollut. 116, 5–32.

Fuentes, J.D., Hayden, B.P., Garstang, M., Lerdau, M., Fitzjarrald, D., Baldocchi, D.D., Monson, R., Lamb, B., Geron, C., 2001. New directions: VOCs and biosphere–atmosphere feedbacks. Atmos. Environ. 35, 189–191.

Grimmond, C.S.B., King, T.S., Cropley, F.D., Nowak, D.J., Souch, C., 2002. Local-scale fluxes of carbon dioxide in urban environments: methodological challenges and results from Chicago. Environ. Pollut. 116, S243–S254.

Gupta, H.K., Gupta, V.B., Rao, C.V.C., Gajghate, D.G., Hasan, M.Z., 2002. Urban ambient air quality and its management strategy for a metropolitan city in India. Environ. Contam. Toxicol. 68, 347–354.

Hendrey, G.R., Ellsworth, D.S., Lewin, K.F., Nagy, J., 1999. A free air enrichment system for exposing tall forest vegetation to elevated atmospheric CO_2. Global Change Biol. 5, 293–309.

Hom, J., Nowak, D., Golub, D., Heisler, G., Grimmond, S., Offerle, B., Scott, S., 2001. Studies on carbon flux in urban forests at the Baltimore Ecosystem Study LTER (P. 1.09.09). Challenges of a Changing Earth. Global Change Open Science Conference, International Geosphere–Biosphere Programme. Amsterdam, Netherlands.

Houghton, J.T., Ding, Y., Griggs, D.J., Noguer, M., van der Linden, P.J., Xiaosu, D., 2001. Climate Change 2001: The Scientific Basis. Cambridge Univ. Press, Cambridge, MA.

ICP (International Cooperative Program on Forests) 2002. The Condition of Forests in Europe. 2002 Executive Report. UNECE and EC, Geneva and Brussels. ISSN 1020-587x.

Idso, C.D., Idso, S.B., Balling, R.B., 2000. An intensive study of the strength and stability of the urban CO_2 dome of Phoenix, AZ. In: Reprints of the Third Urban Environment Symposium, American Meteorological Society, Davis, California, August 2000. American Meteorological Society, Boston, MA, pp. 203–204.

IPCC (Intergovernmental Panel on Climate Change), 2001. Climate Change 2001: Impacts, Adaptation and Vulnerability. Summary for policymakers. Intergovernmental Panel on Climate Change, Geneva.

Jenkins, J.C., Chojnacky, D.C., Heath, L.S., Birdsey, R.A., 2003. National-scale biomass estimators for United States tree species. Forest Sci. 49, 12–35.

Karnosky, D.F., 1997. Impacts of air pollution on forest ecosystems: Implications for sustainable forestry in Eastern Europe. In: Ilavsky, J. (Ed.), Ecological Management of Forests for their Sustainable Development, pp. 51–54.

Karnosky, D.F., 2003. Impacts of elevated atmospheric CO_2 on forest trees and forest ecosystems: knowledge gaps. Environ. Internat. 29, 161–169.

Karnosky, D.F., Gielen, B., Cuelemans, R., Schlesinger, W.H., Norby, R.J., Oksanen, E., Matyssek, R., Hendrey, G.R., 2001. In: Karnosky, D.F., Ceulemans, R., Scarascia-Mugnozza, G.E., Innes, J.L. (Eds.), The Impact of Carbon Dioxide and Other Greenhouse Gases on Forest Ecosystems. CABI Press, pp. 297–324.

Karnosky, D.F., Mankovska, B., Percy, K., Dickson, R.E., Podila, G.K., Sober, J., Noormets, A., Hendrey, G., Coleman, M.D., Kubiske, M., Pregitzer, K.S., Isebrands, J.G., 1999. Effects of tropospheric O_3 on trembling aspen and interaction with CO_2: Results from an O_3-gradient and a FACE experiment. Water Air Soil Pollut. 116, 311–322.

Karnosky, D.F., Percy, K.E., Xiang, B., Callan, B., Noormets, A., Mankovska, B., Hopkin, A., Sober, J., Jones, W., Dickson, R.E., Isebrands, J.G., 2002. Interacting elevated CO_2 and tropospheric O_3 and predisposes aspen (*Populus tremuloides* Michx.) to infection by rust (*Melampsora medusae* f.sp. tremuloidae). Global Change Biol. 8, 329–338.

Karnosky, D.F., Scarascia-Mugnozza, G., Ceulemans, R., Innes, J. (Eds.), 2001. The Impact of Carbon Dioxide and Other Greenhouse Gases on Forest Ecosystems. CABI Publishing, Waalingford, UK.

Karnosky, D.F., Zak, D.R., Pregitzer, K.S., Awmack, C.S., Bockheim, J.G., Dickson, R.E., Hendrey, G.R., Host, G.E., King, J.S., Kopper, B.J., Kruger, E.L., Kubiske, M.E., Lindroth, R.L., Mattson, W.J., McDonald, E.P., Noormets, A., Oksanen, E., Parsons, W.F.J., Percy, K.E., Podila, G.K., Riemenschneider, D.E., Sharma, P., Thakur, R., Sober, A., Sober, J., Jones, W.S., Anttonen, S., Vapaavuori, E., Mankovska, B., Heilman, W.E., Isebrands, J.G., 2003. Tropospheric O_3 moderates responses of temperate hardwood forests to elevated CO_2: A synthesis of molecular to ecosystem results from the Aspen FACE project. Funct. Ecol. 17, 289–304.

Kellomäki, S., Rouvinen, I., Peltola, H., Strandman, H., Steinbrecher, R., 2001. Impact of global warming on the tree species composition of boreal forests in Finland and effects on emissions of isoprenoids. Global Change Biol. 7, 531–544.

Kimmel, V., Tammet, H., Truuts, T., 2002. Variation of atmospheric air pollution under conditions of rapid economic change in Estonia 1994–1999. Atmos. Environ. 36, 4133–4144.

Kolb, T.E., Wagner, M.R., Covington, W.W., 1994. Concepts of forest health: Utilitarian and ecosystem perspectives. J. For. 92, 10–15.

Körner, C., 2003. Carbon limitation in trees. J. Ecol. 9, 4–17.

Krupa, S.V., Legge, A.H., 1998. Sulphur dioxide, particulate sulphur and its impacts on a boreal forest ecosystem. In: Ambasht, R.S. (Ed.), Modern Trends in Ecology and Management. Backhuys Publishers, Leiden, Netherlands, pp. 285–306.

Krupa, S.V., Legge, A.H., 2000. Passive sampling of ambient, gaseous air pollutants: an assessment from an ecological perspective. Environ. Pollut. 107, 31–45.

Krupa, S., Nosal, M., 2001. Relationships between passive sampler and continuous ozone (O_3) measurement data in ecological effects research. The Scientific World 1, 593–601.

Krupa, S., Nosal, M., Peterson, D.L., 2001. Use of passive ozone (O_3) samplers in vegetation effects assessment. Environ. Pollut. 112, 303–309.

Krupa, S., Nosal, M., Ferdinand, J.A., Stevenson, R.E., Skelly, J.M., 2003. A multi-variate statistical model integrating passive sampler and meteorology data to predict the frequency distributions of hourly ambient ozone (O_3) concentrations. Environ. Pollut. 124, 173–178.

Larson, J.L., Zak, D.R., Sinsabaugh, R.L., 2002. Microbial activity beneath temperate trees growing under elevated CO_2 and O_3. Soil Sci. Soc. Amer. J. 66, 1848–1856.

Legge, A.H., Krupa, S.V. (Eds.), 1990. Acidic Deposition: Sulphur and Nitrogen Oxides. Lewis Publishers, Chelsea, MI.

Mäkinen, H., Nöjd, P., Kahle, H.-P., Neumann, U., Tueite, B., Mielikäinen, K., Röhle, H., Spiecker, H., 2003. Large-scale climatic variability and radial increment variation of *Picea abies* (L.) Karst. in central and northern Europe. Trees 17, 173–184.

McDonald, E.P., Kruger, E.L., Riemenschneider, D.E., Isebrands, J.G., 2002. Competitive status influences tree-growth responses to elevated CO_2 and O_3 in aggrading aspen stands. Funct. Ecol. 16, 792–801.

McLaughlin, S.B., Percy, K.E., 1999. Forest health in North America: Some perspectives on potential roles of climate and air pollution. Water Air Soil Pollut. 116, 151–197.

McLaughlin, S.B., Shortle, W.C., Smith, K.T., 2002. Dendroecological applications in air pollution and environmental chemistry: research needs. Dendrochonologia 20, 133–157.

McLeod, A.R., Long, S.P., 1999. Free-air carbon dioxide enrichment (FACE) in global change research: A review. Adv. Ecol. Res. 28, 1–56.

Menzel, A., Fabian, P., 1999. Growing season extended in Europe. Nature 397, 659.

Muller-Starck, G., Degen, B., Hattemer, H., Karnosky, D., Kremer, A., Paule, L., Percy, K., Scholz, F., Shen, X., Vendramin, G., 2000. Genetic response of forest systems to changing environmental conditions—analysis and management. Forest Gen. 7, 247–254.

NPS, 2002. Air quality in the National Parks. National Park Service. 2nd Edition.

Parmesan, C., Yohe, G., 2003. A globally coherent fingerprint of climate change impacts across natural systems. Nature 421, 37–42.

Parson, E.A., Corell, R.W., Barron, E.J., Burkett, V., Janetos, A., Joyce, L., Karl, T.R., Mac-Cracken, M.C., Melillo, J., Morgan, M.G., Schimel, D.S., Wilbanks, T., 2003. Understanding climatic impacts, vulnerabilities, and adaptation in the United States: Building a capacity for assessment. Clim. Change 57, 9–42.

Percy, K.E., 2002. Is air pollution an important factor in forest health? Szaro, R.C., Bytnerowicz, A., Oszlanyi, J. (Eds.), Effects of Air Pollution on Forest Health and Biodiversity in Forests of the Carpathian Mountains. IOS Press, Amsterdam, pp. 23–42.

Percy, K.E., Ferretti, M., 2003. Air pollution and forest health: Towards new monitoring concepts. Environ. Pollut. (in press).

Pitelka, L.F., Bugmann, H., Reynolds, J.F., 2001. How much physiology is needed in forest gap models for simulating long-term vegetation response to global change? Clim. Change 51, 251–257, Introduction.

Poorter, H., Navas, M.-L., 2003. Plant growth and competition at elevated CO_2: on winners, losers and functional groups. New Phytol. 157, 175–198.

Schulze, E.-D., Valentini, R., Sanz, M.-J., 2002. The long way from Kyoto to Marrakesh: Implications of the Kyoto Protocol negotiations for global ecology. Global Change Biol. 8, 505–518.

Sharkey, T.D., Singsaas, E.L., 1995. Why plants emit isoprene. Nature 374, 769.

Shaw, M.R., Zavaleta, E.S., Chiariello, N.R., Cleland, E.E., Mooney, H.A., Field, C.B., 2002. Grassland responses to global environmental changes suppressed by elevated CO_2. Science 298, 1987–1990.

Smith, W.B., 2002. Forest inventory and analysis: a national inventory and monitoring program. Environ. Pollut. 116, S233–S242.

Spiecker, H., Mielikäinen, K., Köhl, M., Skovsgaard, J.P., 1996. Growth trends in European forests. Research Report no. 5. European Forest Institute, Joensuu, Finland.

USDA Forest Service, 2001. Forest facts and historical trends. FS-696. USDA Forest Service Washington, DC.

USDA Forest Service, 2003. Forest Health Monitoring. http://www.fs.fed.us/ne/foresthealth/.

Zak, D.R., Pregitzer, K.S., King, J.S., Holmes, W.E., 2000. Elevated atmospheric CO_2, fine roots and the response of soil microorganisms: A review and hypothesis. New Phytol. 147, 201–222.

Zimmerman, F., Lux, H., Maenhaut, W., Matschullat, J., Plessow, K., Reuter, F., Wienhaus, O., 2003. A review of air pollution and atmospheric deposition dynamics in southern Saxony, Germany, Central Europe. Atmos. Environ. 37, 671–691.

Author Index

461

Subject Index